Vertebrate History: Problems in Evolution

Vertebrate History: Problems in Evolution

BARBARA J. STAHL

St. Anselm's College

McGRAW-HILL
BOOK COMPANY

New York
St. Louis
San Francisco
Düsseldorf
Johannesburg
Kuala Lumpur
London
Mexico
Montreal
New Delhi
Panama
Rio de Janeiro
Singapore
Sydney
Toronto

Library of Congress Cataloging in Publication Data

Stahl, Barbara J
  Vertebrate history.

  (McGraw-Hill series in population biology)
  Bibliography: p.
  1. Vertebrates—Evolution.   2. Vertebrates, Fossil.
  I. Title. [DNLM: 1. Evolution. 2. Vertebrates.
  QL605 S781v 1974]
  QL607.5.S7   596'.03'8   73-13997
  ISBN 0-07-060698-6

Vertebrate History: Problems in Evolution

1234567890MAMM79876543

This book was set in Baskerville by Black Dot, Inc.
The editors were James R. Young, Jr., and Susan Gamer; the designer
was J. E. O'Connor; and the production supervisor was Joe Campanella.
The drawings were done by Vantage Art, Inc.
The printer and binder was The Maple Press Company.

The cover illustration is from J. Millot and J. Anthony:
Colloques Internationaux du Centre National de la Recherche Scientifique
(Editions du Centre National de la Recherche
Scientifique), 1958.

Contents

Preface

Anyone who has written or read a textbook becomes acutely aware of what cannot be included within its covers. The writer sets down an orderly summary of the knowledge in his field and of necessity omits mention of most of the problems with which workers have had to cope in coming to their conclusions. The reader faces an encyclopedia of basic information on the subject and struggles to keep the various facts straight. Although there exist a hardy few who are borne along, fascinated by the facts, many more readers may close the book without sensing the reality of the material or the men and women who have investigated it.

It is especially regrettable if undergraduate students of vertebrate evolution do not know of the problems that currently absorb embryologists, anatomists, and paleontologists, because the chance to wrestle with these problems—the fragmentary evidence and conflicting possibilities that exist—engenders more interest than does a catalog of conclusions. Left without a sense of the excitement of the investigation, students come ultimately to question the reason for having to concern themselves with anatomy and events of the past instead of looking toward the future. Although most students are willing to admit that insatiable curiosity is a human characteristic and that the mere existence of a problem has always been sufficient reason for attacking it, recognition of this fact often does not remove their misgivings. What is required is not argument about the value of basic scientific knowledge but involvement—even vicarious involvement—in the questions that workers are trying now to resolve. This book is written in the hope that those who read it will discover the wide assortment of issues which rarely appear in textbooks but provide the real excitement that holds researchers in vertebrate evolution to the field.

Prefaces, if they are to be read to the end, should be short, but I want to acknowledge the help that I have received from many quarters. I am indebted most of all to Dr. Alfred S. Romer of Harvard University, who encouraged me to undertake the graduate studies that led to my interest in the history of the vertebrate animals. His kindness in permitting me to use his library facilitated my research and allowed me the pleasure of frequenting his office in the Museum of Comparative Zoology in Cambridge, Massachusetts. There I benefited often from the interest of Miss Nelda Wright, who always made sure that I found what I was looking for. I have also been aided immeasurably at the Museum by graduate students and members of the staff who took time to suggest or provide papers that I might never have discovered alone. I owe special thanks to Mrs. Myvanwy Dick because, besides providing references, she sustained me throughout my research by her friendship and countless cups of tea.

I am especially grateful to the men who read parts of the manuscript and, despite schedules crowded with research, teaching, writing, and administrative chores, gave me valuable comment and criticism. Dr. Bobb Schaeffer of the American Museum of Natural History read the introductory chapters and those on fishes; Dr. Ernest E. Williams of Harvard University reviewed the chapter on amphibians; Dr. Romer read the sections covering the reptiles and birds; and Dr. Bryan Patterson of Harvard University scrutinized the chapter on mammals. I have corrected the errors and filled the gaps that these scholars pointed out to the best of my ability. I am entirely responsible for the faults that remain.

Many people gave me permission to reproduce drawings and photographs from their books and papers, and I appreciate their generosity. I should not have been able to reprint the photographs, however, had not their owners taken the time and trouble to obtain negatives or copies of the pictures and forward them to me. For this favor, I am indebted to Mr. Robert Carlisle (Department of the Navy, Washington, D.C.), Dr. Robert H. Denison (retired from The Field Museum of Natural History, Chicago), Dr. Hermann Jaeger (Naturkundemuseum an der Humboldt-Universität zu Berlin), Dr. Zofia Kielan-Jaworowska (Polska Academia Nauk, Zakład Paleozoologii), Dr. Robert F. Normandin (St. Anselm's College), Dr. D. F. G. Poole (Dental School, University of Bristol, England), Dr. Alfred S. Romer (Professor Emeritus, Harvard University), Dr. Morris P. Ruben (Boston University School of Graduate Dentistry), Mr. William H. Tobey (Photographer, *Harvard Alumni Bulletin*), Dr. Dorothy F. Travis (Gerontology Research Center, National Institute of Child Health and Human Development), Dr. Peter Wellnhofer (Bayerische Staatssamlung für Paläontologie und historische Geologie), and Dr. Rainer Zangerl (Field Museum of Natural History, Chicago).

My preoccupation with the research for this book and the writing of it has caused, at times, a degree of inconvenience to those I love best. I want to thank Susan, Nancy, Sarah, and John for their great patience and last, and certainly most, my husband, Dr. David G. Stahl, without whose strong encouragement I could never have brought the book to completion.

Barbara J. Stahl

Vertebrate History: Problems in Evolution

1
Fossils: Getting the Evidence

The work and thought of scientists in many specialized fields stands behind the formation of the theory of evolution in general and of vertebrate evolution in particular. Verification of the actual history of a group of animals, however, depends ultimately upon finding fossilized remains of its members. There is a difference in making the logical assumption that the first vertebrates probably lacked movable jaws and paired fins and in being able to demonstrate that, in fact, this combination of structures was absent. Concrete knowledge of vertebrate evolution advances little by little as new fossils are found, described, and interpreted. Since searching for fossils is often like looking for a needle in a haystack and describing and interpreting the bits and pieces is usually extremely time-consuming work, progress is slow.

Every trace of the physical existence of an extinct animal is considered a fossil and regarded as potentially helpful in determining the history of an ancestral line. Sometimes a fossil hunter is lucky enough to find the whole skeleton of an animal that was covered with sand or mud shortly after death or even an entire body frozen in ice. Far more often he must be content with the broken end of a humerus, a flattened fragment of a skull, some teeth, a spine, or even a mere footprint. Rarely, patches of mummified, scaly skin or impressions of delicate soft tissues are preserved. Coprolites, fossilized contents of the animal's intestine may contribute a clue to the structure of the gut or, preserved after egestion, indicate the diet of the animal that produced them.

Despite the variety of fossils—preserved hard parts, impressions, casts, and molds—the physical record of the vertebrate life of earlier ages is meager in comparison with the great numbers of animals that must have existed. To be preserved, an animal must die in an environment in which its body will be covered immediately and so protected from destruction. Few individuals obliged the future paleontologist, as some saber-toothed tigers did, by drowning in a tar pit. Most dead animals remained on the surface to be eaten by scavengers and rotted away completely by bacteria. Animals that lived in shallow water or near the shore, where they were likely to be buried in the mud, were preserved by the hundreds; other forms—inhabitants of upland forest areas where they lay exposed on the ground after death, or dwellers in deep waters—left no trace.

Some fossils form but are never found. They lie in rocks that are buried far too deep for the paleontologist's pick to reach. The fossils that scientists are able to collect are those in strata brought to the surface as the land is lifted up or the covering rocks are worn away (Fig. 1.1). If no one happens to come upon it, a beautiful and important fossil may be exposed as the rock around it weathers and then, after having existed for millions of years, disintegrate completely in the wind and rain within a few months.

Knowing that many species of animals were not preserved and

Figure 1.1. The history of a fossil. (**a**) Millions of years ago, during the Oligocene epoch, a three-toed horse died near a stream. (**b**) The stream overflowed its banks soon after, covering the remains of the animal with mud. (**c**) After many years, the skeleton lay well buried by sediments. (**d**) As the land in the region subsided, the sediments thickened and became consolidated into rock. (**e**) Eventually, the climate became arid, forces within the earth caused the land to rise, and erosion of the sedimentary rock began. (**f**) Thirty million years later, erosion had created a deep canyon, exposing the stratum containing the fossil and making it accessible to the paleontologist. (From Museum of Comparative Zoology, Harvard University.)

(a)

(b)

(c)

(d)

(e)

(f)

also that a great part of the fossil hoard that has been preserved is either lost or irretrievable, one may look askance at the fossil record as a source of information about vertebrate evolution. In fact, throughout the last hundred years as the modern theory of evolution has been taking shape, some men have done so. Their argument that a quart jar with a pint of water in it is half empty is admittedly true—that the gaps in the fossil record make it impossible to demonstrate the gradual transitions from form to form that evolutionists postulate. It is also true, however, that the quart jar is half full, and paleontologists prefer that view. When a paleontologist finds, interprets, and dates the remains of a vertebrate, he has incontrovertible proof that it existed in a particular place and at a particular time. Accumulation of such data has provided physical evidence of the progressive development of the vertebrate group. In combination with the evidence drawn from the genetics laboratory about the mechanism of change in living organisms, the fossil record firmly supports the contentions of the modern theory of evolution. No paleontologist today feels that the gaps in the fossil record cripple the evolutionary theory so seriously that an alternative explanation is preferable. On the contrary, fossil seekers take to the road in the hope that they will be lucky enough to find the remains of an animal hitherto unknown—a form related to a known one—so that still another sequence in vertebrate descent will become clear.

Searching for fossils

Although an amateur may occasionally discover something of significance by simply looking among the rocks in his own vicinity, the important finds are almost invariably made by the professional paleontologists, versed in the necessary geology, who lay careful plans for their expeditions. Whether they are searching for particular fossils or intend to investigate the fossil content in rocks of a particular age, they must begin by consulting geological maps to determine where promising rocks of the right age are exposed. Their knowledge of geology goes beyond map reading, in fact, to the understanding of rock formations and recognition of different layers of rock in the field. They must distinguish between igneous rocks, solidified from a molten state, and metamorphic rocks, transformed by tremendous stress (materials that never, or almost never, contain vertebrate fossils) and sedimentary rocks, which may. The last are formed when particles of ash, silt, clay, or sand settle and are compacted and cemented into a solid mass. The resulting rocks may be tuff, shale, limestone, mudstone, or sandstone, depending upon the nature and size of the particles of which they are composed. Although some sedimentary deposits, like the tuff from volcanic ash, may form on land, most rocks of this type are laid down under water on the bottoms of lakes, streams, and seas and at the mouths of rivers opening onto continental shelves. Very few fossils are recovered while these deposits are submerged. Prospecting takes place where sedimentary rocks have been lifted out of the water. Repeatedly in the history of the earth, areas have risen, spilling off the water which had covered them and exposing the old bottoms to the forces of erosion. These places are the paleontologist's hunting ground.

The best locations for fossil hunting are often not the most comfortable. Greenland and Spitzbergen, where rocks have been exposed by the melting snow in relatively recent times, can be visited only for a few months in summer because of the cold and the violent storms which occur there during the rest of the year. Digging at more temperate latitudes is not necessarily easier (Fig. 1.2). Exposed rocks, of course, support no protective foliage cover. The paleontologist is usually out in the open under the broiling sun and often in wind as well. In remote areas he must frequently cope with lack of water and an absence of roads. Fossil hunters have exchanged their canteens and horses for more modern equipment where possible, but even today they have to spend a good part of their energy in hiking and hauling supplies. Construction sites—foundation holes and road cuts—in the midst of settled areas afford access to hitherto buried fossils and, compared to other locations, can be relatively convenient places to work.

When he arrives at his destination, the paleontologist must rely on his eyes as the tools he uses first. That he *looks* for fossils may seem too obvious to have to mention, but it should be emphasized that the time spent in looking over the ground can be long indeed. Like a man seeking a four-leaf clover, the paleontologist walks the area, running his eyes over every detail of the surface under his feet. Even though he has a mental picture of the kind of telltale signs which indicate the presence of vertebrate remains, his search may net him nothing. Eventually, if he finds no trace of fossils, he must consider the time and supplies that remain available to him and decide whether or not to move to another place. It is difficult to call a halt to an expedition which has been unsuccessful—to pack and leave an area where there may be important fossils lying by chance unobserved. Even a last-minute find by an assistant at that point is sufficient to raise everyone's morale.

(a)

If fossils are found, another phase of the work begins. Collecting involves certain rules that must be followed if the fossils are to be of use to the student of evolution. Every curator of a natural history museum has received at some time or other a box with a fossil in it sent by an amateur collector who wants to know if his contribution has any importance. If the fossil is unaccompanied by specific data relating to the location from which it was taken and the rocks in which it was found, the curator must usually explain gently that the fossil can be classified only very generally and might be retained by the collector as an interesting souvenir of an extinct form. Fossils collected on expeditions are labeled systematically to provide information for the use of the worker who will attempt to interpret the nature, age, and habitat of the ancient vertebrate.

Although some fossils are collected by merely being picked up from the weathered or unconsolidated sediments in which they lie, most fossils must be extricated from the rocky matrix that encloses them. If the matrix is extremely hard, trying to free the fossil completely in the field would result in severe damage to the specimen. In such a case, the paleontologist cuts out or detaches a slab of the rock containing the fossil and transports it to the laboratory, where a preparator can remove the surrounding rock under conditions that minimize the danger to the specimen. If the paleontologist has discovered a large, well-preserved but solidly embedded skull, the work of removing it is bound to be time-consuming and backbreaking (Fig. 1.3). The exposed portions of the skull are first protected by strips of

(b)

Figure 1.2. Paleontologists in the field. (**a**) In the desert in western Argentina, James Jensen pauses from fossil hunting for a drink of water. (Photograph, *The Harvard Alumni Bulletin*, copyright 1967, Harvard Bulletin, Inc.) (**b**) In chilly Antarctica, William Breed chisels specimens out of the rock at Coalsack Bluff. (Photograph, U.S. Navy.)

burlap soaked in plaster of paris. The water for mixing the plaster often has to be brought from a distance, of course, if the excavation site is in a dry area. The paleontologist and his assistants work with picks, chisels, and hammers to cut and

(a)

separate the block containing the fossil. When the block is free, it is crated and conveyed to the laboratory. If the digging is being done in his own country, and if modern transport facilities are available, the labors of the paleontologist may end when he and his co-workers heft the crate into their truck. Sending fossils home from another country is a more complicated affair. The chief of the expedition has then to reach an agreement with the country in which the digging is taking place giving him permission not only to explore but to export what he finds. No paleontologist wants to risk having his boxes of fossils impounded by a government that suspects him of trying to carry off a national treasure.

Preparing fossils in the laboratory

Most fossils that reach the laboratory must undergo some preparatory work before the paleontologist can study them profitably. The techniques used to expose fossils depend upon the nature of the matrix in which the remains are embedded and the condition of the remains themselves. Resistant structures like teeth, if buried in clay, may simply have to be washed carefully in water, but more often matrices are harder and fossil materials farther removed from their original state. As a rule, animals after death and burial lose not only their soft tissues but also much or all of the organic matter built into their harder parts. Small specimens may be completely carbonized, that is, reduced to a thin, black film of residual carbon, through the gradual transformation of their substance into gases that seep

(b)

Figure 1.3. Recovering and cleaning a fossil. **(a)** In the field, Arnold Lewis covers an exposed fossil with a protective coating before applying a layer of plaster. **(b)** In the laboratory, after removing plaster and much of the matrix, he uses a fine needle to clear particles from the surface of the specimen. (Photographs, *The Harvard Alumni Bulletin*, copyright 1967, Harvard Bulletin, Inc.)

away. Bones which retain their shape are nevertheless sometimes rendered friable through the loss of the organic framework that held their mineral material together. Under certain circumstances, skeletal structures may be hardened during their interment. If ground water carries dissolved calcium, silicon, or iron into their interior, crystals are likely to form, filling every interstice. The bony tissue itself is occasionally petrified as the original salts it contained are replaced by new ones. Extensive petrifaction is not common in vertebrate fossils, however. The preparator usually has to work upon thin or relatively delicate remains that are embedded in rock that is more resistant than they. If he has reason to suspect that preserved elements lie in the interior of a slab, he may use x-rays to detect them, but most often he faces the problem of removing from partially visible fossils as much of the covering matrix as possible.

As every detail of a fossil may be important, every square millimeter that can be exposed represents a gain for the investigator. Rarely, the sedimentary rock containing the fossil is so weathered or unconsolidated that it can be washed and brushed off. More usually, especially with older vertebrate material, the surrounding matrix must be attacked with chisels, small drills, and even needles. Sometimes a small specimen, like the head shield of an ancient jawless fish, is submerged in alcohol or xylol to make fine details visible more clearly. Working on such a fossil, the preparator uses a dissecting scope to lessen further the chance of destroying with his sharp instrument delicate parts of the bony plates or their surface ornamentation. Where chipping or picking proves impractical, careful application of acid may be effective in weakening the rock. Hydrochloric and hydrofluoric acid have both been tried, the former against limy cover and the latter for rock containing a large amount of silica.

Frequently, the preparator knows when he looks at an embedded fossil that removing it from the rock is impossible. This is the case when ancient bone has been infiltrated by mineral material, retaining its distinctive osseous pattern but becoming inseparable from the mass of rock around it. Also, whole skulls are sometimes filled with sediments which, after they have hardened, can never be picked out from the interior. Such fossils were of limited use until W. J. Sollas, a British geologist-paleontologist, devised a new method for their preparation. He resorted to grinding away the entire mass, fossil and rock, thin layer after thin layer. He studied and drew the pattern of preserved bone that appeared on the ground surface before removing each successive layer. Eventually the entire fossil was reduced to a powder, but Sollas had obtained a series of drawings showing the configuration of the skull bones trapped in the rock. One of these drawings, by itself, was of limited significance, as it represented only a section through the skull at one level. From the whole series, however, Sollas was able to reconstruct an accurate model of the original remains. He did so by reproducing the pattern of bone shown in each drawing in a thin layer of plaster and then assembling the plaster layers— usually several hundred of them—in the proper order. In this way he built, from a series of sections, a three-dimensional replica of the fossil skull which had been hidden in the rock. The method used by Sollas was modified by workers who followed him. Although the tedium of making the model could not be avoided, the process was facilitated by using sheets of wax instead of plaster (Fig. 1.4). Now, rather than losing the material of the specimen as it is ground, preparators have found a way to preserve thin sections of it. After grinding away .5 to 2 millimeters of bone and matrix, they etch the surface of the block with hydrochloric acid for less than a minute, wash it, coat it with acetone, and cover it with a sheet of cellulose nitrate. When the preparation is dry, they peel off the sheet of cellulose nitrate. A fine layer of fossil material and sediment adheres to it. Peels, as these sheets are called, can be kept indefinitely and studied and restudied at leisure. Since the particles of bone upon them will accept histological stains, the peels can be colored and projected or viewed under the microscope. Although many workers prefer to prepare specimens in a recalcitrant matrix by using faster methods employing acid, grinding and making peels has become one of the standard techniques for investigating fossils of small size.

Fossils in museums   A small number of the fossils undergo further treatment to prepare them for public display. The bones of land vertebrates, freed entirely from the rock in which they were embedded, are assembled in their proper order and fixed in place. In this way the frames of huge dinosaurs emerge from piles of separate bones to tower over visitors in high-ceilinged museum halls, and skeletons of mammal-like reptiles stand again on all fours in the glass display cases. Fossil fishes and flattened amphibian skulls may be exposed in relief against the rocky matrix, their bones darkened or outlined to make them more visible to the viewer. Whole slabs of rock containing footprints go up on museum walls.

Fossils on display in museums are, for the most part, handsome examples of forms well known to the paleontologist. Drawers and boxes in storage rooms hold, besides additional examples of recognized genera, specimens only cursorily examined, classified in a general way, and labeled with collecting data. These materials wait until someone has the time to

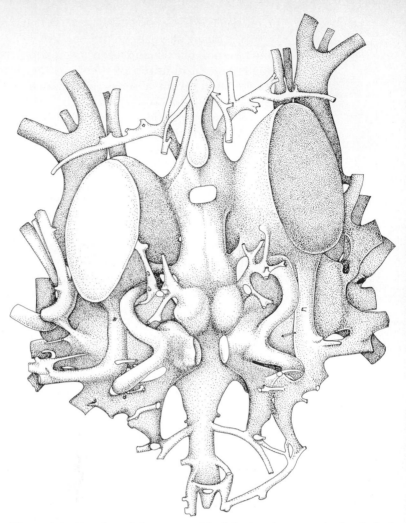

Figure 1.4. Drawing of the wax model of the interior spaces in the skull of *Cephalaspis*, an extinct jawless fish (dorsal view). The model, constructed according to Sollas' method by E. Stensiö, made possible knowledge of the brain and associated structures that could not have been obtained merely by cleaning and observing the fossil. (From Stensiö.)

undertake a thorough study of them. Because fossil hunters bring in material faster than it can be described, all but the newest departments of paleontology have a backlog of specimens on which more work must be done. Paleontologists say, only half in jest, that the key to the solution of several problems in vertebrate history may be lying on the shelf in a laboratory cabinet somewhere.

Studying fossil material

An investigator, as he turns his attention to a particular fossil, begins a four-pronged inquiry. Not only does he try to identify and describe the remains he has in hand, but he wants also to determine the conditions under which the animal lived, the period of its existence, if not already known, and the place of the organism in the evolutionary scheme. The results of his study should provide his colleagues with an accurate description of the specimen and an interpretation of its significance.

Identification and description of fossil material require that the paleontologist draw upon his knowledge of comparative anatomy: by recognizing familiar patterns of skeletal design or arrangement, the researcher is able to assign the fossil to one of the classes of vertebrates and to refer it to an order and family. On the basis of its individual peculiarities, he may give it a new generic and specific name. When a fossil consists of a fairly complete skeleton or the greatest part of a skull, the paleontologist's task is almost always easier than when a fossil is more fragmentary. Small pieces of dermal bone, for instance, he may decide have probably come from the armor of an early jawed fish, but more precise classification may have to wait until he can match the bits of bone with similar bones of a more complete specimen. If he recognizes isolated spines or teeth as having belonged to some sort of shark, he may establish a new genus under the class Chondrichthyes. However, as such fragmentary remains give no further clue to the identity of the fish that bore them, he must list the new genus, not in a particular order of the class Chondrichthyes, but simply under the heading *incertae sedis*, of uncertain seat. When new fossils are found with spines or teeth of the same sort attached to other hard parts, it may become possible to reclassify the genus in a known order or to declare that it represents a new one.

The severest test of the paleontologist's ability to recognize anatomical structures comes when a specimen has been crushed or its parts have become dislocated. It is not unusual to find remains of early fishes and tetrapods that have been disarticulated after death and flattened by the weight of the sediments in which they were buried or broken by the shifting of rock. A first look at the skull of an ancient lobe-finned fish can be discouraging indeed. An untutored eye may see nothing initially but the black shale in which the fossil is embedded. Little by little the outlines of bones appear, but elements project in every direction and are half hidden by one another. Paleontologists have no choice, though, but to work with the specimens they have, and their long study of such fossils has enabled them to discover the complicated patterns of bones that distinguish the various types of early vertebrates.

Type specimens

When a fossil form appears which is different from others that have been described, a new specific name is created for it. The fossil first given the name is known thereafter as the type specimen of that species and serves as a standard of comparison for other fossils thought to be like it. Since the preservation of valuable type specimens is best assured if they remain in one place, paleontologists find it necessary either to visit laboratories and museums to study collections or to work from plaster or latex models of the originals. The use of photographs and drawings is essential, especially in preliminary studies and in general communication, but for the researcher there is no substitute for examination of the fossils themselves. By looking at specimen after specimen in a particular group, the paleontologist sharpens his ability to identify fragmentary or poorly preserved remains.

Reconstructing fossil forms

After describing the fossil that is the subject of his study, a worker tries to obtain some understanding of the appearance of the ancient animal it represents. To this end, he draws a reconstruction—not just a fanciful picture, but one based securely on the anatomical evidence he has (Fig. 1.5). Should the fossil material consist of an entire skeleton, he calculates the bulk of the musculature from the size of the areas of muscle attachment visible on individual bones and draws an outline to encompass both bones and flesh. If the lower jaw is missing, it can be added, its size and position estimated from the length of the opposing upper jaw and teeth and from the site of its articulation with the skull. The orbital opening dictates the size and location of the eye. Although the appearance of the surface of the body is often a matter of conjecture, the preservation of scales or imprints of feathers or "mummified" skin enables the

Figure 1.5. Recreating the appearance of an animal from fossil material. **(a)** The fossil *Protosuchus*, a form related to the crocodiles: (1) in dorsal view; (2) in ventral view as exposed in the matrix. **(b)** Lateral view of the skeleton of *Protosuchus* restored in walking pose. Dotted lines represent missing elements. **(c)** Restoration of *Protosuchus* as the animal may have appeared in life. Restoration by Louise Waller Germann. (*a* and *c* from American Museum of Natural History, *b* from Colbert and Mook.)

a(1)

a(2)

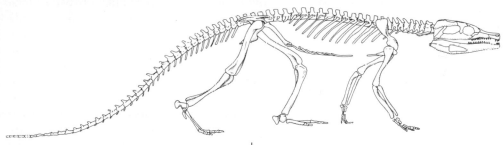

b

c

paleontologist to represent the body covering in some detail. Where parts of the fossil are missing, he may extend his outline in what he assumes is a logical way, but he uses a dotted line to warn the viewer that this part of the drawing is based upon speculation rather than physical evidence.

If a fossil consists only of a bone or two, the investigator can still make some estimate of the nature of the animal from which it came. In making such informed guesses, he again uses his knowledge of comparative anatomy and a method of reasoning developed by George Cuvier early in the nineteenth century in Paris. Cuvier pointed out that all the parts of an animal's body are structurally interdependent and that adaptation for life in a particular environment brings with it the association of a predictable group of structural characteristics. Hence, examination of an isolated femur may allow the paleontologist to estimate the size of the missing leg bones and the posture of the animal as well as its weight. He may also be able to tell from the form of the femur whether its owner was a runner, climber, or a swimmer. Then, from his knowledge of the anatomical characteristics of each of these general types, he can figure out what to expect or look for as he continues his search for more of the fossil material.

A study that brought forth only the reconstruction of an extinct animal would be incomplete. The purpose of identifying fossils is not to compile a picture album of ancient forms but to elucidate vertebrate history from its beginnings to the present.

Assessing ecological conditions of ancient environments

To realize this aim, the paleontologist must also try to determine the physical environment in which each animal lived and the time of its existence. Although he takes his first clue to the habitat of the animal from the adaptive modifications visible in its anatomy, he must find the other ecological clues in the rocks in which the fossil was found. From comparative anatomy, then, he turns to geology for help. Geologists' knowledge of the conditions under which sediments are being laid down today enables them to make statements about the conditions that prevailed when similar types of sediment were deposited in the past. No one has argued against reasoning in this manner since the triumph of the idea, put forth by James Hutton and Charles Lyell in England in the first half of the nineteenth century, that the forces that wreak changes in the surface of the earth have been at work uniformly for hundreds of millions of years. So it is that a geologist can examine the rocks and the beds from which the rocks were taken and make an estimate of the environment of the place in which their materials were laid down. The paleontologist, looking again at the anatomy of his specimen, may suppose that the animal did live in the environment that the geologist describes or if that conclusion seems unreasonable, is forced to speculate how the animal's remains came to rest in such a place. Actually, in determining the nature of ancient environments, geologists and paleontologists find themselves interdependent, for geologists take into consideration the assemblage of fossils in the rocks they are examining. When the chemistry of the rock is not sufficient to distinguish a freshwater deposit from one laid down in sea water, geologists must be able to tell whether the fossils present represent animals characteristic of freshwater or marine habitats. Further, they may even be able to make an estimate of the climate that existed in the earlier period by recognizing ancestors of today's warm- or cold-water invertebrates.

Determining the age of fossil material

To determine the relative age of his fossil material, the paleontologist continues to work in cooperation with geologists. In making assumptions about the age of sedimentary rocks and the remains of organisms contained within them, the men in both fields rely on a fact made clear by William Smith, a countryman and contemporary of Hutton and Lyell. Smith pointed out that since new sediments are laid down on top of old ones, a particular stratum of rock is always younger than the layers below it and older than the ones above. This idea, obvious now but new and striking at the beginning of the nineteenth century, enabled workers to construct a system of relative dating. A certain animal, they could say, lived later than those whose remains were preserved in the rocks below and earlier than the group of animals whose fossils were found at a higher level.

Though Smith's idea was the key to understanding the progress of vertebrate evolution, determining the exact sequence of fossil forms did not prove to be an easy matter. It would have been, if sediments settled everywhere at the same rate and were never disturbed thereafter. In reality, the situation is quite different: not only has the rate and type of deposition varied from time to time and from place to place, but also the resultant sedimentary rocks have been lifted, bent, tipped on end, faulted, and often partially or entirely destroyed by the forces of erosion. It is not unusual for a worker in the field to observe in one place the sequence A, B, C in the rock layers and in another, not too distant, the sequence A, C, due to the wearing away in the second area of layer B before the deposition of the sediments which formed C (Fig. 1.6). Paleontologists who are trying to establish the relative ages of fossils recovered from widely separated sites must contend with even more complex stratigraphical riddles. No overall solutions are in sight—or possible, for that matter. Paleontologists continue to regard the correlation of strata in different areas as being among the most difficult problems in the study of vertebrate evolution.

Paleontologists and geologists have attempted to decipher the history of rock formations one at a time and then to compare series from different localities. Their studies began, as is always the case, with descriptions of the strata which exist and advanced to the search for similar patterns of deposition over a broad area. Following another of William Smith's principles, workers learned to identify contemporaneous rock layers by certain of *Index fossils* the fossils they contain. Index fossils, as these forms are called, are easily recognizable remains of animals that enjoyed a wide distribution for restricted periods of time. They serve as markers in the rock, making it possible to define the temporal relationship of one stratigraphical series to another. Little by little, a general outline has emerged of the succession of sedimentary rocks formed down through the ages and of the living things that have inhabited the earth in the last 600 million years.

As knowledge accumulated, it became possible to draw up a table of geological ages and to associate with each the forms of *The geologic time scale* life which characterized it (Fig. 1.7). The table shows the earth's history divided into five eras. The two oldest, the Archeozoic

Figure 1.6. Diagrammatic cross section of deposits at three widely separated localities, showing how strata are correlated by similarities in their fossil content and how they may be unequally distributed. Stratum D at locality 2 and stratum B at locality 3 are missing; they may have eroded or perhaps never have formed at those places. (From Longwell and Flint.)

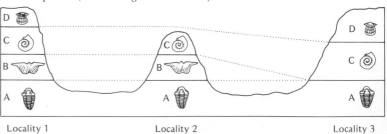

Locality 1   Locality 2   Locality 3

and Proterozoic, are distinguished entirely by physical criteria and are of interest chiefly to the geologist, since few fossils appear in rocks from those times. Paleontologists have given more attention to the last three, the Paleozoic, Mesozoic, and Cenozoic, because it is in the strata laid down then that almost all fossils are found. The names of several of the periods into which these three eras are subdivided reflect the names of places where paleontologists as well as geologists worked to define distinctive layers of rock and fossil forms. The smallest units of time shown on the geological table are the epochs into which the Tertiary and Quaternary periods are divided. Individual rock formations, usually named after the locality where they are first described, are assigned to the proper epoch or period of the geological calendar. A paleontologist who wants to collect remains of bony fishes that lived in the Eocene epoch of the Tertiary period may travel to Wyoming to search among the exposed rocks of the Green River Formation, as the strata in that location are known to have formed at that time.

Radiometric dating   Long after paleontologists began the attempt to discover the age of their fossil material by relative means, it became possible through twentieth-century physics to estimate its absolute age in years. A variety of techniques are in use now which have provided information concerning the duration of the periods and epochs of the geological calendar as well as the approximate age of specific fossils. All the methods of dating involve measurement of the decay of radioactive materials, minute amounts of which exist in both living and nonliving things.

The uranium-lead method   The first of these methods of dating, based upon the decay of uranium and thorium to lead, was proposed in 1905, within a decade of the discovery of radioactivity. Once it was understood that the time can be established, called the half-life of an element, in which half the atoms of the element are transformed from their original state by the spontaneous loss of particles and energy, physicists were able, by calculating the ratio of uranium or thorium to lead in a piece of rock, to estimate the number of years that had elapsed since the rock was formed. The first dates obtained by this method were relatively crude approximations of the age of the material tested, because it was not recognized that besides radiogenic lead the samples contained lead from other sources. Eventually, physicists found that the lead in their samples was a mixture of four isotopes, or forms with the same atomic number but different atomic weights. One isotope was nonradiogenic, two were bred by the decay of different isotopes

Era	Period and epoch, estimated years since each began	Principal animal life and changes	The environment
CENOZOIC — Age of Man	Quaternary Recent 20,000 Pleistocene 2,000,000 to 4,000,000	Rise of modern man; modern animals Primitive man; great mammals extinct at end	Postglacial; warm Alternate glacial cold and interglacial warm
CENOZOIC — Age of Mammals	Tertiary Pliocene Miocene Oligocene Eocene Paleocene 65,000,000	Some mammals decline Mammals at maximum; modern carnivores Rise of higher mammals Archaic mammals disappear; modern orders present Rise of archaic placental mammals and of later birds	Continents mainly above seas; Alps, Himalayas, other mountains rising; warm, gradually cooling; modern plants
MESOZOIC — Age of Reptiles	Cretaceous ("Chalk") 130,000,000	Small marsupial mammals and insectivores; toothed birds; great reptiles specialized, then disappear; first snakes; modern fishes and invertebrates; end of ammonites	Rocky Mountains and Andes forming; great swamps; cooling climates; rise of flowering plants
MESOZOIC — Age of Reptiles	Jurassic 180,000,000	First toothed birds; many giant reptiles; first lizards and crocodiles; modern sharks and bony fishes; many bivalves and ammonites	Continents near sea level; deserts widespread; many cycads
MESOZOIC — Age of Reptiles	Triassic 230,000,000	First mammals; rise of reptiles (ichthyosaurs, plesiosaurs, phytosaurs, pterodactyls, dinosaurs); end of armored amphibians; bony fishes in seas	Large rivers and flood plains locally; climate temperate to subtropical, some aridity; conifers and cycads
PALEOZOIC — Age of Amphibians	Permian 280,000,000	Primitive reptiles; decline of early amphibians; modern insects; last placoderms, trilobites, eurypterids, and other ancient marine life	Continued mountain building; widespread glaciation in Southern Hemisphere; climate various
PALEOZOIC — Age of Amphibians	Carboniferous (Pennsylvanian and Mississippian) 350,000,000	Amphibians dominant; first reptiles; insects suddenly common; brachiopods decline	Land low, sea over much of continents; mountain building at end; warm; seed ferns, lycopods
PALEOZOIC — Fishes	Devonian 400,000,000	First amphibians; many sharks and bony fishes; end of ostracoderms; trilobites decline; many mollusks; ammonites begin; first crabs and land snails	Much land submerged; some aridity on land
PALEOZOIC — Fishes	Silurian 450,000,000	Many ostracoderms; placoderms appear; scorpions and spiders, the first air breathers on land	Much land submerged; mountain building later; first land plants
PALEOZOIC — Age of Invertebrates	Ordovician 500,000,000	Ostracoderms, the first vertebrates; corals and reefs; worms, bryozoans, some uni- and bivalve mollusks, many nautilids, echinoderms, and eurypterids	Great land submergence
PALEOZOIC — Age of Invertebrates	Cambrian 570,000,000	First marine fossils, most invertebrate phyla present; trilobites dominant; many brachiopods; probably no land life	Many lowlands; mild climates
PROTEROZOIC	925,000,000	Probably soft-bodied invertebrates, but fossils scarce; sponge spicules, worm casts	Great sediments, then much volcanic activity, and long period of erosion
ARCHEOZOIC	4,500,000,000	Possibly unicellular plants and animals; no fossils	Much igneous activity, some sediments, later much erosion

of uranium, and one was produced by the decay of thorium. Investigators learned to subtract the nonradiogenic lead in their calculations and to obtain separate ratios for the transformations of each of the isotopes of uranium and thorium. By refining their methods in this way, physicists began to derive more accurate dates for the rocks they tested.

The use of uranium-lead ratios to determine absolute age has been helpful to paleontologists but is not a solution to all their problems of dating. Aside from the many sources of error which still detract from the accuracy of the method, the materials that can be sampled are limited. More often than not, rocks in which fossils are found are either devoid of uranium and thorium or for some other reason are not suitable for analysis. Geologists benefited from calculations based on the decay of another radioactive element, rubidium, to strontium; but rubidium in the proper form is also unusual in fossiliferous rocks. For the paleontologist, the discovery that one of the isotopes of potassium decays into the inert gas argon was an important one. Potassium is a common element, and although age determinations cannot be made on every sample that contains it, the potassium-containing minerals from which accurate dates can be ascertained are relatively numerous. A second advantage for the paleontologist is that the potassium-argon method can be used to measure the age of rocks that are under half a million years old, rocks too young to be dated by uranium-lead, thorium-lead, or rubidium-strontium analyses.

The potassium-argon method

The potassium-argon method is not without its limitations. Since argon is a gas, it may, under certain conditions, escape from the crystal lattice of the mineral in which it was produced by decay. If any of the radiogenic gas is lost, as it may be if the material that contains it is heated, the age of the sample will appear too young. Conversely, if the sample is contaminated by argon from the atmosphere and the contamination is not corrected for, the sample will seem older than it actually is. Materials which have gained or lost significant amounts of argon are among those on which age determinations cannot be profitably performed. Unfortunately, fossil bones and teeth fall into this category, as do most of the minerals in sedimentary rock. It is not impossible to measure the age of sedimentary rock

Figure 1.7. Geologic periods and the forms of life that characterized them. (From Storer.)

directly, however, because glauconite, a form of mica found in such rocks, is suitable for analysis.

Where the presence of glauconite does not render sedimentary materials useful in potassium-argon measurements, the age of the deposits and the fossils they contain can still be deduced from their association with rocks whose age can be determined. It can be assumed, for example, that sedimentary strata which have been penetrated by intrusions of molten rock are older than the intruding masses were when they cooled and became radiometrically datable. Sedimentary layers overlying the datable rocks or containing pebbles worn from them, on the other hand, must be some years younger. Geological conditions are not always conducive to these kinds of calculations, but when radiometric dating can be accomplished even indirectly, crucial information may be gained. By nonradiometric methods, for instance, the earliest members of the family of man found in Olduvai Gorge in East Africa were dated as nearly 1 million years old. When several samples of different minerals from the rocks immediately above and below the fossils proved by potassium-argon analysis to be between 1.6 and 1.9 million years of age, paleontologists learned that the hominid stock was far older than they had believed.

The carbon 14 method

Most fossil material is old enough to require radiometric dating through potassium-argon or uranium-lead analyses of the surrounding rock. Fossils in strata under 50,000 years old, which are too young to be dated reliably even by the potassium-argon method, can be dated directly by measuring the amount of carbon 14 they contain. Carbon 14, a radioactive isotope produced by cosmic-ray bombardment of nitrogen in the atmosphere, exists with other carbon isotopes in carbon dioxide in the air. Living organisms incorporate carbon 14 as they do nonradioactive carbon, so that the relative proportion of the various isotopes of carbon is the same in their tissues as outside. After death, when carbon exchange with the environment ceases, the radioactive isotope in the body decays and is not replaced. Knowing that the half-life of carbon 14 is 5,730 years, workers can calculate the time that has passed since the animal or plant died by measuring the amount of carbon 14 remaining in samples of the fossil tissue. Because materials to be dated by carbon 14 determination must be so recent, the method has been of greater use to archaeologists and anthropologists than to paleontologists. The latter have been able to use it, however, in fixing the time, tens of thousands of years ago, when great mammals like the ground sloths, the mammoths, and the mastodons became extinct.

Paleontologists who have found unknown fossils, dug them from the ground, prepared them for study, and determined the nature, the habitat, and the age of the animals they represent have still other difficult questions to answer. Before they can appreciate the significance of their finds, they must determine the place of the animals in the evolutionary scheme. From what ancestral stock did the animals spring? To which forms did they give rise? The speculations of paleontologists concerning these matters have provided material for all the chapters that follow.

2

The Origin of the Vertebrates

In the southwest part of England and in neighboring Wales, there lie exposed layers of rock that geologists agree were formed some 400 million years ago in Devonian times. Paleontologists know these deposits as the Old Red Sandstone and value them as one of the few formations in which the passage from late Silurian to early and middle Devonian sediments is recorded without major gaps occasioned by erosion. The lower layers were formed as silt and sand settled on the sea bottom. The invertebrate animals preserved in the oldest rocks are recognizable as relatives of modern genera which are still marine in habit. The sediments must have accumulated in the shallows of a subsiding sea, however, because the invertebrate fauna at the next higher level are a group characteristic of brackish, estuarine waters. Finally, the uppermost rocks of the Old Red Sandstone are composed of materials that were deposited in fresh water, dropped perhaps as the current slowed in rivers about to empty into the sea. It was in the rocks of this formation that paleontologists found, only a little over a hundred years ago, fossils belonging to the earliest known group of vertebrate animals.

Early vertebrate fossils in the Old Red Sandstone

The remains of these animals consisted of bony plates that could be retrieved from every level of the Old Red Sandstone. Despite the differences in the microscopic structure of the plates, all the hard material was recognizable as vertebrate bony tissue and quite distinct from the shells and exoskeletons of the invertebrates found in the same deposits. The first fossils found were fragmentary and gave few clues to the general form of the animals. As paleontologists looked elsewhere—in Scotland, Norway, and Spitzbergen and later in other places—they discovered remains that were less broken and scattered. It became obvious that these early vertebrates had been fish-like in shape. Some had been streamlined like free-swimming fishes of today; others were flattened and presumably bottom-dwellers. They varied in size from a few inches to more than a foot in length. All were encased in a bony armor which was so designed as to leave no doubt in paleontologists' minds that the animals lacked movable jaws and paired fins like those of later backboned forms. Because they shared these characteristics, the ancient vertebrates, or jawless fishes, were grouped together as ostracoderms, animals with a shell-like skin.

Continued study of fossil material made apparent the existence of many different types of ostracoderms. E. R. Lankester, who wrote a description of the forms dug from the Old Red Sandstone before 1868, divided the group into the Osteostraci and the Heterostraci. Thirty years later, R. H. Traquair reported the discovery of two additional kinds of ostracoderms, the Anaspida and the Coelolepida. Modern workers in the field continue to find this classification valid and are still investigating the structure of the animals in each of the four subdivisions.

Varieties of ostracoderms

That the Osteostraci are the type about whose anatomy most is known is due largely to the research of E. A. Stensiö. Since the early 1920s, he and his technical assistants in laboratories at the University of Stockholm have picked the rock away from the fossilized armor and scales of numerous specimens and ground hundreds of sections to determine the nature of the internal structures of the head in these forms. In all the Osteostraci the head is covered dorsally with a shield of bone ornamented over its surface by denticles (Fig. 2.1). The shield shows a characteristic pattern of openings and depressed areas: close to the midline in the center of the head or somewhat forward there lie a pair of circular holes through which, in life, the eyes must have stared

The Osteostraci

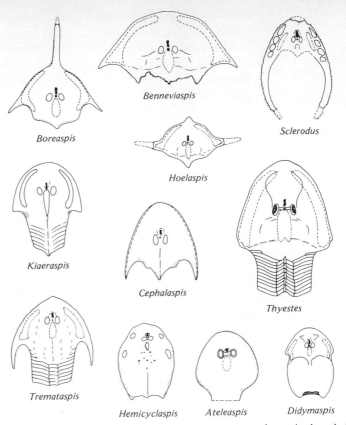

Figure 2.1. Head shields of various osteostracan ostracoderms in dorsal view. (From Gregory.)

directly upward. Between these eyes a small pineal opening is invariably present, and in front of it a single aperture for the naso-hypophyseal canal. Although the purpose of the four openings in the head shield is obvious and proved without a doubt by the anatomy of the underlying internal structures, the significance of the two elongated depressed areas on either side of the shield and of the median one behind the pineal opening is still not certain. There is a pavement of small tile-like scales in the floor of each depression and, beneath, a number of bone-lined channels which lead inward to the ear region of the brain. Stensiö concluded that the depressed areas held electric organs, but other workers have guessed instead that they housed sense-receptor cells which detected disturbances in the water. An aggregation of sensory structures on the dorsal side of the head suggests that the osteostracans may have swum on or near the bottom, and the fact that the underside of the head was flattened reinforces that possibility. The head shield lapped over to protect the edges of the ventral surface, the remainder of the area being covered by little scales fitted tightly together. Between the ventral tip of the head shield and the front of the scaly region lay a very small mouth, surely useless for ingesting any but the most minute particles. On each side of the mouth a curved row of external gill openings stretched to the posterior margin of the shield. In most of the Osteostraci, the shield was prolonged at its lateral or posterior edges into a pair of prongs, which may have helped to stabilize the animal as it swam. The stabilizing effect was enhanced, it seems, by paddle-shaped flaps, which projected behind the rigid prongs in several forms (Fig. 2.2). Posterior to the head shield, the body was covered more flexibly with rows of elongated scales. It is certain that these ancient vertebrates swam as modern fishes do by side-to-side movements of the tail, for not only was the body built to bend, but the caudal fin showed the same heterocercal design that still characterizes a number of fishes today (Fig. 5.8a).

Internal anatomy of the head region

Although paleontologists had early made the assumption that the fossil ostracoderms, so fish-like in form, were true vertebrates, Stensiö's revelations concerning the internal anatomy of the head region of the Osteostraci provided unequivocal evidence that the conclusion was correct. When the Swedish scientist made ground sections of the head shield, he found that the outlines of what was probably a cartilaginous endoskeleton were preserved by a very thin layer of bone. It was possible to discover the configuration of the spaces within the head skeleton and thus to know the shape of the soft organs which had filled them (Fig. 2.3). The brain housed by the cranium was similar to the brain of modern jawless, or agnathous, vertebrates: it was

Figure 2.2. Reconstruction of *Hemicyclaspis*, a member of the Osteostraci. (From Stensiö.)

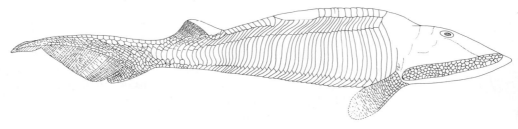

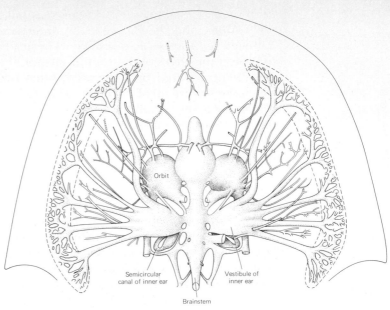

Figure 2.3. Cast of the cranial cavity, orbits, inner-ear regions, and interior canals of *Kiaeraspis*, an osteostracan, in ventral view. (From Stensiö.)

middorsal in position and subdivided into a forebrain, midbrain, and hindbrain. Connections between the forebrain and the eyes, nostril, and hypophysis were evident as well as a cerebellum atop the hindbrain and an internal ear with two semicircular canals on each side. Cranial nerves could be followed to the region of the gill pouches, as they can in any piscine vertebrate. The most peculiar structures were the large tubular passages extending from the cavity of the internal ear outward and upward to the depressed areas in the dorsal surface of the head shield. The endoskeleton that protected the brain was continuous with that which supported the gill pouches. The large size of the pouches, especially the anterior ones, seemed to confirm the idea that the animals must have been filter feeders, sucking water through the small mouth, straining it of organic matter, and expelling it over the gills and through the openings on the underside of the head. Despite the rigidity of the head shield, the ventral patch of fitted scales could have been flexible enough to allow the muscular contractions necessary to maintain such a flow of water. The thin layer of bone which made it possible to learn something about the structure of the gill pouches and the brain did not continue, apparently, posterior to the head shield. As a result, neither Stensiö nor any other investigator has been able to describe the internal skeleton that supported the trunk and tail of the Osteostraci. Doubtless, these animals, like all primitive vertebrates, had an unmineralized but turgid rod-like structure, called a notochord, beneath the nerve cord, but whether they possessed any skeletal elements in the vicinity of the notochord and the nerve cord is a question that cannot be answered at the present time.

Habitat and fate of the Osteostraci

Once the vertebrate nature of the Osteostraci had been determined, paleontologists sought to abstract from the fossil evidence some idea of the habitat and history of these primitive fishes. Even as they drew their conclusions, they remained aware of the small amount of material on which they had to base their thinking. Although distribution of these forms may have been worldwide, collectors have dug fossils only from northern Europe and widely separated sites in North America; and, of course, they can know nothing of the large number of osteostracans that were not preserved or whose fossilized remains eroded away. All the described forms but one seem to have lived in fresh or brackish water. The single marine form, *Sclerodus*, found in the Old Red Sandstone of England, may represent an aberrant subgroup or the only evidence remaining of a marine branch of the Osteostraci whose members swam over the sea bottom in other parts of the globe. The gradual extinction of the osteostracans is proved by the decreasing number of fossils found in middle and late Devonian rocks. They disappeared from the earth's waters at the end of the Devonian, undoubtedly because they were displaced by the more progressive fishes—forms with jaws and paired fins that were more efficient than the osteostracans and the other ostracoderms in swimming and food getting. The Osteostraci, in fact, seem to have been suffering in competition with the burgeoning jawed, or gnathostome, fishes throughout Devonian time. Gradually, they were restricted to relatively undisturbed areas at the bottom of streams and no longer frequented more open waters near the floor of lagoons and lakes. Although later osteostracans show some advance in the development of flaps and spines that would have increased their stability in swimming, there occurred a steady decrease in the length of the head shield that was ominous. A loss in dermal bone has been a sure sign of decline in several vertebrate groups.

Origin of the Osteostraci

The origin of the Osteostraci is much more mysterious than

their disappearance. An investigator who seeks to know the steps by which these forms came into existence finds himself confronting one of the most perplexing problems in the study of vertebrate evolution. The earliest known osteostracans, those found in Silurian deposits, show all the typical characteristics of the group. No one has found any older fossils whose structure is clearly antecedent to the osteostracan pattern. When a vertebrate group seems to burst upon the scene in this fashion, paleontologists can say only that they may possess fossils of the ancestral forms which they have not properly identified or that the progenitors of the group were too few in number or too fragile to leave remains that were likely to be found. The latter explanation is probably the true one for the Osteostraci. Early forms of the head shield may have been thinner than the known ones—so thin that they crumbled away after the death of the animals that produced them. It is also possible that the evolution of the Osteostraci may have taken place relatively rapidly, the oldest recognizable forms having developed from a small ancestral stock which inhabited a restricted area. Although paleontologists may find many specimens of the successfully established forms which gained a wider geographical distribution, there is a strong probability that the resting place of the rarer ancestral types will never be discovered.

The Anaspida   There is only one known group of ostracoderms, the Anaspida, that seems related to the Osteostraci (Fig. 2.4). The forms it includes were contemporaries of the osteostracans and thus are surely cousins rather than parents of the latter. Researchers trying to assess the degree of relationship between the anaspids and osteostracans must base their conclusions upon a comparison of the external anatomy of the two types of jawless fishes, as the internal structure of the anaspids is unknown. The chief reason for postulating kinship between the two groups lies

Figure 2.4.   Reconstruction of *Rhyncholepis*, an anaspid. (From Kiaer.)

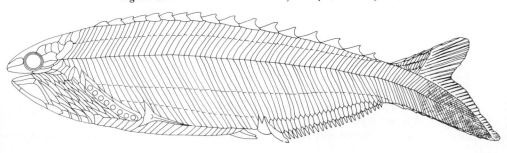

in the anaspids' possession of the dorsally located and character-istically arranged single nostril, pineal opening, and pair of eyes. Although the eyes are somewhat farther apart in the anaspids than they are in the osteostracans, the similarity of the pattern is unmistakable.

Mode of life of the anaspids

The remaining structural characteristics of the anaspids sug-gest that although they too seem generally to have inhabited brackish or fresh water, their mode of life in that environment was quite different from that of the osteostracans. The anaspids were not flattened bottom-dwellers. They possessed a stream-lined body that achieved some stability through a projecting ridge of dorsal spines and in some cases a ventrolaterally placed pair of elongated folds. The caudal fin was peculiarly designed to give an upward thrust to the head as the anaspid swam: the lower of its two lobes was the longer and was probably stiffened by the extension into it of the axial supporting element of the body; the upper lobe, shorter and more flexible, would have driven the tail downward when the fish moved forward. Such a caudal fin may have substituted effectively for a pair of pectoral fins in elevating the head and so in keeping the anaspid away from the bottom. The terminal position of the mouth suggests that anaspids may have remained just under the surface, sucking in algae which floated in the sun-lit water. The body must have been fairly flexible, for fossil remains show it to have been covered by a mosaic of very small plates in front and, behind the downward-curving row of gill slits, by rows of narrow, vertically set, bar-shaped scales. If the muscles were as well developed as the shape of the trunk and tail lead one to guess, the anaspids may have been relatively strong and quick in their movements. Among the anaspids there were forms that showed reduction and even disappearance of the dermal plates and scales over much of the body. With what adaptation or change in habits this loss was associated, paleontologists may never know.

The Heterostraci

The third group of ostracoderms, in contrast to the first two, contained a number of marine forms (Fig. 2.5). In fact, the Heterostraci, as these ostracoderms are called, must have radi-ated widely and lived successfully in a broad range of watery habitats. Many specimens have been recovered from sediments deposited in shallow seas, yet in every family of these jawless fishes there seem to have been freshwater representatives. To what special abilities the heterostracans owed their success is not clear: they were no better equipped for locomotion than other

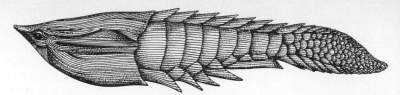

Figure 2.5. Reconstruction of *Anglaspis*, a heterostracan. (From Kiaer.)

ostracoderms. Like the anaspids, they relied upon a hypocercal caudal fin for lifting the head in swimming, but their projections from the body to restrict their rolling while moving forward were even less extensive. Some forms possessed a single, long dorsal spine and a pair of lateral ones; other types had no more than ridges of sharp, elevated scales to steady them. Most heterostracans, from the primitive cyathaspids to the more advanced pteraspids, seem from the cylindrical shape of the body to have been swimmers in open water. Only one subgroup, the Drepanaspidae, prominent in Devonian times, showed the dorsoventral flattening that indicates the bottom-dwelling habit of life. The body in all the Heterostraci was covered in front by a sheathing of dermal plates and in back by large, overlapping or small, closely fitted scales. The presence of a row of scales at the edge of the mouth has caused some paleontologists to speculate that heterostracans may have been able to nibble at food rather than merely to draw in organic bits with the respiratory current. If that were so, the Heterostraci may have enjoyed some slight advantage over other types of ostracoderms in utilizing materials in the environment as sources of food.

Paleontologists agree that the Heterostraci struck out on a separate evolutionary path from that followed by the Osteostraci and the Anaspida. One indication of the divergence is the difference in arrangement of the openings in the head shield. Absent are the close-set eyes with single nostril and pineal aperture between them on the middorsal line. Instead, heterostracans show what may be the first appearance of the pattern that was to characterize later vertebrates: the eyes are situated more laterally, the pineal opening is middorsal but not invariably present, and a single nostril is nowhere apparent. The location of the nasal aperture has not been proved to the satisfaction of all paleontologists, but most researchers are of the opinion that the nostrils were paired and opened at each side of the mouth. Stensiö and his colleague Jarvik insist, however, that the heterostracans retained a single nostril anteriorly on the head. The head shield itself differed in its construction from that of the Osteostraci. The heterostracan covering contained no bone cells and was, in early forms, composed basically of four plates, one dorsal, one ventral, and, between them on each side, a branchial plate covering the gill area. Water passing over the gills emerged from a single opening behind the branchial elements rather than through separate holes in the ventral part of the shield, as in most of the Osteostraci. That there was an evolutionary trend toward reduction in the solidity of the dermal head covering in heterostracans is indicated by the increasing subdivision of the plates in later forms.

The Coelolepida   Similar to the Heterostraci in body form and in lack of the osteostracan-anaspid arrangement of eyes and median nostril are the mysterious ostracoderms, which have been classified as the Coelolepida (Fig. 2.6). Paleontologists know little more about them than that they did exist. Isolated scales have turned up in rocks of late Silurian and early Devonian age as well as a few whole specimens which have revealed nothing beyond the general shape and size of these forms. They were small, apparently flattened, and covered with tiny spined scales. Their fossils come from deposits laid down in brackish waters. Presumably they swam in company with other forms of ostracoderms in bays where fresh water from rivers diluted the salty sea. Since their structure is poorly known, it is almost

Figure 2.6. Reconstruction of *Phlebolepis,* a coelolepid. **(a)** Lateral view; **(b)** dorsal view. (From Ritchie.)

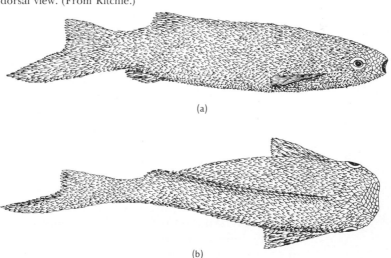

(a)

(b)

impossible to speculate upon their relationship to the other jawless fishes. Although they may simply represent a separate radiation from the ancient base of the ostracoderm stock, some paleontologists prefer to regard the coelolepid group as a degenerate offshoot from heterostracan stock. They emphasize the likeness in shape and the lateral position of the eyes in the two forms and point out that the evolutionary trend toward fragmentation of the heterostracan dermal elements, if continued to an extreme point, could well have resulted in a covering of fine scales like that of the Coelolepida. Westoll, however, suggested that coelolepids might be incompletely ossified young heterostracan individuals.

Origin of the vertebrates an unsolved problem

The possibility of a relationship between the Heterostraci and the Coelolepida is but a small part of the problems that paleontologists would like to solve. What remains to be understood is the early evolution of the entire ostracoderm assemblage and the origin in even more ancient times of the vertebrate line from its nonvertebrate ancestral source. Difficulties of such magnitude exist, however, in the study of these matters that investigators have had to confine themselves to defining the problems and to building theories on the small amount of evidence they do have. Although nonscientists sometimes accept current theories as the answers given by science to these questions of universal interest, students of evolution understand them as starting points for research.

The earliest known ostracoderms

The earliest known evidence of ostracoderm existence consists almost entirely of fragments of dermal bone that show similarities in microscopic structure to heterostracan dermal plates. These plates of bone are present in rocks of Middle Ordovician age in formations exposed in South Dakota, Wyoming, and Colorado. There is one deposit of vertebrate material that is apparently older, a group of small denticles of early Ordovician date found at a site near Leningrad. Since the denticles have no distinctive characteristics relating them to a particular type of ostracoderm, it must remain only an assumption that the vertebrates that produced the dentin and enamel-like material were ostracoderms of some kind. The assumption seems a safe one, however, because the discovery of an impression of a head shield of a heterostracan of Ordovician age has demonstrated the existence of differentiated forms as early as the middle of the period. All the Ordovician fragments have been made the basis of genera: the Russian fossils are classified as *Paleodus* and *Archodus*, the North American ones as *Astraspis* and *Eriptychius*.

These early vertebrate fossils, although they bear witness to the existence of ostracoderms in the Ordovician period, do not afford much help to paleontologists who are trying to discover the origin of the jawless fishes. Although some of the fragments of dermal bone are at least an inch long, they are not large enough to give any idea of what *Astraspis* and *Eriptychius* looked like. If one assumes from the histological evidence that the animals were recognizable as heterostracans, it follows that this Ordovician material, even should more of it be found, can reveal little about the transition to the ostracoderm level from a preexisting lower line.

What the earlier history of the vertebrate group was and why no vertebrate fossils have been found in pre-Ordovician rocks, paleontologists can only guess. It may be that the first dermal ossifications were so thin that none has been preserved in recognizable form. Possibly, dermal bone appeared with comparative suddenness, after the organization of the vertebrate body was well established, as the result of a small but crucial change in the chemistry of the deep layers of the skin: research into the question of hard-tissue formation has shown that ossification will take place in connective tissue when a certain chemical equilibrium is reached. The vertebrate pattern of soft organ systems may have been evolving for millions of years before the relatively modest genetic mutation occurred which was responsible for realizing the biochemical balance necessary for the deposition of bone. If, as this theory suggests, the dermal skeleton was among the last of the ostracoderm structures to evolve, the earliest vertebrates would have been soft-bodied and thus rarely preserved after death. A paleontologist might expect, as a stroke of good fortune, to discover an impression of such a soft animal but no more.

In the absence of known vertebrate fossils from rocks older than those of Ordovician time, it has been concluded that the backboned animals were the last of the great groups to evolve. Representatives of virtually all the invertebrate phyla appear more than 100 million years earlier in Cambrian deposits, in the oldest strata sufficiently undistorted by the forces of heat and pressure to yield significant amounts of fossil material. Although so large a time gap exists between the first appearance of vertebrate and invertebrate organisms, zoologists and paleontologists have assumed that the former must have been derived from the latter and have speculated widely about the most logical or probable nonvertebrate ancestor for vertebrate animals.

Some nineteenth-century biologists, observing what they considered to be similarities between the anatomy of living, adult vertebrates and invertebrates, suggested that vertebrates evolved from annelid worms or from arthropods. Some hypotheses entailed turning the lower forms upside down to explain the dorsal location of the vertebrate nerve cord and the ventral position of the heart. The weakness of these ideas was soon apparent: aside from the improbability of a dorsoventral reversal, it was obvious that the double, solid invertebrate nerve cord was not translatable into the single, hollow vertebrate structure and that neither the annelid nor the arthropod blood-pumping devices could give rise to the vertebrate heart. In fact, every part of these theories of vertebrate origin disintegrated under examination.

Those biologists who turned for clues to the vertebrate characteristics that appear early in the embryo produced more viable theories. In accordance with the general concept of Ernst Haeckel that the developmental steps in the formation of an individual give some indication of the evolutionary history of the group to which it belongs, these biologists focused their attention upon the establishment of the three primary layers of cells in the embryo (the outer ectoderm, the inner endoderm, and the central mesoderm) and the first structures to arise. The discovery of similarities in the origin of the mesodermal layer in echinoderms and vertebrates and of the distinctly different method of its origin in other multicellular animals gave rise to the possibility of an evolutionary link between these two groups. This idea has proved difficult to pursue, however, because all extant echinoderms develop a radial symmetry and a semisessile habit that are far distant from the vertebrate condition. A connection between echinoderms and vertebrates is not impossible on this account, but to maintain it, one must assume that a protovertebrate group diverged from the echinoderm line in pre-Cambrian times before the development of the specialized, radially symmetrical forms.

The vertebrates as chordates

Relationship to two, and possibly three, other groups of nonvertebrate organisms has been postulated on the basis of the appearance in vertebrate embryos of three structures, the rod-like notochord, which develops in the dorsal midline; the tubular medial nerve cord, that forms above it; and a series of paired, laterally placed clefts in the body wall, which allow passage to the outside from the pharyngeal region of the gut (Fig. 2.7). These structures were discovered in the little boneless water-dwelling lancelet *Branchiostoma*, better known as amphiox-

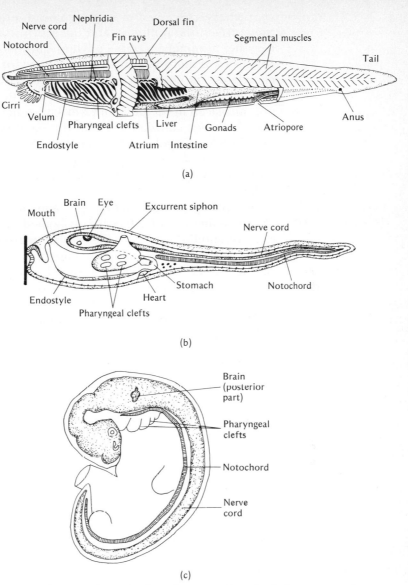

Figure 2.7. The three chordate characters (dorsal tubular nerve cord, notochord, and pharyngeal clefts) as seen in **(a)** *Branchiostoma* (= *Amphioxus*); **(b)** larval tunicate; **(c)** vertebrate embryo (*a,b* from Storer; *c* after Ballard).

us, and also in the larvae of tunicate forms. A suggestion of their presence was found in a third group, the acorn worms and their relatives. The common possession of the three anatomical characteristics seemed of such significance that the vertebrates, the lancelet, the tunicates, and the acorn worms were classified together as members of the phylum Chordata. Since no one has suggested a more logical alliance for any of the nonvertebrate chordates, the group has been maintained in the classification scheme. Although some biologists believe that the acorn worms are misplaced in the phylum Chordata, others see these animals and the other chordates as constituting a natural group whose members have evolved from a common stock. The latter workers have formed a theory of vertebrate origin based on descent from primitive forms related to the lower chordates existing today.

The evolution of the early chordates

According to this theory, the animals at the base of the chordate line must have been soft-bodied, sessile forms of small size, resembling perhaps the minute pterobranch relatives of *Balanoglossus*, the acorn worm. These hypothetical forerunners of the chordates may have trapped food particles (as pterobranchs still do) by waving soft, unjointed appendages in the surrounding water. One group of pterobranchs exhibits a pair of slits in the body wall, through which excess water, ingested with food material, finds its way out of the gut (Fig. 2.8). The existence of related animals some of which have food-collecting appendages and no openings through the body wall, some of which have both, and some of which (like the acorn worms) have only a series of paired clefts has given rise to the thought that a transition from one type of structure to the other may have occurred early in the evolution of the chordate line. With the change in structure would have come a change in the way food particles were collected: in the absence of oral appendages the animals would have had to draw food-laden water into the gut and expel the water, strained of its organic material, through the paired openings in the body wall. This idea is supported by the fact that filter feeding, as this process is called, is characteristic of all the lower chordates and was also used, paleontologists are quite sure, by the ostracoderms.

If the filter-feeding acorn worms are not degenerate but exhibit a very primitive state of chordate organization, one must conclude that the paired openings from the gut to the outside, generally called gill slits, were the first of the chordate characteristics to evolve. The other two, the notochord and the dorsal hollow nerve cord, are poorly expressed in acorn worms. A short rod which stiffens only the proboscis of *Balanoglossus* is hesitantly accepted as representing the former and a concentra-

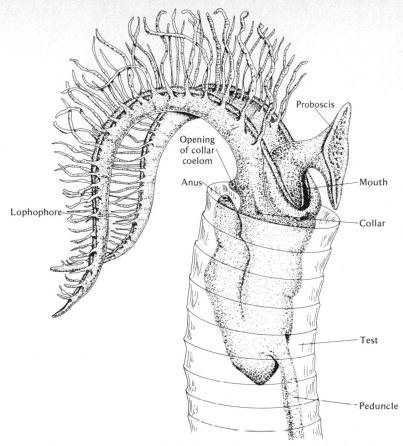

Figure 2.8. A pterobranch, *Rhabdopleura*, showing one member of a colony, connected by a test-encased peduncle to other individuals. In another genus of pterobranchs, there exists one pair of pharyngeal clefts posterior to the collar. (From Torrey.)

tion of nervous tissue in the dorsal region of the anterior collar some development of the latter. The incomplete nature of the typical chordate structures in these animals is of ambiguous significance (Fig. 2.9). Either the acorn worms represent a group that had gone half the distance to full chordate status—and that belief is evidenced in the assignment of the name Hemichordata—or they are actually unrelated forms whose anatomy has developed superficial similarities to that of chordates. If the latter is true, of course, those who suppose the chordates to have arisen from pterobranch-like animals are on the wrong track.

Whether they are correct or not, it does seem possible that gill

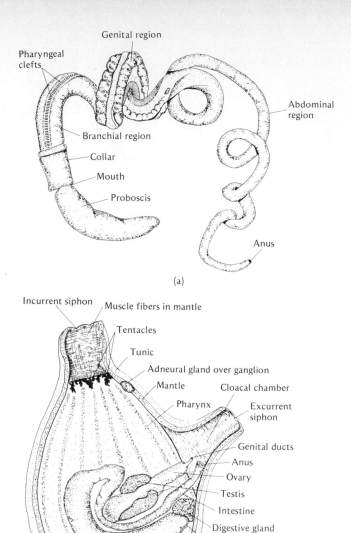

Genital region

Pharyngeal
clefts

Abdominal
region

Branchial region

Collar

Mouth

Proboscis

Anus

(a)

Incurrent siphon — Muscle fibers in mantle

Tentacles

Tunic

Adneural gland over ganglion

Mantle

Cloacal chamber

Pharynx

Excurrent
siphon

Genital ducts

Anus

Ovary

Testis

Intestine

Digestive gland

Esophagus

Stomach

Position of endostyle

(b)

Figure 2.9. Adult invertebrate chordates in which the typical chordate characters are incompletely expressed. **(a)** *Dolichoglossus*, a hemichordate with pharyngeal clefts but a ventral nerve cord as well as a dorsal one and a small
notochord-like structure confined to the base of the proboscis. **(b)** *Molgula*, a
sessile tunicate possessing pharyngeal clefts but only a residual nervous ganglion
dorsally and no remnant of the notochord which exists in the larva. (From
Weichert.)

slits which functioned in the collection of food and so had a vital role in sustaining life may have been the earliest of the chordate structures to become firmly established. In the tunicates, assigned to the subphylum Urochordata, only the gill slits are present throughout the life of the animal. The notochord and dorsal tubular nerve cord in these marine forms exist during the larval period, disappearing during the metamorphosis which produces the sessile or free-floating adult. They compose the bulk of a muscular tail by means of which the larvae may swim some distance from the parent animals before assuming their vegetative existence. Those who regard the urochordates as modern representatives of animals ancestral to vertebrates suggest that the role of the notochord and nerve cord in tunicates is primitive and witness to the fact that these structures arose as adaptations for dispersal of the young. If mobility proved to be a decided advantage, allowing individuals to avoid injury or to follow the food supply, it is possible that animals that retained their larval form longest survived more often to reproduce their kind. In time, a population characterized by members having a prolonged juvenile stage and a relatively short, nonmotile adult life would have evolved. If, then, paedogenesis took place, that is, reproduction when the animals were still in the larval state, the degenerate adult form could have disappeared entirely.

Relationship of amphioxus to vertebrate ancestors

Had the evolution of chordates taken place in this way, the earliest of the animals to possess all three of the typical characteristics as well as the segmented axial muscles responsible for locomotion might have resembled the modern amphioxus. Although amphioxus has had a long history of its own, it may have carried into the present slightly modified versions of the primitive notochord and nerve cord, the former extending from the anterior tip of the body into the tail and the latter having at its anterior end a slight enlargement presaging the brain. The common ancestor of amphioxus and the vertebrates, if it existed as this theory suggests, may have inhabited shallow places, filter feeding in quiet water, and swimming away from invertebrates that chanced to disturb it. Beyond this supposition theorists cannot go. Although they have based their reasoning carefully upon the evidence of comparative anatomy and relied upon the phenomena of variation and natural selection as agents of evolutionary change, they cannot offer the fossil evidence that would substantiate their ideas.

As paleontologists are unable to trace the progression of forms from the invertebrate level through the earliest chordates to the bone and scale-covered ostracoderms, they have tried to determine at least the nature of the watery environment in which the evolution of the jawless vertebrates took place. In this problem, there is some pertinent paleontological and geological evidence. Nevertheless, the question at hand—whether the vertebrates arose in fresh or salt water—is far from settled. By late Silurian time, it appears, ostracoderms were living in both marine and freshwater environments. It is their habitat in the earlier Ordovician period that assumes crucial importance. Among the sedimentary rocks of that age which are now accessible, few were deposited in bodies of fresh water or on floodplains of ancient continents. Waters that fill depressions or run through furrows in land masses are relatively shallow: minor changes in level serve to elevate old bottoms and expose the accumulated sediments to the forces of erosion. The small number of Ordovician continental sediments which have survived continue to be searched, but undoubtedly much fossil material from the period has long since disappeared. Marine deposits of Ordovician age and older are more common than continental deposits. Although many of these rocks have proved to be fossiliferous, the absence of vertebrate remains in them is striking. Paleontologists who maintain that vertebrates originated in fresh water point to the lack of vertebrate fossils in early marine rocks as evidence in favor of their argument. Those on the opposite side in the debate believe that the fossils exist in such deposits but have not yet been found. The advocates of freshwater origin do not accept this reasoning as valid. Why, they ask, should vertebrate fossils have been completely missed in pre-Silurian marine sediments when they are found easily in post-Silurian ones? Possibly, comes the reply, because the first vertebrates were so few in number that traces of their existence are extremely rare: one might expect such fossils to remain undiscovered. The absence of vertebrate remains in material collected from marine beds of early Paleozoic age constitutes a piece of negative evidence. Those who favor the idea of vertebrate origin in the sea are quick to point out that negative evidence is not conclusive in any argument, and their opponents are forced to agree.

Paleontologists who believe that vertebrates developed in fresh water have received backing from another quarter. Homer Smith, a renal physiologist, argued that the vertebrate kidney is basically a water-excreting, salt-conserving organ that arose in a group of animals establishing itself in a freshwater habitat. In such a situation, the environmental water is hypotonic to the

Habitat of the first vertebrates

Smith's argument for the origin of vertebrates in fresh water

Vertebrate History: Problems in Evolution

body fluids of the organism. Vital salts are lost from the tissues by diffusion through the semipermeable epidermis, and water tends always to seep into the body by osmosis. Any animal which lives in fresh water must be able to pump out the excess water and prevent the loss of its salts. Vertebrate kidney tubules are built expressly to perform these tasks. The glomerulus at the head of each tubule is a water-expressing structure, and the cells of the tubule walls are specialized to resorb necessary ions from the dilute filtrate passing outward. Smith was sure that kidney tubules with large glomeruli were primitive and arose with the vertebrate body itself because they are widespread throughout the vertebrate line. To Smith it seemed obvious that the earliest vertebrates must have been freshwater forms. Their descendants migrated to salt water and eventually to land as modifications arose in the structure of the kidney tubules, modifications that restricted the volume of water excreted. Marine fishes, unable to live unless they prevent loss of water to the hypertonic environmental fluid, show tubules without glomeruli. Land animals, whose bodies are exposed to the drying air, either have glomeruli reduced in size or develop kidney tubules with special water-resorbing segments. Smith's ideas do not discourage the workers who continue to postulate a marine cradle for the vertebrate line. They argue that animals living in sea water hypotonic to their body fluids would require, like freshwater forms, kidney tubules which pumped out water and saved salts. They agree with Smith that large glomeruli characterized the primitive kidney but maintain that such a kidney could have evolved in dilute salt water as readily as in fresh.

Rebuttal by advocates of a marine origin

Positive evidence should come from a study of the Ordovician fossils that are available. However, paleontologists and geologists who have analyzed the sediments in which the oldest pieces of dermal bone were found have produced, not a definitive answer to the question, but testimony which has either been disputed or rationalized by theorists on both sides. The presence in the sediments of kinds of invertebrates which live in shallow marine habitats—brachiopods, pelecypods, and gastropods, for example—has led some investigators to suppose that the Ordovician ostracoderms lived in sea water. Others caution against jumping too quickly to this conclusion on the basis of a general inspection of invertebrate fossils present in the rock formation. Seas advance and ebb within relatively few millions of years, leaving behind marine and continental deposits interlayered in a complicated pattern. One must be sure that the marine invertebrates are discovered at the same location and in the same stratum as the ostracoderm bone.

Interpretation of fossil evidence for the habitat of the first vertebrates

Even supposing arbitrarily that the vertebrate fossils were buried in sand under brackish water at the sea edge, however, does not solve the problem. These remains may represent early forms that were newly adapted to marine life and not at all typical of the larger number of vertebrate animals of the period. It is possible, too, that the assumption that the ostracoderm fossils are remnants of marine forms is itself false. The living animals may have inhabited rivers or streams and been washed seaward after death. Advocates of a freshwater habitat for Ordovician vertebrates have proposed this interpretation. It can be put to the test, because the condition of fossil material gives some clue to whether transportation of the dead animals occurred or not. An animal which dies, sinks, and is quickly buried is often preserved with a minimum of disarticulation of its skeletal elements. Dislocated or partial remains, however, do not prove that transportation from another area took place, because dead animals may be broken apart by scavengers and their bones scattered over the bottom. Fossil fragments that have been transported often show broken edges worn smooth or sorting of materials, as larger pieces settle out of a current sooner than smaller ones. The rocks in which such fossils lie sometimes reveal a pattern of ridges which were formed as water currents flowed in ripples over the mud. With these factors in mind, investigators have attempted to make a judgment concerning the source of the Ordovician ostracoderm fossils. They have found the pieces of dermal bone to be somewhat worn but not distinctly sorted. Some workers believe that the large quantity of bone found in the Harding Formation in Colorado makes it unlikely that the material was transported from a distance. Others think that accumulation of bone fragments in shallow bays where rivers emptied should still be considered as a possibility. The discussion goes on. For the present, at least, the environment of the first vertebrates, like the identity of their ancestors, remains unknown.

The fate of the ostracoderms

About the fate of the early jawless vertebrates paleontologists are in agreement. The fossil record shows that the ostracoderms disappeared at the end of the Devonian period. In the last 60 million years of their existence they shared their environmental niche with new forms of vertebrates whose ability to swim and feed was superior to their own. The more advanced fishes which arose in the Devonian possessed hinged jaws and could engulf relatively large objects. Although it is possible that some of them sustained themselves on plant material and small invertebrates, others must surely have preyed upon the defenseless ostracoderms. The jawless forms which survived in the presence of these powerful natural enemies may have adopted habits of concealment or developed protective coloration. Their numbers dwindled, nevertheless, and extinction followed.

Since the ostracoderms were the first vertebrates, it is thought that they must have given rise to the jawed forms which succeeded them. No series of fossils has appeared, however, to testify how the transition occurred. The known ostracoderms seem to have been too specialized in their structure to have served as ancestral stock for any of the primitive jawed fishes. Most of the jawed forms themselves are far from generalized types when they first appear on the scene. In the absence of fossil evidence, paleontologists have been cautious in their speculations: they cast the osteostracans and anaspids out of the direct line toward gnathostomes because the peculiar arrangement of eyes and pineal and naso-hypophyseal openings characteristic of these ostracoderms seems unlikely as a basis for the arrangement of the comparable structures in jawed vertebrates. The Heterostraci emerge as a more probable source, in their minds, only because these jawless forms show eyes located more laterally and are generally presumed to have had paired nostrils situated near the mouth. Since all known heterostracans demonstrate peculiar structural features of one sort or another, paleontologists do not regard any of them as the ancestral form they are seeking. If the ostracoderm ancestor of the jawed fishes ever comes to light, many workers believe that it might prove to be a heterostracan-like form of a very primitive type.

Relationship of the ostracoderms to the first jawed fishes

There is some evidence that, besides giving rise to more advanced forms, the ostracoderms may have left behind a group of jawless descendants. The extant cyclostomes—the hagfishes and lampreys—show, side by side with specialized structures, certain characteristics that are similar to those of the ancient jawless fishes (Fig. 2.10). The ammocoete larvae of the lamprey are filter feeders, sucking water from the stream bottom into the mouth and expressing it through seven pairs of gill slits. At metamorphosis, modifications in the oral anatomy take place which prepare the animals to assume the parasitic habit peculiar to the adult. If one regards the specialized mouth structures which appear late in the development of the individual as relatively recent evolutionary occurrences connected with deg-

Modern descendants of the ostracoderms

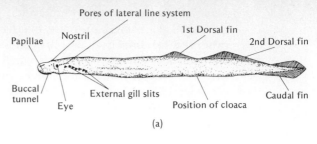

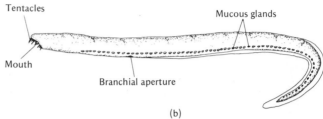

Figure 2.10. Modern jawless fishes: the cyclostomes. (a) The lamprey *Petromyzon*; (b) The hagfish *Myxine*. (From Weichert.)

radation to parasitism, it is possible to suppose that lampreys are derived from ostracoderm stock.

The absence of jaws and paired fins in the cyclostomes is surely primitive: no anatomical or embryological evidence exists to suggest that lampreys or hagfishes are descendants of forms that once had these structures. Between the eyes lampreys have a naso-hypophyseal opening and a pineal organ whose arrangement on the dorsal surface of the head is reminiscent of that of osteostracans and anaspids. In fact, the appearance of these structures and the possession of two rather than three semicircular canals in the ear region have led investigators to suggest that the lampreys are modern representatives of the osteostracan-anaspid line. This reasoning is supported by the appearance of an extraordinary fossil form called *Jamoytius* (Fig. 2.11). It was found first in Silurian rocks in Lanarkshire, Scotland, by an amateur collector who sold his two specimens to the British Museum (Natural History) in 1914. Both consisted only of a carbonaceous film on the surface of rock fragments, but certain important structures were decipherable: the animal, about 6 inches in length, had large eyes set on either side of a terminal oval mouth, a pair of lateral fin folds, and segmental, vertically arranged markings on the body that were assumed at first to represent well-developed but primitively constructed myo-

Jamoytius

meres. E. I. White, observing these features in *Jamoytius* and noting the absence of bony dermal plates, stated that the fossil was evidence at last of the primordial ancestor of the higher chordates. He believed that *Jamoytius* was a conservative descendant of the stock that gave rise to the first vertebrates and the cephalochordate line to which amphioxus belongs.

The finding of new specimens of *Jamoytius* resulted in interpretations of the fossil that differed from White's. Although no one denied that *Jamoytius* had bulky axial musculature, the surface markings that had seemed to White to be segmental myomeres were recognized in the later specimens as scales similar in arrangement to those of anaspids. They were peculiar in being thin, unossified, and flexible. In each segment of the body, one scale extended from the middorsal to the midventral region without subdivision. The arrangement of the scales, eyes, mouth, and lateral fin folds and the structure of the branchial apparatus (which had not been visible in the first two specimens) led Ritchie to identify *Jamoytius* as an anaspid that was allied to a group transitional to cyclostomes. In support of this conclusion, Ritchie remarks that the mouth of *Jamoytius* may even have been partially specialized for the type of feeding characteristic of the Recent jawless fishes. He reasoned that the ovoid mouth was too small to have allowed sufficient intake for the filter feeding of so large an animal; he speculated that it might have served instead as a scraper, enabling *Jamoytius* to remove algae from rocks. Later forms might have progressed from this type of feeding to scraping organisms from the surface of other animals and finally to puncturing the skin and subsisting on flesh and blood, as adult cyclostomes do today.

Although Ritchie observed that *Jamoytius* shows closer affinity to lampreys than to hagfishes, he concluded that the anaspids were ancestral to both groups of cyclostomes. Other paleontologists have taken a different point of view. Stensiö suggested that

Figure 2.11. Reconstruction of *Jamoytius*, a Silurian form which may be an anaspid distantly related to modern cyclostomes. (From Ritchie.)

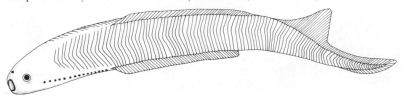

the osteostracan-anaspid line of ostracoderms gave rise to the lampreys alone and that the hagfishes were derived independently from the heterostracans. He based his argument in part on his opinion that the heterostracans, like the hagfishes, had a single median nostril. This theory has found little favor with the majority of workers, who think it more likely that the Heterostraci had a pair of nostrils and were closely related to the forebears of the jawed fishes. The single known fossil cyclostome from the Paleozoic era helps little to settle any of the questions that have been raised: *Mayomyzon* (Fig. 2.12) from the Upper Carboniferous is very similar to *Petromyzon*, the modern lamprey. Zangerl and Bardack, who described *Mayomyzon*, discovered in its structure no new evidence corroborating an anaspid-lamprey transition and nothing that might make clearer the relationship between the ancient lampreys and the hagfishes.

Mayomyzon, a
fossil cyclostome

Figure 2.12. *Mayomyzon*, a lamprey of the Carboniferous period, fossilized in an ironstone concretion. The head is at left with an annular cartilage around the mouth and the dark, round eye clearly visible. Vertical markings immediately to the right of the eye and below it are gill pouches. (From Zangerl and Bardack.)

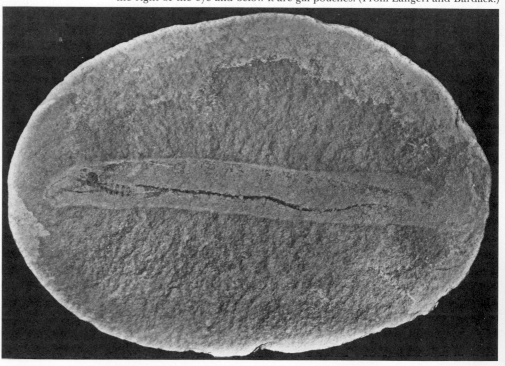

Loss of bone in
the evolution of
modern jawless
fishes

In postulating a phylogenetic connection between the ostracoderms and the cyclostomes, paleontologists are not ignoring the presence of dermal bone in the older animals. The lack of bone and denticles in all living jawless fishes does not preclude their origin from the heavily armored extinct forms. The gradual reduction and loss of bone has occurred in several vertebrate lines as they passed their peak and entered into a decline. The cyclostomes are obviously the last members of a once more numerous group. The basic cartilaginous structures of the braincase and closely associated visceral skeleton characterizing the few existing genera may well be all that remains of the protective skeletal complex housing the brain, sense organs, and branchial pouches within the head shield of the ostracoderms that flourished in early Paleozoic times. Unfortunately, there is no way to verify the theory that has sprung from studies in comparative anatomy, because paleontologists have found, except for *Mayomyzon*, no fossilized jawless fishes in rocks formed between the Devonian period and the current age. They have just enough fossil evidence to establish the nature of the ostracoderms at the time of their greatest success but not enough to ascertain either how they gave rise to later forms or even how, originally, they came into being.

3

Bone and Cartilage in Early Vertebrates

Mineralized tissue in many fossils

Anyone who becomes interested in tracing the course of vertebrate evolution inevitably finds that he must devote considerable study to the nature of the hard tissues developed by the backboned animals. Although some of the evidence that falls into the hands of paleontologists consists of imprints or fossilized soft tissues, by far the largest amount of material is bone, tooth, scale, and spine—structures that were mineralized in the living organism. The collector must be able to distinguish these vertebrate remains from hardened elements produced by other kinds of animals: the invertebrates that shared watery habitats with the fishes left exoskeletons and shells which sank into the sediments and were often preserved in close association with vertebrate materials.

Its structure is important to paleontologists

The distinction between hard tissues of vertebrates and invertebrates is made on the basis of structure. When a whole element is found, a glance at the general design of the piece may be sufficient, but fragmentary remains, especially very small ones, must frequently be studied microscopically before a decision can be made concerning their source. To those who imagine paleontologists as men whose proper tools are measuring tapes rather than microscopes it should be pointed out that the structure of fossils comprises a continuum from gross form to the arrangement of minute crystals. An investigator looks for significant clues to the identity and phylogenetic relationship of a specimen at every structural level and uses whatever he needs to do the job—eyes, lenses, or x-rays.

An increased amount of fossil material and a more thorough study of its hard-tissue components with twentieth-century instruments and techniques have already forced the revision of *An old theory* one long-honored theory of early vertebrate evolution. It was *about hard tissues* believed for many years that the development of the bony *in vertebrates* skeleton was a gradual affair, that the first vertebrates were cartilaginous fishes, and that the ancient bony fishes were the bearers of a new, more advanced type of skeletal material. This idea seems to have originated with those who first ranked the animals in a series from lower, or less perfect, to higher, or more perfect, forms. The biologist Lamarck was merely expressing a thought already current in scientific thinking in Europe at the beginning of the 1800s when he wrote in his book *Zoological Philosophy*, "for among the cartilaginous fishes the softness and cartilaginous condition of the parts intended to stiffen their bodies and aid their movements indicate that it is among them that . . . nature has sketched its [the skeleton's] first rudiments." In early classification schemes, all fishes with cartilaginous skeletons were lumped together and regarded as primitive. Later, when anatomical knowledge of fishes became more sophisticated, some forms, such as the sturgeon, were recognized as being more closely related to the bony fishes than to sharks, rays, and chimaeras. They were regarded thereafter as degenerate forms whose skeletons had reverted to the ancestral type. Cyclostomes, too, were separated from other cartilaginous fishes and placed in their own class at the base of the vertebrate line.

The old idea that cartilage was a more primitive tissue than bone received new support in the last half of the nineteenth century, because it seemed, with other evidence, to bear out a theory made popular in that period by the German biologist Ernst Haeckel. He stated, as was mentioned briefly in the last chapter, that an organism in the course of its embryonic development passes through a succession of structural stages which represent the series of adult forms that appear in the evolution of the group to which it belonged. Haeckel's theory, encapsulated in the phrase "ontogeny recapitulates phylogeny," came at the height of the furor over the Darwinian concept and was attractive to those who were ready to grant that all living things, however complex, are descended from the single-celled state, just as the adult is traceable to the zygote, the cell formed by the fusion of egg and sperm. It seemed obvious that the early appearance of cartilage in the vertebrate embryo signified the primitive nature of that tissue. Just as bone replaced cartilage in the skeleton of the developing individual, the theory ran, so bony skeletons succeeded cartilaginous ones as the vertebrates evolved to higher levels. This interpretation of the history of skeletal materials led authors of textbooks to place the description of cartilaginous fishes before that of bony forms. Students inferred from such an arrangement not only that cartilage was an older, more primitive tissue than bone but that the sharks and their relatives came before the bony fishes in time.

Modern theory concerning the relative antiquity of bone and cartilage

Neither of these ideas is correct. Fossilized sharks have not been found earlier than the late Devonian period. Bony fish are known, however, from older rock layers—those of middle and early Devonian age. Thus, although the antecedents and the evolutionary history of the first sharks are still veiled in mystery, it seems quite certain that they descended from, rather than gave rise to, fishes that possessed bone. In any case, cartilage was not a new tissue when it appeared in the shark line: it is evident that cartilage, like bone, existed in the ancient jawless vertebrates which antedated jawed fishes of all kinds. Stensiö's ground sections of the ostracoderm *Cephalaspis* revealed beneath the dermal bone of the head shield an internal skeleton with a thin bony outer layer and what appears to have been a cartilaginous inner portion. Evidence from the structure of other ostracoderms reinforces the conclusion that bone and cartilage are equally old.

As paleontologists were disproving the contention that cartilaginous vertebrates evolved before bony ones, biologists were knocking the Haeckelian prop from under the idea that cartilage was basically the more primitive tissue. They agreed that too rigid an interpretation of Haeckel's theory of recapitulation led to erroneous conclusions. The evolution of the different classes of vertebrates had proceeded along divergent pathways: a higher vertebrate never passed through a stage equivalent to the adult form in a lower group. One could not say that an embryonic mammal or reptile whose skeleton was being elaborated in cartilage was in the primitive cartilaginous fish stage of its development. It is true only that vertebrates, being more closely related to each other than to animals outside the phylum Chordata, share a distinctive developmental pattern. Vertebrate embryos look most alike when they are youngest and are passing through structural stages common to the group as a whole; later in the ontogenetic process, as they develop the structures peculiar to their class, they resemble each other less and less. If, at an early stage in embryonic development, vertebrate animals all show cartilaginous elements in the skeleton, it means not that their common ancestor was a cartilaginous form but that the formation of cartilage is a fundamental step in the construction of the vertebrate body. Cartilage is to be regarded, then, as an essential embryonic tissue rather than a primitive one.

Significance of cartilage in the embryo

A corollary to this interpretation is that cyclostomes, sharks, and sturgeons do not show reversion to a primitive state but the retention of an embryonic skeletal material. This supposition does not seem implausible to embryologists. They know that each system of organs develops at its own rate—the nervous system more quickly, the skeletal and reproductive systems more slowly—and that variations from the characteristic rates have been observed. Usually, certain parts of the skeleton are laid down in cartilage and then gradually replaced by bone. If the onset of the replacement process was delayed until after sexual maturity, a reproductive adult with an entirely cartilaginous skeleton would result. Should the appearance of bone be delayed indefinitely, the animal would live its whole life as a cartilaginous form. Embryologists have coined the term neoteny for the retention of an embryonic character and have theorized that the process may often have played a part in the evolution of new forms. In many vertebrate lines there is a trend toward less extensive ossification in both the deeper and the more superficial parts of the skeleton. It is possible that any decrease in bone which is evident in animals belonging to declining or degenerating groups is due to retardation in the rate of skeletal maturation rather than to a more radical kind of change.

Adult cartilaginous forms possibly evolved through neoteny

Once examination of Ordovician and Silurian ostracoderm remains had proved that bone characterized the earliest vertebrates, scientists began to speculate about the advantages that might have accrued to the first fishes through their possession of the hard tissue. Homer Smith, in accordance with his belief in the freshwater origin of vertebrates, thought that an outer, dermal, skeleton of plates and thick scales would have prevented

The advantages of bone in early vertebrates

the seepage of fresh water into the hypertonic body fluids of the jawless forms that arose in lakes and streams. A. S. Romer, convinced that the microscopic structure of ostracoderm armor rendered it unsuitable as an osmotic barrier, believed instead that it served to protect the little vertebrates from the claws of the predatory eurypterid invertebrates that inhabited the fresh waters of the time. Bone situated deeper internally must have functioned then as now for support and the attachment of muscles. It is likely that the appearance of bone and cartilage, whose supportive capacities far outstripped those of the simple notochord, triggered the development of the vertebrates as large and active animals. Far from being a mere evolutionary afterthought—an improvement in the skeletal system—the development of bone may have been the crucial innovation in structure responsible for the sudden rise and radiation of the vertebrate group.

Chemistry and structure of bone as a clue to the origin of vertebrates

Those who believe that the appearance of bone and the origin of vertebrates were closely related events have investigated the chemistry and fine structure of ossified material as part of their attack upon the problem of the derivation of the vertebrate line. Although bone is a distinctive tissue peculiar to vertebrates and has not yet been traced to a particular precursor tissue at the invertebrate level, research has shown that it has much in common with hard tissues produced by living things other than vertebrates. Bone, like arthropod exoskeletons, mollusk shells, and the hard parts of even more lowly metazoa, is produced by the mineralization of a matrix or ground substance elaborated by specialized cells. Among vertebrates and invertebrates, cells active in the formation of hard tissues are drawn from both the ectodermal and mesodermal layers of the embryo. The matrices they secrete are alike in that they almost always contain fibers of polysaccharide or polymerized protein. The minerals deposited in the ground substance of the hard tissues of most animals consist generally of calcium carbonate or phosphate in an amorphous form or as crystals of various sizes. Bone is unique in its microscopic structure and in the particular combination of characteristics of its cells, matrix, and mineral content, but the

Figure 3.1. Structural components of hard tissue. **(a)** Chick bone in formative stage. X48,000. **(b)** Exoskeleton of a crustacean (freshwater crayfish). X81,000. (From Annals of New York Academy of Sciences.) Abbreviations: *C*, cell; *Cm*, cell membrane; *F*, fibers; *M*, mineral particles; *Ma*, matrix. (Photographs courtesy of D. F. Travis.)

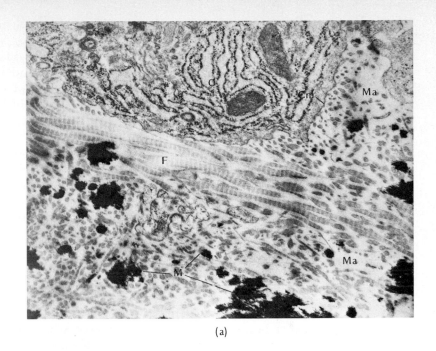

(a)

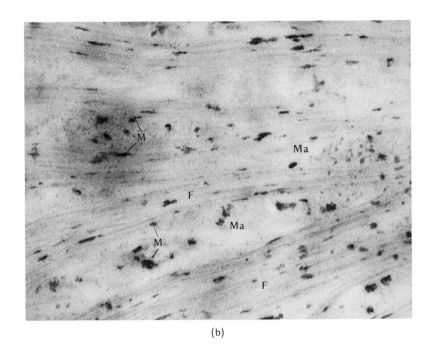

(b)

constituents of the vertebrate tissue existed in other animals before backboned forms appeared. For instance, certain brachiopods, or "lamp shells," have been laying down calcium as hydroxyapatite, the phosphate-rich salt found in bone, since early Ordovician times, and many sponges of great antiquity have produced collagenous fibers like those which appear in the matrix formed by vertebrate osteoblasts. Bone, it seems, did not originate in early vertebrates as a tissue completely new in every sense but simply represented a new combination of materials assembled in an already old and well-established way. (Fig. 3.1.)

Despite the similarities between bone and hard tissues formed by living things other than vertebrates, relatively few investigators have placed emphasis on comparative studies. In fact, by concentrating upon the distinctive nature of bone, vertebrate histologists and medical researchers have, perhaps unintentionally, strengthened the assumption that bone is a unique tissue whose origin might never be explained. Those who view bone in a broader context, however, have found a new line of reasoning that may lead them toward an understanding of how the vertebrate tissue evolved. They have focused their attention on the basic steps in the development of mineralized tissues and tried to determine the factors responsible for the chemical and structural variety of the hard materials produced. Although research has advanced little beyond the descriptive phase, it is already recognized that the nature of the matrix and the fibers it contains is crucial: the deposition of salts occurs only in a matrix that has passed a certain chemical threshold. The type of mineral and the form in which it is laid down depend in large part upon the composition of the receptive matrix and the construction of its fibers. Since the source of substances for both the matrix and its fibers is the cell, the nature of these elements is determined by the special metabolic activities of which the cell is capable. Ultimately, all chemical activity in cells is governed by genes through their control of enzyme formation. It is possible that the genetic prerequisite for the production of bone evolved slowly, step by step, as the chordate line differentiated from some earlier group. The first chordates probably inherited the ability to produce nonmineralized connective tissues. In the course of time, successive mutations would have resulted in the development of a matrix peculiar to the new group. Eventually, a final, even relatively minor, genetic change might have rendered cells capable of secreting a matrix that, in its own particular way, met threshold conditions for mineralization. The hard tissue produced from such a matrix would be different from any other and deserving of a separate name—bone. Such a sequence of events, while not demonstrable, is consonant with the principles of evolution and provides an explanation for the sudden appearance of bone at the base of the vertebrate line.

The variety of mineralized tissues in vertebrates

Often when a new tissue or structure appears on the scene, it develops a wide variety of forms with some rapidity. Subsequently, the less successful versions are eliminated or confined to a small number of organisms. The evolution of hard tissues in vertebrates seems to have followed such a course. Histologists, who divide vertebrate mineralized tissues into four general categories (calcified cartilage, bone, dentin, and enamel), have had to subdivide these categories repeatedly to account for all the variations present among fishes. As the list of distinctive structural arrangements grows longer, the problem of establishing phylogenetic relationships becomes more complex. Sometimes microscopic examination of fossilized hard tissue enables a paleontologist to relate the remains to a known group of vertebrates, but not infrequently he finds a degree of variation which forces him to conclude that he has in hand either the remains of an aberrant individual or material from an animal of still another evolutionary line. (Vertebrate mineralized tissues are shown in Fig. 3.2.)

Dermal armor in ancient fishes

Paleohistology made it clear that the structure of dermal bone in the ostracoderms and in the earliest jawed fishes was much more complicated than it is in water-dwellers of today. In its most elaborate form, the hard tissue that lay beneath the epidermis was built into large plates or thick scales which consisted of three layers. The innermost one was constructed compactly of several superimposed ossified lamellae. Fine blood vessels from the underlying soft tissue penetrated the dense bone, passing through it to ramify in a layer of spongy bone which formed the midregion of the plate or scale. The outermost surface consisted of still another variant of bony material, a thin but densely mineralized layer containing delicate little canals radiating from the base toward the surface. In some early vertebrates, whole pear-shaped cells as well as their processes were included in the material, but in most forms the cell bodies apparently remained beneath in the pulpy areas continuous with the vascular channels of the spongy bone. This histologically distinct material is recognizable as a kind of dentin. In the dermal armor of the ancient vertebrates, it was covered with a very thin, glassy, enamel-like substance (Fig. 3.3). Since some

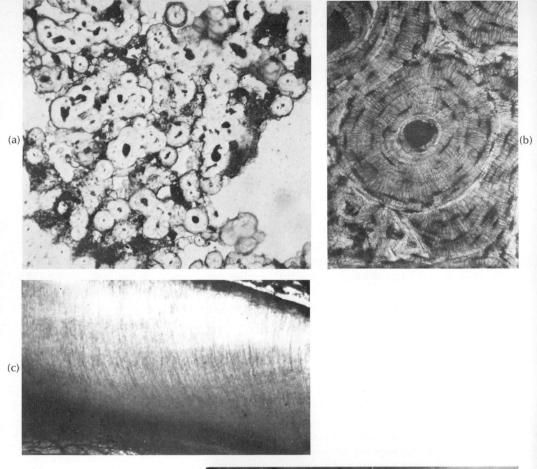

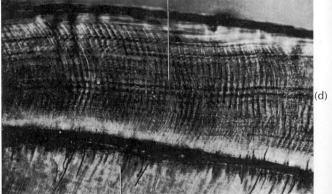

Figure 3.2. Mineralized tissues of vertebrates. **(a)** Calcified cartilage, of a type in which mineralization produces globular structures. **(b)** Bone; cross section through a Haversian system. In the living tissue, central canal is occupied by capillaries and small spaces are filled by bone cells and their processes. **(c)** Dentin; contains tubules penetrated by processes of cells (odontoblasts) whose bodies remain in the adjacent soft tissue. **(d)** Enamel, a highly mineralized acellular tissue. (*a* courtesy of R. H. Denison; *b* courtesy of R. F. Normandin; *c* courtesy of M. P. Ruben; *d* from Peyer.)

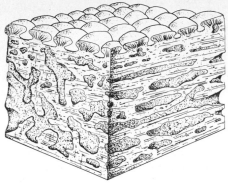

Figure 3.3. Three-dimensional diagram of dermal armor of an ostracoderm, showing basal lamellar bone, intermediate spongy layer, and superficial denticles. (From Ørvig.)

form of dentin overlaid by vitreous material is characteristic of all vertebrate teeth, it is rather certain that the basic mechanism for tooth formation was already operative in the oldest fishes. The superficial layer of the armor usually displayed a surface marked by tubercles called denticles ("little teeth") in direct allusion to their obvious homology to the teeth that, in most vertebrates, became restricted to the oral cavity.

Formation of teeth and tooth-like structures

Embryologists know that teeth (and some fish spines which are tooth-like structures) develop differently from underlying dermal bone. While the latter is laid down entirely by mesodermally derived cells called osteoblasts, teeth arise as the result of a complex interaction between cells of the ectoderm and mesoderm. When stimulated by the presence of neural crest elements that migrate to the site, the ectoderm buds inward to make an enamel organ in the form of a double-layered inverted cup. The cells of the inner layer of the cup, the ameloblasts, are responsible for secreting the matrix which is transformed by mineralization into enamel. Dentin is produced within the enamel cone by the odontoblasts, cells which differentiate from the mesoderm trapped within the enamel organ. Although the process by which the ornamental tubercles of extinct fishes were formed can never be surely determined, there is no reason to believe that it differed essentially from the method by which tooth-like structures develop among vertebrates generally. The variety of shapes assumed by the denticles on the surface of dermal armor can be explained by modification in the shape of the enamel organ. Just as the crowns of mammalian teeth show diversification of form, so tubercles ranged in design from flattened mounds to bent cones and from individual peaks to those which coalesced into ridges. (Fig. 3.4.)

If current theory is correct, then, in the earliest vertebrates two mechanisms were already at work in the differentiation of hard tissues in the organism. Despite their association in the building of dermal armor, both processes were able to function independently. The thin layer of bone which appears in the internal skeleton of Osteostraci like *Cephalaspis* was the product of osteoblasts working alone. In cartilaginous fishes, where osteoblasts do not appear, enamel organs and odontoblasts arise

Figure 3.4. Cross section through lower jaw of a salamander (axolotl), showing a mature tooth at the right and, in the center, two replacement teeth in the process of formation. In the smaller replacement tooth, enamel-forming cells make a cap (dark) over the crown. The bulk of the crown consists of dentin formed by odontoblasts situated at the periphery of the interior pulp (dark). (From Peyer.)

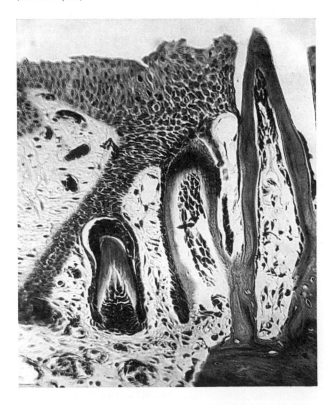

to form the placoid scales that characterize the group. The scales, basically tooth-like in structure, are attached by a bony baseplate (also made by the odontoblasts) and by fibers to the tough but boneless dermis. Each mechanism has produced, in every vertebrate line, distinct varieties of the kind of hard tissue for which it is responsible. The paleontologist and the histologist have discovered many different structural patterns in bone laid down by osteoblasts and an almost bewildering assortment of denticles, spines, and teeth containing several kinds of dentin and enamel formed by odontoblasts in cooperation with ecto-dermally derived cells. If paleontologists were originally drawn into the study of the microscopic structure of these hard tissues through their efforts to identify and classify fossil fragments, they persisted in their investigations because they recognized that they had opened a new path of inquiry into vertebrate evolution. In the teeth and the skeletal system, at least, they could begin to understand how tissues develop and change as well as how modifications in gross morphology follow one another in the course of vertebrate history. (Fig. 3.5 shows placoid scales.)

Those who chose to investigate dental tissues soon found that the hard parts of mammalian teeth, with which they were most familiar, showed considerable specialization and gave no hint of

Enamel and enameloid materials

the broad spectrum of enamel and dentin structure that existed in the vertebrate group. Mammalian enamel, for instance, is composed of highly crystalline rods separated from each other by a small amount of proteinaceous material. The matrix for each rod is laid down by a retreating ameloblast, which also produces an organic sheath. Ensuing mineralization of the matrix with gradual crystal formation does not change the fundamental organization of the tissue established by the work of the ameloblasts: in mature mammalian enamel, the calcium apatite rods extend parallel to one another from the periphery of the dentin to the surface of the tooth. Reptilian enamel, which is surely homologous to the mammalian type, differs struc-turally in not having ensheathed crystalline enamel rods. The enamel of anamniotes is so unlike any form of enamel found in the higher vertebrates that many workers have called it by some other name. Since it appears as a thin layer of homogeneous, glassy material whose boundary with the dentin is often indis-tinct, some have regarded it as a hypermineralized extension of the latter tissue and have labeled it vitrodentin or durodentin.

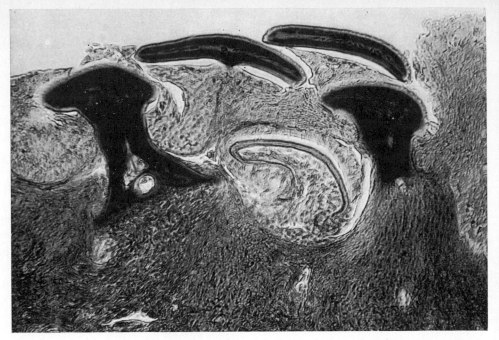

Figure 3.5. Placoid scales. Section through several functional placoid scales of a shark (*Scyliorhinus*) and one replacement scale in the process of formation. Note the structural similarity between these scales and the teeth shown in Fig. 3.4. (From Peyer.)

Whether the enameloid covering of lower vertebrate teeth is a variety of enamel or dentin could be settled, theoretically, by determining its embryonic origin, because the former tissue is derived from ectoderm and the latter from mesoderm. In practice, some investigators have come away from the laboratory convinced that the shiny covering of fish and amphibian teeth is produced by mesodermal tissue; others have declared with equal conviction that ectodermal cells are surely involved, since the enamel-like material appears only over areas of the tooth that are covered initially by epidermis. It is possible that the difference of opinion will disappear as tooth development in a larger number of anamniote embryos is described. (Fig. 3.6.)

Dentin   Although dentin appears in as many varieties as enamel, there is no doubt about its homology in all vertebrates. No matter what structural arrangement the tissue finally assumes, the matrix for it is always secreted by mesodermally derived odontoblasts. In mammals, the odontoblasts surround the embryonic pulp and

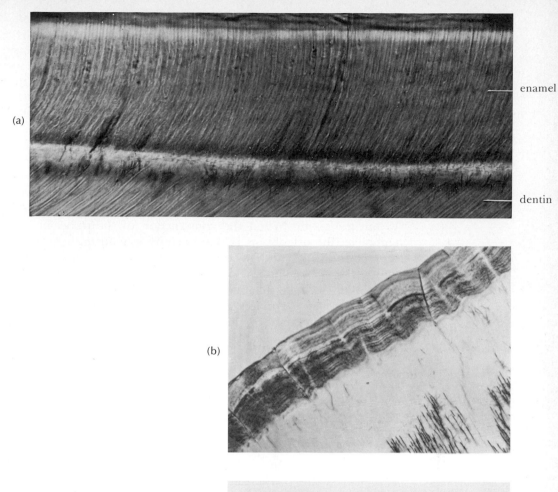

(a)

enamel

dentin

(b)

(c)

Figure 3.6. Enamel and enameloid material at the tooth surface. **(a)** Enamel from the tooth of a mole, showing enamel prisms characteristic of mammals. **(b)** Enamel from tooth of a lizard, lacking prisms but showing incremental growth lines and a distinct boundary between enamel and subjacent dentin. **(c)** Enameloid surface layer, or vitrodentin, from tooth of a cartilaginous fish; the boundary between enameloid and dentin is indistinct, and fibers are visible in the enameloid substance. (*a* from Peyer; *b,c* from Poole.)

retreat inward as they produce the dentin layer, their long cell processes remaining behind in fine tubules. The resulting tissue is known as circumpulpar orthodentin because of its position and the straight course of the parallel dentinal tubules. Orthodentin is recognizable in lower vertebrates, too, although in some of them the dentinal tubules may be more freely branching and the matrix less well mineralized than in mammals. Among fishes, other varieties of dentin occur, some of which show a certain structural similarity to bone. In one form of the material, osteodentin, the two tissues are associated intimately. In the development of osteodentin true osteoblasts appear first in the developing pulp. They lay down a network of delicate bony bars, or trabeculae, as the initial supportive framework within the tooth. When the construction of the trabeculae is complete, the osteoblasts in their vicinity transform themselves into odontoblasts and produce a layer of dentin against the bone. The new material covers the trabeculae and lines the remaining vascular channels that twist and turn through the interior of the tooth. As the dentin becomes thicker, the vascular channels narrow. The resulting structural pattern bears a strong resemblance to primary osteons in bone. Since the hard tissue which encroaches on a vascular channel is truly dentin, with typical tubules for the odontoblast processes, the structural unit in this tissue has been called a dentinal osteon or a denteon. Denteons, when they rise straight through the crown parallel to one another and separated by columns of enameloid substance, as they do in the plate-like teeth of several Recent and extinct fish, are a component of still another type of piscine dentin called tubular dentin.

Relationship between dentin and bone

In fishes, the relationship between dentin and bone seems provocatively close. Both tissues are produced by cells whose processes radiate through fine channels, or canaliculi, into the mineralized matrix. The cells of bone stay in place as they prepare the matrix and so are surrounded as ossification occurs; odontoblasts retreat from the area of matrix and mineral deposition, their cell bodies remaining in soft tissue. The difference in location of osteoblasts and odontoblasts and the distinctive patterns made by their processes extending through the hard matter, being immediately visible under the light microscope, have been accorded importance and emphasized repeatedly. The student of hard tissues in fishes soon perceives, however, that there are morphological intermediates between the "typical" bone and the "typical" dentin described in elementary textbooks (Fig. 3.7). The material known as semidentin,

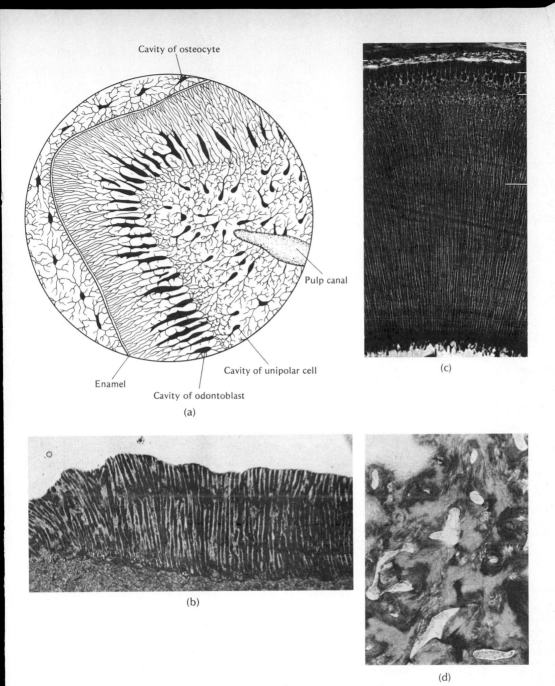

Figure 3.7. Dentin in fishes. **(a)** Semidentin, drawn from a denticle on the armor of *Sedowichthys*, an arthrodire. **(b)** Tubular dentin from tooth of *Sagenodus*, an extinct lungfish. **(c)** Orthodentin from tooth of the modern angler fish, *Lophius*, a teleost. **(d)** Osteodentin from toothplate of the modern chimeroid, *Hydrolagus*, a cartilaginous fish. (*a* from Bystrow; *b* from Denison; *c* from Peyer; *d* from Stahl.)

for example, found in the earliest jawed fishes, has cells that resemble odontoblasts in sending processes into the canaliculi from only one end of the cell body; but the cell bodies are enclosed in the hard tissue along its inner edge instead of lying just beyond in the adjacent soft layer. That dentin in fishes should include a variety which approaches bone so nearly in the distribution of its cells makes it evident that there is little distance between bone and dentin in the spectrum of mineralized tissues. The resemblance of circumvascular denteons to osteons and the possibility that osteoblasts convert to odontoblasts in the development of osteodentin suggests strongly that dentin evolved as a highly specialized and divergent form of ossified tissue. Even in mammals, where dentin has assumed morphological characteristics which distinguish it clearly from bone, the chemistry of the matrix in the two tissues is similar.

Varieties of bone

Unlike dentin, which forms only where mesoderm and ectoderm meet, bone arises in areas of the mesoderm that are free from ectodermal influence. Whether it appears superficially in the dermis or deeper within the body, bone is built either in compact layers (lamellae) or as a network of trabeculae through which fine blood vessels wend their way. Compact bone is the stronger of the two types but also the heavier. It develops at the periphery of individual bones, where it sustains the greatest weight or the direct tensile stress from tendons and muscle fibers. Spongy, or cancellous, bone, still remarkably strong because of the orientation of its trabeculae parallel to the lines of force projected through it, fills the interior of skeletal elements where the weight of compact bone would be intolerable. Both types of bone occur in the armor and scales of ostracoderms as well as in the skeletons of more advanced vertebrates. (Fig. 3.8.)

By studying bone that was formed from Ordovician to Recent times, the paleohistologist has obtained some idea of the evolutionary changes that have taken place in the structure of the tissue. He has found that although most bone exhibits lacunae for the cell bodies of the osteoblasts, these spaces are absent in fishes of two widely separated groups. In the ancient heterostracan ostracoderms and most of the modern teleosts, the bone-building cells are obliterated as mineralization proceeds, leaving an ossified tissue which is acellular. Several workers have speculated that animals with acellular bone enjoy a metabolic saving in not having to sustain living osteoblasts, but no one has explained why the same peculiar modification should have arisen in vertebrates so far apart on the evolutionary scale.

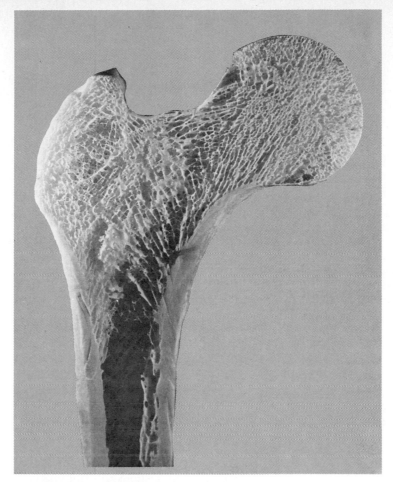

Figure 3.8. Upper end of the human femur cut longitudinally to show compact bone and spongy bone. The compact bone forms a thick layer in the shaft around the marrow cavity and a thin layer over the spongy bone. Trabeculae in the spongy bone are oriented parallel to the lines of stress projected through the femur. (Photograph courtesy of R. F. Normandin.)

As well as distinguishing between cellular and acellular bone, investigators have been able to tell the difference between primary bone, produced initially in the growth of a part of the skeleton, and secondary bone, which results from the rebuilding that goes on within an element. Under the microscope, secondary bone presents a distinctive pattern of juxtaposed Haversian systems, each consisting of several concentric bony lamellae surrounding a narrow canal through which one or two fine blood vessels pass. The earliest vertebrates laid down primary bone but had probably little ability to resorb and replace it. The apparent absence of secondary bone in ostracoderms has led paleontologists to wonder how these animals were able to grow. The head shield which encased the anterior end of the jawless fish shows no evidence of sutures separating bony components which could have enlarged by adding primary bone at their edges. There is no evidence, either, of incompletely built head shields that could have belonged to growing juveniles. Fully formed head shields of the same design (and thus attributable to the same species) have been found in a variety of sizes within a certain characteristic range. Paleontologists who accept these head shields as full-grown must conclude either that adult size showed extreme variation within a species of ostracoderms or that the different sized individuals belonged to a different species. Neither possibility seems likely. Those workers who think that they have detected areas of bone resorption in a few forms are confident that smaller specimens represent younger individuals whose head shields were growing in a manner as yet unrecognized.

It is possible that the production of secondary bone in ostracoderms, if it occurred, was very limited. Examination of bone from many kinds of jawed fish, fossil and Recent, has shown that even among the more advanced fishes secondary bone is not highly developed or constant in its appearance. Haversian systems become common only at the tetrapod level. In primitive amphibians and reptiles they appear scattered in an irregular way. Only in dinosaurs and the advanced mammal-like reptiles do the Haversian systems develop densely. Closely packed Haversian systems are not characteristic of the bone in all the highest vertebrates, however. Avian bone lacks them, as does bone in several types of mammals. The variation in occurrence of Haversian tissue among amniote vertebrates seems not to be of great significance. The important events in the evolution of bone are surely the early appearance of the primary tissue in its compact and cancellous forms and the increase in growth of secondary material as tetrapod vertebrates arose and gained the land. (Fig. 3.9.)

Early studies of the vertebrate skeleton
Although studies of the microscopic anatomy of hard tissues have provided new insights into the evolution of vertebrates, paleontologists continue to rely upon examination of the gross structure of the skeleton in describing new specimens and trying to place them correctly on the family tree. Since ancient times, of

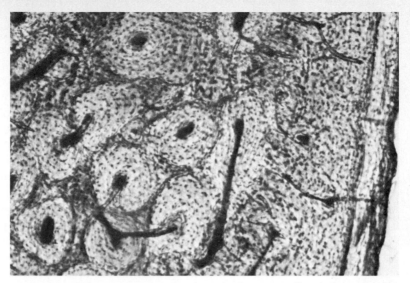

Figure 3.9. Primary and secondary bone in the mammalian skeleton. Layers, or lamellae, visible at the right are primary bone, having been formed at the surface as the skeletal element enlarged. The concentric rings of bone surrounding blood vessels at the left constitute Haversian systems and are secondary tissue, having been laid down in place of older bone which has been resorbed. (Photograph courtesy of M. P. Ruben.)

course, men have been making comparisons between animals on the basis of characteristics visible to the naked eye, but not until the nineteenth century did workers concentrate in earnest upon a detailed study of the osteology of the vertebrates. It was obvious even to foes of the evolutionary theory, like Cuvier, that the common plan of the skeleton of the backboned animals set them apart from other forms whose supportive structures were completely different. The earliest evolutionists assumed that animals with like skeletons were closely related. Their supposition was later substantiated by the geneticists, who linked common characteristics to common hereditary material and defined relationship in terms of the possession of like genes.

Even though they were unaware of what the science of genetics would eventually reveal, biologists in the 1800s accumulated a vast amount of osteological data and forged the method of study which paleontologists have used ever since. Rather than drawing descriptions vaguely or fancifully, as had been done in the past, these men made careful observations of the specimens that lay before them and compared skeletons part by part, bone by bone. Whether they thought they were proving evolutionary relationships or merely improving the classification of creatures created by God, they recognized that comparisons were meaningless unless they involved elements that were the same in terms of their relative position within the skeleton. Thus, the concept of homology was born. Determining which units of the skeletons in two or more animals were homologous, however, was not always an easy task. The skeleton, like the soft organs, often underwent specialization to such a degree that individual parts changed radically in their appearance or location. Difficulties which arose in the laboratory were challenging rather than discouraging, for evidence was accumulating from studies of living forms that the concept of homology was indeed a sound one. Embryologists were proving that many apparently dissimilar adult structures were comparable by retracing their development to early stages at which their essential sameness was obvious. The fledgling comparative anatomists, quick to grasp the significance of the embryologists' demonstrations, tied their science firmly to that of embryology by accepting the ultimate definition of homologous structures as those which arise from the same germ layers in the same relative location in the embryo.

The principle of homology

Paleontologists, who use the techniques of comparative anatomy in studying the skeletons of extinct forms, cannot confirm predicated homologies by investigating embryonic material. They must rely entirely upon the structural characteristics of a fossil bone or calcified cartilage and its position in the body. They can demonstrate the homology of a highly specialized element only by finding a series of fossils which link it to a more generalized form of the structure whose homology is easier to ascertain. From its position at the distal end of the appendage, the single-hoofed foot of the horse was adjudged homologous to a digit, but verification of that conclusion was not possible until O. C. Marsh assembled the series of fossil horses extending backward from *Pliohippus*, a near-modern form, to *Eohippus* (now called *Hyracotherium*), the little Eocene animal whose hind feet had three toes and front feet, four. It became plain that the foot of the modern animal is indeed a single digit and has evolved from the basic pentadactyl appendage by a process of reduction that has been in progress for millions of years.

Every paleontologist dreams of finding a series of fossil skeletons so complete that he can follow the evolution of the vertebrates in that particular line step by step. Since nearly complete sequences are a rarity, workers almost always find themselves puzzling over scattered specimens that, despite certain similarities, are obviously not ancestral one to another. Even though evolutionary lines cannot be traced neatly from this kind of fossil evidence, much of value can be learned nevertheless. By painstaking analysis of the anatomy of isolated forms, paleontologists have been able to put together a general picture of the great groups of vertebrate animals that have come into existence since the Ordovician period. Further, they have been able to identify certain trends in the evolution of the skeleton itself which serve as a guide in drawing the vertebrates' family tree.

Dermal bone in the vertebrate skeleton

One such trend is evident in the history of dermal bone. Dermal ossification occurs more extensively in lower vertebrates than in higher ones and, especially among the various groups of fishes, is usually more elaborate among earlier than later forms. The jawless ostracoderms possessed a heavier dermal covering than any of their more advanced descendants. Their bony armor, the three-layered structure of which was described earlier, encased the body from snout to tail. Even though this well-developed dermal skeleton characterized the ostracoderm group as a whole, a close look at its nature and the changes that took place in it reveal a pattern that persisted throughout the vertebrate line. The deposition of bone was generally heavier anteriorly: in the region of the head, the ostracoderms exhibited massive bony shields, or at least plates that were larger than mere scales. In each of the three major ostracoderm groups, continuing evolution brought a diminution of the dermal skeleton. If current estimates of the relative ages of the different ostracoderms are correct, the head shields of the Heterostraci, originally composed of a few large plates, eventually exhibited a greater number of smaller ones; osteostracan shields, which once covered a part of the trunk as well as the head, became more and more restricted in their length; last, and most extreme, among the anaspids, whose armor showed subdivisions into scales, there appeared at least one late form that had lost almost all its dermal bone.

In jawed vertebrates, dermal bone remained more concentrated in the area of the head and shoulder and tended to disappear more posteriorly. The dermal bones in the skull and shoulder girdle of fishes constitute the largest part of the dermal skeleton in these animals. Over the trunk and tail, dermal bone exists only in the form of scales. A survey of the structure of scales in ancient and modern fishes provides evidence of the gradual disappearance of dermal bone in the region behind the shoulder girdle. Reduction in the thickness of scales had occurred among the earliest ancestors of the majority of present-day bony fishes. The ganoid scales of these animals still covered the body with a pavement of scales fitted edge to edge, but the middle layer of spongy bone, present in the armor of older forms, was almost gone, as were the elaborate, ornamental denticles. The lamellar base of the scale was covered superficially by a dentinous cosmoid layer and a hard substance called ganoine (Fig. 3.10). As time went on, all trace of vascular bone and cosmoid layer disappeared, leaving thin scales of much reduced, acellular, lamellar bone arranged in overlapping rows. In many families of existing fishes, these scales are minute or have finally disappeared altogether. The same tendency toward reduction of dermal ossification appeared in the ancient line of bony fishes which gave rise to tetrapod vertebrates. The first members of this group possessed a type of scale described as cosmoid. Heavy with lamellar and spongy bone, it too underwent reduction as millions of years passed. The earliest tetrapods carried with them onto the land only a remnant of these scales, which they eventually lost. Ossification in the dermis behind the shoulder girdle is totally absent in almost all modern land vertebrates. The only tetrapods that show it are the reptilian forms which develop rib-like dermal bones in the skin under the belly and some animals, like the turtle and the armadillo, whose powers of dermal ossification have revived to produce new types of protective armor. Except in these forms, dermal bones are found only in the skull and anterior limb

Figure 3.10. Three-dimensional diagrams of portions of scales of archaic bony fishes. **(a)** Ganoid scale found in early ray-finned fishes. **(b)** Cosmoid scale, found in early lobe-finned fishes. (From Romer.)

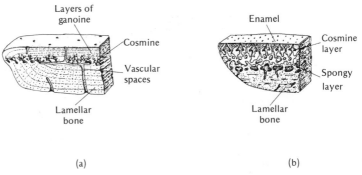

(a)   (b)

girdle of land animals, and even in these parts of the skeleton a trend toward reduction is apparent.

Endochondral bone in the verte-brate skeleton

As the dermal skeleton diminished, bone which arose in the deeper region of the body increased in importance. Called endochondral, this bone replaced elements of the skeleton first formed in cartilage in the embryo. As they lost their protective covering of plates and scales, the vertebrates developed an endochondral skeleton which, while still shielding vital organs, would afford support with minimal weight and allow great mobility. The evolution of the endochondral skeleton was surely one of the key factors in preparing the vertebrates for a successful invasion of the land: endochondral vertebral elements which permitted the undulations of the tail that moved fishes through the water developed into the flexible bridgework from which the weight of the tetrapod body was hung. The girdle bones that initially provided sockets for appendages which steered and stabilized the moving fish became parts of stout bracing rings that transmitted force from the limbs to the backbone. The simple endochondral bones supporting blade-like fins were ancestral to the elements which allow birds to fly and mammals to run. Had the trend toward increased endochondral ossification not occurred, land vertebrates either would not have evolved or would have developed a shape and mode of behavior that are difficult to imagine.

Paleontologists have found it somewhat easier, as a rule, to ascertain homologies among endochondral bones than among dermal ones. Especially among lower vertebrates, dermal plates or bones of the skull vary widely in number and arrangement. It is often not possible to determine surely which bones are comparable in the skulls of two fossil fishes that are not closely related. Some dermal bones of the head in teleost fishes have been assigned names applied to bones in the same region of the tetrapod skull, but no worker pretends that the use of the same terms is anything more than a temporary expedient. Changes in the pattern of endochondral bone, on the other hand, have been more conservative and easier to follow. Elements tend to stay in place and to become adapted by a change in form to fulfill a new function. Vertebrae, usually spool-like with separate neural arches in fishes, develop heavy, solid centra and interlocking neural arches in land animals whose backbones bear more weight. The size, shape, and number (rather than the basic pattern) of bones in the limb vary in animals which swim, run, dig, climb, or fly. Even in changes which are not quite so obviously adaptive, endochondral bones maintain their original spatial relationships to one another: the three mammalian middle-ear bones retain the association which appeared first among primitive fishes, where the homologous elements formed the jaw articulation and its brace against the skull.

Problem of establishing relationships on the basis of osteological evidence

It is true that investigators who are in possession of whole bones or of pieces large enough to be identified are in a better position to discover evolutionary relationships than workers who have in hand only very small fragments. Yet even when nearly complete skeletons are available, difficulties still arise in determining the mutual standing of two forms in an evolutionary line or in placing an isolated specimen in its proper group. The paleontologist must assess the degree of difference between a particular bone or complex of bones and homologous structures in similar vertebrates. On the strength of what he can see, he has to speculate upon the closeness or distance of a relationship and whether one form is more likely to have been the ancestor of another or a collateral descendant from the same stock. Although what he is attempting to establish, actually, is the hereditary or genetic sequence in vertebrate evolution, he is forced to gauge it indirectly by the appearance of bones whose development is governed by the genetic material. He is well aware that his subjective estimate of the amount of change in the design of a bone may not correspond with the dimensions of the genetic mutation responsible for it. In one case, a small modification in appearance may be the result of many individual mutations accumulated in the course of long separate evolutionary development. In another, what appears to be broad skeletal change could have stemmed from a relatively few genetic rearrangements that affected skeletal structure profoundly. The individual in the first case would seem less divergent from the basic stock than it really was in terms of its genetic history; the individual in the second case would appear to be more so.

Despite the impossibility of studying the genetic material of extinct forms, the paleontologist cannot but gain insight from information uncovered in the future about the interrelationship of genetic and morphological change in living animals. His interpretation of the significance of skeletal change may be affected, too, by clarification of the influence of environmental factors upon genetic mutation. Although modern paleontologists are generally Darwinist in their thinking, many of them utilize Lamarckian language in relating changes in structure to changing environmental demands. Whether the subject is the development of limbs and increased skeletal support in the earliest tetrapods, the appearance of flippers in animals that reentered the sea, or the formation of teeth especially suited for a particular diet, authors often imply a connection between the need for a structure and its appearance. Since the occurrence of random mutations is usually assumed by writers rather than stated explicitly, the casual reader may come to regard evolutionary change in vertebrates as a straight-line progression toward greater complexity and more perfect adaptation, effected through some inner initiative on the part of the animals and an environmental force. Those who enter more seriously into the study of evolution soon recognize that they must confine themselves to the interpretation of events for which there is fossil evidence without making groundless assumptions about the specific causes of change. They are aware, though, that while they hunt, dig, clean, and compare the bones of ancient animals, studies are in progress in genetics and biochemistry which may help to dispel the darkness in which they work. Although Lamarckian theory in its original form is dead, geneticists are defining subtle factors in the environment that do affect the hereditary material within the cell. Biochemists, who have begun to determine the way in which genes direct cellular activity, are attempting to understand genetic control of the differentiation of tissues. It is possible that continuing research in these areas will provide clarification of interrelationships among environment, heredity, and morphological change. If new solutions emerge to the old problems of environmental and genetic control of change in structure, paleontologists should be able to give a better explanation of the evolution of the vertebrates that they are documenting through their studies of bones, scales, and teeth.

4
The First Fishes with Jaws

The advent of jawed fishes

The vertebrates, covered with bony plates and scales, had come late into waters already populated with other kinds of living things. Nevertheless, they established themselves so successfully that within a few tens of millions of years they differentiated in form and swam nearly everywhere. Continuing evolution of the group in Silurian and Devonian times produced not only the ostracoderms described in Chap. 2 but also varieties of fishes that represented a higher level of vertebrate structure. These fishes had jaws and paired appendages: presumably, they set the pattern of voraciousness and mobility that has characterized backboned animals ever since. Because these earliest jawed forms were contemporaries in the Devonian period, gradually replacing the ostracoderms and then being themselves replaced by even more advanced fishes, they are generally placed together in the class Placodermi and introduced to students as a cohesive group. Actually, these forms were strikingly diverse. Paleontologists are still studying and restudying their fossils, trying to figure out where they came from, how they were related to each other, and what their connections might have been to the Devonian ancestors of modern fishes.

Although no one relishes making the account of vertebrate history more complicated, it is clear that two groups within the old designation Placodermi are morphologically quite dissimilar, surely represent different evolutionary lines, and ought not to be classified under the same heading. Accordingly, it is becoming common practice to confine the term "placoderm" to the fishes known as arthrodires and their relatives and to place those termed acanthodians in another category.

Acanthodians

Of the two kinds of early jawed forms, the acanthodians, or "spiny sharks" as they are sometimes called, are the older (Fig.

Figure 4.1.   Restoration of two acanthodians in lateral view. **(a)** *Parexus*, a Lower Devonian form. **(b)** *Ischnacanthus*, a form from the Lower and Middle Devonian. Abbreviations: *Anal*, anal fin; D^1, D^2, dorsal fins; *Hy*, upper end of the hyoid arch; I^1, I^2, I^3, intermediate paired spines; *LL*, lateral line; *PFR*, pectoral fin rays; *Pect*, pectoral fin; *Pect Sp*, pectoral fin spine; *Pelv*, pelvic fin. (From Watson.)

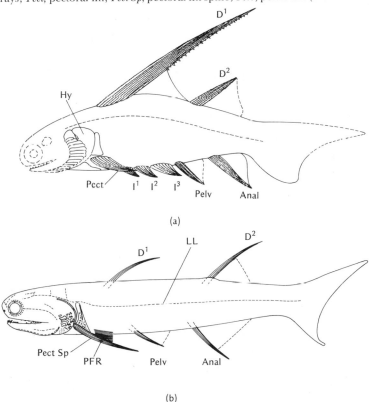

4.1). Their fin spines and isolated scales have been recovered from Silurian deposits in company with heterostracan remains. The first known acanthodians, though they may have swum in the same waters with ostracoderms, must have looked little like their more primitive contemporaries. Less fragmentary acanthodian fossils from Devonian rocks show the fishes to have been small and somewhat like sharks in shape, with prominent spines forming cutwaters in front of membranous median and paired fins. Their heterocercal caudal fin was not distinctive, but the hinged lower jaw that bordered the terminal mouth and the exceedingly large eyes placed far forward gave the acanthodian head a unique appearance. Behind the head, the gill slits were partially covered by an operculum that projected posteriorly to the shoulder girdle. The dermal armor over the branchial region consisted of little plates, but elsewhere the acanthodians were protected by a close-fitted mosaic of small, tile-like scales.

Compared to the number of solitary spines that have been found, the more articulated acanthodian skeletons are few. From D. M. S. Watson's description of the best-preserved remains, however, it would seem that the acanthodians were a relatively conservative group. Genera varied to a limited degree in the shape of the body, the number of paired fins, the dimensions of the fin spines, and the way the jaws ossified. The changes in the fins are of special interest to paleontologists who

Figure 4.2. Pectoral fins of acanthodians, showing supporting elements. **(a)** *Climatius*, an early Devonian form with broad spine supporting the fin membrane and no known endoskeletal elements. **(b)** *Acanthodes*, a later acanthodian with a narrow, deep-set spine anteriorly and supporting elements within the fin web. (From Jollie.)

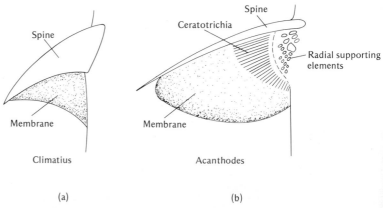

Vertebrate History: Problems in Evolution

are curious about the establishment of appendages in the vertebrate line. The older acanthodians like *Climatius* had what must have been immobile or nearly immobile fins, each held anteriorly by a short, stout spine that was anchored against the body wall. Besides one anal and two dorsal median fins, there was a ventrolateral row of paired fins. Of the latter, the first and last, comparable in location to typical pectoral and pelvic appendages, were the largest (Fig. 4.2). Those between were unequal in size and probably auxiliary in function. As millions of years passed, the acanthodians exhibited a tendency to lose the intermediate paired fins and to retain the pectoral and pelvic ones. During the same period, the fin spines became thinner and more deeply set into the body wall. The development of basal elements beneath them, providing surface for attachment of muscle fibers, suggests the achievement of at least some mobility. There is evidence, too, that by the Middle Devonian some acanthodians possessed skeletal support within the fin membrane itself: several fossils show basal and radial elements and even ceratotrichia.* By the time of their extinction the acanthodians had evolved a kind of fin structure similar in principle to that which appeared in other groups of fishes. Their ability to swim must have been a factor in their long survival: although they enjoyed their greatest success in the Devonian, a few genera outlived other types of early jawed fishes, existing through the Carboniferous period and even into the beginning of Permian time.

Unlike the acanthodians, the placoderms developed a wide variety of forms. Lower Devonian rocks yield fossils of several different types of these fishes. Although a common stock for all of them must have been present in Silurian waters, paleontologists have as yet found no trace of it. They do know that the placoderms, despite their structural plasticity, were short-lived: sediments laid down after the Devonian are virtually bare of placoderm remains.

Because of their diversity, the placoderms are not so easy to characterize as the acanthodians. Their chief similarity lay in the design of their dermal armor: the head and the anterior part of the trunk were protected by separate shields made of bony plates. The two shields articulated with one another behind the gill region, often through a ball-and-socket joint. A pair of

---

*Ceratotrichia: unjointed fibrous rods that support the outermost part of the fin in cartilaginous fishes and certain primitive forms (see Figs. 4.2 and 4.13).

strong spines, or remnants of such structures, generally appeared on the anteroventral corners of the trunk armor. The shape of the body and the nature of the fins and jaws varied widely from group to group within the placoderm assemblage.

Arthrodires

The best known of the placoderms are the arthrodires. The earliest ones, the Arctolepida, seem to have been small, freshwater types that hugged the bottom. Their eyes peered forward and upward through orbit holes in the front part of the flattened head shield. The shield, marked by sensory canals in a characteristic arrangement, covered a braincase that was often well ossified. The thoracic armor, fitted close behind, extended far down the trunk, allowing that part of the body little flexibility. An enormous spine projected from each side of the shoulder. Presumably, there was a fin web posterior to it, for the trunk shield shows a gap behind the spine through which blood vessels and nerves could have passed outward to the soft tissue of a pectoral appendage. The small size of the gap indicates that the fin had a narrow base and was probably not very big.

Continuing evolution of the arthrodires led to larger forms, classified as the Brachythoraci, many of which were surely marine (Fig. 4.3). The dermal armor, especially in the trunk region, diminished in solidity and extent. In some of the later arthrodires, the thoracic shield became so short that it could be accurately described as a kind of dermal pectoral girdle. Its pair of spines was reduced and in some genera disappeared completely. The open space in the armor behind the spines enlarged sufficiently to allow the growth of greatly expanded pectoral fins more broadly attached to the body wall in these fishes. These later arthrodires, like *Coccosteus* and *Hadrosteus*, were probably more active animals than the arctolepids. In all but a few forms, the head and shoulder shields fitted together less tightly and had a well-developed ball-and-socket joint between them. This arrangement and the cervical articulation present between the braincase and the anterior end of the vertebral column doubtless permitted freer movement of the head, useful not only in respiration but also in catching food. The shape of their teeth—curious spikes and sharpened blades jutting from the edge of the jawbones—is evidence that they were predaceous. No one looking at the remains of *Dunkleosteus*, with its huge mouth gape and powerfully armed jaws, could doubt that this arthrodire must have cleared the seas of many smaller fishes.

Although the best known of the brachythoracids were mobile, aggressive forms, there is evidence that during the middle and

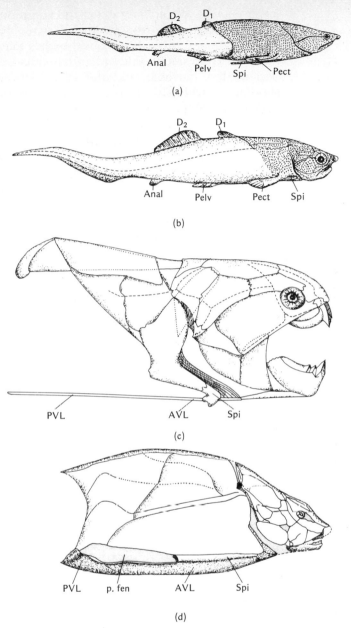

Figure 4.3. Restorations of four arthrodires in lateral view. **(a)** *Arctolepis* of the Lower Devonian, an arctolepid. **(b)** *Coccosteus*, a Middle Devonian brachythoracid. **(c)** *Dunkleosteus*, a highly predaceous Upper Devonian brachythoracid, showing dermal elements of the skull and shoulder shield. **(d)** Head and shoulder shield of *Holonema*, an Upper Devonian brachythoracid adapted for bottom dwelling, perhaps as a scavenger. Abbreviations: *AVL, PVL,* plates in the ventral trunk armor; *p. fen,* opening in the trunk armor for the pectoral fin; *Spi,* spinal plate of trunk armor; other abbreviations as in Fig. 4.1. (*a,b* from Stensiö, *c* from Heintz; *d* from Miles.)

late Devonian the advanced arthrodires underwent a radiation that produced some groups adapted for a different mode of life. In two families, the fishes evolved bodies that were deeper and flattened from side to side rather than dorsoventrally. The head became less movable, and the space between the head and shoulder shields shrank progressively. Among these animals, which probably swam closer to the surface than most arthrodires, one form finally appeared in which the head and shoulder shields fused. With reduced mobility of the head, the mouth opened less widely, and in some genera the snout elongated. Arthrodires of this type were surely restricted by their narrower mouth gape to a diet of smaller animals, invertebrate as well as vertebrate, which they may not have had to pursue very vigorously.

Another family of small-mouthed brachythoracids appeared in the middle Devonian, perhaps as a branch from the arthrodiran stock just approaching the brachythoracid level. These fishes, the holonematids, retained the bottom-dwelling habits and long trunk shield of their ancestors but possessed the broad-based pectoral fins, reduced spines, and elaborate articulations between head and trunk that characterize the advanced members of the group. In comparison to the trunk shield, the head was extremely short. The high, posterior part of the armor covering it pitched sharply downward to the edge of the upper jaw, which bore short, minimally developed tooth plates. Because the structure of the tooth plates was relatively weak and the space for the muscles which closed the mouth was small, it appears that the holonematids may have been scavengers, maintaining themselves on dead organisms or soft-bodied ones that they found as they moved over the mud. (See Fig. 4.3d.)

Ptyctodonts

The bottom waters evidently offered a number of adaptive zones to the placoderms, because by late Devonian times several different groups existed in that environment (Fig. 4.4). In the marine shallows, where clams and other hard-shelled invertebrates abounded, there appeared forms called ptyctodonts, placoderms with short jaws, heavy tooth plates, and strong muscles to give power to the bite. The reduction of the armor in these fishes had gone far beyond that of the late Devonian brachythoracid arthrodires. The ptyctodonts retained only a narrow band of bone in the shoulder region and little of the head shield except the plates above and behind the eyes. Despite the loss of dermal bone, their classification as placoderms is not in dispute. The ptyctodonts still exhibit the characteristic joint

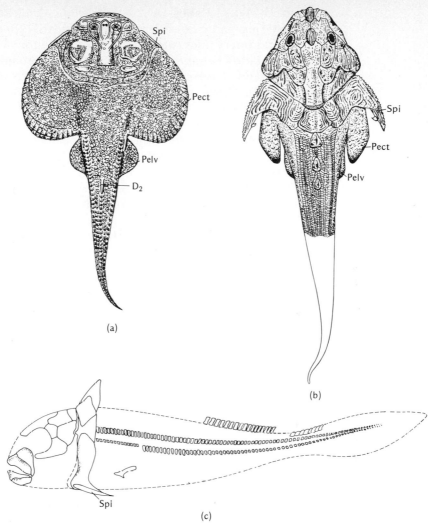

Figure 4.4. Restorations of various placoderms. **(a)** The rhenanid *Gemuendina* (dorsal view). **(b)** The petalichthyid *Lunaspis* (dorsal view). **(c)** The ptyctodont *Rhamphodopsis* (lateral view). Abbreviations as in Fig. 4.1. (*a,b* from Stensiö; *c* from Watson.)

between the armor of the head and shoulder, and at least one form retained typical pectoral spines.

Other kinds of placoderms

Besides the ptyctodonts, a variety of flattened forms possessing the dermal head and shoulder shields inhabited the offshore sea bottoms. Some were even more depressed than the early arthrodires. Fishes such as *Homostius* and *Phyllolepis*, whose differing patterns of dermal armor identify them with separate evolutionary lines, were highly specialized for swimming close to the muddy or sandy undersurface. The rhenanids like *Gemuendina* and *Jagorina* developed a body form similar to that of modern skates and rays, with broad-based pectoral fins that spread widely on either side of a blunt head and tapering tail. Still another placoderm group related in some way to the arthrodires consisted of fishes, the petalichthyids, that retained fins of a more conservative size than those of the rhenanids and a body more arthrodiran in shape. They cannot be considered closely akin to the arthrodires, however, for the arrangement of the bony plates in their armor differs from that in the arctolepids and brachythoracids.

The antiarchs represented an offshoot of placoderm stock that remained in fresh waters. Though these fishes were like the early arthrodires in having their head armor fitted closely to a long thoracic shield, their peculiarly foreshortened head and jointed spike-like pectoral appendages gave them a distinctive and bizarre appearance (Fig. 4.5). There was no exposed pectoral fin flap: all the soft tissue was enclosed within a long sheath of dermal bone that was movably articulated with the thoracic shield. The sheath, made of many bony plates, was sharply pointed at its distal end and (in all known forms but one) jointed once in the middle of its length. These strange appendages may have served as props as the antiarch half glided, half crawled over the muddy lake bottom in search of food.

The efforts of paleontologists to learn the history of the early placoderms are frustrated by lack of fossil material. The deficiency will not be easy to overcome, for the Silurian freshwater deposits that might have received the remains of the first placoderms have either disappeared or become inaccessible.

Figure 4.5. Restoration of the Middle Devonian antiarch, *Pterichthyodes*. (From Romer.)

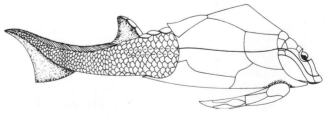

Some workers suspect that these rocks held not only the earliest freshwater arthrodires and ancestral antiarchs but the series of forms that led toward the first jawed invaders of the sea. If they are right, evidence of the evolution of acanthodians as well as placoderms may remain unavailable and our understanding of the origin of jaws and paired fins continue unimproved.

The evolution of paired appendages

How paired appendages developed in the vertebrate line has long been the subject of speculation. Despite the gaps in the fossil record of the acanthodians and placoderms, hypotheses concerning the appearance of pectoral and pelvic fins have become steadily more sophisticated as investigators take into consideration evidence from a wider range of biological studies. In the nineteenth century, Gegenbaur put forth a theory based solely upon morphological similarity: he fancied a resemblance between the arrangement of the skeletal elements in the branchial arches and those in the appendicular girdles and suggested that the former might have given rise to the latter. The fin skeleton could have been derived, he thought, by elaboration of the extrabranchial elements that ran outward from the arch to support the gill filaments. Except as an illustration of the folly of building an hypothesis upon too slight a base, Gegenbaur's theory is valueless today. It was demolished by the combined efforts of embryologists and anatomists. They demonstrated the development of the branchial and appendicular skeletons from different germ layers in the embryo and their control by dissimilar muscles. It is not likely that the branchial skeleton, derived from neural crest ectoderm and associated with special muscle of visceral origin, could have given rise to appendicular elements which differentiate from mesoderm in conjunction with somatic musculature.

Balfour's theory

An alternative suggestion for the appearance of paired fins has proved more viable than Gegenbaur's theory because it is less at variance with developmental and anatomical principles. Balfour proposed that paired fins be regarded simply as outfoldings from the body wall similar in structure to the unpaired fins and that the origin of the two should be sought in common (Fig. 4.6). The tendency of different kinds of Devonian fishes to produce midline and ventrolateral fins could have been inherited from the unknown first vertebrates in which a generalized pattern of fin folding was established. Some adherents of Balfour's theory traced the source of such a pattern to the ancestral protochordates of which amphioxus is a remnant. They assumed that the protochordates from which vertebrates

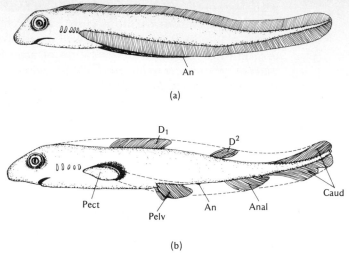

Figure 4.6. Diagrams illustrating Balfour's theory of the origin of the paired fins. **(a)** Early vertebrate with continuous fin fold paired ventrally anterior to the anus. **(b)** More advanced vertebrate having discrete paired and median fins as a result of loss of parts of the original fin fold (shown in dotted lines). Abbreviations: *An*, anus; *Caud*, caudal fin; other abbreviations as in Fig. 4.1. (After Wiedersheim.)

descended had, like amphioxus, a continuous fin, unpaired in the dorsal midline and around the margin of the tail but dividing in two as it extended anteriorly on the ventral side of the body. Paired fins could have arisen from the ventrolateral folds of such organisms through the retention of certain parts of the folds and the suppression of intermediate areas. The acanthodians with a series of paired fins are cited as examples of early fishes that show a pattern transitional between continuous ventrolateral fin folds and their reduction to two pairs of appendages.

Closer scrutiny reveals difficulties that stand in the way of this apparently logical solution of the problem. First, the paired metapleural folds of amphioxus end at the atriopore, an opening in the midline some distance anterior to the anus: immediately in front of the anus, where vertebrate pelvic appendages develop, the finfold of amphioxus is unpaired. This difference in the pattern of outfolding lessens the likelihood that the arrangement in amphioxus resembles an ancestral one in the vertebrate line.

Second, the transitional status of the acanthodian fins is questionable. A survey of the ostracoderms, more primitive vertebrates than the acanthodians, brings forth no forms with an arrangement of median and paired fins that seems intermediate between the theoretical continuous fin fold and its interruption to produce an acanthodian-like condition. Although it is true that the ostracoderms from which the acanthodians presumably came are not known, there is little fossil evidence to suggest that any of the ostracoderms evolved from ancestors with uninterrupted outfoldings from the body wall. One of the anaspids, *Endeiolepis*, had elongated fin folds both dorsally and ventrolaterally, but *Endeiolepis* was a late Devonian form and degenerate in several respects. In older anaspids, like *Birkenia* from the Upper Silurian, there is no hint of long fin folds, paired or unpaired. It is probable that lateral fin folds appearing in such anaspids as *Pterygolepis* were developed within the group rather than inherited from a protochordate or even a primary vertebrate ancestor.

In this century, the search has abated for a single ancestral group with structures that could have been the precursors of all the paired appendages that were to follow. The likelihood that a single source existed diminished as paleontologists discovered the wide variety of folds, flaps, and spines which were present in ostracoderms. When the diverse types of paired fins in acanthodians, arthrodires, and antiarchs are also taken into consideration, it seems more probable that paired appendages arose *A modern hypothesis: parallel evolution* independently in several different vertebrate lines. This hypothesis accords with genetic and evolutionary principles. Since the emergent vertebrates were more closely related to each other than to animals of other groups, they shared a significant number of genes of like structure in which similar mutations could take place. For this reason, it is not beyond the realm of probability that genetic changes of the same sort occurred several times to produce, in different lines of fishes, folds in the body wall or enlarged spines in the dermal armor. If paired appendages arose in this way, their origin is an example of the phenomenon known as parallel evolution. The term is used to describe the development of new characters of a similar nature in organisms descended from the same stock but belonging to separate evolutionary lines. Structures which appear through parallel evolution are homologous only in the broadest sense. Since they are not descended from a common antecedent, they fail the more rigorous test of homology that requires the derivation of homologous characters from a single preexisting structure.

In primitive fishes, appendages which arose through mutation would have been preserved by natural selection for their adaptive value. Although many of the early fishes hugged the bottom, their axial musculature made freer, self-directed locomotion possible. Unlike the jellyfishes and smaller planktonic animals that were adapted to live as the currents carried them, the muscular vertebrates never attained the open waters by developing forms that could be tossed about at random. Their success depended from the first upon the appearance of appendages which restricted the aberrant motions that any body moving through the water incurs. In every line, continuing evolution would have brought about a combination of fins and spines that gave maximum stability to the animal while interfering as little as possible with the streamlining necessary for successful swimming. Dorsal and ventral keels in the midline would have lessened the tendency of the body to roll around its axis; paired appendages, while contributing to the control of rolling, would also have reduced pitching vertically and yawing from side to side. The better developed these mechanisms of control became, the more successful the fishes were likely to be as swimming forms.

The function of fins in fishes

Another factor which would have increased the natural selection for survival of those vertebrates which developed paired appendages was the necessity of obtaining lift in swimming away from the bottom or maintaining an upward course through the water. Among the lower fishes discussed so far, it is significant that flexible paired appendages appear in combination with a heterocercal caudal fin—a tail fin whose larger upper lobe elevates the posterior part of the body and depresses the head. *Cephalaspis,* among the osteostracans, had such a caudal fin and not surprisingly had pectoral flaps, which, held at the proper angle, doubtless prevented the head from being driven into the mud. Whether the laterally projecting spines borne by many ostracoderms helped in lifting the head would have depended on the extent of their surface and the plane in which they were set relative to the body. In forms where such spines were absent or not effective in producing lift, the caudal fin seems to have been of the reversed heterocercal type, one with a larger lower lobe which elevated the head instead of depressing it.

The first jawed fishes lived under the same environmental pressures as the ostracoderms and thus competed for survival more successfully as they developed their own mechanisms for control and lift. The paired fins and spines of placoderms and acanthodians differed in design but surely exerted a similar stabilizing effect. Since the heterocercal caudal fin seems to have been widespread in both groups, the same requirement for elevation of the anterior end of the body may have influenced the evolution of pectoral appendages as the dominant paired projections from the trunk. The complexity of the paired appendages of the early gnathostome fishes eventually surpassed that of the ostracoderms. For the filter-feeding ostracoderms there may have been little adaptive advantage in more elaborate structures. If any appeared, they were not selected for. In the case of the acanthodians and placoderms, however, improved appendages that contributed to better swimming were vital in the development of the predatory habit the hinged jaws allowed. The appendages showed progressive change, not as a result of any mysterious orthogenetic trend, but because they were among a number of characters whose simultaneous development produced an increasingly effective underwater form.

The evolution of fin shape

There is still disagreement among paleontologists concerning the sequence of changes in the paired fins of the earliest gnathostomes. Although few now seek to prove the derivation of the fins from a pair of continuous fin folds, the idea lingers that appendages broadly attached to the body wall are primitive. From such fins, useful as stabilizers and elevators, goes a broadly accepted hypothesis, those with a narrower base evolved. The latter, supported by a more specialized endoskeleton and manipulated by more complex musculature, would increase the maneuverability of the fish. Evidence to support this version of the evolution of fins has been drawn from fossils of higher fishes: changes from broad- to narrow-based fins took place in both the shark and bony-fish lines. Trends in the evolution of acanthodian fins are not incompatible with this principle. Even though acanthodian fins remained broad at the base, it seems that they became more mobile as the anterior spine slimmed, the web enlarged, and basal and radial supporting elements appeared to serve as areas of muscle attachment.

Evolution of paired fins in placoderms

It is by no means so clear that such a sequence of evolutionary change occurred in the placoderm line. Among these varied fishes both broad- and narrow-based paired fins are known. Breadth of attachment ranged from the skate-like pectoral- and pelvic-fin expansion of the highly specialized rhenanids to the slender tribasic fins of *Ctenurella*, an equally specialized ptyctodont described by Ørvig. Within the arthrodire group, where fossilized remnants of fin skeletons are infrequently found, it is possible to estimate the extent to which the fin was attached to the body wall by the size of the opening in the trunk armor behind the pectoral spine. In the arctolepids, the gap between the anterior lateral and anterior ventrolateral dermal plates was small and would have allowed the emergence of a small fin with a constricted base. In the brachythoracids, the postspinal gap is larger: in what appear to be the most advanced members of the group, not only do the anterior lateral and anterior ventrolateral plates fail to meet, but the posterior lateral and posterior ventrolateral plates behind them are widely separated as well.

A difference of opinion exists among paleontologists concerning the interpretation of the fossil evidence. T. S. Westoll has expressed one idea, E. Stensiö and E. White quite another. Westoll, who regards the Lower Devonian arctolepids as more primitive arthrodires than the middle and late Devonian brachythoracids, reasons that the earliest arthrodire fins were narrow-based (Fig. 4.7). As they appeared in the arctolepids, the pectoral fins were housed at first within the hollow pectoral spines, Westoll believes, their blood vessels and nerves passing outward from trunk to fin through foramina in a bony partition across the base of the spine called the prespinal lamella. The development of a freer fin began as the posterior edge of the fin emerged from behind the spine through the small gap in the trunk armor. As the brachythoracids evolved, the spine diminished in size, the prespinal lamella disappeared, and the free fin became more and more broadly attached to the body wall. According to Westoll, the reduction of the trunk armor not only permitted the development of freer fins but necessitated it, as the liberation of the trunk musculature would have increased the vigor of the swimming movements to a point at which survival without increased control would have been impossible.

Stensiö and White reject the idea of evolution from a narrow-based to a wide-based type of fin. They insist upon the broad-based fin as the ancestral condition, holding that the brachythoracids have preserved it to the end of the Devonian period. The arctolepids, they think, with their large pectoral fin spines and small area of fin attachment, constitute a group that has diverged far from the base of the arthrodiran line. Stensiö and White view the arctolepid fins as becoming more concentrated at the base and the spines as evolving after the fins, enlarging to cover the endoskeleton of the pectoral girdle, of which the prespinal lamella they believe formed a part. To the objection that their interpretation of the fossil evidence ignores the

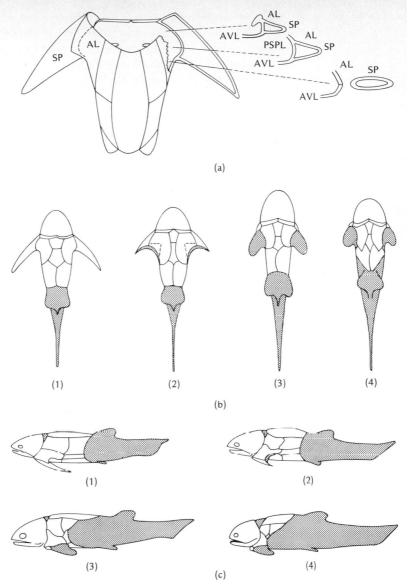

Figure 4.7. Westoll's theory of the evolution of the pectoral fins in arthrodires. **(a)** Trunk armor and pectoral spine of an arctolepid arthrodire in dorsal view with cross sections cut at intervals to show position of the prespinal lamella. **(b)** Ventral and **(c)** lateral views of (1) an arctolepid; (2) *Phlyctaenaspis*; (3) *Coccosteus*; and (4) *Dunkleosteus* (= *Dinichthys*), showing gradual extension of the fin web, reduction of the pectoral spine, and shortening of the trunk armor. Abbreviations: *AL, AVL, SP,* anterolateral, anteroventrolateral, and spinal plates of trunk armor, respectively; *PSPL,* prespinal lamella. (From Westoll.)

succession of the arthrodires established by stratigraphical data, Stensiö and White answer that the really small number of known fossils may represent a skewed sample of the living forms. Since the gaps in the record are large, they argue, there are no grounds for dismissing their interpretation of fin evolution, logically based on principles of structure, merely because it conflicts with present ideas of arthrodire chronology.

Pectoral fins of antiarchs

The evolution of the armored, spine-like pectoral fins of the antiarchs constitutes a special problem and one that is far from being resolved. When the antiarchs appear in the Lower Devonian, they possess, fully formed, the peculiar appendages that W. Gross has dubbed "arthropterygia" because of their superficial resemblance to arthropod limbs. No fossil has been found bearing appendages recognizably transitional between those of antiarchs and any other group of placoderms. Most paleontologists agree that the antiarchs must have diverged from arthrodiran stock very early in the history of the jawed fishes. Little progress has been made in understanding how the fins might have developed from the arthrodiran pattern. Gross theorized that the dermal plates that covered the fin were derived in some manner from the pectoral spine. Westoll accepted Gross's hypothesis and attempted to explain the development of the articulation that made the antiarch fin movable about its base. He thinks that the helmet process borne on the anterior ventrolateral dermal plate which serves as the proximal articular surface has been derived from the arctolepid prespinal lamella. The helmet process is described further by Watson as receiving an endoskeletal element in a pit at its center and dorsal and ventral dermal arm plates on its periphery (Fig. 4.8). Stensiö has suggested that the endoskeletal member was derived from a central endoskeletal axis that he believes existed in ancient placoderms like *Paleacanthaspis*. White has found the ideas of both Westoll and Stensiö unsatisfactory but has not proposed an alternative explanation.

Evolution of jaws

While the evolution of paired appendages among the placoderms and acanthodians was foreshadowed by a number of structures at the ostracoderm level, the appearance of jaws in these fishes was not. The mouth in ostracoderms was edged by dermal bone in such a way that its aperture could be changed very little; in the first gnathostomes the hinged lower jaw existed in a fully functional condition, making possible the enlarged gape and forceful closure that provided at once a new mechanism for food getting, defense, and offense. There is no

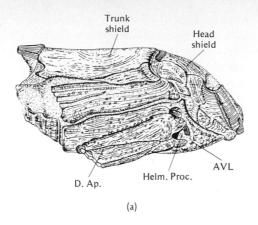

Trunk
shield

Head
shield

AVL

Helm. Proc.

D. Ap.

(a)

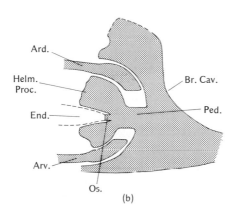

Ard.

Helm.
Proc.

Br. Cav.

End.

Ped.

Arv.

Os.

(b)

Figure 4.8. The base of the antiarch fin skeleton as reconstructed by Watson. **(a)** Head and trunk armor of the antiarch *Bothriolepis* in right lateral view, showing upper part of pectoral appendage articulating with helmet process on the anterior ventrolateral plate of the trunk shield. **(b)** Diagram of the structure of the articulation. Abbreviations: *Ard., Arv.,* dorsal and ventral dermal plates of the appendage; *AVL,* anterior ventrolateral plate of trunk shield; *Br. Cav,* branchial cavity; *D. Ap,* dermal plates sheathing appendage; *End.,* endoskeletal element supporting appendage; *Helm. Proc.,* helmet process; *Os.,* ossified articular facet; *Ped.,* pedicel. (From Watson.)

evidence in the fossil record of stages transitional to the jaw apparatus that proved indispensable then and thereafter to the success of the vertebrates.

Although paleontologists have not been able to trace the evolution of the jaws, anatomists and embryologists have made their source clear. In his textbook written in the 1870s, Gegenbaur stated a fact obvious to every student: the palatoquadrate, or upper, and mandibular, or lower, jaw elements are serially homologous to parts of the gill arches just behind them. Embryologists substantiated the point by demonstrating that the endochondral portions of the jaws are derived, like the branchial skeleton, from ectomesenchyme of neural crest origin (Fig. 4.9). Their musculature and its innervation is, they found, an extension of the special visceral system associated with the gill arches. The identification of the jaws is sure, having been confirmed by numerous studies. After the discovery of the ectomesenchyme by Platt in 1895 and its derivation from the neural crest by Landacre in 1921, the development of the jaw cartilages and the gill arches from this common material was observed in every class of vertebrates.

Still hopeful of finding relevant fossil material, paleontologists have proposed certain hypotheses which they expect new evidence to support. Since osteostracans and anaspids show more than the six pairs of branchial arches generally characteristic of gnathostomes, it has been suggested that more than one anterior pair became involved in the development of the jaws. Sewertzoff and Jaekel postulated that two or three arches lay before what is now considered the first, or mandibular, arch, while Allis, De Beer, and Jarvik concluded that there was a single premandibu-

Figure 4.9. The basic structure of the branchial skeleton in a cartilaginous fish and the position of the branchial elements in relation to the braincase and anterior part of the vertebral column. The palatoquadrate element in the first, or mandibular arch, and the hyomandibula in the second, or hyoid arch, are serially homologous to the epibranchial elements of the more posterior arches. Meckel's cartilage and the ceratohyal are homologous to the posterior ceratobranchial elements. (From Weichert.)

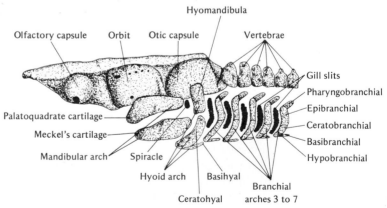

lar pair. Jarvik believes that he has recognized traces of that premandibular arch, not in placoderms, but in living jawless fishes of the genus *Petromyzon* and in an early crossopterygian gnathostome, *Eusthenopteron*. His studies of cephalaspids convince him that the premandibular arch contributed, even before the advent of jaws, to the base of the endocranium, as anterior branchial elements do in extant forms.

Extrapolating from the branchial skeleton of ostracoderms to its homologues in the first jawed fishes is hazardous because its anatomy in the extinct agnaths is not well known. Gill arches like those of the cephalaspids, consisting of uninterrupted elements continuous with each other and with the cranium, constitute problematic antecedents for the jaws and other parts of the branchial skeleton in gnathostomes composed of discrete units movably articulated with one another. Some workers have sought to resolve this difficulty by supposing that joints could have evolved in the unbroken skeletal framework supporting the osteostracan gills. No example of such a change in any part of the vertebrate skeleton is known, however, and a mechanism for the process is hard to imagine. It is possible that the condition of the osteostracan branchial arch system is not ancestral to that of the jawed fishes but a specialization that survived, perhaps, only in cyclostomes. The branchial skeleton of the heterostracans, when it becomes known, may prove to be quite different from the osteostracan model and a more feasible precursor for the jaws and gill arches of acanthodians and placoderms.

The search for primitive jaws in fossil fishes

It would have been helpful to those attempting to understand the evolution of the jaws had there appeared among the first gnathostomes an arrangement of the jaw elements that was obviously primitive. Watson believed that in both placoderms and acanthodians such a condition did exist. Modification of the anterior branchial arches had produced the jaws, he thought, but had not yet transformed the dorsal part of the second, or hyoid, arch into a suspensory element—one that braced the jaw joint in front of it. The gill slit of normal size that he found between the mandibular and hyoid arch seemed to him proof of his contention, for the opening in that position is reduced to a spiracle or lost as the dorsal hyoid element, the hyomandibula or epihyal, becomes associated with the jaw articulation. On the basis of this interpretation of the jaw apparatus, Watson erected the category Aphetohyoidea and included within it all the early gnathostomes whose hyoid arch was, he supposed, in a primitive state. In his opinion, this characteristic united the acanthodians, arthrodires, antiarchs, petalichthyids, and rhenanids.

It appears now that Watson was wrong, and his Aphetohyoidea has been abandoned. Aside from the fact that acanthodians and placoderms are too different to be housed in the same group on the basis of a single characteristic, the structure of the jaws in both kinds of fishes has proved to be not so primitive as Watson thought. Stensiö found, through his extensive studies of the head in placoderms, that considerable variety in the form and suspension of the jaws existed among these fishes. In some, the jaws are propped against the hyoid in hyostylic fashion; in others, the palatoquadrate is fused to the cranium, the hyoid having returned secondarily to the nonsuspensory condition. None of the placoderms shows clear evidence of the full gill slit between mandibular and hyoid arches that Watson believed was present. Among the acanthodians, study of the branchial skeleton has been possible only in *Acanthodes*. The dorsal part of the hyoid arch in this fish is not highly modified, but it does extend far forward to participate in the jaw suspension. In none of the known lower gnathostomes, it seems, are clues forthcoming to the earliest stages in the history of the jaws.

Although the crucial changes in the evolution of the gnathostome mouth were those involved in the transformation of anterior branchial arches into hinged jaw elements, other innovations contributed to the functional efficiency of the new apparatus. Dermal bones developed in several lines to brace or cover the endochondral bars, and teeth appeared generally at the margins of the jaws. As in the case of the jaw suspension, no arrangement of dermal plates and teeth now known among the acanthodians and placoderms is more primitive than the others. The various kinds of tooth and bone development are rather the diversified end products of adaptive changes in the divergent gnathostome groups.

The appearance of teeth

The evolution of teeth or analogous structures in nearly all the early jawed fishes can be attributed to the universal advantage of having barbs on the jaws to prevent the escape of live food. Variety in the structure of the teeth and jaws in different groups resulted from the interplay of mutation and natural selection as acanthodians and placoderms spread into different watery environments (Fig. 4.10). Placoderms, like the antiarchs and petalichthyids that scavenged on the bottom, had short jaws bearing nipping tooth plates. The rhenanids, also bottom-feeders, evolved a protrusible mouth, toothless except for small

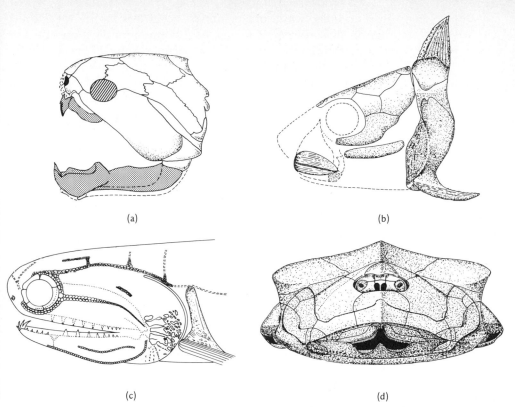

(a)　　　　　　　　　　　　　　(b)

(c)　　　　　　　　　　　　　　(d)

Figure 4.10. Teeth in early jawed fishes. **(a)** Head of the brachythoracid arthrodire *Dunkleosteus*, with jaws bearing bony prongs and blades. **(b)** Head of the ptyctodont *Rhamphodopsis*, showing tooth plates. **(c)** Head of the acanthodian *Ischnacanthus*, showing small marginal teeth and symphyseal tooth whorl. **(d)** Head of the antiarch *Bothriolepis*, in frontal view, showing small bony plates edging the mouth. (*a,b* from Miles; *c* from Watson; *d* from Stensiö.)

tubercles mounted on the palatoquadrate. The seagoing brachythoracids, which were predaceous, grew prongs and cutting blades on long jaw elements, the upper of which were braced by the heavy dermal bones of the cheek. The jaws in ptyctodonts lacked dermal reinforcement but supported tooth plates which enabled their owners to utilize food unavailable to fishes with less massive dental structures. Slender, slightly curving teeth developed in acanthodians but disappeared from the palatoquadrate in some genera and from both jaws in others. The endochondral elements were flanked by dermal scales in the acanthodian fishes and in *Acanthodes* were supported by a narrow submandibular dermal bar. The teeth and jaws in acanthodians, while not primitive, seem less highly specialized than in placoderms. The lack of extreme modification of these structures in connection with a peculiar habit of feeding may have contributed to the long survival of the acanthodian group.

For investigators working on acanthodians and placoderms there is an unsolved problem even broader in scope than the origin of jaws and paired appendages. The phylogenetic relationships among the first gnathostomes are poorly known, and everyone interested in these fishes is bent on their further elucidation. Students of every category of vertebrates are faced with the task of determining how the members of the group are related to each other, but for forms that lived 400 million years ago ascertaining evolutionary connections is enormously difficult. The meagerness of the fossil record, especially the lack of placoderm material, will support little more than the most general speculations. These speculations are worthwhile, however, when they are made in accordance with the principles of modern systematics.

Systematics: its use in the study of evolution

Since the concepts of systematics are applicable to the study of the evolutionary history of all the vertebrates, it would be well to expand upon them at least briefly before considering specific problems of interrelationships among the placoderms or any other group. Systematics may be described as the study of diverse organisms from a broadly comparative point of view. The systematist weighs and synthesizes evidence from several scientific disciplines to produce as accurate an estimate as possible of the relationships exhibited by the forms under investigation. He may express his conclusions diagrammatically as a system of classification. Though the methods of systematics are applicable far beyond the realm of biology and classifications can be variously based, students of evolution use hereditary relationships to order organisms into natural groups. Natural groups are those, then, whose members are descended from a common ancestor. Insofar as a system of classification can be built of natural groups of vertebrates, it will accurately reflect the progress of their evolution. The inclusion of artificial aggregations of animals that have one or more traits in common but are not genetically close is an error that subtracts from the validity of the scheme.

In estimating hereditary relationships of extinct vertebrates, modern paleontologists integrate information from a wider range of sources than earlier workers used. Technological advances, responsible for new tools of investigation in the field and in the laboratory, have increased the amount of data streaming in from every direction. Reinterpretation of old conclusions is constantly necessary as new knowledge accumulates. If, as a result of their work, paleontologists change their minds about the position of a group of animals, they modify the existing classification. The emphasis is on modification rather than addition. The classification scheme resembles not so much a highway being extended over new ground as a road which is continuously being repaired. The reparations rarely produce a clear passage, especially where fossil evidence is fragmentary, because workers in the same area often disagree about the changes that should be made. Men of different opinions hold to different versions of the classification, and, as a result, those using the system must discriminate among parallel paths.

The significance of structural similarity

In spite of their adherence to the principles of systematics, paleontologists still face pitfalls that cause them to make errors in identifying evolutionary lines and determining the order of descent within them. Structural similarity is easy to misinterpret. When organisms share characteristics, there are three possible explanations: one animal may have been derived from the stock to which the other belongs; the two may belong to different stocks that have arisen from a common ancestor; or the forms may represent individuals of widely separated lines in which similar traits have developed convergently. To ascertain which explanation is correct in a particular case, it is necessary to analyze the distinguishing characteristics of the organisms in question and to know the animals' relative age. If structural differences are of a superficial nature and extremely divergent modifications in the older form are few, it is possible that the animals are representatives of a single evolutionary line. The presence, beside structures showing minor differences, of a significant number whose specialization seems to be proceeding in separate directions usually means that the specimens belong to groups which radiated independently from a common stock. If fossils of the ancestral forms are known, it is often possible to determine whether similar characters in the descendants were carried from an ancient antecedent or developed afterward through parallel evolution. The appearance of like traits through convergence is the explanation to be preferred if the distinctive characters of the animals differ greatly and the similarities, few in number, involve structures that are strongly adaptive, like teeth or general body form.

In practice, recognition of the true relationship between

animals with morphological similarities is a more complicated affair than it might seem at first. For extinct forms, the relative age of the fossil remains is a significant factor but not a sure criterion for making interpretations of evolutionary status. Despite detailed stratigraphical studies, it is usually not possible to correlate closely the ages of fossils dug from rocks in widely separated locations. Assigning several different placoderms to the Lower Devonian masks what may be a spread of several million years among the various groups at a time when these early jawed fishes were evolving at a rapid rate. Under these conditions, the age of the fossils cannot be used to determine whether forms that show a degree of similarity are in the same line or the result of parallel evolution. Where change has occurred at a fast pace to produce a number of divergent groups, it is frequently impossible to tell which groups are more closely related and thus to trace the pattern of the evolutionary branching from the basal stock.

Age of fossils as a factor in determining relationships

When a difference in age can be discerned, it contróls the interpretation of the significance of similarity between two forms. If, for instance, two animals resemble each other markedly and one is known to have preceded the other in time, it can be argued that the stock of the older may be ancestral to that of the younger. The argument holds especially well if the earlier form shows characteristics that seem primitive or that could have been transitional to structures exhibited by the later specimen. Should the later specimen appear to be in many ways the less advanced of the two, the resemblance between the animals must be interpreted differently. It would be unlikely in that case that the animals belonged to the same line. The younger might be a late representative of a conservative branch whose members had changed little from the ancestral stock. The earlier but more divergent form could have belonged to a group that differentiated from the same base at a faster rate. It is also possible that both forms were equally removed from their common ancestor but that the animals of the younger type had reverted secondarily in several salient characters to what appears to be a simpler level of structure.

The argument that collateral lines may evolve at different rates has been used by paleontologists to support interpretations which run counter to theories based on chronology. An example of such reasoning has already been seen in Stensiö and White's interpretation of the evolution of arthrodire appendages. The establishment of the arctolepids as the older and the brachythoracids as the younger forms has not deterred Stensiö and White from accepting the brachythoracid broad-based fin as the more primitive type. They regard it as having been retained conservatively into Middle and Upper Devonian times, long after the arctolepid appendage had changed to the narrow-based and long-spined form visible in the Lower Devonian specimens.

Where a large difference in age exists between fossil forms of similar appearance, the likelihood of convergence is strong. Entrance by successive groups into the same kind of environment, development there of adaptive characteristics, and eventual extinction leaves a record of animals with similar structures at widely spaced stratigraphical levels. Although resemblances between some of these forms may be quite startling, extended studies usually reveal widely different evolutionary histories for the forms in question. Convergence explains the remarkable resemblance between the Devonian rhenanids and modern skates. Although the shape of the body is strikingly similar in the two kinds of fishes, paleontologists know that they are far apart on the evolutionary scale. Convergence is not always the answer in cases of morphological likeness between two specimens of distinctly different age, however. It is within the realm of possibility that such forms may even belong within the same line. Stensiö suggested long ago that the cyclostomes, known in fossil form until recently only from Tertiary rocks, were modern descendants of the Silurian and Devonian ostracoderms. Evidence supporting Stensiö's contention was presented in 1968 by Zangerl and Bardack, in the form of a fossil cyclostome from the Carboniferous period. (See Fig. 2.12.) This specimen, found in Pennsylvanian shale, is a lamprey little different from the modern one. It is apparent now that the transition from armored to unarmored fish took place during the Devonian and that the surviving forms have descended through 300 million years virtually unchanged. Paleontologists pondering similar specimens of widely disparate age may think first of convergence, but the discovery of the probable link between the ostracoderms and cyclostomes serves as a reminder that other interpretations are possible.

Geographical distribution is important to the systematist

Geographical distribution has a bearing on the interpretation of relationships between morphologically similar forms and therefore must be considered by the systematist. For example, he cannot propose a phylogenetic connection between two organisms that were separated spatially throughout their entire history. Nor can he ignore the support given by propinquity to the possibility of the direct descent of one form from another when structure and relative age suggest it. Evidence drawn from geographical distribution is especially crucial in determining relationships among land animals. Such forms do, in time, migrate over vast distances within continents and between those that are connected by land bridges. Though many kinds of land animals spread gradually through archipelagos by island-hopping, broad expanses of water confine whole populations to a particular land mass. A study of the phylogenetic relationship between two groups is not complete without an assessment of their geographical range. To make such an assessment the investigator must familiarize himself with the location and extent of land and water areas on the earth during the lifetime of the groups in question. Since the continents themselves probably drifted in addition to changing their shape as the level of the ocean rose and fell, determination of paleogeography is often a complicated matter. Changes in climate have to be considered as well as those in topography, because temperature barriers may have prevented movement of populations where the lay of the land did not.

The peregrinations of ancient fishes are even harder to trace than those of land animals. Barriers to the free passage of underwater vertebrates certainly existed in the form of temperature, salinity, and pressure differences in former times, as they do now, but the patterns they formed and the migratory channels they allowed are impossible to discover. Moreover, the physiology of fishes seems remarkably adaptive. Recent species evolve varieties that extend the range of the group, and fossil evidence shows that many extinct fishes overcame at least the salinity barrier by developing both freshwater and marine forms. Since paleontologists have found that representatives of the major taxa were geographically widespread, they have been able to base few arguments concerning possible relationships, especially among marine fishes, on peculiarities of distribution.

Definition of monophyly and polyphyly

When sufficient information is ascertainable about the structure, age, and geographical distribution of similar fossil forms, it may be clear that the animals descended, in all probability, from a common ancestor and thus constitute a natural or monophyletic group. The aim of the systematists is to separate eventually all the vertebrates—fishes and land forms—into monophyletic aggregations. While they have succeeded in eliminating the extremely artificial groups of the early classifiers (such as the Unguiculata of the seventeenth-century naturalist

John Ray that included as clawed animals mice, monkeys, elephants, and camels), they are still far short of their goal. Groups like the Teleostei that embrace most of the modern bony fishes contain forms with a set of characteristics which distinguish their members from other animals but which have almost certainly evolved in a parallel fashion in families derived from different precursors. Such an association is polyphyletic: its members have reached approximately the same level or grade of development but do not share a common origin. Several taxa that now appear in the classification are suspected of being of this type. The placoderms, for instance, may be polyphyletic, since they do show structures, like the ball-and-socket joint between the head and shoulder shields, whose diverse design precludes their having evolved from a single source. Polyphyletic groups will remain in the classification until systematists sort out the separate lineages of their members. Since the evidence required for this task accumulates very slowly and in some cases may not come to light at all, the ideal scheme will probably never be attained. Systematists are not preoccupied with the ultimate possibilities, however. They are satisfied if their work results in improvements in the classification currently in use.

Management of the classification system

The participation in the study of vertebrate relationships of an ever increasing number of scientists from different fields has made management of the classification scheme a serious problem. The difficulty that stems from the existence of several versions of the classification for the same group has already been mentioned. The student who undertakes an investigation in depth must face a fearsome and sometimes confusing array of terms. Complication of this kind has to be tolerated as unavoidable in an area of active research, but a lack of clarity in the application of names does not.

To suppress the chaos that began to appear in the naming of newly discovered forms as the study of animal life burgeoned in the nineteenth century, a code was drawn up, the International Rules of Zoological Nomenclature. It is designed to prevent the use of more than one name for a single taxon and to maintain a uniform content for a category designated by a particular name. The code has been especially effective in controling terminology at the level of the genus and species. Although nothing can prevent two workers from discovering the same form and naming it differently, the rules give priority to the name which is published first. Strict application of the law of priority has occasionally produced some problems of its own: sometimes a name appears first in a journal with a small circulation and is not brought into general view until long after a later name has established itself in the literature. Having to abandon, for reasons of priority, a familiar name like *Amphioxus*, coined by Yarrel in 1836, for *Branchiostoma*, which was, someone discovered, published two years earlier by Costa, is a painful process. An international commission meets from time to time to review the rules and to grant exceptions to the law of priority where it works excessive hardship. The commission cannot cover all the cases that arise, however, and as a result some argument concerning names goes on continually.

The use of type specimens

To guard against the possibility that different organisms would be classified under the same name, the use of type specimens was introduced. Under this system, a worker who is describing a newly discovered organism devises a name for the species and assigns it to a particular specimen. That specimen becomes known as the type for the name. Other workers who wish to use that specific designation must compare the organism to which they want to assign it with the type specimen. Since the details on which such comparisons are based are usually very fine, an investigator often prefers to see the type specimen rather than to rely on verbal descriptions or photographs of it. He must arrange to borrow it or, since the owner may be loath to risk its loss in transit, travel to the museum or laboratory where it is kept.

In using the type system to control the content of a category in the classification scheme, scientists have to treat the type specimen as a standard for the terminology alone and not as an ideal representative of the species itself. The distinction, though subtle, is a most important one, as it reflects the change that has taken place in the concept of the species during the last century. Those who see in the type specimen the perfect model of a particular kind of animal are reverting to the old philosophic idea of archetypes—symbolic forms that embodied the essence of the species they represented. The archetype of an organism was its epitome, changeless and perfect. The concept of the archetype was useful to the eighteenth-century classifiers who labored mightily to describe and put in order the known forms of life. By defining separate archetypes of living things and arranging them in a *scala naturae*, a ladder of natural organisms, these early workers made themselves aware of the spectrum of life from the small forms to the large, from the simple to the complex.

Although the possibility of a serial arrangement of living things emerged through the definition of archetypes, the transformation of the static *scala naturae* into the dynamic, multibranched evolutionary tree did not and could not take place until the idea of the archetype as the expression of the species was discarded. The archetype was a metaphysical concept, one which led biologists toward philosophical and theological interpretations of natural phenomena. By the middle of the nineteenth century, when the schism widened between science and philosophy, it became obvious to observers of nature that archetypes did not exist in the field. A species turned out in reality to be a population of organisms whose characteristics varied about a mean. Darwin made this discovery for himself as he traveled through South America and the Galapagos Islands in the 1830s; Wallace observed the same thing in Malaysia two decades later. It was no coincidence that these were the men who placed forcefully before the world the idea that species were variable and over long periods of time could change their nature entirely. Once the species was established as an aggregation of mutable, interbreeding individuals rather than organisms that approached an archetypical form, the way was cleared for all the work that led to the modern theory of evolution.

Drawing taxonomic boundaries

The variation shown by species, both geographically and in time, sets still another difficulty in the way of the systematists. Since populations undergoing gradual change fit imperfectly into a scheme consisting of discrete taxonomical units, workers often disagree about the best way to classify related organisms. Although the problem of the confines of the species has been argued interminably, it is controlled, at least in principle, by the assertion that members of the same species must be able to interbreed either directly or indirectly through an intermediate population. Where information concerning breeding is unavailable (as is the case with fossil organisms), the decision about the likelihood of a reproductive relationship between contemporary groups rests on deductions from their morphology, ecology, and geographical distribution. If the last two are not distinctive, paleontologists must decide whether the degree of morphological variation that can be observed is evidence of mere individual peculiarity or is an indicator of a difference in specific rank. Usually a disparity in age is accepted as proof that the organisms were not members of the same breeding group, although it is conceivable that this reasoning has resulted in dividing a continuously evolving line into a succession of separate species.

Above the species, there is no principle that governs the content of the group besides the stricture that its members should trace their ancestry to a common source. Whether an aggregation under study should be maintained in a single genus, segregated further into subgenera, or cut apart into several genera is a matter for the investigator to decide. His choice will depend upon the significance he attaches to the differences among the organisms in the aggregation. The specialist, thoroughly familiar with a circumscribed group, is likely to create an elaborately divided classification scheme, since minute differences that exist among similar forms are important to him. A worker whose view is broader tends to minimize the number of subdivisions. G. G. Simpson has characterized systematists as "splitters" or "lumpers" on the basis of their approach to classification. Splitters and lumpers may agree on the general way in which a group of vertebrates has evolved, but the classification schemes they construct will differ in the number of categories they contain.

Both splitters and lumpers have found classification of the first gnathostomes troublesome, because there has been so little evidence with which to work. After the failure of Watson's attempt to unite acanthodians and placoderms in the same class, the acanthodians were recognized as an independent and undoubtedly monophyletic group. The argument centered, in the case of these fishes, around the question of the higher taxon to which they belonged. The placoderms posed a knottier problem: besides the matter of their affiliation, systematists had to consider the classification of the various specimens within the group. Although the distinct head and shoulder shield and pectoral spines are common characters for the assemblage, the diversity of jaw and fin structure and the differences in pattern of the dermal armor obscure the nature of the interrelationships among the different orders.

Evolution of dermal armor in placoderms

The evolution of the dermal armor of placoderms has been much discussed because it is one of the keys to understanding the history of the group. It is evident that the arthrodires, the antiarchs, and the petalichthyids developed different patterns of plates in the head and shoulder shields as their lines diverged, but the way in which their armor changed from the ancestral state is unknown. Since paleontologists have not recognized any fossil specimens as forms that could be directly antecedent to the placoderm fishes, they can only guess the nature of the dermal covering that was primary for the group. Although it is possible

that "eoarthrodires" (Stensiö's name for the hypothetical ancestors of the placoderms) had a solid dermal shield which frag-
mented into separate plates, most investigators think it more
probable that these animals bore tessellae, or small, tile-like
units, which tended to fuse (Fig. 4.11). L. B. Halstead-Tarlo
contends that the presence of a tessellated exoskeleton in
protoplacoderms would be consonant with the macular pattern
of dermal derivatives which he suggests is fundamental in
vertebrates. The existence of tessellae in the dermal covering of
forms like *Radotina* does not weigh in favor of either argument,
because there is no way of telling whether the minute plates
were derived secondarily through fragmentation or held over
from a primary mosaic arrangement.

As crucial an event in the evolution of the placoderms as the
development of the dermal armor, and as poorly understood, is
the appearance of the articulations between the skeleton of the

Appearance of the
cranio-vertebral
joint

head and trunk regions. All the placoderms seem to have had an
endoskeletal joint between the cranium and the vertebral column, and the most advanced and least flattened forms had, in

Figure 4.11. The head shield of *Astraspis* in dorsal view, showing tessellated
structure of the armor. (From Halstead-Tarlo.)

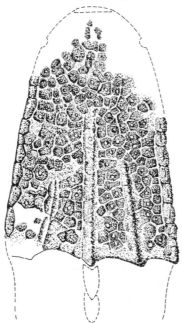

addition, a pair of intricate articulations between the head and shoulder shields. Flexibility in the skeleton at these points allowed the animals to move the head through several degrees in the vertical plane. The motion, limited as it was, contributed to the maintenance of equilibrium, the production of respiratory currents, and the enlargement of the mouth opening in predaceous forms.

Since the cranio-vertebral joint is always present and less variable in structure than the articulation between the dermal shields, White supposed that it evolved earlier even than the dermal armor. R. S. Miles has disagreed with this conclusion. His own hypothesis, that the evolution of the two was interdependent, he based on functional as well as morphological considerations. From his observations of extant fishes, he had learned that mobility of the head results from the flexibility of the anterior vertebrae or, in forms like holocephalians, skates, and rays, whose anterior vertebrae are fused, from a cranio-vertebral joint. He reasoned that the ancestors of the placoderms developed the cranio-vertebral joint as their evolving thoracic shield gradually restricted movement of the trunk vertebrae. Although Miles cannot explain the reason for the evolution of the heavy dermal shield in early placoderms (the advantage of having such a structure is not clear, since it diminished in many successful advanced forms), he believes that its appearance induced selection for the articulation between the cranium and the first vertebra.

Are placoderms monophyletic or polyphyletic?

Because of the common possession of a cranio-vertebral joint, unusual in fishes, and the development of the cephalic and thoracic shields, the placoderms have been long regarded as a monophyletic group. The possibility of polyphyly has been raised, however, because of the structural differences apparent in the paired joint between the two shields and in the pattern of the dermal plates over the head. Although these differences could signify that fishes descendent from more than one precursor group reached the placoderm grade, most workers are unwilling to accept the idea in the absence of positive fossil evidence of multiple ancestry. The placoderms could still be monophyletic, it is felt, their distinctive characteristics appearing as their lines diverged. Radiation from a common source would have been gradual, of course: formation of the thoracic armor, establishment of the cranio-vertebral joint, and movement of the branchial skeleton forward under the back of the skull would have been well advanced as the eoarthrodire group began to subdivide. Final expansion of the cephalic shield and completion of the articular facets between comparable head and shoulder plates could have occurred as parallel developments in the several lines of placoderms that became genetically isolated from one another. Inheritance of a common pool of genes from a single ancestral stock would explain the tendency to evolve parallel structures, while accumulation of unshared mutations would account for the variability in the pattern of plates in the head shield and the different design of the dermal articulations (Fig. 4.12). Miles believes that the ball-and-socket form of these joints has evolved at least three times: in the arthrodires, where the socket is on the head shield; in the petalichthyids, where a differently constructed socket is in the same place; and in the antiarchs, where the socket is borne on the thoracic armor.

Disagreement concerning interrelationships among the placoderms

Investigators trying to clarify interrelationships among the placoderms have to face the puzzling array of similarities and differences that existed among these fishes. Not surprisingly, their interpretations of the way in which the evolutionary tree has branched are not the same. There is general agreement, for instance, that the arctolepids and the brachythoracids are closely connected but no consensus on the nature of their relationship. While Westoll thinks that the former gave rise to the latter, Stensiö (believing that broad-based fins are primitive among the placoderms) reverses the order, and White suggests that the two had a common ancestor but diverged from one another. J. A.

Figure 4.12. Head shields of (*a*) a brachythoracid arthrodire and (*b*) an antiarch in dorsal view, showing difference in structure of facet articulating with trunk shield. **(a)** *Coccosteus*, with socket for reception of knob-like process on trunk armor. **(b)** *Asterolepis*, with trochlear process for projection into concavity in front of trunk armor. (*a* from Dean; *b* from Stensiö.)

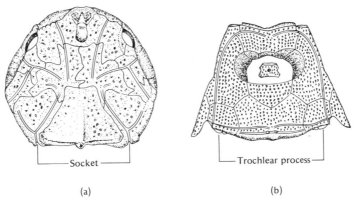

Socket

Trochlear process

(a)

(b)

Moy-Thomas placed rhenanids and apparently similar forms, the stensioellids, on a common stem in the order Stegoselachii; a dozen years later, White labeled the Stegoselachii "a systematic dust-bin for arthrodires of uncertain position" and set the rhenanids and stensioellids far apart in his evolutionary scheme. Romer followed Westoll in this separation, but the Russian D. W. Obrutchev has recently returned to the idea that the two are closely associated. As a last example of the confusion that reigns at present in the matter of placoderm evolution, it is Romer's opinion that the petalichthyids were a primitive group that diverged early from the line of fishes leading to the arctolepid arthrodires; White thinks that the two emerged from a single stock—the petalichthyids surviving into late Devonian times and the arctolepids becoming extinct before the middle of the period.

Relationship of placoderms and acanthodians to more advanced fishes

Paleontologists are equally unsure of the relationships of the placoderms and the acanthodians to other Devonian fishes. As has been seen, the closest scrutiny of the structure of the appendages and the jaws has not enabled investigators to connect the first gnathostomes with any one of the known jawless forms that lived earlier. Similarly, inspection of the characteristics of placoderms and acanthodians has not made clear how these animals are related to the osteichthyan and chondrichthyan forms that made their appearance in the Devonian. There are two points of view on the subject: in the past it was widely held that certain of the earliest known gnathostomes gave rise to the first bony fishes and sharks through a gradual transformation of structure; more recently, as fossil gnathostomes have been restudied and their anatomy more thoroughly deciphered, the acanthodians and placoderms have been interpreted as being out of the direct line to the more successful fishes. They have been regarded instead as side branches in the evolutionary scheme. Adherents of the former theory must assume that the structures of the earlier forms were converted to those of the later fishes or lost. Those who believe that acanthodians and placoderms have developed collaterally with the bony and cartilaginous fishes are spared the necessity of accounting for every anatomical character in this way. They may remark upon the resemblances of acanthodians and placoderms either to bony fishes or sharks but are free to regard the more peculiar features of the first-known gnathostomes as having evolved after these animals were established as a separate line.

Stensiö's Elasmo-branchiomorphi

On the basis of his own studies and those of Holmgren and

Jarvik, Stensiö has associated the acanthodians, the placoderms, and the chondrichthyan fishes under the heading Elasmobranchiomorphi. The placoderms and early sharks resemble each other, he finds, in several aspects of their endocranial anatomy and in the tendency to develop prominent spines (Fig. 4.13). He suggested, after making detailed comparisons, that there may be special affinity between petalichthyids and sharks and between ptyctodonts and holocephalians. Although he is less specific about particular relationships, he finds ample ground for tying the acanthodians into the assemblage. They

Figure 4.13. Structural similarities between acanthodians and sharks. **(a)** The pectoral skeleton in (1) *Acanthodes* and (2) the shark *Squalus*. **(b)** The symphyseal tooth whorl in (1) *Gomphodus* and (2) the Carboniferous shark *Edestus*. In *Edestus*, only a part of the whorl is shown. (*a*1 from Watson; *a*2 from Jollie; *b*1 from Gross; *b*2 from Romer.)

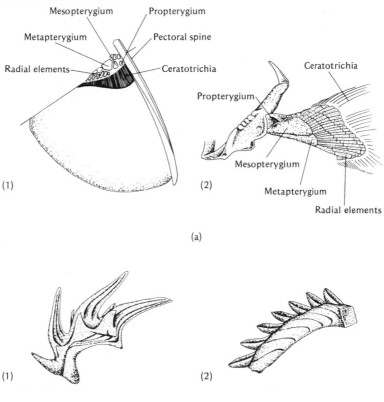

produce spines anterior to the fins, also, and have pectoral skeletal elements similar to those of elasmobranchs. In *Acanthodes*, for instance, there is a large scapula, a ventral coracoid, and three basals, a series of radials and ceratotrichia, extending into the fin behind the deep-set spine. *Climatius* seems to have had separate openings from the gill slits, as sharks do, and *Ischnacanthus* had a whorl of teeth at the symphysis of the lower jaw which, in its position and spiral form, was like that borne by the edestid sharks of later Paleozoic time. Besides the readily observable similarities that led to the acanthodians' being called "spiny sharks," Holmgren listed others of a more technical nature gleaned from his studies of the cranium and visceral arches. Those who recognize the Elasmobranchiomorphi as a natural group point out that the acanthodians, placoderms, and chondrichthyans are in all probability more closely related to each other than to the osteichthyans, since they were well established in marine waters hundreds of millions of years before the bony fishes reached the sea.

Not everyone agrees with Stensiö that the members of his Elasmobranchiomorphi constitute a natural group. Watson has attacked Stensiö's arguments point by point, favoring instead a relationship between the petalichthyids and the elasmobranchs; Moy-Thomas has found considerable support for the idea that holocephalians are a branch of the shark line rather than an offshoot from ptyctodont stock. There has been even less enthusiasm among paleontologists for associating the acanthodians with the cartilaginous fishes. In placing the acanthodians in the Elasmobranchiomorphi, Stensiö dismissed as having little significance the resemblance they bear to the earliest ray-finned bony fish. Other workers have found the likeness too strong to ignore. In his nineteenth-century study of fossil fishes, L. Agassiz bracketed the acanthodians and the first actinopterygians as ganoid fishes because both were covered with close-fitting, rectangular scales similar in their ganoine surface and deeper structure. Although the acanthodian body was admittedly shark-shaped, the appearance of the head, with its short pre-orbital region and large eyes, was remarkably like that of the bony fish. The anatomy of the visceral arches further strengthened the possibility of a relationship between the acanthodians and the actinopterygians: the dorsal elements of each arch, the pharyngobranchials, and the ventral hypobranchials were directed forward in acanthodians, as they are in actinopterygians, instead of backward, as in elasmobranchs. As a result, the acanthodian and actinopterygian arches are similarly $>$ -shaped and distinct from the elasmobranch $\lessgtr$ -shaped structures.

Miles' association
of acanthodians
with actinopterygi-
ans

In his comments on the relationships of the acanthodians, Miles argues forcefully that they share a common ancestry with the bony fishes rather than the elasmobranchs. Although he acknowledges that the resemblance between the head of acanthodians and that of actinopterygians may be due to convergence, he is convinced that the structural similarities in the branchial skeleton spring from common heredity rather than coincidence. The separate dorsal ossification in the hyoid arch of *Acanthodes* is homologous, he believes, to the laterohyal element that arises in the same location in the ray-finned bony fish. This dorsal portion of the hyoid in both kinds of fishes is lateral to the vein which drains the head instead of medial to it, as is the hyomandibula of elasmobranchs. He extends his argument by speculating that the space found between the jaw joint and the ventral end of the epihyal element in *Acanthodes* may have been occupied in life by a separate ossification equivalent to the symplectic bone of actinopterygians. He questions Watson's theory that the gill cover in acanthodians was distinct from that of bony fishes in developing from the mandibular rather than the hyoid arch. Miles, like Stensiö and Jarvik, interprets the fossil evidence as supporting the existence in acanthodians of an ordinary hyoidean operculum which may have covered the gill chamber completely, as it does in the bony fishes. Miles discounts Watson's restoration of the gill cover in *Climatius* and other early acanthodians as a small structure that left exposed, dorsally and posteriorly, shark-like separate gill openings. Watson's drawing was determined not so much by the configuration of the fossil material, in Miles's opinion, as by the prevailing idea that acanthodians resembled elasmobranchs. Any such resemblance seems less and less significant, however, to paleontologists who oppose Stensiö's classification, as the list of characteristics found in both acanthodians and actinopterygians grows longer. Miles rests his case for the association of the acanthodians and the bony fishes not only on the nature of the branchial arches but also on similarities in the cranium. It shows ossification in both and shares a number of structural features: the parasphenoid bone is similarly shaped and bears impressions of the spiracular pouches in the same location, the upper jaw articulates with the base of the braincase in the same way, and the inner ear contains compact otoliths unlike those of cartilaginous fishes.

Even the strongest advocates of an alliance between acanthodians and actinopterygians recognize that the older fishes were not directly ancestral to the later forms. The acanthodians show no trace of the dermal shoulder girdle that was already prominent in the first bony fishes, nor do they possess even the beginnings of a dermal palate beneath the braincase. The teeth in acanthodians, when they were present, were borne on the endoskeletal upper and lower jaws, not on dermal bones that edged the mouth, as the teeth of early bony fish were. No hint of the spines characteristic of acanthodians is found in any of the known actinopterygians. Their presence in the acanthodians may be the clearest evidence, in fact, that these fish were evolving in a line independent of other Devonian vertebrates. The development of these structures at the anterior margin of every appendage is unique: placoderms have only pectoral spines (and an occasional dorsal spike), elasmobranchs produce spines only in the midline, and early osteichthyes are devoid of spines altogether from their first appearance. The special characters of acanthodians and the early appearance of the group prove to Miles that these fishes split away from the base of the osteichthyan line. He finds no grounds in their deviation from the actinopterygian pattern, however, for allying them more closely to the placoderms and cartilaginous fish, as Stensiö does. D. Heyler, another investigator of early fishes, reconciles the similarities of the acanthodians to both cartilaginous and bony fishes by suggesting that acanthodians evolved simultaneously with sharks from a proselachian stock and later gave rise to the first actinopterygians. Heyler is aware of the objections that can be made to his hypothesis. In formulating it, he has assumed not only that acanthodians were directly ancestral to the more advanced bony fishes but also that the origin of the elasmobranchs occurred far back in the Silurian period. That fossil sharks have been found no earlier than the Middle Devonian he dismisses as an obstacle to his theory by supposing the remains of the first sharks to be still undiscovered.

Fate of the acanthodians and placoderms

While it is not possible to determine which of the several theories concerning the relationships of the primitive gnathostomes is most nearly correct, it is clear that acanthodians and placoderms competed unsuccessfully with the newer, more advanced swimmers. Dwindling numbers of fossils in Upper Devonian rocks attest to the gradual reduction in the population of the archaic fishes. It is safe to assume that their decline, slow at first, eventually became more rapid as survivors were left in isolated, inbreeding groups. The fossil record doubtless ends at this stage in the downhill progression of each line when the number of individuals fell low enough to make the recovery of remains highly unlikely. Extinction of the fishes would have occurred subsequently as their increasing rarity resulted in their meeting less and less regularly to reproduce.

Although it is never possible to ascertain exactly what structural and functional disadvantages were responsible for the disappearance of a group of animals, it is easier to guess their nature in some cases than in others. It seems most logical, for instance, that the ostracoderms lost the battle for survival because of their inferior feeding apparatus and swimming ability. The reason for the extinction of the acanthodians and the placoderms is not quite so obvious. These early gnathostomes had radiated into a wide variety of habitats from fresh waters to salt and had evolved a range of adaptive types from free swimmers to forms that stayed close to the bottom. It is curious that, despite their plasticity, they failed in competition everywhere and became the only major aggregation of vertebrates to leave no surviving remnant of their line.

Paradoxically, the acanthodians continued in existence for the longest time, though their spined fins and sparse teeth must have been less effective than the comparable structures of contemporary bony and cartilaginous fishes. The brachythoracid arthrodires, with more maneuverable fins and knife-sharp tooth blades, died out much sooner. Their disappearance is the more mysterious since they varied in size from a few inches to many feet and must have swum in water of different depths and relied on different organisms for their food supply. The extinction of the heavily armored antiarchs and arctolepids is easier to understand, since their archaic and highly specialized structure was probably associated with inflexible feeding habits and a relatively restricted habitat. What is amazing, really, is not the failure of one or the other of these groups but the loss of them all without exception. In the absence of a known environmental change that would have discriminated selectively against acanthodians and placoderms, paleontologists must suppose that the first jawed vertebrates were overwhelmed everywhere by representatives of the newer groups of fishes.

5

The Rise of the Modern Fishes

*Early appearance
of modern classes
of fishes*

Those who regard evolution as a process through which a group of organisms is gradually advanced or improved may find it paradoxical that several of the most successful vertebrate lines were established in early Devonian times, long before the appearance of land forms. The fishes that supplanted the archaic ostracoderms and placoderms have maintained supremacy in their environment for 400 million years and even now show no sign of having passed their peak (Fig. 5.1). The bony fishes, or Osteichthyes, are represented today by the conservative lungfishes, the burgeoning ray-finned teleosts, and a handful of forms left over from once more numerous groups. The cartilaginous fishes, an equally tenacious though less diversified assemblage, survive in the sharks, the batoids, and the holocephalians. Though these fishes are set lower than the tetrapods in diagrams of vertebrate evolution, they are certainly not primitive forms. They have developed as broad a range of adaptive specializations as their terrestrial contemporaries. If their structural, physiological, and behavioral diversity is not widely recognized, it is because they are less well known than the four-legged vertebrates. Men discriminate more finely among animals that share their environment than among those that live, largely unobserved, in another.

Tracing the history of modern fishes is a complex task

The availability of living forms has not made the evolution of the higher fishes easier to understand than that of the ostracoderms and placoderms. Recent bony and cartilaginous fish have left behind them a long trail of extinct forebears whose remains must be found, cleaned, interpreted, and fitted into the record before the whole story is known. The presence of living representatives has, in effect, doubled the task: ichthyologists work continuously to describe and classify extant species while paleontologists labor over fossil material. Each group is striving to define monophyletic assemblages and to link them with each other. So far, ichthyologists and paleoichthyologists have met to drive few golden spikes. The ichthyologists have sorted living fishes into ever more narrowly drawn categories, but they have much yet to learn about the deep anatomy of the many forms whose external characteristics they have described. Their most recent classification schemes contain some groups whose affinities are unclear or whose members may be less closely related than they seem. To resolve their problems, ichthyologists look to comparative anatomy and physiology, but studies in these areas, as extensive as they have been, are far too few to provide the information required. Some fishes have been examined in detail, some only cursorily, and others not at all. The amount of work that remains to be done on living fishes is enormous and grows steadily larger as new forms are discovered in previously unexplored parts of the ocean and in the fresh waters of hitherto inaccessible continental interiors.

The problems of the paleoichthyologist are different but no less perplexing. The fossils of higher fishes, like those of archaic forms, are delicate, often crushed, and hard to interpret. Dermal bones and scales frequently hide endoskeletal elements, and, especially in Paleozoic specimens, the rostral portion of the skull or the postcranial skeleton may be missing (Fig. 5.2). Since cartilage, even when calcified, does not preserve very well, unossified braincases and vertebral elements usually disappear entirely. Ancient sharks are reduced to a scattering of teeth, spines, and dermal denticles. As with any group of extinct organisms, fossilization in certain environments and not in others, destruction of fossils by geologic change or weathering, and inaccessibility or nondiscovery of existing specimens com-

Vertebrate History: Problems in Evolution

bine to produce a discontinuous record of the evolution of the higher fishes. Because of the slimness of the record, even after he has determined the age of his fossils, the paleoichthyologist may still founder upon the question of the relationship of the forms they represent. Since analysis of structure does not always enable an investigator to distinguish between parallel or convergent development and true genetic affinity, he must extend his research to include the nature of the environment and the geographical distribution of the fishes he is attempting to relate. To establish the possibility of the descent of one form from another it may be necessary to correlate the age of sediments in different locations, an undertaking of which the difficulty has already been described. In short, the paleoichthyologist who attempts to trace the evolution of modern bony and cartilaginous fishes encounters the same obstacles that impede the study of any group. His problems are multiplied, however, by the great length of the history of the higher fishes. No other classes of vertebrates have left so complex a trail of fossils over so long a period of time.

Importance of paleoecology in study of fossil fishes

Although the evolution of the fishes of any period is usually summarized in terms of structural change, an appreciation of the vitality of ancient forms and the selective forces to which they were subject comes only from a consideration of the animals as inhabitants of a particular environment. By reconstructing the environment and visualizing the fishes that he is studying as part of a biological community, the paleontologist can understand the functional value of their characteristic structures. The adaptive aspect of certain modifications becomes apparent, and characters that he would otherwise have described separately he can identify as functionally interrelated. Though the causes of specific genetic mutations remain as obscure as ever, he can at least supply a rationale for the association of certain body forms, fin positions, and jaw mechanisms. Further, he gains a basis for speculating upon the nature

Figure 5.1. Various kinds of living fishes, not drawn to the same scale. (*a, b, c, d*) bony fishes (Osteichthyes); (*e, f, g*) cartilaginous fishes (Chondrichthyes). **(a)** *Neoceratodus*, the Australian lungfish, one of the few remaining lobe-finned fishes. **(b)** *Beryx*, a ray-finned fish of the advanced teleost type. **(c)** *Lepidosteus*, the freshwater garpike, a surviving ray-fin of the archaic holostean type. **(d)** *Scaphyrhynchus*, the sturgeon, a degenerate form called chondrostean, akin in basic structure to the most primitive Paleozoic ray-fins, the paleoniscoids. **(e)** *Harriotta*, a holocephalian. **(f)** *Raia*, a batoid. **(g)** *Carcharodon*, a shark. (*a,c,d* from Dean; *b* from Cambridge Natural History; *e,f,g* from Garman.)

(a)

of structures that are not preserved. From observation of living fishes in similar environments, he may make what he hopes are reasonable assumptions about respiratory and excretory organs, for instance, or about the development of the swim bladder.

These assumptions are supported by the uniformitarian principle that earlier fishes met the same forces and coped with them in much the same way as fishes do today. Belief in the continuity of physical phenomena also underpins the paleontologist's reliance on the facts of modern ecology as a guide to understanding ancient environments. In order to reconstruct the habitats of millions of years ago he has to know in detail the conditions that exist undersea and in fresh waters at the present time. The study requires an imaginative mind. It is not sufficient to review merely in a general way the differences in salinity, oxygen content, temperature, light, and pressure that may occur in various waters. To grasp fully the diversity of the submerged world, it is necessary to pass before the mind's eye the almost endless series of specific niches that house swimming vertebrates. The watery medium interconnects them all, but the different areas constitute distinctly separate realms. Because each demands for survival a unique combination of structural and physiological characteristics, passage by fishes or any form of life from one kind of environment to another is usually accompanied by significant evolutionary change.

A description of the many different habitats into which the higher fishes radiated during the course of their evolution is beyond the scope of this book. However, the importance of ecological criteria in the interpretation of fossil material can be made clear by an example of paleontologists' use of them in

(b)

Figure 5.2. Two examples of the fossil material with which paleoichthyologists work. **(a)** A Triassic coelacanth, *Chinlea sorenseni*; the fossil has been photographed under ultraviolet light to render the bones fluorescent and thus more easily distinguishable from the matrix. **(b)** The skull of a Triassic ray-finned fish, *Redfieldia*, X 5. The skull appears, crushed, in ventral view. The left and right dentary bones of the lower jaw (*dent*) are visible on either side of the median gular plate (*gu*), which protected the throat. Flanking the dentary bones are the tooth-bearing upper-jaw elements (*mx*) and, above them, various bones of the snout and face (*ros*, rostral; *ant*, antorbital; *adn*, adnasal; *dsph*, dermosphenoid). (From Schaeffer.)

furthering their understanding of extinct fishes. Without an analysis of its surroundings, a specimen like *Dorypterus* would have remained a complete enigma to the men who described its structure. This bony fish was a paleoniscoid, an early ray-finned form, but a very strange-looking one. Unlike most of the other Permian paleoniscoids, which were fusiform, carnivorous fishes, *Dorypterus* had a deep body that was laterally compressed and jaws that were ill-suited for predatory habits (Fig. 5.3). Though the upper lobe of its caudal fin was stiffened by an extension of the vertebral column in primitive fashion, the large size of the lower lobe reduced the lifting force of the tail. Since *Dorypterus'* caudal fin did not pitch the head downward, the pectoral fins were not required to act as elevators. They had abandoned their customary ventral position and had climbed high on the side of the body. These fins and the pelvic pair, which were set far forward beneath the gill chamber, must have been used solely as braking and steering devices. Rolling of the deep, slender body was prevented by the long dorsal and anal fins, which bordered the upper and lower margins of the caudal region. The dorsal

Figure 5.3. Reconstructed skeleton of the Permian paleoniscoid fish *Dorypterus*. The living fish was about 5 inches long. (From Westoll.)

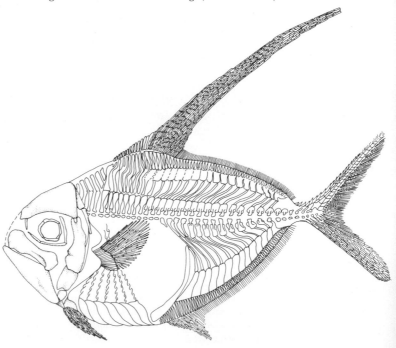

fin, prolonged anteriorly into a tall blade, increased the stability of the fish still further. Fishes with a similar body design appeared much later among the more advanced ray-finned forms, but the presence of *Dorypterus* in the company of more conventional paleoniscoids in the Permian period constituted a special problem for the paleontologists.

There was no doubt that the habitat of *Dorypterus* was marine. The English Marl Slate and German Kupferschiefer strata in which the fish was found have given up an assortment of sea-floor invertebrates. Interpretation of the other fishes found in these deposits as saltwater forms presented no difficulty. Besides paleoniscoids, collectors had retrieved sharks and coelacanths structurally adapted for the rapid swimming that survival in open water requires. Despite its superior ability to maneuver, the 5-inch *Dorypterus* could not have escaped the attacks of these larger, faster fishes. Unless paleontologists could conceive of a different ecological niche for *Dorypterus,* its coexistence with these forms was inexplicable. Since fishes of *Dorypterus'* shape and size inhabit modern coral reefs, where crevices and labyrinthine channels provide refuge from pursuit, the suggestion was made that *Dorypterus* and two similar but less highly modified deep-bodied Permian fishes were early exploiters of such a habitat. This idea was abandoned when geologists found no trace of reef formation in either the Marl Slate or the Kupferschiefer deposits. There was little possibility that *Dorypterus* had lived protected in fresh water and drifted out to sea after death. The good condition of dorypterid fossils argues against their having been transported any distance, and the similarity of the German and the English forms makes it unlikely that they were evolving separately in the fresh waters of two different areas.

Guided by the paleontological evidence and ecological principles, Westoll proposed a solution to the dorypterid problem. He concluded from the presence of fragmentary land plants and reptilian bones in Marl Slate and Kupferschiefer rocks that the areas of deposition must have been near shore, where the water was not deep. He imagined calm lagoons that received the outflow from marsh-bordered streams and opened seaward into deeper, less protected waters. The sunlit shallows would have supported a heavy algal growth, in which *Dorypterus* and other deep-bodied fishes could have lived, safe from attack by predatory paleoniscoids. He could not prove that algal jungles existed in the Marl Slate and Kupferschiefer lagoons because soft plants are not preserved, but he assumed their presence for two reasons. First, they appear in modern environments similar to the one he is proposing, and, second, they would provide the kind of food for which *Dorypterus'* jaws seem to be specialized. *Dorypterus'* mouth was tilted upward and edged on each side by a strong, toothless maxilla and mandible that closed like scissors blades. Westoll supposed that these fishes grazed on the plant matter, darting agilely away from coelacanths that occasionally penetrated their green cloud, while the fusiform sharks and ray-finned fishes swam outside through the clear water in the deeper part of the lagoon.

Adaptive traits of the haplolepid fishes

Like other paleontologists who try to explain the succession of extinct forms, Westoll has had to visualize not only the conditions which existed in various habitats but also the ecological changes that occurred in particular regions over long periods of time. His account of the Haplolepidae, an aberrant family of Carboniferous paleoniscoids, would have been incomplete had he not attempted to describe the alterations in the environment that led to the substitution of this group for older, more generalized fishes. Less obviously divergent in body form than the dorypterids, the haplolepids had nevertheless certain peculiar traits that set them apart (Fig. 5.4). They had lost the prominent rostrum that projected over the mouth as well as the broad cheek and opercular bones that characterized the earlier fishes. Though they retained the primitive paleoniscoid caudal fin, the haplolepids were stouter than their predecessors and somewhat flattened ventrally in the trunk region. The fossil record indicates that they lived in an area populated also by a distinctive group of ancient amphibians. The earliest known paleoniscoids in the same location had been associated with an entirely different fauna. They had formed part of the community of fishes which existed in lakes between the mountains newly raised at the end of Silurian time.

Figure 5.4. *Haplolepis ovoidea*, a haplolepid fish from the Pennsylvanian of Linton, Ohio. (From Westoll.)

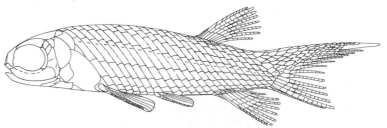

According to Westoll's interpretation, these lakes persisted for millions of years, receiving the rainwater that rushed down from high ground loaded with sand and silt. The water became shallower as sediments accumulated on the lake floors and then deeper again as the underlying rock sank beneath their weight. Eventually, the mountain peaks wore away. Their substance formed sloping piedmonts, and the lakes gave way to streams, which deposited their finer sediments on lower marshy ground. The lake fishes never colonized the muddy, stagnant pools that formed in depressions in the swamp. The haplolepids, Westoll thinks, were better adapted for such an environment. The shape of the rostrum and the reduction in the size of the gill chamber under the small opercular cover may have been related to the poor aeration of the water. Since gills serve effectively as respiratory organs only when they are exposed to a flow of well-oxygenated water, their reduction in the haplolepids would not have been a disadvantage. On the contrary, if the fishes used the air bladder as an auxiliary respiratory structure, a decrease in the surface area of the gill filaments would have been favored by natural selection. Since blood which has gained oxygen in the wall of the air bladder goes directly through the heart and to the gill capillaries, restriction in the number of these vessels prevents the loss of oxygen by diffusion into the stagnant environmental water. That an air bladder or some kind of supplementary respiratory organ was present is not unlikely. The absence of an overhanging rostrum would have allowed the haplolepids to approach the surface, lips first, to suck in bubbles of air. Westoll thought of these fishes as cruising slowly in the upper water of the swamp above the mud made toxic by hydrogen sulfide and other products of decay. He believed that they lived in this narrow niche until the advancing Permian seas obliterated it. While his reconstruction of the circumstances that governed the evolution and survival of the haplolepids may not be entirely correct, it is founded carefully upon available evidence and transforms these fishes from abstractions in stone to the real animals they once were.

Coexistence of fishes of different adaptive lines

Although a change in the environment may cause the extinction of existing species, there seem always to be some forms that can maintain themselves under the new conditions until physical modifications occur that substantially improve their chances of survival. When a sufficient number of these adaptive traits have appeared, the population enlarges rapidly and leaves its imprint in the fossil record. Because the beginning of the adaptive process is generally hidden from the investigator, the earliest known members of a group always show a degree of specialization for the kind of life they led. The higher fishes, when they appear in the Devonian period, have already acquired the characteristics that identify them as belonging to one or another of the major assemblages of bony or cartilaginous forms.

The subsequent history of these Devonian fishes was one of simultaneous diversification of a myriad of species that lived side by side with the remaining archaic forms. The necessity of describing each category of fishes separately has often prevented the newcomer to the field from appreciating the variety of the underwater vertebrates at this stage of their evolution. He is apt to see the different groups as fishes succeeding one another like floats in a parade or, worse, as forming serial parts of the line leading to the first tetrapods. Examination of the fossil fishes collected from any Paleozoic stratum of mid-Devonian age or later soon erases such impressions. In the Old Red Sandstone deposits in northern Scotland, for instance, every known group of fishes is represented except the sharks. The mid-Devonian freshwater lakes in which the Old Red sediments settled supported the newer bony fishes as well as populations of antiarchs, free-swimming and bottom-living arthrodires, acanthodians, and jawless forms. At this time, when *Cephalaspis* was still in existence, the bony fishes were already differentiated into ray-finned paleoniscoids, lungfishes, and two lines of lobe-finned rhipidistian fishes. Not long afterward, the coelacanths appeared in fresh water, and sharks joined other swimmers in the sea.

Origin of the modern orders of fishes is obscure

The origin of all these fishes is obscure. As has been shown earlier, it is not possible to demonstrate unequivocally the descent of any group of the higher fishes from a specific stock of placoderms or acanthodians. Although the bony and cartilaginous forms have been classified separately as Osteichthyes and Chondrichthyes respectively, paleontologists lack fossil evidence to prove that either of these assemblages is a natural group. Stensiö, in arguing that the Holocephali stem from ptyctodonts and the sharks from a different group among the Elasmobranchiomorphi, has treated the Chondrichthyes as polyphyletic and omitted the term from his classification scheme. In the case of the bony fishes, the differences which exist among the early actinopterygians, rhipidistians, coelacanths, and lungfishes make it possible that these types are not derived from a common ancestor and should not be lumped in a single class. Paleontologists who believe that all the bony fishes evolved from acanthodians still debate whether they may therefore be regarded as monophyletic. In the narrow sense, if the different groups arose from different acanthodian lineages, the bony fishes might still be considered a polyphyletic assemblage.

Since the first bony fish appear in Lower Devonian rocks, paleontologists speculate that the transition from archaic to more modern forms began before the close of the Silurian period. Whether it continued well into the Devonian to produce independently the groups of higher fishes that enter the record later, or whether the process was completed in a shorter time, producing basal stocks of higher fishes that enlarged their populations at different rates, is a question to which there is at present no answer. Propounding one would require a better understanding than is now available of the factors that determine the rate of evolution. The nature and number of these factors and the complexity of their interrelationship have combined to prevent the formation of any rules that would guide paleontologists in divining precisely the speed of the transition from lower to higher fishes without additional fossil evidence.

Measuring the rate of evolution

The variability of evolutionary rates through time and among members of different taxonomic groups is evident not only to students of fishes but to those of vertebrates at every level. Discussion of the pace of change within a particular line is hampered, however, by the difficulty of measuring it. In rare cases, where a relatively complete series of fossils exists, the rate of evolution can be gauged by quantitative changes in one or more of the animals' structures. Measurements of the length of the skull or of individual bones serve, when charted along the geologic time scale, to give the viewer some idea of the rate of transformation that is taking place. If a sufficient number of characters are analyzed in this way, the investigator can make a general statement about the progress of change. While such an approach is helpful in calculating the evolutionary rate of animals, like the horse, that have left numerous, comparatively well-preserved remains, it is hardly useful when applied to fishes. Even if enough fossils were known in a single lineage of bony or cartilaginous fish, their usually incomplete condition would make it impossible to find a significant number of homologous structures that could be measured throughout the series.

To quantify the course of morphological change in the evolution of lungfishes, Westoll devised a more flexible system.

Instead of measuring specific structures, he assigned graded numerical values to characteristics at different stages of their development. He obtained a score for each fossil by adding the points it merited on each trait. Since he chose higher ratings for characters in their more primitive state, the lower its total score, the more advanced or differentiated a lungfish was considered to be. By using Westoll's method, Schaeffer calculated the rate of character change per million years for both lungfishes and coelacanths and so was able to compare the rate at which the two groups of lobe-finned fishes evolved (Fig. 5.5). Although such a comparison is undoubtedly meaningful, purists are quick to point out that there is a large subjective factor in the analysis. It is, after all, the investigator who chooses the structures to be included in the survey and who passes judgment on their level of differentiation. This shortcoming and others, like the impossibility of weighting numerical values according to the relative importance of various traits, Schaeffer admits, but he still believes that his use of Westoll's method reveals a pattern in evolutionary rates that is significant and informative.

The difficulties inherent in measuring homologous structures or in some other way quantifying morphological transformation can be avoided by calculating rates of change on a taxonomic basis. The assumption which underlies the taxonomic approach is that the number of genera (the category most frequently used) in any group of vertebrates is an index of its rate of evolution. One method of taxonomically graphing the progress of change

Figure 5.5. Rate of structural change in coelacanths and dipnoans from the Devonian period onward. The vertical scale represents the number of suprageneric character changes per million years; the horizontal scale indicates geological periods. (From Schaeffer.)

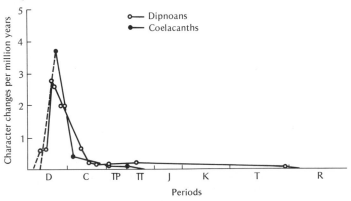

Vertebrate History: Problems in Evolution

is to plot the number of new genera (or orders, or families) that appear per million years. As Simpson makes clear, however, this sort of analysis has its own pitfalls. It blurs distinctions that should be made: a line which has produced three successive genera in a million years each of which is more specialized or more advanced than the one before is said to be evolving at the same rate as a line that adds, in the same amount of time, three genera that are variants at the same level of development. The method is no less subjective than any other. The erection of taxonomic categories, especially genera, is a process upon which workers often disagree. The rate of evolution would certainly be faster for splitters than for lumpers. There is no way of achieving uniformity in separating organisms into genera, even if classification of the lines to be compared were reviewed by one man, because it is impossible to base the separation on an equal degree of genetic change. If the amount of genetic change that has taken place in the successive evolution of five genera of fish is actually greater than that which has occurred in the same time to produce five new genera of birds, the rate of evolution is not really the same. Rather than being predicated upon accumulation of mutations, however, generic distinction necessarily rests upon observable differences. Collectively, these are hard to evaluate as indicators of the rate of evolution. Organisms are usually a mosaic of conservative and progressive characters: a line of fishes may be divided into several genera on the basis of change in a few rapidly differentiating structures while a large number of other traits are carried along for millions of years virtually unmodified.

The fragmentary nature of the fossil record prevents the paleontologist from using the taxonomic approach or any other method for calculating evolutionary rates in specific phyletic lines. Since he is likely to have available a scattered assortment of fossil forms of whose exact relationship he is unsure, he is restricted to charting the course of change in broadly inclusive groups. He bases an estimate of the evolutionary rate of ray-finned fishes on all the known actinopterygian fossil genera and then assumes that his sample reflects the composition of that large assemblage reasonably closely.

Measurements of the rate of evolution, despite all their deficiencies, have given investigators some idea of the temporal pattern of change in the vertebrate line. Simpson was able to demonstrate, by figuring the origination rate of various categories of these animals, that a new class projected itself first by a burst of ordinal groups (Fig. 5.6). Diversification within each order followed, producing a peak in the number of new genera per million years about 50 or 60 million years later. The rapid radiation of generic forms, when it occurs, gives the impression of a sudden population explosion within the class, although, in truth, the increase represents the high point of a geometrical progression that began perhaps 100 million years earlier. The subsequent history of different groups varies. In the less successful classes, the rate of appearance of new forms dwindles, never to rise again. The origination rate of the more dynamic aggregations declines and then peaks once or twice more as radiation takes place at new adaptive levels.

Relating biological and geological transformations

Because his taxonomic data were plotted on the geologic time scale, Simpson could comment upon the suggestion, often made, that the appearance and proliferation of new groups of

Figure 5.6. First appearances of known orders and genera of cartilaginous fishes (Chondrichthyes) and bony fishes (Osteichthyes). Orders: ---□---; genera: ____•____. Note that the peaks in the appearance of new orders are generally followed by a rise in the number of new genera. Vertical scale: new orders (*left*) and new genera (*right*) per million years. Horizontal scale: geological periods. (Modified from Simpson.)

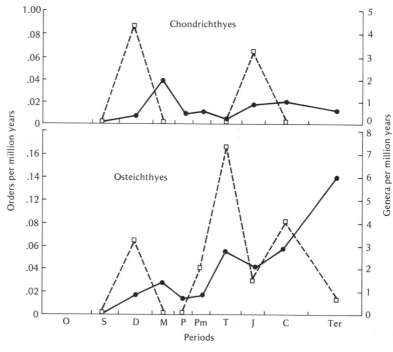

Vertebrate History: Problems in Evolution

vertebrates was related in some way to major disturbances of the earth's surface. To many paleontologists it seemed possible that the evolution of early forms was spurred by the mountain-building episodes that occurred periodically. Examples of association between the biological and geological phenomena were not lacking. Establishment of the various groups of fishes and the earliest amphibians took place as the Caledonian orogeny reshaped northern Europe and the Acadian changed the eastern part of North America in Silurian and Devonian time. The radiation of the amphibians and the rise of the reptiles followed during the Carboniferous and Permian, while the Hercynian mountains rose in Europe and the Appalachians appeared in continental North America. The diversification of mammals in the New World at the beginning of the present era coincided with the vast Laramide revolution, which raised the Rockies and the Andes chain. The simultaneous occurrence of geological and biological transformations, however, does not prove the dependence of organic evolution upon topographical change. Simpson believes that the timing of the appearance of new groups is determined by prior events in the evolutionary process itself. Although the environmental changes brought about by movements of the earth's crust exert strong selective pressure, they cannot produce new forms unless preliminary genetic mutations have already occurred. There is no evidence that the rate of mutation is altered by gradual change in the nature of the environment, and, in any event, Simpson points out, it is unjustifiable to suppose that the beginning of a series of mutations which culminated in the establishment of a new class or order coincided with a particular geologic disturbance.

Variation in the rate of evolution among fishes

Although the factors that govern the rate of evolution can be discussed in a general way, they cannot be analyzed precisely enough to explain why some groups change more quickly than others. Even closely related vertebrates show wide variation in the pace at which they evolve. Among the bony fishes, the actinopterygians have progressed rapidly from paleoniscoid to holostean to teleostean grade, proliferating numerous genera at each level. None of the lobe-finned forms produced such a rapid succession of forms. The rhipidistian lobe-fins gave rise to the first amphibians late in the Devonian but then faded away without undergoing great diversification. The coelacanths and lungfishes have been conservative in their evolution for most of their history. Neither has produced many more than two dozen known genera or has changed much in the 350 million years that

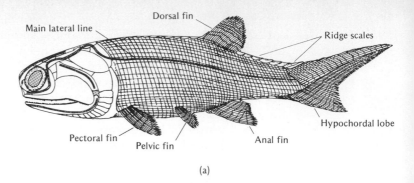

Main lateral line

Dorsal fin

Ridge scales

Pectoral fin

Pelvic fin

Anal fin

Hypochordal lobe

(a)

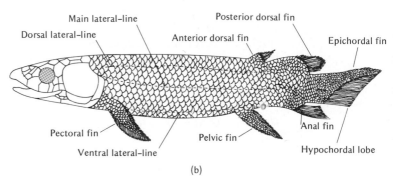

Main lateral–line

Dorsal lateral–line

Posterior dorsal fin

Anterior dorsal fin

Epichordal fin

Pectoral fin

Ventral lateral–line

Pelvic fin

Anal fin

Hypochordal lobe

(b)

have elapsed since the end of the Devonian period. Among the vertebrates they are the best examples of forms whose evolution has been extremely slow or bradytelic. The cartilaginous fishes included in the Chondrichthyes show a rate intermediate between that of the actinopterygians and the lobe-finned forms.

It is possible to demonstrate without going beyond the higher fishes that the rate of evolution within a line is not directly related to its survival. The coelacanths and lungfishes have outlived many forms that during their existence differentiated more rapidly. It is true that lungfishes escaped vigorous competition with marine ray-fins by retreating to fresh water early in their history, but the coelacanths endured the contest for eons before leaving open waters for the protection of the deeper sea. The progress of these forms hardly augured their future: had paleontologists been set down in the Carboniferous period to survey the extant fishes, they would not have marked for longevity two retrograde types that were even then showing evidence of skeletal degeneration and specialization.

Paleontologists tracing the history of these persistent animals

Devonian lobe-finned fishes: lungfishes, coelacanths, rhipidistians

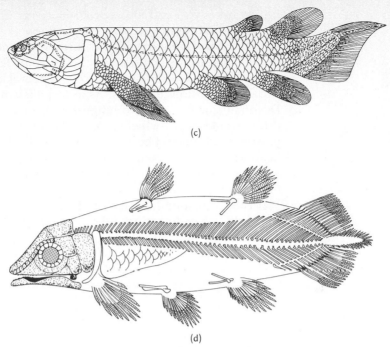

Figure 5.7. The external form of a primitive ray-finned fish (*a*) and that of early lobe-finned fishes (*b,c,d*), compared. **(a)** The paleoniscoid *Moythomasia.* **(b)** The lungfish *Dipterus valenciennesi.* **(c)** The rhipidistian *Holoptychus.* **(d)** The coelacanth *Undina.* Note in the lobe-finned forms two dorsal fins rather than one, the somewhat smaller eye, the more sloping snout, and the flesh-covered axis extending into the fins. (*a,b,c* from Moy-Thomas and Miles *Palaeozoic Fishes,* Chapman and Hall, Ltd.; *d* from Smith Woodward.)

have found that their earliest representatives were similar in form to the Devonian lobe-finned fishes known as rhipidistians. Unlike contemporary ray-fins, Devonian rhipidistians, lungfishes, and coelacanths had two dorsal fins, small eyes, and a forward-sloping snout (Fig. 5.7). The paired fins were supported by endoskeletal elements different from the parallel radials of the paleoniscoids. In rhipidistians and lungfishes, small bones formed an axis that projected a considerable distance into the fin and underpinned the muscle responsible for the lobed appearance of the fin. Although no remains of these ossifications have been found in extinct coelacanths, the shape of the fin bases implies their presence. The caudal fin showed a tendency to lose its heterocercal form in the lobe-finned fishes, as it did in the progressive ray-fins, but the change occurred differently in the two groups (Fig. 5.8). Whereas symmetry was attained in the former by enlargement of the upper lobe, it was achieved in the latter by lengthening of the lower lobe and withdrawal of the vertebral column from the expanded portion of the fin. As in rhipidistians, the body of the Devonian coelacanths and lungfishes was covered with scales of the cosmoid type, which lacked the heavy ganoine layer of paleoniscoid scales. The arrangement of the dermal bones of the head in the three groups of lobe-finned fishes, though difficult to compare, was distinctly different from that in the ray-finned forms.

It is hard to lengthen the list of characteristics which unite all the early known lobe-fins and differentiate them from paleonis-

Figure 5.8. Tail region of ray-finned (*a,b*) and lobe-finned (*c,d*) fishes, showing modification of caudal fin. **(a)** *Moythomasia*, showing asymmetrical caudal fin with vertebral column extending into the upper lobe (heterocercal type). **(b)** *Chirodus*, with caudal fin that has become superficially symmetrical through enlargement of the lower lobe (homocercal type). **(c)** *Osteolepis*, a rhipidistian, with heterocercal caudal fin. **(d)** *Gyroptychius*, also a rhipidistian, with caudal fin which has acquired symmetry through enlargement of the upper lobe (diphycercal type). (Modified from Miles.)

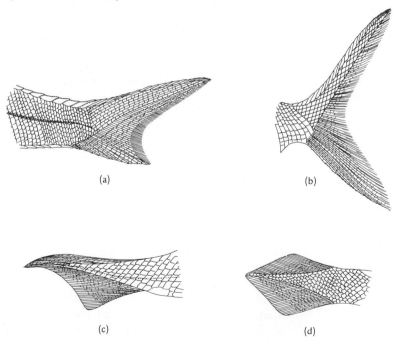

(a)

(b)

(c)

(d)

coids, because Devonian lungfishes, coelacanths, and rhipidistians had been evolving divergently for many millions of years. There are likenesses between two of the three groups, however, which strengthen the argument that these fishes are related in some special way. Paleontologists point out that the endocranium of rhipidistians and coelacanths is unique in its division into two parts (Fig. 5.9). The anterior division, which develops from the embryonic trabeculae, becomes hinged instead of fused to the posterior portion derived from the otic capsules and parachordal cartilages. Although the earliest ray-finned fishes seem to have had a divided endocranium, the separation did not correspond to that in the braincase of the lobe-fins and soon disappeared. The lungfishes, even the oldest known forms, showed no evidence of a jointed endocranium, but they shared with certain rhipidistians another characteristic not exhibited by the paleoniscoids: they produced a mineralized cosmine layer that covered the outer surface of dermal bones over areas in the head region. The material was apparently laid down and resorbed in a seasonal cycle that allowed periodic growth of the skull.

Relationships and classification of the lobe-finned fishes

While it may seem that the rhipidistians, lungfishes, and coelacanths are more closely allied to each other than any of them are to the ray-finned fishes, there is some disagreement among paleontologists on this point. In his discussion of the affinities of lungfishes, Westoll emphasized their similarity to the rhipidistians and suggested that the two groups might have had a common ancestor. Jarvik, reviewing the same paleontological evidence interpreted it differently. He attacked Westoll's reasoning, point by point. The superficial cosmine layer, he maintained, is not peculiar to lungfishes and rhipidistians, since ostracoderms like *Cephalaspis* also had it. He questioned the assertion that the arrangement of the dermal bones and sensory lines in the head region is comparable and described what he believed are important differences in the anatomy of the snout in the two types of fishes. He denied the lobed fins as strong evidence of common ancestry: their occasional presence among ray-finned fishes and sharks shows that they could have arisen independently in several separate lines.

Jarvik also argued against the contention that rhipidistians and coelacanths are closely united. Having disposed of their fleshy fins as evidence of relationship, he sought to prove that the presence in both fishes of a jointed endocranium was not significant of a connection between them. He regarded the

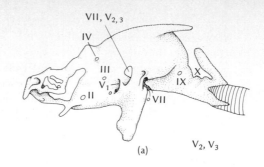

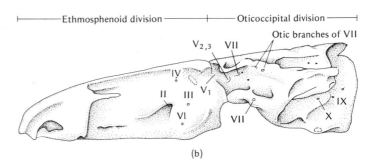

Ethmosphenoid division — Oticoccipital division

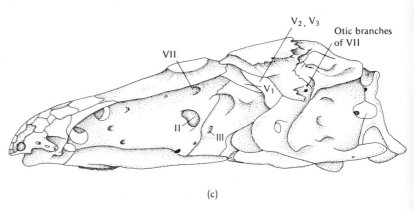

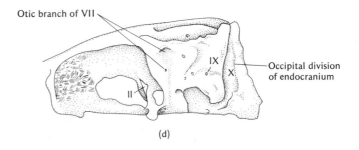

intracranial articulation merely as an extremely ancient vertebrate characteristic that happened to survive in two different groups of fishes. This conclusion derives from his opinion that the posterior part of the braincase, which encloses the front of the notochord, represents the cranial division of the vertebral column and its joint with the anterior region of the braincase a primitive articulation of the axial skeleton. After looking in vain for structures that distinguish rhipidistians and coelacanths from all other bony fishes, Jarvik ceased to consider them a natural group and rejected the practice of bracketing them as Crossopterygii in the classification scheme.

Not all paleontologists have gone as far as Jarvik in dividing the bony fishes. Romer set the ray-finned fishes apart in the subclass Actinopterygii, but, taking note of the characteristics common to the fleshy-finned fishes, he grouped those forms in the subclass Sarcopterygii. Within the latter category he united the rhipidistians and coelacanths as members of the order Crossopterygii but gave separate status to the lungfishes under the ordinal heading Dipnoi. Schaeffer, after investigating the possibility of an independent origin for the Dipnoi, returned to the view that dipnoans and crossopterygians are closely related. He therefore retained the sarcopterygian assemblage as a monophyletic group. R. H. Denison agrees with Romer and Schaeffer about the close relationship of the dipnoans to the other lobe-finned fishes. His examination of *Uranolophus,* a Lower Devonian lungfish that he regards as possibly the oldest known member of the Dipnoi, has convinced him that the first lungfishes were very similar to rhipidistians in postcranial structure. Since he believes that all the lobe-finned fishes had a common ancestor, he would not divide them systematically quite as sharply from each other as from the actinopterygians.

Paleontologists inquiring into the interrelationships among the higher fishes focus their attention upon the earliest and least specialized members in each group. Fossils of the later fishes become important when investigators try to determine the

Figure 5.9. The endocranium (internal head skeleton enclosing the brain and sense-receptor organs) in various fishes, in left lateral view. **(a)** The lungfish *Neoceratodus*, showing absence of any transverse division in the endocranium. **(b)** The rhipidistian *Eusthenopteron*, showing division of the endocranium anterior to the otic region. **(c)** The coelacanth *Latimeria*, showing division at the same level. **(d)** The paleoniscoid *Pteronisculus*, showing division posterior to the otic region. Roman numerals signify cranial nerves which issue from foramina indicated. (From Jollie.)

subsequent course of evolution within each separate line. In the case of the lungfishes, a survey of specimens in chronological order proved that these animals diverged from more generalized predecessors among the bony fishes and belonged, in fact, to that group rather than to the Amphibia. Biologists of the nineteenth century assigned the lungfishes to the latter class because the living African *Protopterus* and South American *Lepidosiren* bear a superficial resemblance to salamanders, like *Amphiuma,* which have a long, cylindrical body, grossly undersized paired appendages, and a diphycercal caudal fin. Its large scales notwithstanding, *Neoceratodus,* the Australian representative, was also regarded as a monstrous salamander by those who described it first. Near the end of the last century, the Belgian paleontologist L. Dollo reoriented the view of workers studying the dipnoans by demonstrating that the peculiarly shaped modern forms had evolved step by step from Devonian fishes with a deeper body and fins like those of rhipidistians. He argued that a series of fossil forms shows the gradual fusion of the second dorsal, anal, and caudal fins to produce the symmetrical fringe around the tail characteristic of lungfishes of the Carboniferous and later periods (Fig. 5.10).

Before the body of the lungfishes had lost its rhipidistian appearance, important modifications took place in the region of the head. Devonian lungfishes had already undergone changes that were surely associated with adaptation to a diet of hard-shelled animals of some sort. The endocranium was rigidly constructed and the upper jaw fused rather than movably jointed to it. As in many fishes that crush invertebrate shells or exoskeletons, the gape was short and the bite presumably powerful. Low plates, denticles, or ridges of tubular dentin mounted on the palate and opposing part of the lower jaw took the place of the sharp marginal teeth present in other fleshy-finned fishes. The special design of the cranium and jaws was one from which there was no retreat. Post-Devonian lungfishes kept the same endoskeletal framework, although their ability to ossify it degenerated rapidly.

The dipnoan skull was strengthened by a dermal bone cover which diminished with the passage of time but never quite disappeared (Fig. 5.11). In its early form it resembled that of other bony fishes in consisting of a large number of small plates enclosing the cranium, the branchial region, and the lower jaw. Despite the fragmentary nature of lungfish fossils, paleontologists have been able to follow the changes that took place in the

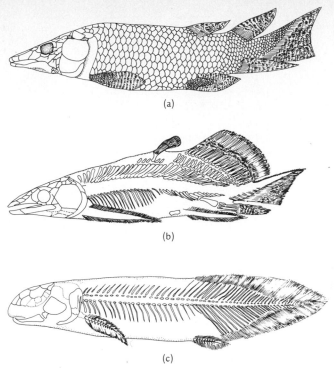

(a)

(b)

(c)

Figure 5.10. Restorations of lungfishes showing stages in the evolution of the diphycercal caudal fin. **(a)** *Griphognathus*, an Upper Devonian form with a heterocercal caudal fin and dorsal fins similar to those of rhipidistians. **(b)** *Fleurantia*, a more specialized Upper Devonian lungfish, showing a broadening of the posterior dorsal fin. **(c)** *Conchopoma*, of Upper Carboniferous and Lower Permian time, in which a diphycercal caudal fin has formed and become continuous with the dorsal and anal fins. (From Moy-Thomas and Miles.)

complex of dermal elements. The amount of bone decreased, as it had in the endoskeleton, but at a slower rate. The dermal bones, which had been thick and cosmine-coated in the earliest lungfishes, lost their cosmine layer, thinned, and sank deep beneath the surface in later forms. The disappearance of dermal elements from the margins of the jaws and palate, additional evidence of the general decline in ossification, coincided with the change in the jaw mechanism and the emergence of various dental structures on more medial bones. In the absence of anterior teeth, further reduction occurred in the exoskeleton of the snout. That area, which had been shielded initially by a rhipidistian-like mosaic of small bones covered with cosmine, eventually lost all its hard tissue.

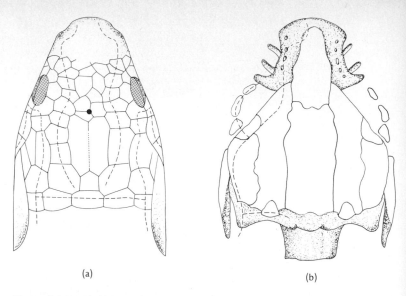

(a)   (b)

Figure 5.11. Skulls of lungfishes in dorsal view, showing dermal bones of the roof. **(a)** *Dipnorhynchus*, an early Devonian form. A mosaic of large and small plates covered the skull roof; a superficial layer of enameloid material over the snout obliterated subdivisions in the armor of that region. **(b)** *Neoceratodus*, the modern Australian lungfish. The snout is devoid of dermal elements; the few large plates over the skull roof (unstippled) have sunk beneath the surface. (From Jollie.)

Formation and variation in dermal bones of the fish skull

Although many changes in the dermal bones of the head region can be attributed partly to the dipnoans' diminished powers of ossification and partly to the special jaw mechanism they developed, paleontologists believe that modifications in the pattern of the dermal plates of the skull roof are less closely related to these factors than to changes in the cephalic part of the lateral-line system. This system, universally present in fishes, consists of special sensory cells that are stimulated by turbulence in the water around the animal. The cells, called neuromasts, lie in a row at the base of pits, grooves, or canals that extend along the length of the body and form a characteristic pattern over the head. Embryologists have shown that the neuromasts differentiate early in development and attract bone-building cells about them. Consequently, the sensory cells aid in determining the arrangement of dermal plates in their vicinity. Where a row of neuromasts is well developed, there is likely to be a series of dermal bones to house them. When such a row degenerates, the bones that contained it tend to disappear unless they have become an integral part of a stress-resistant area of the dermal shield. The stability of the dermal elements associated with the lateral-line system has been immensely helpful to paleontologists seeking homologous bones in different specimens. Bones which form between sensory lines show greater variation in number and arrangement. Since these bones, called "anamestic" by Westoll, may differ in members of the same genus, they cannot be used as reliable criteria of evolutionary change.

In dipnoans, the mosaic of little bones over the dorsal part of the head can be resolved into a series that underlies the supraorbital sensory canal, another that encloses the infraorbital one, and a group of intervening anamestic elements. Before they could analyze the changes that occurred in these bones, paleontologists had to face the problem of terminology. Assigning to the dipnoan plates names used for roofing bones in other fishes would have implied homologies of which investigators were unsure, while inventing a special set of terms for the large number of elements present in Devonian lungfishes would have increased the linguistic burden excessively. When a system of identification by letter and number proved workable, paleontologists were able to communicate with one another about their observations. They agreed that the progressive reduction in separate bones evident among dipnoans was not completely a random affair. Although early forms showed variety in the fusion of adjacent plates, a degree of order seems to have been imposed by the sensory lines. Late Devonian lungfishes exhibited a tendency toward fusion of certain bones in the series under the supraorbital canal, of others under the infraorbital canal, and of postorbital bones from both series after the two canals evolved a connection behind the eye. Posterior to the point at which the lines joined, the supraorbital canal soon degenerated into a row of isolated pits, and in later lungfishes the bones that had enclosed it were lost. The anamestic plates that filled the space between the supraorbital series became fewer and larger as the sensory-line elements changed, but the distinction between the two groups of bones never disappeared.

Because lungfishes, like rhipidistians, showed a mosaic of small bones that underwent reduction, paleontologists once hoped to recognize homologous elements in the two kinds of fishes. Despite Westoll's attempt to compare the posterior elements of the supraorbital series in dipnoans with the parietals of the rhipidistians and the X and Y bones of the former with the supratemporal series of the latter, it was soon evident that such homologies were highly speculative. Rather than postulating homology, Schaeffer attributed the similar alignment of bones to adaptation to common kinds of stress associated with feeding and respiration. He extended this explanation to cover the reduction of dermal bones that occurred also in other types of bony fishes, surmising that all of them were derived from ancestors with numerous roofing bones and had undergone parallel evolution under the influence of similar selective forces.

Dermal bones in the skull of coelacanths

Like the dipnoans, the coelacanths appeared in the Devonian with a pattern of dermal bones that was already distinctive. Because the large paired elements in the middorsal part of the skull in these fishes did resemble certain bones in rhipidistians, paleontologists used the same names for them. Their acceptance of a common nomenclature did not mean, however, that they were certain of the homologies which in this case they were willing to propose. The row of supraorbital plates in coelacanths was lateral to the corresponding sensory line rather than directly beneath it, as in rhipidistians, and the bones which carried the infraorbital line were elongated with unusually large openings for communication between the sensory canal and the outside. The rest of the skull was covered with bones whose arrangement could also be compared only generally to that of rhipidistians (Fig. 5.12). The long snout in several genera was shielded at the

Figure 5.12. Skull of the coelacanth *Rhabdoderma elegans* in lateral view. The skeletal elements are named, and the sensory lateral-line canals are shown in the restoration. (From Moy-Thomas and Miles.)

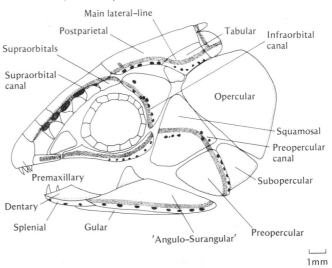

tip by a group of small rostrals and edged by a pair of toothed premaxillaries. Posterior marginal teeth were missing in all forms so that the mouth was bordered dorsally by the bones containing the infraorbital canal. The lower jaw was completely enclosed by a variable number of dermal elements, and, behind it, the corner of the cheek anterior to the opercular plate was covered by bones which have been called the squamosal and the preopercular.

Even though they recognize their inability to establish homologies beyond a doubt, paleontologists continue to discuss the manner in which the coelacanth pattern of dermal bones might have appeared. Stensiö has contended that the larger plates developed by fusion of smaller ones; others argue that expansion of certain bones and the loss of others could have brought about the same arrangement. Unfortunately, there is little hope of settling the question until geneticists and embryologists learn more about the factors which govern the formation of dermal bone. The unsolved problem is a crucial one, not only for the study of coelacanth relationships, but for those of other vertebrates as well. Whether investigators are attempting to trace lines of descent within the coelacanth group or to construct the family tree of ray-finned fishes, rhipidistians, or tetrapods, they must review comparatively the patterns of the dermal bones of the skull. Ignorance of the precise mechanisms controlling development and change in these elements hinders paleontologists in their efforts to determine whether forms with similar arrangements of dermal bones could have been derived one from another or were more likely to have been related in some other way.

Evolution of the coelacanths

Most paleontologists felt that they were on firm ground in likening the dermal bones of the skull in coelacanths to those of rhipidistians because the two groups of fishes exhibited a number of other structural similarities that suggested close relationship. Besides the lobed fins, both possessed an endocranium jointed in the same way between the trabecular and parachordal portions. Although the flexible braincase retained much of its embryonic cartilage in post-Devonian coelacanths, in earlier forms it ossified in rhipidistian fashion. The basisphenoid bone, one of the few elements in the cranium which did persist, demonstrated progressive change from the rhipidistian model, losing a common facet for articulation with the upper jaw and evolving a special one. The retention of some rhipidistian characters and the loss of others has led investigators to conclude, over the objections of Jarvik, that the coelacanths may have split away from rhipidistian stock late in the Silurian period and evolved divergently thereafter.

As has been already noted, structural change in coelacanths was limited and slow after their first appearance in the fossil record. The singular caudal fin, with its equal-sized dorsal and ventral lobes separated by a fringed extension of the axial skeleton, characterized *Nesides* and *Diplocercides* in the Devonian and remained as a hallmark of the group. Though the depth of the body increased in later forms, all the coelacanths were basically similar in shape. Their two lobed dorsal fins projected upward and the anal and paired fins downward from a heavy, almost ungainly, body behind a broad head and sloping snout. For reasons that are not clear, their peculiar shape was an extraordinarily successful one. Unchanged, it allowed the coelacanths to survive in a wide range of habitats. They maintained their place in Devonian seas and in Carboniferous and Permian rivers, swamps, and lakes in competition with the much more changeable paleoniscoids. Whereas the latter underwent countless metamorphoses in adapting to different niches, the coelacanths accompanied them virtually unmodified. In the Triassic period, as the ray-fins were advancing to the holostean level, the coelacanths enjoyed their greatest radiation. They swam then in salt and sweet waters almost worldwide. Paleontologists have found them in the rocks of continental Europe, the British Isles, Madagascar, Spitzbergen, Greenland, and North America.

The fossil record of the coelacanths ends abruptly in the Cretaceous period. Every one assumed that the versatile coelacanths had finally failed, overwhelmed either by the rapidly spreading teleost ray-fins or environmental changes to which they could not adapt. Then, in 1938, a South African fisherman

The modern coelacanth Latimeria

hauled in a live one. Its identity as a coelacanth was unquestionable, since it differed little in appearance from the Cretaceous *Macropoma* or the Jurassic *Undina*. The fish rotted before it could be transported to a laboratory, but scientists saved the skin and alerted fishermen to watch for a second specimen. Ichthyologists and paleontologists waited a long time before their dream of examining the soft tissues of the coelacanth was realized. The second fish was not caught until 1952 and also underwent partial decay. Its loss spurred the more organized efforts to find and preserve specimens that resulted in the taking of nearly a dozen *Latimeria* in the vicinity of the Comoro Islands within the next three years (Fig. 5.13).

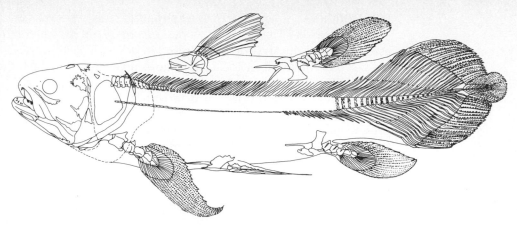

Figure 5.13. The modern coelacanth *Latimeria chalumnae*, showing the anatomy of the skeleton. (From Millot and Anthony.)

The coelacanth (named for Miss Courtenay-Latimer, the museum curator who had called attention to the first specimen) proved to be a strong fish rather than a delicate relict. J. Millot, who with J. Anthony and other assistants undertook the description of the anatomy of *Latimeria,* reported that the fish was capable of fighting for hours when turned out of the net onto the deck of a fishing vessel. The phosphorescent eyes and silvery-blue color of freshly caught specimens suggested that the coelacanths live in water 200 to 500 meters deep, where they lunge at passing fishes from concealed places among the rocks. Finding a whole fish in the stomach of one individual attested to these predatory habits.

Although the description of the internal structure of *Latimeria* has not yet been completed, Millot and Anthony have found that the modern form is much like the extinct ones. The arrangement of the dermal bones covering the skull is substantially unchanged, and the braincase beneath it still retains the intracranial articulation. Study of the postcranial skeleton has confirmed reconstructions made on the basis of fossil evidence: the body is supported in primitive fashion by a notochord unconstricted by ossified centra. Bone is confined to dermal elements in the pectoral girdle, a thin layer encasing each vertebral spine and fin ray, and an endochondral plate at the base of each fin.

The anatomy of the viscera as well as the structure of the skeleton negates the possibility that the coelacanths could have given rise to a known line of land animals, as the rhipidistians undoubtedly did. Despite the fact that *Latimeria* can presumably use the paired fins to prop its body and to propel it forward over the bottom, the modern coelacanth shows no evidence of having paralleled other rhipidistian descendants in evolving internal organs preadapted for use in a terrestrial environment. The outpocketing of the gut that serves as a lung in land animals is present but vestigial in *Latimeria*. The vein that drains its wall returns blood, not to the left side of the heart, as in tetrapods, but to the sinus venosus at the back of the heart, as it does directly or indirectly in all osteichthyans except lungfishes. The heart is characteristically fish-like in showing no sign of division into left and right sides, and the gut, with its spiral-valved intestine, is of a type common to all fishes except the most advanced ray-fins.

At one time, paleontologists thought that coelacanths, like air-breathing tetrapods, had nasal passages that opened into the mouth cavity, but dissection of *Latimeria* disproved that idea. Despite their fleshy fins, the coelacanths were no nearer the ancestral stock of land vertebrates than the dipnoans, fishes in which internal nares, or choanae, were also shown to be nonexistent. The discovery that internal nares were absent in coelacanths as well as dipnoans removed the basis for union of these fishes with the rhipidistians in the subclass Choanichthyes. The category was abandoned and the rhipidistians recognized as the only fishes that evolved the structural arrangement which appears in tetrapods.

The rhipidistian fishes

Of the fleshy-finned fishes, the rhipidistians are the only ones that have left no surviving species. The group is known to paleontologists only through a relatively small number of fossils, many of which show little more than dermal bones and scales. Although anything is possible, it seems unlikely that a living form will be discovered, because the rhipidistians had declined precipitously before they were swept away with many other kinds of vertebrates at the end of the Permian period. The reason for their disappearance is not easy to ascertain. They established themselves firmly in Devonian fresh waters as vigorous predators, probably as fearsome as the modern pikes, and developed no specialized structures that would have limited their ability to adapt to changed conditions. The rhipidistian body seemed designed for strong if not rapid swimming: robust in some genera and more fusiform in others, it carried dorsal, anal, and pelvic fins far back near a heterocercal or diphycercal tail. The head, heavily armored by dermal plates, bore small eyes and a large mouth. Any fish caught between a rhipidistian's jaws would have been fixed there by the sharp marginal teeth and punctured by spikes that projected from the palate. The powerful bite with which a rhipidistian closed upon its prey can be inferred from the arrangements within the skull that permitted shock absorption. The upper jaw articulated firmly with the anterior part of the braincase behind the eye, but movement at the intracranial joint would have prevented transmission of undamped pressure waves backward through the rear of the skull and body. It may have been that the jaw mechanism, however powerful a closure it allowed, was ultimately disadvantageous to the rhipidistians in their competition with freshwater paleoniscoid fishes. Although the ray-fins did not evolve their extraordinary maneuverability until post-Paleozoic times, they had shown by the Permian period a tendency toward variation of the mouthparts which assured them as a group a broader diet than the rhipidistians could consume.

Retention of an old jaw mechanism and a relatively invariable form of the body and fins does not in itself explain satisfactorily the failure of the rhipidistian fishes to survive. The existence in modern fresh waters of ray-finned forms like the ancient gar, *Lepidosteus,* and the bichir, *Polypterus,* demonstrates the viability of long-bodied predators and fishes with lobed fins. Although the structure of the rhipidistians may have put them at some disadvantage, their disappearance might have been due more directly to their inability to reproduce as rapidly as the actinopterygians. It is possible that as the ray-fins developed the immense egg-laying capacity for which existing forms are famous, the rhipidistians continued the perhaps more primitive habit of producing a smaller number of unprotected eggs. Paleontologists who discuss the derivation of amphibians from rhipidistian stock cite the preadaptive value of the appendicular skeleton and the incipient lungs, but it may well be that evolution of a protective gelatinous covering for their fertilized eggs was a crucial factor in the survival of animals crossing the rhipidistian-amphibian boundary.

Rhipidistians as amphibian ancestors

It has long been obvious that of all the Devonian fishes the rhipidistians were the most likely to have produced the first vertebrates that could live for substantial periods on land. Although most of the bony fishes possessed a diverticulum of the gut through which they could extract oxygen from the air, only the rhipidistians evolved a skeleton that could support the body out of the water. Unlike the lungfishes and coelacanths, the rhipidistians maintained their ability to lay down bone in the endoskeleton and eventually developed ossified centra which reinforced the notochord. The ray-finned fishes showed the same tendency toward increased ossification of the endoskeleton, but the arrangement of bones in the fin did not favor the evolution of strong, elongated appendages. The series of stubby bones that extended into the paired fins of the rhipidistians provided a better basis for the skeletal pattern that emerged in the amphibian limb. The rhipidistians had paralleled the ray-fins and other bony fishes in losing dermal bones from the mosaic that covered the skull, but as anamestic bones disappeared from the snout and elements fused along the path of the sensory canals, their dermal skeleton alone retained its solidity and approximated the amphibian design.

Although the relationship of the rhipidistians to the amphibians will be discussed in greater detail in the next chapter, it should be said here that none of the known fishes is thought to be directly ancestral to the earliest land vertebrates. Most of them lived after the first amphibians appeared, and those that came before show no evidence of developing the stout limbs and ribs that characterized the primitive tetrapods. While paleontologists hope to find remains of the rhipidistian line in which these structures evolved, they have no intention of neglecting the history of the other members of the group.

Classification of the rhipidistians

The investigators who have studied these fishes agree that the rhipidistians comprise several lineages, but the extent to which they diverged is in dispute. Jarvik, who has described the available fossil material in minute detail, recognizes a division of the rhipidistians at the ordinal level. To one group, the Porolepiformes, he assigns *Porolepis, Holoptychus,* and their allies, fishes distinguished outwardly from other rhipidistians (if *Holoptychus* is representative of forms whose postcranial skeletons are missing) by their relatively deep body and elongated pectoral fins. The more streamlined forms like *Osteolepis* and the later *Eusthenopteron* Jarvik places among the Osteolepiformes, the larger of his two divisions. In support of his broad separation of the two groups, Jarvik lists numerous differences in the skeletal anatomy of the porolepiforms and osteolepiforms. He finds divergence in the structure of the snout, the nasal cavities, and the palate as well as in the pattern of the dermal bones over the cranium and cheek. Romer and K. S. Thomson, who disagree with Jarvik, do not challenge his anatomical descriptions but argue against his interpretation of them. In their opinion, the differences that exist among the rhipidistians are not great enough to require the erection of two orders. Instead, Romer distinguishes two superfamilies, the Osteolepidoidea and the Holoptychoidea. The latter category houses the same types as the Porolepiformes of Jarvik but is named for a representative that is more completely preserved than the earlier *Porolepis*. During the Devonian period, diversification occurred within each of the major groups. Both Romer and Jarvik separate the large, slim, especially predaceous-looking rhizodonts with cycloid scales, like *Eusthenopteron*, from their more conservative relatives, like *Osteolepis*, that bore a covering of closely fitted rhombic scales. Thomson believes that the rhizodonts and osteolepids are distinct enough to merit assignment to different superfamilies. The classification of the rhipidistians has been complicated still further by the discovery of a very small form, *Strunius* (Fig. 5.14), that does not seem to belong to any of the established groups. H. Jessen, who described the fish, suggested the creation of a separate order for it and *Onychodus*, a little-known, possibly marine rhipidistian that appears to be related to it. The Struniiformes, as Jessen named the taxon, has been received with some hesitation, since the forms on which it is based are not yet well understood.

Classification of the ray-finned or actinopterygian fishes

While paleontologists working on the classification of the rhipidistians trace their troubles to the paucity of fossil material and the lack of living forms, those who study the ray-finned fishes find their task complicated by the abundance of both. Usable descriptions of modern species have been accumulating since Rondelet and Belon wrote their treatises in the sixteenth

Figure 5.14. *Strunius walteri*, a Devonian form of uncertain relationships. (From Jessen.)

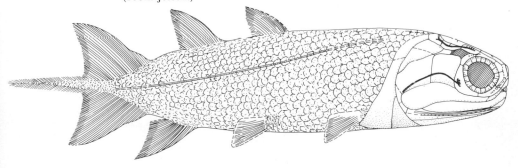

century, and the number of fossil forms has increased steadily since Agassiz published the first volume of *Recherches sur les poissons fossiles* in 1833. The first attempt to categorize the fishes was based upon study of the living forms since their anatomy was known far better than that of extinct ones. Ichthyologists distinguished from the many modern species (which they called teleosts) a few fishes that seemed more archaic in structure. Of these, they assumed that *Lepidosteus* and the bowfin, *Amia*, represented the stock from which the current swimmers arose and the sturgeon, *Acipenser*, and the spoonbill, *Polyodon*, being cartilaginous, even more primitive forms (Fig. 5.15). They fitted these fishes into the classification scheme as the Holostei and the Chondrostei respectively. When fossil evidence became available, it appeared that the earliest ray-fins were bony rather than cartilaginous forms. Since the term Chondrostei was obviously inappropriate for animals with a skeleton of bone, the ancient relatives of the sturgeon and the spoonbill, though listed as

Figure 5.15. Surviving archaic ray-finned fishes. **(a)** *Polyodon*, the river spoonbill of southern United States. *Polyodon* and a related genus native to China, the sturgeons (see Fig. 5.1*d*), and the bichir, *Polypterus*, are the sole surviving members of the Chondrostei. **(b)** *Amia*, the freshwater bowfin. This form and *Lepidosteus* (see Fig. 5.1*c*) are the only living representatives of the Holostei. (*a* from Dean; *b* from Cambridge Natural History.)

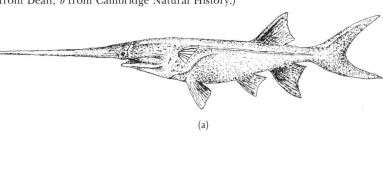

(a)

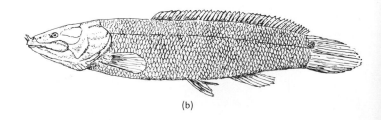

(b)

chondrosteans, were distinguished as paleoniscoids. The name was retained even after it was discovered that not all the fishes in the group belonged to the genus *Paleoniscus.* The paleoniscoids had diversified far more widely, it seems, than any one had imagined.

Structure and origin of paleoniscoid fishes

The earliest known paleoniscoids shared certain characteristics with the rhipidistians but had clearly evolved separately from them for a long time. Although the braincase, jaws, and branchial chamber of the paleoniscoids were encased in bone, the pattern of the sensory canals was unique in these fishes and the arrangement of the dermal plates distinct from that of lobe-finned forms. While the dermal part of the pectoral girdle was similarly constructed in both types of bony fishes and early representatives of both had an unconstricted notochord that extended from a point behind the basisphenoid in the skull to the tip of the heterocercal caudal fin, fishes in the two groups differed in the structure of their fins and scales. With rare exceptions, the paired fins of the paleoniscoids were broad-based and articulated with the girdle by a row of basal elements that did not project far outward into the fin. The dorsal fin was single, as has been noted; there were never two of them, as in rhipidistians. Scales of the ganoid type grew in paleoniscoids by enlargement at the edges and sometimes by superimposition of dentin tubercles. The paleoniscoids did not evolve a mechanism for absorption and redeposition of the hard tissue of skull and scale, as certain rhipidistians and dipnoans did in Devonian time.

There seems to be as little chance of tracing the origin of the paleoniscoids as of any other kind of fishes, but investigators speculate on the matter, nevertheless. They debate not only the relationship of the first ray-fins to the acanthodians but also the type of habitat in which the line developed. As was seen in the last chapter, close comparative study of the acanthodian and paleoniscoid fossils resulted in several conflicting hypotheses concerning the ancestry of the later group. The theories about the environment in which the fishes appeared have been discussed to a lesser degree, because there is no positive evidence to support them. Westoll and Romer believe that the ray-fins originated in fresh water, since the oldest remains of these forms are found in freshwater deposits rather than in contemporary marine strata. They point out that fossils of fishes transitional to the paleoniscoid level may never be found because of the rarity of Silurian and Lower Devonian freshwater beds. White and

Gardiner argue in opposition that ray-fins must have evolved in salt water even though Devonian shallow marine deposits do not seem to contain their bones. The broad geographical distribution attained by the Devonian paleoniscoids is explicable, they feel, only if these fishes had ancestors that could cross the sea. Since migration from continent to continent overland through freshwater routes is a slow and hazardous process, the existence of marine predecessors for the first known paleoniscoids would seem logical. It should be mentioned, however, that if Europe, North America, Antarctica, Australia, and Africa once formed a single land mass, as the proponents of the theory of continental drift assert, the presence of Devonian paleoniscoids in the lakes and rivers of these lands would not have required passage through a marine environment. Although it appears that continental drifting may be proved or disproved within one or two decades by bouncing laser beams off the reflector on the moon, the paleontological evidence necessary to determine the earliest environment of the paleoniscoids may never be available.

Later history of the paleoniscoids

The later history of the paleoniscoids is not in dispute. They became the most successful of the freshwater fishes, exceeding the crossopterygians and dipnoans in number and variety. During the Carboniferous and Permian periods, some of them ventured into the sea, where they survived and evolved advanced forms despite the presence of the fiercely predaceous ancient sharks. Their fortunes declined as the Paleozoic era came to a close: of the thirty-nine families that Gardiner lists, only twelve had representatives that lived into the Triassic period. The fishes to which the paleoniscoids lost their place in the sea were more progressive members of their own line. These forms, the holosteans, supplanted the marine paleoniscoids entirely by the time the dinosaurs walked unchallenged on the land. The last of the old ray-fins to become extinct were those which reinvaded fresh water. Triassic freshwater deposits in eastern and southwestern parts of the United States have yielded assemblages of paleoniscoids—both archaic types and those which approached but did not quite attain the holostean level—that remind paleontologists of older communities of ray-finned fishes. Isolated in separate systems of lakes and rivers, the remaining paleoniscoids evolved a variety of indigenous types and finally disappeared almost completely. Just as a freshwater niche protected the lungfishes from total extinction, so it allowed the survival of a few degenerate remnants of the paleoniscoid group—the sturgeon, the spoonbill, and the African bichir, *Polypterus*.

Paleontologists have found it easier to characterize the history of the paleoniscoids than to clarify the descent of the many genera they have described (Fig. 5.16). In assigning all the chondrosteans except the living species and their immediate ancestors to the order Paleonisciformes, Romer emphasizes the common structural ground plan of the fossil forms and implies that they have radiated from a central ancestral stock. B. G. Gardiner agrees that the paleoniscoids were a close-knit group, but, being more of a splitter than Romer, he restricts the membership of the Paleonisciformes more severely. Extremely specialized forms like the frill-finned *Tarrasius* and late, advanced types such as *Dorypterus* and the almost holostean-level *Redfieldia* he places in separate orders. Even within the narrowly defined order Paleonisciformes, Gardiner observes a strong tendency toward diversification. The paleoniscoids which appeared in the Devonian period belonged to four families so distinct that Gardiner supposes each to have been evolving divergently for some time. Although these Devonian forms were primitive in certain respects, none seems to have been ancestral to the Carboniferous members of the order. Neither the acanthodian-like, minute scales of *Cheirolepis* nor the divided cranium and pineal foramen in the skull of *Kentuckia* was carried onward in the paleoniscoid line. In Gardiner's opinion, the twelve families that included the bulk of the Carboniferous forms did have a common Devonian ancestor but one that is still unknown. The later paleoniscoids were derived from several of the Carboniferous families rather than a single one, but here again actual relationships are obscure. Since a number of groups showed evolutionary tendencies of a similar kind, it is often impossible to discriminate between parallel development and true descent. By the Triassic period, fishes of more than one family possessed shorter jaws than the older paleoniscoids as well as thinner scales, a reduced number of fin rays, and a more symmetrical-looking caudal fin.

These advanced characteristics are the ones by which paleontologists recognize the holostean ray-fins that dominated the waters of the Mesozoic era. When investigators discovered that the structural pattern of these fishes had emerged more than once, they put aside the old theory that the central holostean types—the semionotids, amioids, and pholidophoroid fishes—had arisen from a single paleoniscoid stock. The Holostei constituted, they came to believe, not a monophyletic group but an assemblage of fishes which had arrived independently at a similar level or grade of structural advancement. To paleonis-

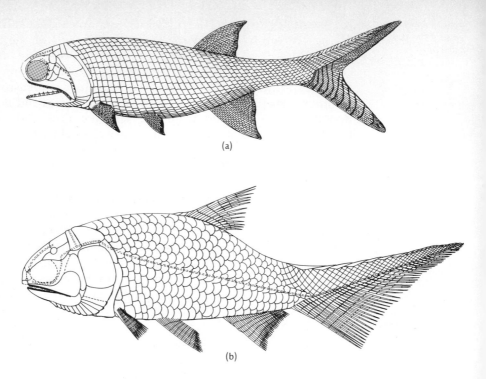

(a)

(b)

coids that had evolved some but not all of the holostean traits, J. Brough applied the term "subholostean." That the ray-finned fishes had undergone evolution from the paleoniscoid to sub-holostean and holostean grades proved a key idea for workers studying the group. They recognized that the same sort of parallel development continued among the actinopterygians throughout the Mesozoic era to produce several lines of fishes that approached the teleost level.

Despite the availability of subholostean fossil material, paleontologists have not yet succeeded in firmly establishing the origin of each holostean group (Fig. 5.17). The elongated aspidorhynchids and the deep-bodied pycnodontids meet the anatomical criteria for inclusion in the Holostei, but they have evolved a special character which is not foreshadowed by any of the earlier forms. Although the other three holostean groups are not quite so divergent, only one has been linked directly to a particular subholostean source. After comparing the pattern of dermal bones sheathing the head, Brough claimed that the first amioids could have been derived from the subholostean parasemionotids which had lived before them in the Triassic period.

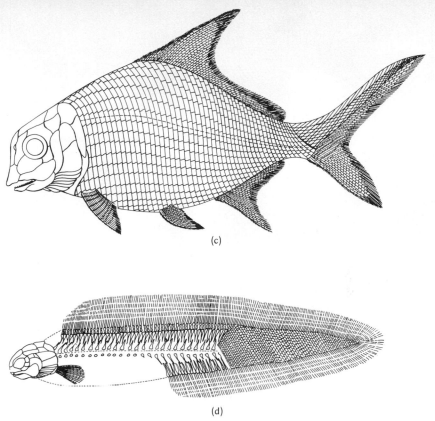

(c)

(d)

Figure 5.16. Paleoniscoid fishes. **(a)** *Dicellopygae.* **(b)** *Sphaerolepis.* **(c)** *Mesolepis.* **(d)** *Tarrasius.* (*a* from Brough, *b* from Gardiner, *c* from Traquair, *d* from Moy-Thomas.)

His theory has been widely accepted but not without reservation. The dermal bones in parasemionotids and the Mesozoic forebears of the modern bowfin, *Amia*, do not resemble each other in every respect: the parasemionotids lack the series of suborbital bones that appears in the amioids. Paleontologists who believe that the amioids descended from the parasemionotids rather than from subholosteans which did have these bones must explain the discrepancy.

Another kind of problem faces those who seek the origin of the semionotids, the holosteans to which the living garpike is allied. Schaeffer and Dunkle find no subholosteans more suitable as ancestors for these short-jawed forms than the parasem-

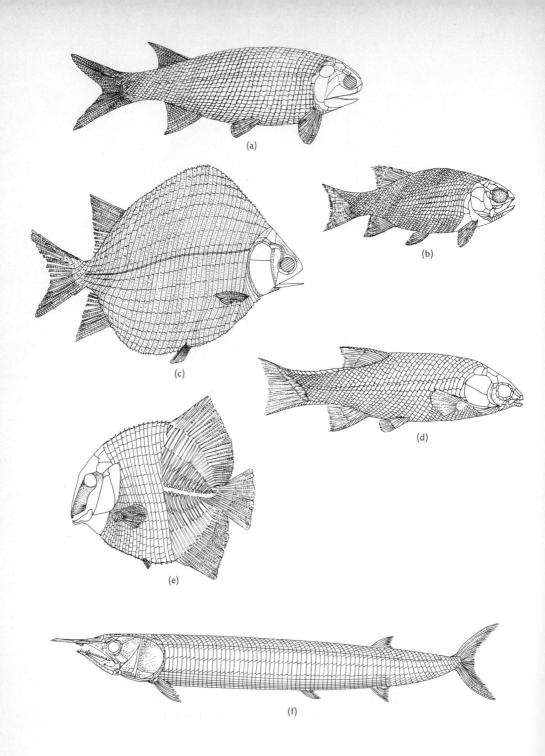

ionotids, but they appear later in the fossil record than *Acentro-phorus*, the first known member of the semionotid assemblage. While other paleontologists have barred a linear relationship between the parasemionotids and the semionotids because of the chronological discordance, Schaeffer and Dunkle mention the possibility that unknown parasemionotids might have lived long enough before the semionotids to have sired the group. Even if an early Permian parasemionotid were discovered, however, it is probable that the question of semionotid ancestry would remain unsettled. It would still be necessary to demonstrate that the special cheek and jaw structure of *Acentrophorus* and its relatives could have been derived from the parasemionotid pattern.

It is, after all, the problem of structural transformation which is the most difficult one. Although paleontologists can reasonably assume that the fragmentary nature of the fossil record results sometimes in misleading conclusions concerning the temporal limits of various groups, they cannot overstep incongruities of structure quite as easily. For the ray-finned fishes, as for the coelacanths, investigators have in hand a number of skulls with variations in the arrangement of the dermal elements. Not knowing enough about mechanisms of development and change in the dermal skeleton, they have few biological criteria to guide them in their search for phyletic sequence. When structural similarity is not close, paleontologists cannot carry their speculations very far. For this reason, the origin of the semionotids can only be debated, and the source of the pholidophorids, holostean forerunners of the teleosts, can hardly be supposed until subholosteans more like them are found.

Relative superiority of the early actinopterygians

Besides attempting to determine the correct interrelationships among extinct forms, paleontologists have used their studies of comparative structure to explain the functional changes that made the extensive radiation of the ray-finned fishes possible. It is obvious that the success of the actinopterygians depended in great part upon the development of locomotor and food-getting mechanisms which were more efficient than those of other kinds of fishes. Although the paired fins and hinged jaws that arose in the placoderms and acanthodians sustained those vertebrates

Figure 5.17. Subholostean and holostean fishes. Subholosteans: **(a)** *Redfieldia* (= *Catopterus*), **(b)** *Parasemionotus*, and **(c)** *Cleithrolepis*. Holosteans: **(d)** *Acentrophorus*, **(e)** *Mesodon*, a pycnodont, and **(f)** *Aspidorhynchus*. (*a,c* from Brough; *b* from Romer; *d* from Moy-Thomas; *e* from Smith Woodward; *f* from Assman.)

for millions of years and, with certain modifications, served the sharks for an even longer period of time, the structures in the ray-fins achieved a level of complexity that enabled those forms to swim and eat with unparalleled finesse. In becoming increasingly well adapted for engaging in these activities underwater, the ray-finned fishes lost the generalized body shape which enabled the rhipidistians to engender a line of land forms. Confined by the course of their evolution to the submerged parts of the earth, the actinopterygians spread throughout their realm and made themselves masters of it.

Swimming ability of the paleoniscoids

At first, the paleoniscoids must have been no better swimmers than other Devonian fishes. Although they presumably had an air bladder outpocketing from the pharyngeal region, their thick, ganoine-covered scales and well-armored head made them still much heavier than the water they displaced. As in other primitive fishes, the lift provided by the heterocercal caudal fin kept them from sinking as they swam, and the low-set, broad-based pectoral fins countered the tendency of the snout to plunge downward. They could change direction by contracting the fusiform body strongly on one side to pull the head around, but stopping suddenly must have been almost impossible. Even if the paired fins could have been twisted outward to act as brakes, they would not have served effectively. Since they were located below the fish's center of gravity, they would have acted as a pivot around which the body would have tumbled tail over snout. Despite the limited ability of the early ray-fins to turn and stop quickly, the design of their body and the placement of the fins should have made them fair swimmers in open water.

Adaptations for more efficient locomotion in ray-fins

Even in Paleozoic times, however, modifications in structure produced ray-fins that could maneuver with greater agility than the ordinary forms. These changes occurred not in one line but in several. From the Carboniferous period onward, families of fishes can be found that show the improved design to a varying degree. Alterations that contributed to locomotor efficiency continued to appear and to be preserved by natural selection until the most modern fishes in the group, the Recent spiny-finned teleosts, became the most adept swimmers ever to exist. In analyzing the progressive structural changes that improved the swimming ability of the ray-fins, paleontologists have been guided by the work of Gray and Harris on the locomotion of living fishes. Their demonstration of the forces that come into play as fishes of certain shape move through the water has enabled paleontologists to estimate the functional significance of the structures they observe in fossil forms.

One of the earliest and most universal signs of change related to swimming efficiency in ray-fins was the gradual disappearance of the heterocercal caudal fin. Its transformation to a symmetrical shape increased its forward drive at the expense of the component of lift that had been essential in keeping ancient fishes afloat. Since the newer forms were not adapted for bottom dwelling, it was plain to paleontologists that the specific gravity of the fishes must have decreased sufficiently to allow them to stay in midwater without sinking. The thinning of the scales that characterized more advanced ray-fins and the loss of scales and dermal skull bones that occurred in numerous families had surely made the fishes lighter; and it is possible that by Carboniferous time the air bladder had begun to function as a hydrostatic organ, further reducing the weight of the fish relative to the water around it. The new equilibrium and the disappearance of the heterocercal tail made it unnecessary for the pectoral fins to remain as hydrofoils below the head. Especially in deep-bodied forms, they moved upward, closer to the center of gravity, where they could brake the forward motion of the fish without causing it to somersault. In this position, one fin held at right angles to the body could effect a sudden turn with a smaller expenditure of energy than bending the whole body required. The reason for the migration forward of the pelvic fins in forms with high-set pectorals was clarified by Harris's experiments on living fishes: he found that elevated pectorals impart an upward, and pelvic fins a downward, motion to the body when they are extended; if the pelvics remained in a posterior position, the concerted action of the paired fins in braking would cause the fish to twist head up and tail down as it slowed. The closer the pelvic fins are to the raised pectoral pair, the less tendency there is for the fish to rotate when it stops.

The shape and length of the fish changed, as well as the form and position of the fins, as the actinopterygians evolved their extraordinary swimming skill. The use of the pectorals for braking and turning was coordinated, as noted above, with the development of a short, deep body. For the head and tail to pivot quickly around an outspread pectoral fin, there had to be minimum resistance of these areas against the adjacent water. The shorter and more compact these parts of the body became, the faster the fish was able to turn. Forms inhabiting environments where maneuverability was more essential than speed evolved a streamlined disk shape, which was stabilized in the vertical plane by prominent dorsal and anal fins. Swimmers in open water retained a more elongated body to accommodate the bulkier axial musculature they required, but even among these fishes the most advanced types show compression or reduction in the number of vertebrae in the trunk and tail region.

Modifications of the endoskeleton in teleosts

Although the structural changes which took place as the actinopterygians progressed must be described one by one, it is important to remember that they appeared together as a functionally interrelated complex. As the body underwent some degree of shortening in front of and behind the raised pectoral fins, modifications occurred in the endoskeleton that allowed the remaining muscles to pull upon it more effectively. The unconstricted notochord of paleoniscoids and holosteans was replaced as the chief axial support in teleosts by a series of ossified centra, each of which fused with the neural arch above it and, in the caudal region, with the hemal arch below. Besides providing stronger anchorage for the musculature, the bony vertebrae, sculptured and drawn into processes, afforded greater surface area for the attachment of muscle fibers than the smooth notochordal sheath had. As the spiny-finned, or acanthopterygian, teleosts evolved, the proportions of the spinal column changed in a way that favored the most effective distribution of the axial musculature. Although the total number of vertebrae decreased, the caudal elements which bore the muscles principally responsible for propulsion eventually exceeded those of the less muscular trunk region. The shape of the skull in these fishes changed, too, allowing many more fibers to insert on the back of the head. In the disk-shaped forms the occipital region rose high above the passage for the spinal cord to receive the heavy epaxial musculature, and even in longer-bodied species the supraoccipital bone, to which much of the dorsal muscle mass attached anteriorly, enlarged its surface by growing forward between the parietals and developing a tall crest.

Diversification of feeding mechanisms in ray-fins

While the back of the skull changed in relation to the depth of the epaxial musculature, the anterior part underwent modifications that were associated with the development of new feeding mechanisms. Early actinopterygians, like contemporary Devonian rhipidistians, had long, toothed jaws that were splendid equipment for predation but unsuitable for any other kind of food gathering. The remodeling of the mouthparts that occurred in many later ray-fins increased the diet of the group as a whole and made possible the adaptation of one or more of its members to almost every imaginable underwater area. Like the structural innovations that resulted in improved locomotor ability, those that produced different techniques of food gathering appeared in many lines and developed to a different degree in each. At every level some forms evolved with shortened jaws and a smaller mouth gape. The change, which occurred independently many times, was brought about by a swinging forward of the bones through which the jaw joint is braced against the otic region of the skull. The suspensorium, as the elements are called collectively, was directed posteriorly in early paleoniscoids so that the upper and lower jaws hinged well behind the eye. First in aberrant paleoniscoids like the haplolepids and later in many subholostean forms, the suspensorium became more vertical, setting the articulation of the jaws farther forward beneath the orbit and reducing the size of the mouth concomitantly. The semionotids and a number of teleosts carried the trend to an extreme by developing a suspensorium which slanted anteriorly to support jaws that were even shorter than those of other fishes.

By studying the arrangement of the jaws, the size of the gape, and the nature of the teeth, paleoichthyologists can follow the development of new feeding habits among the diversifying ray-fins. Beside the predatory paleoniscoids that grew sharp teeth on their long jaws and patches of conical denticles more medially within the oral cavity, there arose the deep-bodied platysomids, whose smaller mouth armed with low crushing or pointed teeth fitted them for eating chitin-covered invertebrates or chipping out polyps from ancient coral reefs. Before the end of the Paleozoic era, forms had evolved that could make use of the abundant vegetation that floated in still waters: the toothless scissor-like jaws of *Dorypterus* mark it as specialized for grazing, and it is not unlikely that other freshwater subholosteans subsisted at least partly on plant material. The elaboration of the mouthparts increased rapidly with the advent of the holosteans and teleosts. In the latter group the structure of the jaws and teeth proved especially plastic. Scavenging along the bottom became possible for fishes that evolved a mouth beneath the snout; nibbling at floating plankton, for others whose jaws tilted in an upward direction. Forms with tube-like mouths sucked in diminutive organisms, while others, still long-jawed, continued the old predaceous habit. Teeth varied in shape and location (Fig. 5.18): although their hard tissues remained characteristically fish-like, marginal teeth in different forms developed as fangs, multiple ranks of shorter spikes, or minute sharp blades fused, hinged, or even set in sockets along the bones at the boundaries of the mouth. When the marginal teeth were small

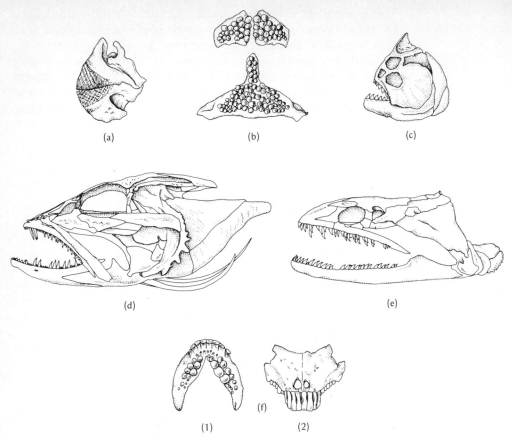

Figure 5.18. Teeth in teleost fishes. **(a)** Jaws of the parrot fish *Pseudoscarus*, showing strong, knife-sharp plates formed by fusion of individual teeth as adaptation for nibbling coral. **(b)** Upper and lower pharyngeal bones of the wrasse *Labrus*, with knob-like teeth for crushing vegetable matter. **(c)** Skull of the piranha, *Serrasalmus*, showing short, sharp teeth for tearing flesh. **(d)** Skull of the mail-cheeked fish *Ophiodon*, showing mouth with fine, conical teeth at the margin and large fangs. **(e)** Skull of the carnivorous muraenid *Thyrsoidea*, with long jaws bearing large, back-curved marginal teeth. **(f)** Teeth of the sparid *Sargus*: (1) dorsal view of lower jaw, showing knob-shaped teeth on inner surface; (2) front view of upper jaw, showing incisiform marginal teeth. (*a,c* from Norman; *b,d,e* from Cambridge Natural History; *f* from Goodrich.)

or absent, batteries of denticles grew on the upper and lower elements of the more posterior gill arches. Often massed in patches, these pharyngeal structures macerated food as it passed backward from the mouth.

*Improved effi-
ciency of the jaw
apparatus in ray-
fins*

The teleost fishes were successful not only because of the wide variety of their oral structures but because there had been real improvement in the efficiency of the jaw mechanism as the ray-fins advanced from grade to grade. The jaws and dermal bones of the paleoniscoids had been very rigid. Hinged at its posterior end, the mandible was pulled upward by adductor muscles that were confined at their origin to a small area enclosed by the maxilla, preoperculum, infraorbital, and palato-quadrate bones of the skull. Contraction of the adductor fibers, which inserted in a depression along the side of the mandible, raised the lower jaw against the immobile maxillary and rostral elements. The mouth in the paleoniscoids was a mere snapping device: the fishes lowered the mandible, engulfed their prey and brought the jaw straight up with as much force as the closely housed adductor muscles could produce.

With the advent of the holosteans, changes occurred that laid the groundwork for an entirely different kind of jaw action (Fig. 5.19). In discussing the evolution of the mouthparts in those fishes and more advanced forms, Schaeffer and Rosen have emphasized change in the maxilla as an essential first step in the development of the new mechanism. The forward movement of the suspensorium in holostean ray-fins brought with it the separation of the maxilla from the adjacent bones of the cheek and palate. The adductor muscles, no longer enclosed in a bony box, grew more massive and more complex. A portion of the fibers inserted upon the newly raised coronoid process of the mandible so that the lower jaw was twisted as well as lifted upward. For the first time the movement of the lower jaw was able to induce some motion in the upper one. Since the coronoid process was attached by a ligament to the relatively free, posterior part of the maxilla, the two moved forward together as the mandible was depressed. The fold of skin that must have existed around the maxilla would have been stretched taut and the oral cavity enlarged slightly. There was undoubtedly a distinct advantage in the stronger, freer movement of the jaws, because the new arrangement of the mouthparts was retained as the ray-fins advanced to the teleost level.

Although the most primitive teleost fishes retained the loosened maxilla as a marginal bone of the upper jaw, there was a tendency in several lines toward its elimination from the border of the mouth. Observation of fossil and Recent forms suggests that the maxilla was relegated to its new position by the gradual enlargement of the premaxilla. That bone, which in lower

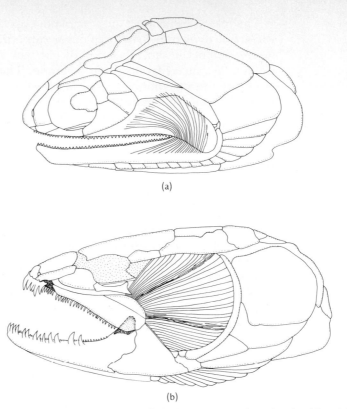

(a)

(b)

Figure 5.19. Comparison of the jaw apparatus in paleoniscoid and holostean fishes. **(a)** *Pteronisculus*, a paleoniscoid with adductor muscles restored. The muscles which lifted the mandible were restricted in volume by the small space in which they were enclosed. **(b)** *Amia*, a holostean, showing enlargement and differentiation of the adductor muscles made possible by the separation of the bones of the cheek and upper jaw. (From Schaeffer and Rosen.)

ray-fins was a small dentigerous element adjacent to the midline, developed processes extending dorsally and laterally as teleost evolution continued. The lateral process grew downward in front of the maxilla and gradually displaced it from the margin. In teleosts in which the maxilla was left, toothless, behind the upper jaw, it became important in effecting increased protrusion of the mouth. The sudden extension of the mouth as it opened created suction that aided significantly in drawing food inward. The superiority of this mechanism of ingestion can be inferred from the independent evolution of a highly protrusile mouth in several lines of teleost fishes. The arrangement of the bones differed in the various groups, but the structural principle was always the same. Mandible, maxilla, and premaxilla were tied together so that the maxilla rode forward, pushing the premaxilla before it when the lower jaw was pulled downward. The degree to which the mouth could be protruded depended upon the nature of the attachment of the premaxilla to the rostrum. In modern teleosts, like the cod, in which the medial part of the premaxilla is anchored to the front of the cranium, only the lateral process of the bone is movable, and protrusion is limited (Fig. 5.20). In the carp and its relatives, the premaxilla is connected by a ligament that allows the entire bone to slide forward as the fish opens its mouth to feed. The acanthopterygians have evolved an apparatus that is as protrusible as that of the carps but even more firmly braced. In these fishes the premaxilla has developed a long ascending process that slips up and down along the dorsal midline beneath strong transverse ligaments. In some forms the ascending process moves through a groove in the medial edge of the maxilla, and that element articulates in turn with the bones of the palato-ethmoid region. This arrangement allows great extension of the mouth with little risk of lateral displacement of its parts.

Origin of the tele-osts

The teleost fishes, beautifully adapted for feeding and swimming, increased so rapidly that investigators have had difficulty in following their evolution. They appeared first as rare forms in the Triassic period, offspring perhaps of the pholidophoroid holosteans. Whether they are a monophyletic or a polyphyletic group is still an open question. It has been maintained by some paleoichthyologists that the teleost assemblage arose from the mid-Triassic leptolepids, unspecialized forms whose overlapping scales bore only a vestige of the old armorial ganoine. Other workers point out that though the earliest teleosts share the traits characteristic of the grade, they show variations in cranial structure which suggest a separate origin from different holostean (or even subholostean) ancestors. Lack of adequate fossil material makes it impossible to decide which hypothesis is correct. As is usual with swimming vertebrates, a series of forms transitional to members of established groups has not been recognized. Paleontologists suspect that the forerunners of the highly diversified teleost population of Cretaceous time may not have arisen in the shallow seas under which the few known Triassic and Jurassic forms were preserved but in fresh water or the deep parts of the ocean from which sediments are no longer available.

Since the history of the earliest teleosts is obscure, paleontolo-

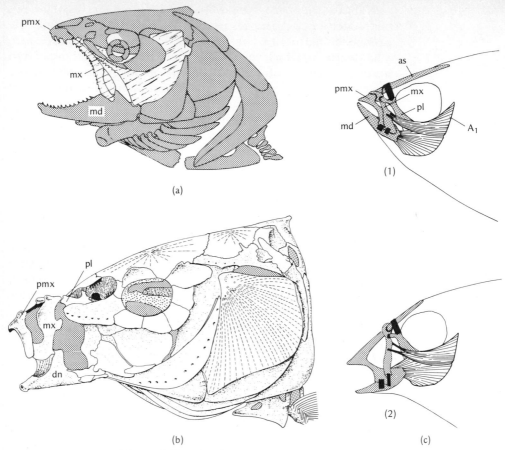

(a)

(1)

(b)   (c)

Figure 5.20. The jaw mechanism in teleost fishes. **(a)** *Salmo*, the salmon, showing a nonprotrusile jaw in open position. The gape is large because the jaws are long. The premaxilla is fixed in position, and the maxilla constitutes part of the margin of the upper jaw. **(b)** *Cyprinus*, the carp, showing its protrusile jaw in open position. The premaxilla has been forced forward by depression of the mandible but is restrained by ligaments secured to median, posterior elements. **(c)** *Gerres*, a perch-like fish, showing a protrusile jaw different from that of the carps: (1) mouth closed; (2) mouth opened, showing premaxilla drawn forward, its ascending process restrained by ligaments and a process of the maxilla. Abbreviations: A_1, adductor muscle (external part); *as*, ascending part of premaxilla; *dn*, dentary bone of mandible; *md*, mandible; *mx*, maxilla; *pl*, palatine bone; *pmx*, premaxilla; ligaments in black. (*a,c* from Schaeffer and Rosen; *b* from Gregory.)

gists have been unable to define the interrelationships among the many types that had all but replaced the older holosteans by the beginning of the Cretaceous period. The numerous fossils of this age indicate that more than half a dozen major groups of teleost ray-fins dominated the scene for a time, only to be eclipsed by the appearance of the acanthopterygian teleosts at the end of the period. These new spiny-finned fishes radiated widely in marine and freshwater environments beside surviving members of the more primitive orders until, by the Eocene epoch of the Cenozoic era, the assemblage of teleosts had assumed its modern character.

Classification of the teleosts

Because most of the principal families of Cretaceous and early Cenozoic times have living representatives, investigators tried first to determine the relative position of the various forms by examining extant fishes. Both E. D. Cope and T. N. Gill used this method in the 1870s to classify the teleosts. The scheme they produced, revised and refined over the years by a succession of eminent ichthyologists, became a standard reference for students of these ray-finned fishes. Despite its place in the literature, however, the classification has not satisfied workers who are interested in establishing phylogenetic sequences. Even in L. S. Berg's carefully drawn version of 1940, the most primitive-appearing teleosts—the herrings, the tarpons, and the salmon and trout—are lumped in the single order Isospondyli although they may not be derived from the same stock, while fishes that are possibly part of a single lineage, like the beryciformes and the perches, are listed in separate groups. The arrangement serves to segregate the different types of teleosts that are known but does not project any evolutionary scheme for them.

In an effort to replace the typological classification with one that does reflect the phyletic relationships of the teleosts, Greenwood, Rosen, Weitzman, and Myers published in 1966 a new analysis of the advanced ray-fins based on fossil as well as living genera (Fig. 5.21). They postulated independent attainment of the teleost level by three or possibly four groups of holostean descendants. Although fossil evidence for their theory is slim, they think that the salmonoids, the elopoid ancestors of the tarpon, the forerunners of the African elephant fishes, and perhaps the clupeoid precursors of modern herrings and sardines were already evolving divergently early in the Mesozoic era. The subsequent history of these fishes was very different. The clupeoids produced successful forms without progressing further. The elopoids also survived as a low-level teleost group but gave rise to the highly specialized eels, with which they share at present little but a unique larval stage. The line to which the elephant fishes belong declined from a vigorous tribe that included monstrous marine specimens like the 12-foot-long *Portheus* to a remnant confined to tropical fresh waters. The

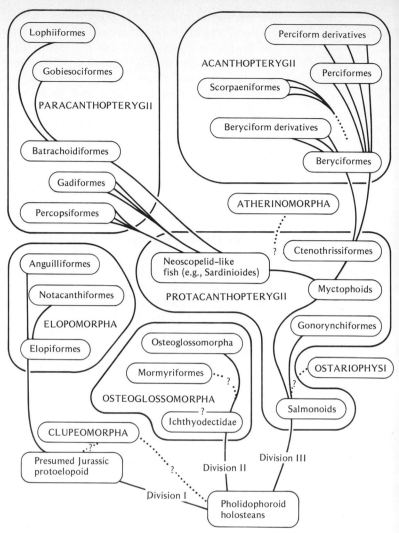

Figure 5.21. A suggested scheme of the interrelationships among modern teleost fishes. (From Greenwood, Rosen, Weitzman, and Myers.)

salmonoids proved the most versatile group of all. In addition to maintaining themselves as primitive teleosts with toothed maxillae and low-set paired fins, they provided the stock from which, in the opinion of Greenwood and his colleagues, the whole assemblage of more advanced and more specialized teleosts was derived. According to these authors, there was in this division of the teleosts a general tendency to progress. Fishes of several lineages traceable to the salmonoids developed characteristics that brought them closer and closer to the acanthopterygian level.

In recognition of their evolutionary course, the authors called the salmonoids and their near relatives Protacanthopterygii and found among them groups which seem to have been immediately ancestral to the most advanced of the ray-finned forms. They follow C. Patterson in designating as predecessors of the first true acanthopterygians the ctenothrissiforms, a Cretaceous order which sprang, they think, from the myctophoids, an earlier and more central group of protacanthopterygians represented today by the lantern fishes. Both the ctenothrissiforms and the myctophoids diversified as they evolved: the former produced possibly more than one of the genera that paleo-ichthyologists have placed among the beryciforms at the base of the Acanthopterygii, and the latter, besides siring the ctenothrissiforms, gave rise indirectly to another, larger aggregation of fishes, the Paracanthopterygii, which developed in parallel with the spiny-finned forms.

Until much more fossil evidence is accumulated, it will probably be impossible to segregate the teleosts into narrower phyletic sequences. The Paracanthopterygii, while distinguished by certain characteristics from other advanced ray-fins, includes such diverse forms as the cod and haddock, the clingfishes, the toadfishes, and the weird-looking deep-ocean anglerfishes. Since not all these forms share an immediate common ancestor among the protacanthopterygians, the superorder in which they are housed is less than satisfactory as a category in the classification scheme. The Acanthopterygii may also be polyphyletic, despite the fact that its members are all alike in possessing stiff spines instead of soft rays in the anterior part of the dorsal and anal fins. The order Perciformes, in which most of the spiny-finned species are classified, almost surely comprises families that are traceable to different beryciform precursors, and, as noted above, the beryciforms, relatives of the modern squirrelfish, may have descended in turn from different members of the Cte-nothrissiformes. Other acanthopterygians, like the sculpin and the sea robin, included in the order Scorpaeniformes, have not been linked to any of the ctenothrissiforms and may not have come from among the protacanthopterygians at all.

Besides including categories that are probably polyphyletic, the new classification scheme lists as separate entities two groups of fishes equal in rank to the three superorders discussed above.

Although they cannot be surely placed in the evolutionary sequence, they seem allied with the salmonoids or their close relatives in some way. One of them, the Atherinomorpha, is poised on the boundary between the Protacanthopterygii and the spiny-finned forms. The fishes that belong to this group, the teleost garfishes, the killifishes, the halfbeaks, and the flying fishes, show a mosaic of primitive and advanced characters: they are as advanced as the acanthopterygians in the structure of the jaws but are not at that level in the construction of the fins. Spines do appear in the median fins of many fishes in the group, but they are never developed to the extent that they are in the beryciforms and their descendants; and despite the elevation of their pectoral fins, the pelvic fins of the atherinomorphs remain in the position they occupied in lower teleosts. It has not been possible, because of the sketchiness of the fossil record, to relate these fishes to any of the protacanthopterygian forms or even to determine whether they are in fact a natural group.

The second superorder with uncertain affinities, the Ostariophysi, is more certainly monophyletic. All the fishes in this category—the characins and electric eels, the carps, and the catfishes—have an arrangement of small bones between the swim bladder and the inner ear so peculiar that it is unlikely to have evolved more than once. Derived from parts of vertebrae, the bones, called Weberian ossicles, lie beside the anterior end of the spinal column and transmit vibrations from the air sac behind to the walls of the fluid-filled otic cavities enclosed within the skull. The fishes with this unusual auditory mechanism are almost all freshwater forms that are in other ways rather primitive teleosts. Although no Cretaceous fossils related to the Ostariophysi have been found, Greenwood and his co-authors believe that the group may have originated from salmonoid stock before the end of the Mesozoic era.

Geographical distribution of the teleosts

Lack of fossil material has prevented investigators not only from ascertaining phyletic relationships but also from following the geographical distribution of many of the modern teleosts. As oceanographic research has increased and the study of freshwater fauna intensified, ichthyologists have mapped more and more accurately the areas occupied by the different living forms, but without adequate fossil evidence they are unable to speak surely about the way in which the fishes reached their present habitat. Of all the problems concerning geographical distribution, the most difficult has been that of explaining the establishment of similar types of fishes in widely different regions. In accounting for the scattered occurrence of certain forms, scientists have had to rely on theories based solely upon evidence from the ecology of the fishes in question and from paleogeography and paleoclimatology. These theories have been most satisfactory in the case of fishes that are marine or, during their life cycle, spend some time in salt water. No one doubts, for instance, that pelagic ray-fins, like the mackerel, which appear on both shores of the North Atlantic migrated across the open ocean or that closely related forms on opposite sides of the Panamanian isthmus differentiated from a common ancestor which swam east and west when the land bridge was submerged. Since there were broad connections between the Atlantic, Indian, and Pacific oceans when advanced teleosts were radiating rapidly at the beginning of the Cenozoic era, ichthyologists find it probable that modern marine ray-fins achieved broad distribution at the outset of their history and later, when the areas of land and water changed, were confined in separate basins.

The explanation conceived for the exclusively northern distribution of the salmon rests strongly upon the adaptation of that form for life in cold water, salt and fresh. Unable to withstand the warmth of tropical oceans, it has never crossed them to invade the cool parts of the Southern Hemisphere. It lives in the upper reaches of both Pacific and Atlantic, traveling as far south as the cold waters to which it is habituated extend. The salmon in each ocean follow long-established migratory routes through the salt water and then swim up rivers to spawn. Because the populations in the Atlantic and Pacific have differentiated sufficiently to be placed in different genera, investigators are quite sure that the paths of the fishes have not overlapped for millions of years. Each of these groups of salmon has related to it landlocked forms, which were trapped in fresh waters cut off from the sea, and fluviatile varieties, which ceased to go downstream all the way to the ocean. Apparently, the tolerance of the salmon for a wide range of salinity allowed it to adapt with relative ease to an entirely freshwater habitat. In most cases, freshwater salmon have been obviously derived from populations in the nearest body of salt water. The presence of salmon in European rivers which empty into the warm waters of the Mediterranean was initially a puzzle, however, since no marine varieties occur in that sea. An explanation for the existence of the isolated European tribes suggested itself when changes in the climate of the region were taken into consideration. Although the theory cannot be proved until relevant paleontological evidence appears, there is a possibility that the Mediterranean was cold enough during the glaciation of Europe to be inhabited by salmon and that the fish which swam there then were the ancestors of the modern inland forms.

Distribution of freshwater forms

It has been easier to explain the distribution of marine fishes and of anadromous forms like the salmon than to account for the spread of ray-fins that are restricted to a freshwater habitat. Ichthyologists have not been able to discover, by studying changes in climate and geography, how certain fishes came to exist in the lakes and streams of widely separated drainage systems or how related forms reached the fresh waters of distant continents. Observation of the movements of living fishes has produced no helpful clues. Unlike marine ray-fins, which can extend their range through hundreds or even thousands of miles, freshwater fishes are often confined within a small area where special conditions prevail. A form adapted for life in a mountain brook cannot survive in warmer, muddy waters downstream, and fishes accustomed to a river current rarely colonize the quiet lakes to which they have access. As larvae, freshwater fishes are no better able to go beyond the boundaries of their narrow ecological niche. They hatch from eggs that are usually prevented by weight or mode of deposition from float-ing about and grow to maturity within a short distance of their adult home. That fishes which spend their entire lives in fresh water are unable to spread easily from place to place seems borne out by the failure of species in northern South America to migrate through the Isthmus of Panama in large numbers. Since the land bridge last arose from the sea a million years ago, only a few forms have passed through Central America to reach the southern border of the United States.

In spite of their restricted movement, freshwater fishes have appeared widely dispersed in every age. Carboniferous haplo-lepids have been found in scattered continental deposits from Ohio to Czechoslovakia, and members of several Triassic genera have been recovered from both eastern and western North America. Paleontologists have attempted to explain the distribu-tion of these and other freshwater forms by suggesting a number of factors which might have facilitated their migration. They point out that passage between different drainage systems would have been possible if changes in the course of streams or the development of swamps had afforded connecting waterways. Temporary recession of fingers of salt water from lowland channels might have left rivers by which fishes could reach tributaries that were formerly separate streams. If these rivers rushed to the sea with sufficient force, their outflowing fresh waters might have carried hearty varieties of ray-fins along the coast to other rivers which led upstream to distant regions of the continent.

Since shallow or narrow bodies of sea water have appeared and disappeared in the course of time, paleontologists are not obliged to regard them as absolute barriers to the passage of freshwater fishes. The oceans, however, have separated the continents either for the duration of vertebrate history or (if continental drift has occurred) at least since the latter part of the Mesozoic era. Their continuous presence around the major land masses has made it especially difficult to understand the distribution of freshwater fishes in the Southern Hemisphere. The three continents in that part of the world share a remarkable number of genera that exist nowhere else at the present time and whose origins are incompletely known. It was once thought that a land bridge had served as a corridor for the exchange of South American and African forms, but those who supported this theory had little to back it except wishful thinking and the legend of a lost Atlantis beyond the Pillars of Hercules. The idea was abandoned when studies of the ocean floor showed no evidence for the rise or fall of any intercontinental terrain. Further research suggested that there had been instead a different kind of connection between Africa and South America. The presence of a mid-Atlantic ridge from which the ocean floor was spreading away both east and west strengthened the possibility, discussed since the seventeenth century, that the two continents had once been part of a single block called Gondwanaland and were being moved away from each other as the Atlantic widened between them. Although geologists have demonstrated the good fit of Brazil under the bulge of West Africa and have found the rocks on the opposing shores to be of corresponding age, vertebrate paleontologists, including men like Schaeffer, who have studied the freshwater fishes in both areas, have been unable to produce evidence of their own for continental drift that they regard as conclusive. While acknowledging the similarity of Cretaceous fishes from the Gabon basin in Africa and the Brazilian region, Schaeffer points out that older African freshwater forms like the bichir, *Polypterus*, and the elephant fishes are not represented in the South American fauna. In his opinion, the absence of these fishes, which could conceivably have spread throughout the fresh waters of a single

Intercontinental distribution of freshwater forms of the Southern Hemisphere

land mass, is possible evidence against the theory that pre-Cretaceous Africa and South America, with Antarctica, India, and Australia, formed one large unit. While this paleontological evidence is not strong enough, particularly in the face of new geological discoveries, to compromise the concept of a Gond-wanaland, Schaeffer believes that the distribution of freshwater fishes is explained as well by theories which do not necessitate contact between the continents of the Southern Hemisphere.

There are two such theories. According to one, the fishes that live in the lakes and rivers of modern South America and Africa are descendants of marine forms that swam freely from one coast to the other. The second proposes that the ancestors of the living fishes were distributed widely in the Northern Hemisphere and made their way south through waterways in the Asian land mass and in the Isthmus of Panama. Fossil material supporting the second hypothesis is meager: it consists chiefly of a North American Eocene osteoglossid, *Phareodus*, which advocates of the theory suggest may be a remnant of a population allied to the modern South American forms. They reason that if the osteoglossids lived in fresh water north of the equator in Mesozoic time, they must have reached the southern continents by migrating along routes overland. The same logic suggests the arrival of the lungfishes by a land route. Since no marine dipnoans are known after the Devonian period, it is likely that their dispersal took place through fresh water in Paleozoic and Mesozoic times. Extinction of the northern genera finally left populations isolated in South America, Africa, and Australia.

The speculation that freshwater fishes of the Southern Hemisphere entered continental waters from the ocean is founded on evidence that is somewhat more direct. The characins, which now, like all the Ostariophysi are confined to a freshwater habitat, have left scales in Cretaceous marine deposits of western North and South America, and other types of fishes peculiar to the African and South American continents exhibit even closer ties to forms in the sea. The siluroid catfishes common to both lands are members of the semimarine family Ariidae, and although the freshwater forms which appear in each continent are probably not descended from that stock, they may be derived from related marine ancestors. The spiny-finned cichlids of Africa and South America are as a group even more tolerant of salt water than the catfishes. Since they include among their number forms that live in estuaries where the water is brackish, it is not unreasonable to postulate their descent from marine fishes that entered rivers and gradually gave rise to landlocked species. What is needed, obviously, to cut through the thicket of speculations is more fossil evidence. At the present time, extinct fishes in Africa and South America are known only from the Triassic and late Cretaceous periods. The earlier freshwater fauna consisted of forms that were to be replaced by the modern teleosts. In the later one, the advanced fishes appear established substantially as they exist today. Clues to the origin of the current assemblage should be forthcoming from intermediate Jurassic and early Cretaceous continental deposits. Unfortunately, fish-bearing layers in such deposits, if they have endured, lie still undiscovered.

The teleosts, like the holosteans and the paleoniscoids before them, displaced the inhabitants of the regions into which they moved as they distributed themselves through continental and oceanic waters. As has been seen, the vigorous and adaptable ray-fins were largely responsible for the extinction of the placoderms and acanthodians as well as the decline of the dipnoans, crossopterygians, and less progressive members of their own kind. Only one group of fishes survived the onslaught

The sharks

of the actinopterygians: the sharks that arose beside the paleoniscoids during the Devonian period claimed a place in the sea which they never relinquished. There they evolved as the ray-fins did, passing from one grade of organization to another and diversifying at each level. Although from time to time a few forms dwelled in lakes and rivers, the sharks seem not to have seriously challenged the bony fishes in fresh water. They showed no tendency to develop the high degree of maneuverability and buoyancy that enduring success in such an environment requires. Their broad bodies and low-set paired fins fitted them for powerful swimming in open water or for cruising along the bottom but made them less capable of the abrupt stops and quick turns necessary in the confines of a lake or stream. The boundless spaces of the sea were more suitable for the sharks and other cartilaginous swimmers of the class Chondrichthyes that, lacking a gas- or air-filled diverticulum of the gut, were heavier than water and thus obliged to move continuously to keep from sinking. Their inability to float made it impossible for sharks to conceal themselves as bony fish often do by remaining motionless in midwater among the weeds or behind rocks. Besides reducing their buoyancy, the absence of an air bladder deprived the sharks of an auxiliary respiratory organ. Without the structure which serves many bony fishes in an emergency as a crude lung, sharks are potentially less well adapted to survive in fresh water, where, on a warm or windless day, the amount of dissolved oxygen may fall below the minimum required for adequate functioning of the gills.

Ancestry of the sharks

Surprisingly little is known about the origin of the sharks or the interrelationships among the different kinds of cartilaginous fishes that had emerged by the Carboniferous period. The fossil record is poor because dead sharks were almost never preserved in recognizable form. Their skeletons rotted with the rest of the soft tissue, and spines, teeth, and scales tumbled separately through the shifting sediments. Studying these isolated structures has not enabled paleontologists to associate the earliest sharks with any particular group of archaic fishes. Since the acanthodians seem allied to the bony fishes, investigators have speculated that the ancestral stock of the cartilaginous forms should be sought among the placoderms. Narrowing the field, they rejected the antiarchs and the arctolepids as too specialized and the short-spined Middle Devonian arthrodires as too late to have served as the source. Disregarding aberrant forms like *Holonema* and *Homostius,* they have cautiously suggested that the ancestors of the sharks might have been petalichthyids or forms related closely to them. These fishes, inhabitants of the sea in early Devonian time, seem to have been somewhat dorsoventrally flattened but otherwise not unlike sharks in their general shape. The tubercles that ornamented the dermal plates of the head could possibly have been precursors of the placoid scales characteristic of chondrichthyans. Both structures are built of dentin, a hard tissue that forms independently of bone. Since petalichthyids shared the arthrodiran tendency toward diminished ossification, they might have lost the deep parts of the dermal plates, retained the superficial denticles, and so given rise to forms that were, in fact, the first sharks.

Although the descent of the cartilaginous fishes from petalichthyids is far from proved, paleontologists are sure that sharks were derived from vertebrates with bone. The possibility that the members of the class Chondrichthyes were related directly to vertebrates with a primitively unossified skeleton was virtually abandoned when the antiquity of bone was recognized. Finding residual bone in the scales and tooth bases of the most ancient sharks made the secondary nature of the cartilage in these fishes even more certain. It became increasingly obvious as research progressed that the sharks were not in any sense primitive forerunners of the more advanced bony fishes, as was

Sharks are not forerunners of bony fishes

once thought. They enter the paleontological record later—not a spine or tooth remains from the Lower Devonian, when paleoniscoids were already in existence—and appear with structures which can by no stretch of the imagination be considered transitional to those of any of the bony fishes. In the Carboniferous period, the sharks diversified into forms that had no parallel among the osteichthyans. Even though most of the fossils are fragmentary, paleontologists have detected representatives of no fewer than five separate lines. Being unsure of the phylogenetic relationships among these early forms, they have listed them tentatively as cladodonts, xenacanths, bradyodonts, edestids, and hybodonts. To distinguish them from the cartilaginous holocephalians which apparently had a separate history, systematists have housed the Paleozoic sharks and their descendants in the subclass Elasmobranchii.

Cladodonts, the earliest sharks

The marine cladodonts appear to be the earliest and most generalized fishes of the elasmobranch assemblage. They are known in part from isolated teeth, calcified jaw cartilages, braincases, spines, and fin radials that have been found in rocks of Middle Devonian to Permian age. In addition, and most fortunately, complete specimens have been discovered in the Upper Devonian Cleveland shales. The latter are remains of sharks which lived near the edge of the sea that covered central North America for much of the Paleozoic era. In the region that is now Ohio, numerous individuals sank after death to the muddy bottom and were encapsulated in rapidly hardening sediments. They were preserved in their entirety before scavengers or bacteria had a chance to destroy them. The commonest form in the Cleveland shales is *Cladoselache* (Fig. 5.22). This fish, like its fellow cladodonts, had a long body supported

Figure 5.22. Restoration of *Cladoselache*, a cladodont shark from the Upper Devonian Cleveland shale. Inset: cladodont teeth. (From Schaeffer.)

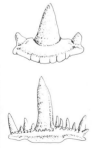
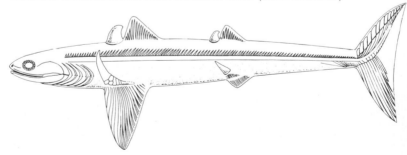

by an unconstricted notochord surmounted by a row of carti-laginous neural arches. Arches and notochord extended into the upper lobe of an outwardly symmetrical caudal fin. At least the anterior of the two dorsal fins was preceded by a broad spine. The hard structure acted undoubtedly as a cutwater and a stabilizer and may have assumed the defensive function that it serves in modern squaloid sharks. The paired fins of *Cladoselache* were extremely broad-based. The large pectorals and smaller pelvics were stiffened to their margins by radial elements that articulated proximally with a long row of basal cartilages embed-ded in the body wall. The anterior basals of the pectoral fin abutted a girdle that had already developed the high-rising scapular process characteristic of later elasmobranchs. The braincase, too, had evolved a general form that was to last, with minor changes, for more than 350 million years. It showed no intracranial subdivision but sheathed the brain as a single entity. Open behind the rostrum, where the cartilaginous roof was incomplete, it enclosed the olfactory organs in bulbous nasal capsules. Midlaterally, antorbital and postorbital processes pro-jected to embrace the eye, and a narrow suborbital shelf extended outward from below. The posterior part of the braincase, which housed the inner ear, was shorter in *Cladosel-ache* than in some of the other cladodont sharks but still provided a resting place for the hyomandibular element that intervened between skull and upper jaw. Since the rostrum was short and the jaw cartilages long, the mouth opened at the end of the snout rather than beneath it. As in other cladodonts, the palatoquadrate in *Cladoselache* was bound tightly to the neuro-cranium. Besides being braced against the otic region through the hyomandibula, it articulated with the postorbital process and, farther forward, was attached by ligaments to the subor-bital shelf. The teeth mounted on the palatoquadrate and mandibular cartilages were those of a carnivorous animal. Each had a tall cusp flanked fore and aft by lower ones. Teeth of this sort, specialized for impaling, cutting, and holding live prey, characterized all the marine cladodonts, including the earliest, Middle Devonian forms of which they are the only remains.

Since cladodont sharks with teeth like those of *Cladoselache* shared a number of common characteristics, it was assumed for a long time that tooth form was a reliable guide to relationships among the cartilaginous fishes. Paleontologists studied the varie-ty of elasmobranch teeth discovered in Carboniferous deposits in Europe and in the comparable Mississippian and Pennsylvanian rocks of North America and, on the basis of differences they noted, raised separate genera and subdivisions among the sharks of this period. To a limited extent, the use of dental anatomy as a basis for classification proved valid. The finding of skeletal remains in association with teeth not of the cladodont type showed that the possessors of these teeth were distinct from *Cladoselache* and its relatives in several respects. Multicusped teeth different from the cladodont kind in that the central prong was small and the flanking ones larger belonged to freshwater sharks that surely represented an independent evolutionary line. The xenacanths, or pleuracanths as these sharks used to be called, gave evidence of cladodont ancestry but had several peculiar traits. *Xenacanthus* (Fig. 5.23), whose skeleton is well known, grew a long spine from the rear of the skull that extended backward over the branchial region. Behind it, a low dorsal fin edged the trunk to the base of the tapering diphycercal tail. The pectoral fins of *Xenacanthus* were unique in having a central axis with anterior and posterior radials. This archipterygial design may have evolved through the separation from the body wall of a cladodont-like row of basal cartilages and the development thereon of accessory radials. The transformation, if it did take place, seems not to have proceeded as far in the pelvic fins. The basal axis projected from the body in those appendages but bore no radials on its posterior edge. The paddle-like structure of the paired fins, the elongated dorsal fin, and the diminished, diphycercal tail suggest that *Xenacanthus* was not a strong swimmer. It could have glided through the water, undulating its long body and dorsal fin almost imperceptibly, steering or stopping with a sudden thrust of the broad pectorals. The adaptive value of the strange, double anal fin is not clear. In this character, as in the development of an axial skeleton with calcified centra, hemal arches, and short ribs, *Xenacanthus* paralleled sharks of a more advanced type than the cladodonts.

Figure 5.23. Restoration of *Xenacanthus* (= *Pleuracanthus*), a freshwater shark of the latter part of the Paleozoic era. Inset: a xenacanthid tooth. (From Jaekel.)

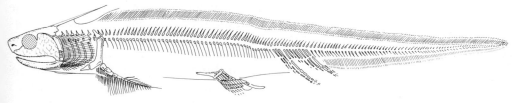

Besides cladodont and xenacanthid teeth, Carboniferous rocks gave up several other kinds. Some of them were so broad and plate-like or had been so deeply rooted in the jaws that they were clearly permanent structures of the mouth rather than teeth that were replaced periodically. Although their form varied, the crowns of these teeth consisted of tubular dentin, the hard tissue described earlier (page 64) in which parallel denteons separated by hypermineralized material rose like columns at right angles to the biting surface. Believing that the similarity of the dental tissue signified phylogenetic relationship, paleontologists grouped all the genera based upon these massive teeth in one taxon, the Bradyodonti. As skeletal elements associated with bradyodont teeth were found, however, it became evident that the fishes to which they belonged had been quite unlike each other (Fig. 5.24). *Janassa*, a form with petal-shaped teeth ranked close together, was flattened as the placoderm rhenanids had been millions of years before and as the skates and rays were to be later. *Chondrenchelys* and *Helodus* were specialized in a different way. *Chondrenchelys*, with its palatoquadrate fused to the cranium and its archipterygial pectoral fins, was an aberrant shark. Perhaps degenerate, it lacked spines and had developed the continuous dorsal fin and diphycercal tail that often mark fishes which have retreated to quiet waters. *Helodus*, though it also had a fused palatoquadrate, showed none of the other peculiarities of *Chondrenchelys*. A stout spine preceded its first dorsal fin, and in the pectoral fins radial cartilages projected from only one side of the basal elements, as is usual in sharks. The diverse nature of the few bradyodonts known from skeletal remains led paleontologists to doubt that the sharks with slowly growing, permanent teeth were as closely related as was once believed. Thus, when L. Radinsky demonstrated that tubular dentin evolved independently in many kinds of fishes that ate hard-shelled organisms, investigators welcomed his argument that tooth structure was not a good criterion for recognizing monophyletic groups among the fishes. The category Bradyodonti is used now for convenience rather than out of conviction that the fishes included in it were descended from a common ancestor. If sufficient fossil material comes to light, it may one day be demonstrated that the different families of bradyodonts arose separately from several groups at the base of the shark line or even below it.

Bradyodonts, sharks with permanent teeth

While the dismantling of the Bradyodonti has multiplied the problem of the origin of the Carboniferous sharks, it has given a

Edestids, sharks with tooth whorls

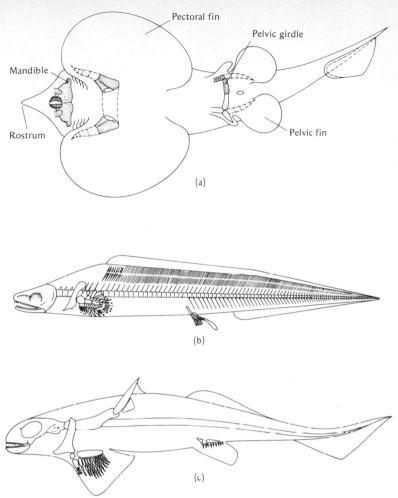

Figure 5.24. Restorations of three Carboniferous fishes with bradyodont teeth. **(a)** *Janassa* (ventral view). **(b)** *Chondrenchelys* (lateral view). **(c)** *Helodus* (lateral view). (From Moy-Thomas and Miles.)

freer hand to workers classifying certain of them known as the edestids (Fig. 5.25). These fishes were distinguished by a series of teeth which spiraled outward from the midline of the lower jaw to form a large whorl that the animal carried throughout its lifetime. Since the teeth in the whorl as well as those at the sides of the mouth contained tubular dentin, the edestids were first identified as a family of bradyodonts. The discovery that the palatoquadrate elements in two forms, *Fadenia* and *Sarcoprion*,

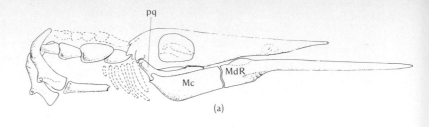

(a)

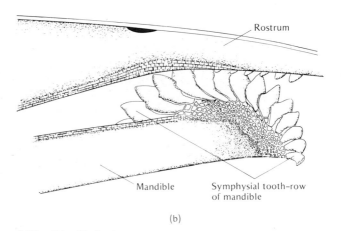

(b)

Figure 5.25. Edestid sharks. **(a)** Restoration of skull and part of postcranial skeleton of *Ornithoprion*. Symphysial whorl of teeth (not shown) was mounted upon the mandibular rostrum (*MdR*) immediately anterior to its articulation with Meckel's cartilage (*Mc*). Because of the shortness of the cartilages of the upper jaw (*pq*), upper teeth with which the symphysial whorl occluded must have been mounted upon the underside of the braincase in this form. (From Zangerl.) **(b)** Restoration of the anterior part of the skull of *Sarcoprion*, showing teeth in place. (From Moy-Thomas and Miles.)

were fused to the braincase, as in *Helodus* and *Chondrenchelys*, seemed to confirm the relationship. No sooner had the edestids been settled among the bradyodonts, however, than reasons for questioning the arrangement became apparent. The palato-quadrate cartilages of the edestids *Helicoprion* and *Ornithoprion* proved to be articulated with the cranium rather than fused to it, and, in several other characters, members of the group known from the most extensive skeletal material seemed more similar to progressive Paleozoic sharks than to those labeled as brady-odonts. Paleontologists hesitated to separate the edestids from the bradyodont assemblage, finally, only because the microscopic structure of their teeth allied them to the group. When the presence of tubular dentin was shown to be adaptive and not necessarily indicative of common ancestry, investigators were able to set the edestids apart from other sharks which had been classified as bradyodonts and seek a new association for them.

Since the postcranial skeleton of the edestids is little known and the skull and jaw cartilages are often found crushed or incomplete, paleontologists have been cautious in revising the position of these sharks in the evolutionary scheme. Bendix-Almgreen divided the group, leaving edestids with palato-quadrate cartilages fused to the braincase in the same taxon as *Chondrenchelys* and *Helodus*; Zangerl postponed his decision until more work on newly discovered fossils is done; and Romer, impressed by the selachian characters of edestids, placed them with the hybodonts, an advanced group (discussed below) presumably derived from cladodont stock. Whatever their source, the edestids were certainly specialized deviants from the main line of shark evolution. Zangerl's reconstruction of the head of *Ornithoprion* shows the fish to have had a long, pointed rostrum that extended far beyond the reduced palatoquadrate cartilages of the upper jaw. The mandibles converged anteriorly to meet an unusual medial element that bore the tooth whorl and thrust forward like a spear in front of the rostrum. The tooth whorl opposed teeth that were evidently set into the neurocranium under the snout, an arrangement extraordinary even among the edestids. Since the snout and spear-like mandible were both protected by bone that formed in the fused bases of the scales, Zangerl speculated that *Ornithoprion* may have probed the bottom in coastal waters for buried mollusks, which it could crack open with its strong anterior teeth. The habits of *Ornithoprion*, like those of other, less aberrant edestids, are not surely known, however, and the function of the tooth whorl has been a matter of dispute for years. The specialized dentition of the edestids, as advantageous as it must have been initially, may have eventually proved a handicap, for the fishes that evolved it did not survive into Mesozoic time.

Hybodonts, ancestors of the modern sharks

The only sharks of the varied Paleozoic assemblage that did survive the stresses of the Permian period were the hybodonts (Fig. 5.26). The key to the continuing success of these fishes, which arose early from cladodont stock, seems to have been the

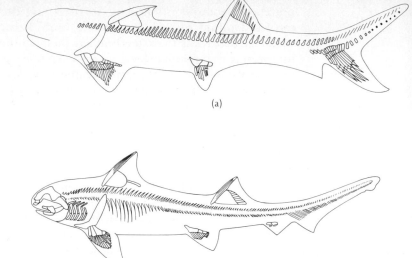

(a)

(b)

Figure 5.26. Restorations of two hybodont sharks. **(a)** *Ctenacanthus.* **(b)** *Hybodus.* (From Schaeffer: *a* after Moy-Thomas; *b* after Smith Woodward.)

improved design of their fins and teeth. The basal cartilages in the pectoral fins had been reduced to three, the radials shortened and jointed, and the distal part of the fin supported by rays of toughened connective tissue. Appendages of this kind could have been twisted more readily than the broad-based fins of the cladodonts and hence would have been more effective as steering devices. The pelvic fins in male hybodonts were equipped with claspers, intromittent organs derived from modified radial cartilages. Their presence indicates the establishment of internal fertilization among the hybodonts. Whether direct transfer of sperm took place in the more primitive cladodonts is debatable (for conflicting opinions, see Romer 1966 and Schaeffer 1967), but its occurrence in the hybodonts can be counted as another of the advantages traceable to the specialization of their fins.

The teeth in many hybodont sharks were high-cusped at the front of the jaw and low-crowned in the cheek region. The differentiated dentition made it possible for these animals to feed upon smaller fish and also to grind up the shells and exoskeletons of invertebrates. The wide range of organisms they were able to utilize as food undoubtedly gave them an advantage over other Paleozoic sharks when, at the end of the era, many species habitually eaten by the cartilaginous fishes became extinct. Although hybodonts were perhaps superior to their contemporaries in other ways not discernible from the fossil record, their omnivorous habits may have been crucial to their survival. Alone, the evolution of narrow-based fins seems not to have enabled a group of cartilaginous fishes to persist. Xenacanths and those of the bradyodonts that had such fins in combination with a uniform dentition failed before the Permian period had run its course. Among the Carboniferous sharks that showed a tendency to develop narrow-based fins, only the hybodonts were prepared by the variety of their teeth to withstand the disruption of long-established feeding patterns.

The hybodonts emerged from the cladodont assemblage at the base of the Carboniferous period, maintained themselves through the remainder of the Paleozoic era, and then, in Mesozoic time, diversified to produce some kinds of fishes whose descendants have survived to the present day. That the ancestors of the hybodonts were sharks similar to *Cladoselache* is suggested but not proved by inspection of fossils belonging to the genus *Ctenacanthus*. The fishes of this group, related through the possession of distinctive dorsal fin spines, include a species contemporary with *Cladoselache* that had cladodont teeth and broad-based fins as well as later forms that showed typical hybodont appendages. Although the skull of most of the ctenacanthids is unknown, the hybodont braincases and jaw cartilages which have been preserved are similar in certain ways to those of cladodonts. In both groups of sharks, the palatoquadrate cartilage of the upper jaw articulated with the neurocranium in front of and behind the eye and extended forward to the tip of the snout. As hybodonts diversified in the Mesozoic era, the most progressive forms tended to lose these cladodont characters. The postorbital articulation between palatoquadrate and neurocranium disappeared, the preorbital contact loosened, and the hyomandibula assumed the most important role in the suspension of the jaws. Although the mouth remained terminal in two surviving genera of hybodont ancestry, in all other modern selachians descended from the Mesozoic sharks the ends of the jaws fall far short of the tip of the rostrum and the mouth opens, as a result, on the undersurface of the head.

The hybodonts that flourished in the Jurassic period were replaced before the end of the Cretaceous by their most successful offspring, the first of the modern elasmobranchs. Only a few hybodonts that lagged in developing advanced characteristics escaped extinction. The descendants of these

Aberrant descendants of the hybodonts

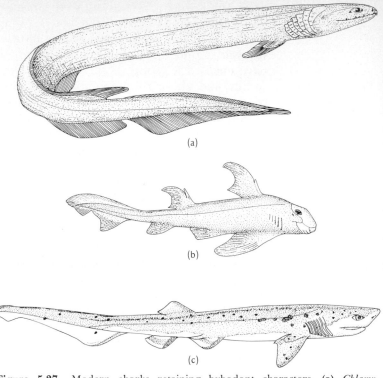

Figure 5.27. Modern sharks retaining hybodont characters. **(a)** *Chlamy-doselachus*, the frilled shark. **(b)** *Heterodontus* (=*Cestracion*), the Port Jackson shark. **(c)** *Heptanchus* (=*Heptranchias*), the seven-gilled shark. (From Dean.)

forms, which still exist, exhibit a mosaic of ancient, modern, and specialized structures (Fig. 5.27). The shark *Heterodontus*, as its name suggests, retains the differentiated dentition which appeared at the hybodont level but has evolved centra, components of the vertebral column lacking in the older fishes. The mouth of *Heterodontus* opens at the end of the snout, as in archaic sharks, and the upper jaw is firmly held, but the braincase of this relict form has developed traits that distinguish it from the hybodont skull. *Chlamydoselachus*, another odd survivor, inherited a different complex of characters. This shark kept the terminal mouth and relatively fixed palatoquadrate but also conserved the ancient uniformity of tooth structure and the supportive notochord. The elongated body, frill-like anal fin, and diphycercal tail are indications that *Chlamydoselachus* has ceased to compete with the strong swimmers and, like the lungfishes, has found for itself a restricted but somewhat sheltered habitat. A third line of selachians whose members progressed a bit beyond the hybodont level before veering off on their own course comprises the hexanchids, sharks which have a shortened palatoquadrate and more than the normal five pairs of gill arches. The presence of six pairs in *Hexanchus* and seven in *Heptranchias* is presumed to be the result of a secondary increase rather than the retention of a primitive number of arches. Since *Chlamydoselachus* also has six pairs and shows other structural similarities to the hexanchids, paleontologists suspect some sort of relationship between the two aberrant types of sharks. The specialization that has occurred among the former and the changes that have taken place in the jaw structure of the latter seem to prove, however, that the two groups have been evolving separately for a long time.

Structural innovations in modern sharks

The structural innovations which characterized the more progressive sharks and rendered them superior to the hybodonts were similar in principle to those which enabled the Cretaceous teleosts to displace their ray-finned forebears. Curiously, in both groups of fishes, increased mobility of the mouthparts and the development of a firmer axial support for the body produced improvements in feeding and locomotion that allowed the rise and radiation of modern forms. In all the sharks that reached the most advanced elasmobranch level, the articulation behind the orbit between the upper jaw and the braincase disappeared, and the more anterior point of contact was reduced to a mere bracing of the orbital process of the palatoquadrate against the chondrocranium wall. The jaws, now anchored against the skull primarily through the movable hyomandibula, could be protruded by special muscles as the shark prepared to close upon its prey. The protrusibility of the mouth made possible the success of the new, shorter jaws. Although reduction in the length of the palatoquadrate and mandibular cartilages produced a more powerful closure, it also brought the mouth opening back beneath the snout. In order to function effectively in this position, whether the shark was feeding upon fish in open water or upon shelled invertebrates from the bottom, the rim of the mouth had necessarily to project beyond the undersurface of the body. The simultaneous appearance of protrusibility and more powerful jaws, rather than the evolution of one or the other of these traits, was responsible, therefore, for the significant improvement in the feeding mechanism.

Schaeffer suggests that the changes which occurred in the

structure of the vertebral column may also have contributed to the effectiveness of the modified jaw apparatus. The hybodonts had relied for support upon the notochord alone, but the newer sharks developed cartilaginous centra around that structure. The centra, which became partially calcified, constricted the notochord and fused with adjacent neural and hemal arches. The vertebral column, made firmer in this way, was able to withstand the shock waves projected through the body when the jaws snapped closed. The column of the sharks was even better braced than that of the teleosts, in which, at the same time, centra appeared independently: in the cartilaginous fishes, the space between successive arches was closed—in the hemal series by broadening of the existing arches and in the neural row by the growth of a new, intercalary arch in each segment. The flexible but firm axial skeleton produced by these modifications provided a more secure anchorage for trunk and tail muscles than the unconstricted notochord and unbraced neural and hemal arches of the earlier fishes.

Variety of modern elasmobranchs

The modern elasmobranchs that exhibit the advanced characters just described can be divided into three groups: the batoids, the squaloids, and the galeoids (Fig. 5.28). Of these, the batoids are the most specialized. The rays and skates, which constitute this assemblage, are extremely flattened, bottom-dwelling forms with vastly expanded pectoral fins and a much reduced tail. The head, depressed and linked with the huge fins, has on its underside a small mouth, transversely set and armed with flat-crowned, crushing teeth. Behind the mouth, five pairs of gill slits open to allow the escape of water that enters the pharynx through a spiracle behind each eye on the dorsal side of the head. In squaloid and galeoid elasmobranchs, the form of the body is less extraordinary. Swimmers in the upper water, they are fusiform with gill slits that perforate the side of the head in front of pectoral fins of normal size. Despite the similarity of their appearance, the two kinds of sharks certainly belong to different lines. They differ consistently from one another in anatomical details of the braincase, jaws, and fins and in the pattern of calcification within the vertebral centra. It is unnecessary to wield a scalpel to distinguish between them: the squaloids produce a strong spine anterior to each dorsal fin and lack an

Figure 5.28. Modern elasmobranch fishes. **(a)** *Carcharhinus*, a galeoid shark. **(b)** *Centrophorus*, a squaloid shark. **(c)** *Rhina* (= *Squatina*), the angel shark. **(d)** *Pristis*, the sawfish, a batoid. **(e)** *Rhinoptera*, a ray, also a batoid. (From Garman.)

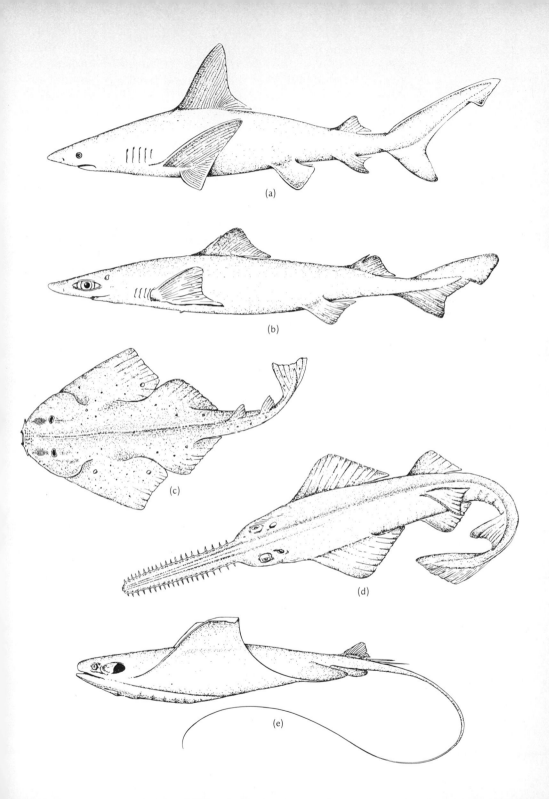

(a)

(b)

(c)

(d)

(e)

anal fin; galeoids have an anal fin but do not develop the dorsal spine. Squaloids, galeoids, and batoids all appeared during the Jurassic period. Whether they diverged from a common hybodont precursor or arose separately from different hybodont stocks is still in dispute. The advanced characteristics they share are not proof of their origin from a single source, since like structures may develop through parallel evolution in lines emanating from separate but related forms.

Holocephalians   Besides the elasmobranchs, another group of modern cartilaginous fishes can be traced back to Jurassic forebears. The holocephalians, scaleless, large-eyed fishes with a whip-like tail, of which there are at present only a few genera, descended from forms akin to those which lived in the sea covering Europe 150 million years ago (Fig. 5.29). Long considered relatives of the sharks, holocephalians resemble those fishes in the architecture of the skeleton and in the structure of their soft organs. Until recently it was assumed that their divergent characteristics proved that they originated, not from hybodonts, but from older, more aberrant shark stock. Holocephalians, living and extinct, have large, slow-growing tooth plates rather than separate, replaceable teeth and a palatoquadrate that is fused to the neurocranium rather than suspended from it, as in elasmobranchs. These two characteristics seemed to ally them to the Paleozoic bradyodonts, and for a long time paleontologists searched within that group for forms that might have been close to the ancestral line. While the relationship to bradyodonts was being explored, new discoveries and new concepts in paleoichthyology allowed the proposal of a different source for the holocephalians. Watson and, later, Ørvig described arthrodiran ptyctodonts as having already evolved tooth plates, a compact skull, a stout dorsal spine, and claspers not unlike those of holocephalians. Since the ptyctodonts had lost most of the heavy dermal armor characteristic of more typical arthrodires, paleontologists deemed it not impossible that those fishes could have given rise to forms whose skeleton was completely cartilaginous. Recognition of the degradation of the skeleton in many vertebrate lines and the convergent evolution of plate-like teeth in unrelated fishes had weakened the case for the derivation of holocephalians from bradyodont sharks. As a result, the idea that holocephalians might be descended from ptyctodonts was worthy of consideration.

While the possibility of a relationship between holocephalians and ptyctodonts was being debated, new fossil finds from

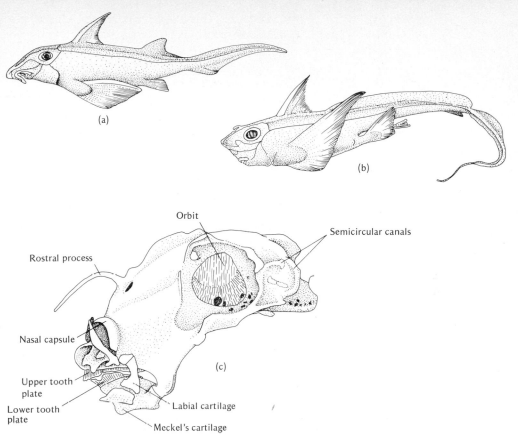

Figure 5.29. Holocephalian fishes. **(a)** *Callorhynchus.* **(b)** *Chimaera monstrosa.* **(c)** Skull of *Hydrolagus* (=*Chimaera*) *colliei*, showing tooth plates and fusion of upper jaw to braincase. (*a,b* from Dean; *c* modified from Jollie.)

Carboniferous shales in central North America revealed something more about the early history of the holocephalians and put in doubt their affinity to any known group of placoderms. Among the many hitherto unknown and as yet undescribed cartilaginous fishes that lived in the shallow sea covering the midportion of the North American continent, there were forms which Zangerl, who has reviewed them, believes are undoubtedly early holocephalians. These fishes possessed the palatoquadrate fused to the neurocranium, the elaborate hook-bearing claspers, and the general body proportions characteristic of the Holocephali. Instead of tooth plates, however, these fishes, which Zangerl has called iniopterygians, bore shark-like rows of

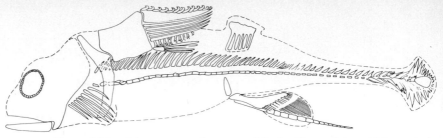

Figure 5.30. *Iniopteryx*, one of a group of Carboniferous fishes which may have been closely related to the holocephalians. (From Zangerl and Case.)

teeth, and their skin was almost entirely bereft of scales. If the teeth are primitive and if the few dermal denticles on the body are representative of the skin-covering of holocephalian ancestors, the holocephalians may have sprung from the same stock of early jawed fishes as the sharks, and not from tooth-plated ptyctodonts with residual armor. The iniopterygians give no surer clue to the identity of chondrichthyan forebears than the Carboniferous sharks do, unfortunately, because they had evolved a complex of specialized characters, the strangest of which were great spined pectoral fins set high against the shoulder that may have beat the water like wings. A fossil fish recently found in Montana, older and less bizarre in appearance than the iniopterygians described by Zangerl, seems to be intermediate in structure between those fishes and the holocephalian group that includes the modern forms. It is further evidence of the antiquity of the holocephalians but no appreciable help in pinpointing their source (Fig. 5.30).

Remarkable success of the modern fishes

Whatever the origin of the holocephalians, it should be obvious that they, like modern elasmobranchs and contemporary bony fishes, cannot be considered more primitive animals than vertebrates which have survived on land. Living fishes of all kinds have behind them an evolutionary history just as long and complicated as that of successful tetrapods. Since their appearance in the Devonian period, fishes in every lineage have undergone progressive changes in structure, older forms giving way to new ones that are increasingly sophisticated in their adaptation for life underwater. If a group of animals is judged, as it should be, on the basis of its continuing refinement for a particular habitat rather than upon the number of anatomical characteristics its members share with mammals, the modern fishes merit a very high place on the evolutionary scale. The only basis for referring to them as lower vertebrates is their retention of the original vertebrate habitat and the physiological mechanisms that made life in the water possible for the earliest gnathostomes. There is a distinction between lower animals and primitive ones, however. Unlike primitive members of a group which are superseded by more elaborately developed forms, the fishes have maintained themselves and diversified even in the face of competition from vertebrates of other classes that have reinvaded the sea. Rather than being remnants of early vertebrate stock that enjoyed their triumph long ago, the fishes have been the most important backboned animals over the submerged three-quarters of the globe in every era. The Age of Fishes, which began in Devonian times, endured throughout the ascendancy on land of the amphibians and reptiles and, during the present Age of Mammals, continues still.

6

The Amphibians: Gaining the Land

The origin of the amphibians was a momentous event, since attainment of the land opened vast new territories to the vertebrate animals and permitted them to evolve their most advanced forms. Despite the importance that terrestrial verte-brates were to have, however, their initial evolution was not in any way unusual or spectacular. The amphibians were not the last survivors of a lesser class but one of a number of new forms produced as the early bony fishes diversified rapidly in the Devonian period. At their first appearance, they gave the impression less of a revolutionary new group than of fishes peculiarly adapted for special habits of life. Outwardly, except for their legs, they resembled the rhipidistian fishes from which they sprang. Very likely, they continued to swim in the shallows, as their sharp-toothed forebears had, preying upon the abun-dant placoderms and early paleoniscoids to be found there. Paleontologists are quite certain of the relationship between the rhipidistians and the amphibians even though they have not discovered the animals intermediate between the finned and limbed forms. The remains of the oldest tetrapods in their collections leave no doubt about the derivation of the axial skeleton from fishes of the rhipidistian group. Since the fossil material provides no evidence of other aspects of the transformation from fish to tetrapod, paleontologists have had to speculate how legs and aerial breathing evolved and why a group of fishes produced forms that habituated themselves little by little to life on land.

Since the evolution of any group of living things is a progressive response to the demands of the environment, investigators have discussed at length the conditions that prevailed at the time of the rhipidistian-amphibian transition. By pooling interpretations based upon geological, botanical, and zoological data, they have been able to visualize the earth as it was then and to guess what might have driven the incipient amphibians on their new *The Devonian en-* course. In the Devonian period, they reason, primitive green *vironment* plants were advancing tentatively beyond the dampest ground at the edges of freshwater lakes and streams. Harbingers of the great tree-fern and horsetail forests of Carboniferous time, these early land plants sheltered the many-legged arthropods which were also acclimating themselves to life in dry air (Fig. 6.1). Beyond the water's edge, the land stretched away unprotected by vegetation. Rain fell and ran off in torrents, crumbling igneous and metamorphic rock into pebbles, sand, and clay. In some areas, the wind picked up the finer particles and blew them into dunes, while in other places loads of sediment were washed into the water and layered over the bottom. The red color of much of this transported material has provided scientists with an important clue to the climatic conditions that existed in the Devonian years. Geologists know that the ruddiness of soils is due to hematite, a compound containing iron in ferric, or highly oxidized, form. Iron in this state is produced when sediments containing ferrous silicates are laid down in warm and moist areas where the surface is subject to drying from time to time. Ferric iron is stable once it appears, and deposits containing it are compressed into layers of rock called red-beds.

As long ago as 1916, Barrell pointed out that the existence of extensive Devonian red-beds signified widespread semiarid conditions during the period when the amphibians arose. He believed that lungs evolved as an adaptation to the droughts that intervened between weeks or months of steamy rainfall. Lull concurred and emphasized the role of natural selection in preserving a population of vertebrates that was developing the

Behavior of the earliest amphibians

ability to sustain itself out of water. Romer described graphically how the earliest tetrapods might have saved themselves from death by leaving a drying pond and hitching themselves overland to a deeper, better-filled pool. Critics of Romer's theory challenged it on several bases. Inger questioned the idea that the Devonian climate was surely semiarid, stating that red soils are being formed today in parts of the world where tropical rains are uninterrupted. Even if dry seasons did exist, he maintained, it was hardly likely that the first amphibians behaved as Romer thought they did. Since they were dependent upon small fish for food, they would have lived in permanent waters rather than in ponds which became periodically uninhabitable. A drought severe enough to destroy the pond or lake in which they had established themselves would certainly evaporate similar bodies of water within a radius of many miles. An attempt to migrate any distance under such conditions would result in the animals' dying of desiccation on the bare ground. Inger imagined that lungs were selected for, not because their possessors ventured

Figure 6.1. Restoration of a late Devonian landscape. Primitive vascular plants were present in the form of leafless psilophytales, tall sphenopsids related to the modern horsetails, and early tree ferns. The root systems of these plants were not extensively developed, and the soil around them was relatively unprotected against erosion. (From a mural by C. R. Knight, courtesy of the Field Museum of Natural History, Chicago.)

onto dry terrain, but because these organs enabled fishes in warm, oxygen-poor water to survive by breathing air at the surface. He thought that the animals' first excursions beyond the water's edge were prompted by increasing population pressures in old communities; during humid nights, individuals driven by lack of adequate food or space might have squirmed over the wet earth in search of less crowded ponds much as the fighting fish *Betta* and the catfish *Clarias* do today in tropical Borneo.

Goin and Goin seconded Inger's suggestion, reporting that this kind of behavior can also be observed in certain modern frogs. These amphibians do not attempt to migrate in the dry season. Sharp reduction in the water level of a pond, in fact, causes the inhabitants to gather in the center, where the remaining water is deepest. Landward movement occurs when rain wets the ground but has not yet restored the pond to its former size. Young frogs leave, then, and colonize less densely populated pools nearby. The Goins agree further with Inger that protoamphibians, impelled to quit their home waters under similar conditions, probably began to eat the newly terrestrial arthropods as a supplement to their regular diet of fish. Once accustomed to a terrestrial food supply, the air-breathing former fishes were launched on their land career.

Romer insists that amphibians remained aquatic animals much longer than Inger and the Goins suppose. Although he holds to his idea that the earliest amphibians traveled over land to escape their drying pools, he believes that well into the Carboniferous period their normal life was carried on in the water. Not until that time, when insects and land plants radiated with amazing rapidity, would there have been sufficient food material on land to sustain a population of vertebrates. The climate in Devonian years was a far harsher one, he still thinks, than Inger envisions. That seasonal droughts were the rule rather than uninterrupted humidity seems proved by the mud cracks and evaporites that appear in the rocks of the age. Transition to terrestrial feeding and a more or less terrestrial habitat would have been unlikely during a period when desert dryness occurred in alternation with earth-soaking rains. Though the first tetrapods were able, in his opinion, to survive an occasional forced march to new waters, the rigorous conditions which necessitated their migration would have precluded their dallying to catch any scorpions or other arthropods that might have crossed their path. During their early history the amphibians probably stayed in the water, protected from the extremes of climatic change, and developed their taste for invertebrates by eating the aquatic insect larvae which hung upside down from the surface of the pond. Romer adduces more direct evidence for his contention that amphibians emerged from the water relatively late by turning to the fossil record; the oldest members of the class retained the sinuous body of the swimmer and had, with few exceptions, small legs that would hardly have allowed easy locomotion on land. Not until the end of the Carboniferous period did a number of amphibians appear whose strong limbs and stout body were adapted for terrestrial rather than aquatic life.

Why the amphibians left the water and when constitute only a part of the puzzle to be solved by those who are studying the origin of the first tetrapods. Paleontologists tracing amphibian history have also to analyze the structural modifications necessary to transform a fish into an animal viable on land before they can explain the steps by which rhipidistian forms evolved into the earliest terrestrial species. It is possible to draw such guidelines, because observation of living animals makes patent the different demands on the vertebrate body of watery and aerial environments. Underwater, animals are supported by the medium, kept moist, and supplied with dissolved oxygen. Once they come out into the dry air, they are faced with the necessity of holding themselves up and moving in a much less buoyant substance, of extracting oxygen from a gaseous mixture, and of conserving the water that forms the basis of their protoplasm. The physical forces operative in the terrestrial environment make certain mechanical arrangements a necessity and others an impossibility if the animal is to succeed in these tasks. Determining the time and the order in which the required structural changes took place is a more speculative matter. Although many of the modifications were correlated and so must have occurred simultaneously, paleontologists are not certain of how much alteration took place as protoamphibians were acclimating themselves to land life and how much occurred preadaptively at an earlier time when the animals were still wholly aquatic.

Although several Devonian fishes possessed lungs, it is not at all clear that elaboration of these structures was an early event in amphibian evolution. Ever since its initial appearance as an outgrowth from the pharyngeal region of the gut, the lung seems to have functioned as a respiratory membrane. Its usefulness to aquatic forms confined in waters of low oxygen content surely explains the perpetuation of the organ in several lines of bony fishes. Although much later in vertebrate evolution the lung itself would increase in complexity as advances in the design of internal organs made possible a higher rate of metabolism, a simple internal respiratory sac not too different from that of the fishes would very likely have sufficed for the cold-blooded protoamphibians. A land vertebrate could not survive, however, without some mechanism for assuring the passage of air from mouth to lungs. In fishes this transport may be a passive process: a fish can gulp air and then plunge head downward, causing the bubbles to rise through the pharynx and so pass backward into the lung. A land animal which remains horizontal must have some way of forcing or drawing air through its respiratory tract. Certain structural innovations

which appear in early tetrapods could have arisen in conjunction with the requirement for a respiratory current. The increased length and stoutness of the ribs characteristic of the first
known amphibians afforded added surface for the attachment
of muscles which could have produced a rise and fall in the body
wall, rhythmically changing the pressure around the lungs. Air,
sucked into the respiratory sacs when their internal pressure fell
below that of the outside, would have been expressed when the
pressure on the lungs rose. This kind of respiratory mechanism,
which survived in higher vertebrates, was surely not the only one
which evolved in the ancient amphibians. The gradual broadening of the head in many forms can be interpreted as evidence for
the existence in Paleozoic times of the force-pump apparatus
that remains in the ribless amphibians of today. In these
animals, the floor of the mouth is lowered with the nostrils open
to draw air into the mouth and then raised with the nostrils
closed to press the air backward into the lungs. The wider the
floor of the mouth, the larger its underlying muscle sheets and
the greater the efficiency of the mechanism that depends upon
their action. The tendency toward flattening and widening of
the head evident in several lines of Paleozoic amphibians could
be explained by postulating use of the force-pump process in
these animals and the continuing selection of types whose skull
structure made possible its more effective operation.

In the fossil record there is evidence to suggest that reliance
on either one of these respiratory mechanisms developed slowly,
both phylogenetically and in the life of the individual. The
young of the early amphibians, like the young of modern forms,
grew gills first and then developed functional lungs as they
matured. During their metamorphosis, the animals almost certainly depended to some extent upon the thin, moist skin (if they
were not heavily armored) and the mouth lining as auxiliary
respiratory surfaces, as existing amphibians do. Whether the
older amphibians were able to continue to respire through the
skin as adults is not clear. Although fossil evidence is scanty, it
seems that many, if not all, of them retained a covering of bony
scales that would have limited gaseous exchange at the surface
of the body. The scales over the back and flanks were thinner
than those of the ancestral rhipidistians, however, and possibly
not present everywhere, so that some cutaneous respiration
might have persisted throughout life in these animals.

One paleontologist believes that the appearance of legs facilitated respiration as well as locomotion in terrestrial vertebrates.

I. I. Schmalhausen points out that a fish lying on its side on the ground bears the weight of the body wall upon its viscera despite the presence of ventral ribs. When the internal organs are compressed in this way, the fish can force air through the pharynx only with great difficulty. For air to pass easily into the lungs, the trunk has to be propped up so that the respiratory organs hang suspended in the body cavity. A rhipidistian fish could have used its muscular fins to lift its body from the surface temporarily, but in the absence of legs the body would have dragged or flopped when the animal moved. Although there might have been a rhipidistian that could expend the large amounts of energy necessary to make the lungs work under these conditions, a form with a fish-like trunk would have been better adapted for living on land if the body was kept clear of the ground.

Elevating the body was also desirable as preparation for easy locomotion over dry ground. A fish moves against the water which buoys it up, generating a minimum of friction, but a tetrapod in contact with the substratum would be considerably hindered by the scraping of its body along the uneven surface unless, like the snakes, it was specially modified for that kind of locomotion. Raising the trunk and tail eliminates at once the necessity of slithering along every rise and fall in the terrain and the danger of scraping the epidermis to shreds in doing so. The

Changes in the skeleton associated with terrestrial life

transformation of the fins from steering devices to piers for the suspension of the body involved changes in every part of the appendicular skeleton. Besides reorienting at least a portion of the limb in a vertical direction and exchanging for the fin rays feet that could be planted flat on the ground, the relationships of the girdles to the axial skeleton had to be substantially modified (Fig. 6.2). In fishes, the pelvic girdle always consisted of a pair of bony plates embedded in the ventral body wall. A tetrapod limb, pressing upward against such a structure, would cause it to sink inward and compress the soft organs in the posterior part of the body cavity. If the hind legs were to hold up the animal and push it forward, there had to emerge a rigid connection between the pelvic girdle and the vertebral column. With such an arrangement, the thrusting force of the foot against the ground could be transmitted with little loss to the main axis of the body.

A union of the appendicular and axial parts of the skeleton in the pectoral region already existed in fishes: from earliest times, the dermal shoulder shield had articulated firmly with the back

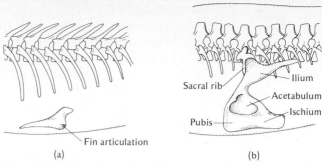

Figure 6.2. The relationship of the pelvic girdle to the vertebral column in fish and early tetrapods. **(a)** Left lateral view of the column and pelvic girdle of a fish, showing the girdle embedded in the musculature of the ventral body wall. **(b)** The same view of the column and girdle in a primitive amphibian, showing the existence of a sacroiliac articulation through which stress can be conveyed from the appendage to the axial skeleton directly. (From Romer.)

of the skull. This association, while it was undoubtedly advantageous for fishes, was unsuitable for a tetrapod (Fig. 6.3). The forelegs are a sort of landing gear in a terrestrial animal: they receive the weight of the body as it is propelled forward by the hind limbs. To absorb the shock generated by the descent of the trunk, the pectoral girdle has to be disconnected from the head. Even if an amphibian had been able to withstand the tension on the skull that would have been created at every step so long as the articulation between girdle and skull remained, another difficulty would have ensued. Because the scapulocoracoid element which receives the limb bone abuts the cleithrum of the shoulder shield along its vertically oriented medial surface, pressure from the leg would set up a shearing stress between the endochondral part of the girdle and the rigidly held dermal portion. Since the skeleton is most vulnerable to shearing forces, the resulting weakness of the suspensory apparatus would limit the weight, and thus the size, of terrestrial vertebrates. Transformation of the girdle into one strong enough to support a walking animal of any mass demanded a loss of the connection of the dermal bones with the skull and a concomitant expansion of the endochondral part to which the legs are attached.

Freeing the skull from the pectoral girdle surely brought with it other advantages for an animal making its way on land. For the first time, a vertebrate would be able to turn its head and to lift it above the level of the body without tilting its tail downward. This ability made it possible to see beyond small obstacles in the immediate environment and so to choose the most convenient path or to discover enemies lurking nearby. Since the skull, separated from the girdle, remained cantilevered from the front of the vertebral column, its elevation above the ground required its development as an inflexible structure with space available on the posterior surface for attachment of the strong neck muscles that hold it up. Specifically, the braincase would have to become more solid than it was in the ancestral fishes and the occipital region broader.

Although many fishes flourished with little ossification of the vertebral column, supporting a body in the aerial environment

Figure 6.3. The relationship of the pectoral girdle to the skull in fish and early tetrapods. **(a)** Skull and pectoral skeleton of a rhipidistian fish, showing abutment of the dorsal part of the pectoral girdle against the rear of the skull. (From Gregory and Raven.) **(b)** Skull and pectoral skeleton of an early amphibian, showing separation of the bones of the shoulder from those of the head. (From Romer.)

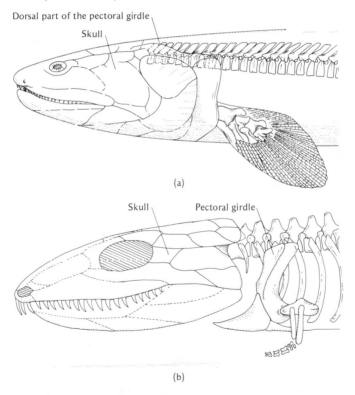

(a)

(b)

necessitated increasing the amount of bone in that part of the skeleton. While the long-bodied early tetrapods still needed the kind of flexibility conferred by the unjointed but pliant notochord, that structure, reinforced as it was in most ancient fishes by cartilaginous elements, never would have sustained the weight of the whole animal out of water. Evolution of a stronger column was made possible by the appearance in the rhipidistians of small bones that hugged and constricted the notochord: anteriorly in each segment an intercentrum embraced the notochord from below, and, just behind, a pair of nubbins called pleurocentra occupied a dorsolateral position (Fig. 6.4). Selected for at first, perhaps, because the intercentrum provided secure footing for the high-spined neural arch, the bones could later expand to share and then entirely accept the stresses projected through the vertebral column of the land animal. Contact between these central elements was apparently insufficient to produce stability, however, for no tetrapod evolved without an accessory articulation more dorsally between the neural arches. The outgrowth fore and aft of paired zygapophyseal, or yoking, processes from the roof of each arch resulted in the development of gliding joints which added strength to the column but interfered little with its flexibility.

Protecting the body from desiccation

The structural modifications which enable an animal to support itself and to respire in an airy medium would be of no benefit unless the animal could protect its body from desiccation. Since vertebrates, like all living things, are composed of watery protoplasm, this is a formidable task. Eventually, in the tetrapod line, it was to be accomplished in a way similar in principle to that utilized by all organisms adapted for terrestrial life: the body became enclosed in a nonliving coat impervious to water, which exposed moist tissues in as few places as possible. Whereas other forms of life manufactured cork and cuticle, shell, slime, and exoskeleton to retard the loss of water, tetrapods evolved a layer of dead, keratin-filled cells at the surface of the epidermis which confined internal fluid reasonably well. Gills, which have to remain moist and exposed if they are to function, had to be abandoned, and respiratory membranes developed in more protected parts of the body. The formation of internal nares was preadaptive for animals that had to restrict contact between the dry air of the environment and their respiratory organs. With a pair of narrow passages that led from the exterior to the front of the oral cavity, protoamphibians could admit in two fine streams the air necessary for respiration rather than exposing the moist interior of the mouth broadly by

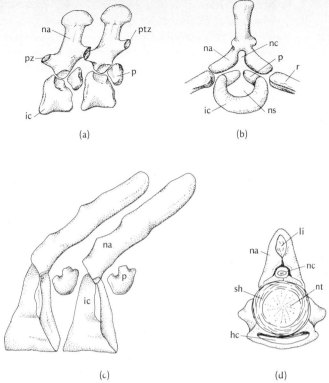

Figure 6.4. The structure of the vertebrae in early amphibians and archaic fishes. **(a)** Left lateral view of two vertebrae of *Archegosaurus*, a Permian amphibian, showing intercentra, pleurocentra, and neural arches, the latter articulating with one another through zygapophyseal processes. **(b)** A vertebra of *Archegosaurus* in anterior view. **(c)** Two vertebrae of the rhipidistian fish *Eusthenopteron*, showing the condition antecedent to that found in the earliest amphibians. Note the absence of articulation between adjacent neural arches. **(d)** Transverse section through vertebral column of *Acipenser*, a modern sturgeon, showing notochord, unconstricted and strengthened by a fibrous sheath, acting as the primary supporting element. The notochord, surrounded wholly or in part by cartilaginous elements, provided the major support in the majority of ancient fishes. Abbreviations: *hc*, haemal canal; *ic*, intercentrum; *li*, ligament; *na*, neural arch; *nc*, neural canal; *ns*, space for notochord; *nt*, notochord; *p*, pleurocentrum; *ptz*, postzygapophysis; *pz*, prezygapophysis; *r*, rib; *sh*, fibrous sheath. (*a,b,d* from Goodrich; *c* from Romer.)

parting the jaws. In higher tetrapods, protection of the internal tissues would be increased by further isolating the respiratory stream within the oral cavity and lengthening the distance through which air had to travel before reaching the lungs.

If the condition of modern amphibians is representative of

the level to which the first tetrapods had brought their water-conserving abilities, it is obvious that the bare beginnings of protection against desiccation were sufficient to allow vertebrates to gain the land. In living amphibians, the skin exhibits a keratinized layer but one thin enough in most species to be permeable to water under certain circumstances. The delicacy of the protective layer enables it to serve a double function: covered with mucus, it slows the loss of water when the animal is exposed to the air but permits fluid to enter when the animal submerges itself after a terrestrial excursion. The increase in cutaneous permeability shown on the latter occasion is under hormonal control. Although it is not known whether endocrine activity of this sort is a specialization in modern amphibians or a legacy from ancient forms, it is safe to suppose that physiological mechanisms must have appeared early to supplement the structural alterations which adapted vertebrates for land life. If hormonal regulation of cutaneous permeability did exist in the first amphibians, it would have been effective combined (as it is in extant forms) with endocrine control of the kidney. If the emergent tetrapods secreted hormones that acted to conserve water by inhibiting filtration of fluid from the bloodstream into the urinary tubules or by encouraging the recovery of water by resorption through the tubule wall, the ability of these animals to survive in the atmosphere would have been enhanced.

Scales disappeared, and other structural defenses against drying out remained limited among the amphibians. These animals assured their viability by behavioral adaptations rather than by paralleling the higher tetrapods in the development of increased physical resistance to desiccation. They frequented moist areas adjacent to streams or lived in marshes, where the heat of the sun made the air steamy rather than dry. When drought was inescapable, they burrowed or lay dormant underground. Undoubtedly, the earliest amphibians fertilized their eggs externally, as many living forms do, laying them in the water, where they were safe from drying if not from the depredations of hungry fishes. Surviving embryos, unprotected by extraembryonic membranes, became gilled larvae that passed through an obligatory aquatic stage before they could step out on land. Those later amphibians which became more terrestrial evolved a variety of special reproductive habits designed to accommodate or abbreviate the larval requirement for water. Their eggs are laid in damp moss, gelatinous froth, or temporary puddles or fertilized and held for a time within the body;

the embryos may mature with unusual speed, acquiring adult characteristics before they leave the egg. None of these peculiar inventions permitted the degree of independence from fresh water enjoyed by higher vertebrates, which enclose their embryos in a fluid-filled sac and protect them further with a shell or by housing them internally until they can survive in dry air.

Modifications of the sensory organs in protoamphibians

Although tetrapods were able to spend time on land before they evolved a truly impervious skin and a way of reproducing in the absence of water, they could hardly have managed if their sensory organs had not soon become adapted to functioning in terrestrial surroundings. The eyes of bony fishes, their chemoreceptors, and their lateral-line organs for sensing vibrations in the environment were designed to work in a watery medium. In air, the surface of the lidless eyes would dry in a short time, as would the soft tissue of the lateral-line canals. The olfactory epithelium, usually folded loosely in fishes, would quickly lose its moisture and shrivel. In addition to glands that secreted mucus over the skin, the first vertebrates which were even partially terrestrial required others that continuously bathed the exposed parts of the sense receptors. The nasal lining, recessed within the respiratory passage, was protected sufficiently by this method alone. The eye acquired a lid that could be passed over the surface periodically for instant rewetting of the cornea. With these minor alterations and certain others in the shape of the cornea and lens, the olfactory epithelium and the eye became immensely valuable for the new vertebrates. Since light travels and substances diffuse more rapidly through air than through water, tetrapods could sense changes in their environment at a much greater distance than fishes ever could. That they relied increasingly upon these sensory organs can be inferred from the gradual enlargement of the cranial spaces for the olfactory and optic lobes of the brain, to which impulses from these receptors pass.

No structural modification, however, could turn the lateral-line system to advantage in a land animal. Its action depended upon the passage through superficial canals of a current of water to deform a gelatinous coating around the cilia projecting from its sensory cells. Even if the gummy coating could have been kept wet, it is unlikely that pressure waves in air would have been forceful enough to disturb it. Because sensitivity to aerial vibrations allows an animal to detect the presence of enemies it cannot see or smell, it was to be expected that a substitute for the lateral-line system, if one appeared, would be selected for. Actually, no new organ arose in the protoamphibians, but by a series of changes in old structures, an arrangement was achieved whereby cells in the inner ear closely related to the sensory cells of the lateral-line system were made responsive to external vibrations. Of chief importance in this renovation was the change that took place in the hyomandibular bone (Fig. 6.5). In its position behind the spiracular gill slit in ancient fishes, this element of the branchial skeleton had served as a prop for the jaws, bracing the quadrate bone against the otic region of the braincase. When, in the evolution of the protoamphibians, the upper jaw became attached firmly to the skull, the hyomandibula relinquished its articulation with the quadrate and expanded in a lateral direction. A process that already extended outward to the operculum in rhipidistians broadened and, after the disappearance of the gill cover in the earliest tetrapods, came to rest against a membrane that spanned the space left between the skull roof and cheek. At its medial end, the hyomandibula remained pressed against the cranial bones that enclosed the inner ear. When eventually the cranial wall became membranous instead of bony where the hyomandibula touched it, the old suspensory element was transformed into a functional stapes or columella. Enveloped by an extension of the spiracular

Figure 6.5. The transformation of the hyomandibular bone into the stapes of the middle ear. **(a)** The hyomandibula acting as a suspensor of the jaws in a rhipidistian fish. **(b)** Hypothetical condition in protoamphibian: the ligamentous connection between the hyomandibula and the jaws becomes more tenuous as the tympanic process elongates and comes in contact with the exposed tympanic membrane. **(c)** The condition present in primitive amphibians: the hyomandibula, now the stapes, no longer braces the jaws but crosses the cavity of the middle ear to abut the dorsally placed tympanic membrane. Abbreviations: *CH*, ceratohyal cartilage; *HMD*, hyomandibular bone; *MC*, Meckel's cartilage of the lower jaw; *PQ*, palatoquadrate element; *pr.ty*, tympanic process; *S*, stapes; *ty*, tympanic membrane. Ligaments are stippled, and the cavity of the middle ear is shaded. (Modified from Westoll.)

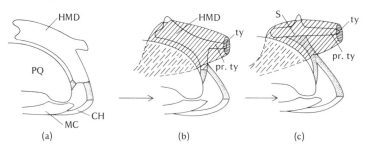

Vertebrate History: Problems in Evolution

pouch, now the cavity of the middle ear, the bone was free to vibrate when the external membrane, or tympanum, against which it rested, was disturbed by vibrations in the air. The new stapes, as it moved, transmitted pressure waves through the membrane at its medial end to the fluid around the inner ear. There, the sensory cells were stimulated, as they always had been, by the agitation of the liquid in which their gelatin-covered cilia were bathed.

Although it is not possible to tell from the remains of the first known amphibians how far evolution of the ear had progressed, it is clear that by late Devonian times a number of other structural adaptations for land life had been realized. Pale-ontologists have found in what is now eastern Greenland fossils of vertebrates that were even then robust tetrapods. These animals were not adapted to withstand the arctic conditions that currently prevail in that part of the world. The climate at the end of the Devonian period was far milder. For most if not all of the year, rain water that fell in the highlands ran freely down to the sea, filling the land with streams, shallow backwaters, and rivers that housed a rich variety of lobe-finned fishes. It was an environment conducive to fossilization: not uncommonly, dead fish were carried along with the current until they were buried in sediment at the bottom, often in estuarine areas, where their bones rested side by side with brackish-water placoderms, acan-thodians, and sharks. The amphibians that appeared in eastern Greenland were transported downstream in the same way. Alive, they shared the freshwater habitat of their rhipidistian cousins, leaving it from time to time to make forays out onto solid ground.

The ichthyostegids, as paleontologists named these early tetrapods, already had well-developed legs. The limb skeleton exhibited, fully defined, the pattern that was to remain charac-teristic of land vertebrates. A single, large bone extended through the upper part of the leg to meet, at knee or elbow, two elements which supported the lower limb. These bones articu-lated distally with a number of small ones that gave flexibility to the wrist or ankle region. The five-toed foot stretched out flat, its metapodial bones bracing those of the digits. Although the proximal bone extended laterally from the side of the body like the fin of a fish, the lower part of the leg was turned in a vertical direction in such a way that it could lift the body from the ground.

The girdles of the ichthyostegid amphibians were as completely adapted for bearing weight as the limbs. The dermal part of the pectoral girdle had lost its connection with the back of the skull: the bones through which the attachment was made in ancestral fishes, the posttemporal and supracleithrum, had disappeared entirely. Instead of being braced against the skull, the remaining dermal bones on each side abutted a new bone formed in the ventral midline, the interclavicle. The dominant element in each half of the pectoral girdle was now the scapulocoracoid, with which the upper limb bone articulated. It rose dorsally to form a strong blade for the attachment of muscles which bound the girdle against the trunk. The pelvic girdle had evolved the rigidity and the contact with the axial skeleton that allowed it to serve as a firm anchor for the hind legs. The two loosely associated ventral plates of fishes had given way to a structure built of left and right halves almost immovably joined to each other in the ventral midline. In its basic design, this girdle was like that of later tetrapods. The three bones of each side met laterally in the acetabulum, the depression into which the head of the proximal limb bone fits. Two of them, the pubis and the ischium, spread from this point ventromedially to meet their counterparts in the midline, affording a surface for the origin of muscles that moved the leg downward and back and forth. The third bone, the ilium, extended upward to reach the sacral rib and vertebra, to which it was securely tied. Besides bracing the girdle against the vertebral column, this bone provided anchorage for the muscles which raised the limb. A posteriorly projecting process of the ilium in ichthyostegids apparently gave attachment to muscles less important in terrestrial locomotion than the levators of the leg, for it appeared in only one later group of amphibians before vanishing forever.

Although fossil evidence for terrestrial adaptation of soft organs is not direct, as it is for the changes in the appendicular skeleton, it is possible to make some inferences about the status of the respiratory organs in the ichthyostegids. The structural requirements for pulmonary breathing were present: internal nares existed at the edge of the palate, and broad ribs framed a capacious chest cavity, in which the lungs could have expanded freely. That the animals may have relied to a great extent upon their lungs is suggested by the limited space which remained for the gills. The branchial chamber, squeezed behind the long jaws, no longer extended posteriorly under a large opercular cover. Its opening apparently fitted under the vestigial subopercular bone, the only part left in ichthyostegids of the movable gill shield of fishes. It seems certain that at this stage in the evolution of amphibians the skin was important as a respiratory organ. The presence of lateral-line canals attests to its having been kept wet, and the reduced scale cover implies exposure of a vascular surface through which the exchange of oxygen and carbon dioxide could have taken place.

Behavior of the ichthyostegids

The persistence of the lateral-line system in ichthyostegids signifies that these animals still lived primarily in the water. They retained a strong resemblance to fishes and continued to swim and to feed in the same manner as their ancestors. Though heavy-bodied, they were elongated and streamlined for movement below the surface. From the remains of the vertebral column, paleontologists can tell that it was supple enough to undulate as the axial muscles pulled upon it. The large notochord, reinforced but apparently not restricted by the adjacent pleurocentra and intercentra, could still be flexed laterally despite the development of neural arches which articulated with each other in tetrapod fashion (Fig. 6.6). If any doubt existed about the ability of the ichthyostegids to swim, it would have been dispelled by a glance at the form of the tail. It was flattened from side to side like that of any fish and supported by radial elements that were braced against the neural and hemal arches. Such a structure acts only to propel an animal through the water and does not appear in vertebrates which rely primarily upon their legs for locomotion. The ichthyostegids glided about in the shallows as easily as their finned relatives, no doubt, and like them lived on small fishes which they snapped up and impaled upon their conical teeth.

Structure of the skull in ichthyostegids

Outwardly, the head of these early tetrapods looked not very different from that of the contemporary lobe-finned fishes. Under the skin, however, the skull had assumed the form which was to remain basic for later terrestrial vertebrates (Fig. 6.7).

Figure 6.6. Skeleton of *Ichthyostega*, a late Devonian amphibian. The oldest known tetrapod, *Ichthyostega* measured about 3 feet in length. (From Jarvik.)

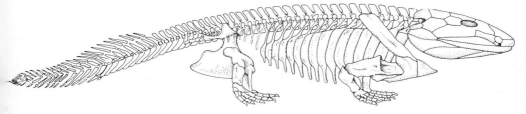

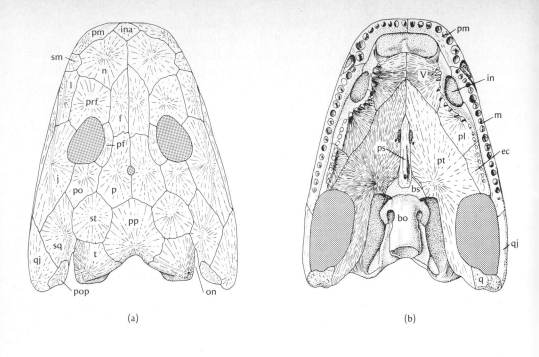

(a)

(b)

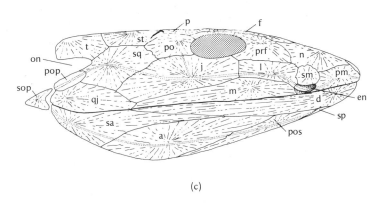

(c)

Figure 6.7. Skull of *Ichthyostega* in **(a)** dorsal, **(b)** palatal, and **(c)** lateral view. Abbreviations: *a*, angular; *ar*, articular; *bo*, basioccipital; *bs*, basisphenoid; *c*, coronoid; *d*, dentary; *e*, epipterygoid; *ec*, ectopterygoid; *en*, external naris; *eo*, exoccipital; *f*, frontal; *fm*, foramen magnum; *in*, internal naris; *ina*, internasal; *it*, intertemporal; *j*, jugal; *l*, lacrimal; *m*, maxilla; *n*, nasal; *on*, otic notch; *op*, opisthotic; *p*, parietal; *pa*, prearticular; *pf*, postfrontal; *pl*, palatine; *pm*, premaxilla; *po*, postorbital; *pop*, preopercular; *pos*, postsplenial; *pp*, postparietal; *pr*, proötic; *prf*, prefrontal; *ps*, parasphenoid; *pt*, pterygoid; *q*, quadrate; *qj*, quadratojugal; *r*, rostral; *s*, stapes; *sa*, surangular; *se*, sphenethmoid; *sm*, septomaxilla; *sop*, subopercular; *sp*, splenial; *sq*, squamosal; *st*, supratemporal; *t*, tabular; *v*, vomer. (After Jarvik.)

The front and rear parts of the braincase joined each other immovably behind the recess for the pituitary gland, and the ensheathing dermal bones of roof and palate exhibited an arrangement so similar to that of other Paleozoic tetrapods that the homology of the corresponding elements is unquestionable. The head was protected by a cover of plate-like bones, which joined each other tightly at sutures. As in later tetrapods, the dorsal region was shielded by a series of paired elements which met at the midline, equivalents, most paleontologists agree, of the nasal, frontal, parietal, and postparietal bones. Beside them were the orbits, surrounded by the prefrontal, postfrontal, postorbital, jugal, and nearby lacrimal plates, five ossifications which persisted in land vertebrates of many lines. Anterior to the lacrimal, a small bone, the septomaxilla, bordered the nasal aperture dorsally, and the premaxilla edged the front of the upper jaw. The latter element bore marginal teeth, as did the maxilla which extended from the nostril back to the quadrato-jugal bone over the corner of the mouth. The side of the head behind the jugal and postorbital bones was sheathed by squamosal, tabular, and supratemporal elements, the first two of which, with a small preopercular plate, rimmed the otic notch, the cleft in the posterior margin of the skull that was closed in the living animal by the tympanic membrane. The undersurface of the skull was as heavily armored as the roof and cheek. The dermal palate that lay beneath the braincase within the arch of the upper jaw consisted of a lateral series of paired bones, the vomers, palatines, and ectopterygoids, all of which bore teeth, and a more medial pair of large pterygoid elements that fitted closely against a central parasphenoid bone. Each pterygoid was prolonged posteriorly into a curved flange that formed the inner side of an opening through which muscles passed to insert on the lower jaw. The bones of the lower jaw—a narrow element bearing teeth and several that supported it from below—were also easily comparable to those of later tetrapods.

Skulls of ichthyostegids and rhipidistians compared

The similarity of the arrangement of the dermal bones in the skulls of ichthyostegids and rhipidistians is one of the strongest pieces of evidence linking the amphibians to that particular group of fishes. No other Devonian vertebrates exhibited a pattern from which that of the tetrapods could have been derived. The elements in the skull of the rhipidistians were not identical to those of the ichthyostegids in every way, but the differences do not seem great enough to preclude a phylogenetic connection between the two kinds of animals (Fig. 6.8). The snout region of the rhipidistian, much shorter than that of the

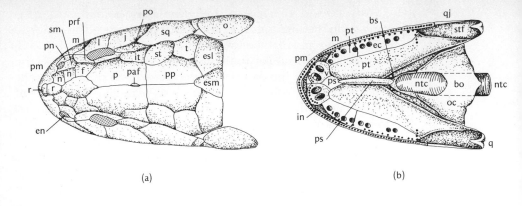

(a)

(b)

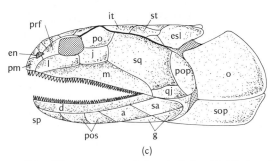

(c)

Figure 6.8. The structure of the skull in rhipidistian fishes. **(a)** Dorsal, **(b)** palatal, and **(c)** lateral view. Abbreviations: *esl*, *esm*, lateral and medial extrascapulars; *g*, gulars; *ntc*, notochord; *o*, opercular; *oc*, otic capsule; *paf*, parietal foramen; *pn*, postnasal; *por*, postrostral; *stf*, subtemporal fossa. For other abbreviations, see Fig. 6.7. (From Romer.)

ichthyostegid, was covered by a number of small bones rather than a single pair of broad nasals. The two large bones of the tetrapod could have resulted from fusion of the little plates present in the fish, however, since the tendency to combine neighboring dermal bones in the skull roof is common among vertebrates. The ichthyostegids had left in front of the paired nasal bones an internasal nubbin, which could have been a last surviving remnant of the mosaic of small plates that once existed in that area. Loss of bones is another well-documented kind of evolutionary change which could account for discrepancies between the skulls of rhipidistians and ichthyostegids. The tetrapods lost the large opercular of the fishes as well as a row of extrascapular elements at the rear of the skull and the small intertemporal bone which in rhipidistians lay anterior to the supratemporal.

The apparent difference in the location of the opening for the median eye proved the most formidable obstacle for those who would relate the ichthyostegids to the rhipidistians. In amphibians and other tetrapods that retained the rudimentary eye in the midline, light reached it through a foramen between the parietal bones. In rhipidistians the foramen existed, but it appeared at first to lie farther forward, between the frontal elements. Paleontologists of Stensiö's school, convinced as they all were that the rhipidistians were ancestral to the tetrapods, suggested that the foramen had changed its position relative to the bones of the skull roof. Unlike fusion or loss of bones, however, this sort of change would have been an extraordinary one. Despite Jarvik's contention that movement of the parietal foramen had occurred in certain instances, other workers remained skeptical. Finally, Westoll proposed what he considered a more probable explanation: the medial eye grew upward from the brain beneath the same pair of bones in the rhipidistians and their tetrapod descendants, but the bones themselves seemed differently placed because of a change in the proportions of the skull (Fig. 6.9). In the short-snouted rhipidistians, the bones between the orbits that flanked the median eye were really the parietals, not the frontals as Jarvik thought. As the tetrapods evolved, the bones in the front of the skull fused and elongated. The true frontals extended posteriorly between the orbits, and the parietals, with the foramen between them, retreated over the skull roof. In support of his hypothesis, Westoll pointed out that in *Elpistostege,* a Devonian animal that he regarded as transitional between the rhipidistians and ichthyostegids, the size and position of the frontal and parietal bones were intermediate. Whether Westoll is correct in using the arrangement of the bones in *Elpistostege* as evidence for his theory is uncertain, since only a partial skull of this form is known and its relationship to the ichthyostegids is not surely established. Support for Westoll's idea has come from other research: Romer has substantiated the possibility of a gradual lengthening of the skull anterior to the orbits and a shortening of the region behind by demonstrating that this trend continued beyond the amphibian level through the first reptiles and into the mammal-like reptile line.

The advantage of the posterior extension of the parietal bones that accompanied this change in early tetrapods may have been that the skull was strengthened thereby. In rhipidistians, the boundary between the parietal and postparietal bones constituted a division of the dermal roof that coincided with the intracranial articulation of the braincase below. The fishes were

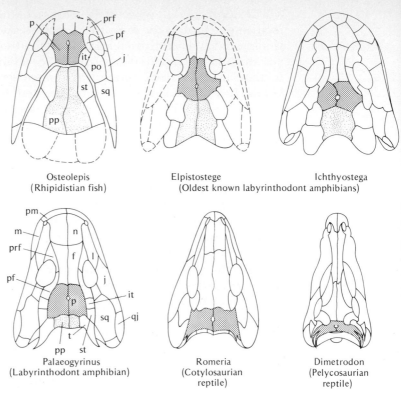

Figure 6.9. Skulls of a rhipidistian fish and several Paleozoic tetrapods in dorsal view, showing posterior migration of the dermal bones of the roof and the progressive shortening of the postparietal elements. Abbreviations as in Fig. 6.7. (From Young after Westoll.)

able, therefore, to raise the entire anterior part of the skull as the mouth was opened. Whatever advantage there was in an enlarged gape was apparently overshadowed by the necessity for rigid construction of the head in an animal that supported itself out of water. In ichthyostegids, the intracranial joint was immobilized, and the overlying dermal bones interlocked in such a way that the roof was no longer effectively transected. Continued posterior displacement of the parietal bones brought them to a position directly above the suture between the two halves of the braincase. The solid dermal cover across the old intracranial division, coupled with the expansion of the parasphenoid bone beneath it, reinforced the interior part of the skull significantly.

If solidification of the skull occurred before tetrapod status was attained, it must have evolved only in the rhipidistians that were directly antecedent to the ichthyostegids. By contrast, internal nares, structures which also seem to be adaptations for land living, characterized rhipidistians generally and constitute another basis for recognizing these fishes as ancestors of the first amphibians. Schmalhausen attributes the presence of openings from the nasal passages into the front of the mouth, not to incipient terrestrial habits on the part of all rhipidistians, but to the advantage these openings conferred upon the rhipidistians underwater. In developing his thesis, Schmalhausen points out that these fishes swam and hunted among the weeds, where their sense of smell was especially important in detecting their prey. He supposes that at an early time in their evolution rhipidistians relied upon currents, as other fish did and still do, to carry water in and out of blind-ended nasal sacs. Like other bony fishes, they must have had at least two external nostrils on each side of the snout: one for entrance and the other for exit of the water. Their olfactory acuity depended upon the rapidity with which the environmental water could be made to flow over the sensitive nasal epithelium. Since the rhipidistians were not built for speedy swimming which would force a steady stream of water through the nostrils, they depended upon their respiratory current to change the water in the nasal cavities. According to Schmalhausen, water traveled more quickly through the nasal sac and out at the back as the posterior nostril became displaced toward the edge of the upper jaw, where suction created by water entering the mouth was strongest. The spread through the population of the genetic factor responsible for the new location of the posterior nostril was encouraged by another advantage besides increased powers of olfaction. Once the nostril had crossed the maxilla to open into the mouth through the palate (and that it did so Schmalhausen believes is indicated by forms like *Panderichthys*, in which the nostril seems to straddle the bone), the fish could take water into the oral cavity without moving the jaws. When a rhipidistian waiting in the weeds to ambush its prey was able to dispense with the rhythmic opening and closing of its mouth, it no longer produced the turbulence that must have betrayed the presence of its ancestors to many of their intended victims. Although Schmalhausen's speculations concerning the hunting habits of the rhipidistians can never be confirmed, his supposition that the internal nares, or choanae, served first for the passage of water and only later for the intake of air is a reasonable one. Internal nares, like many other vertebrate structures, may have been selected for because they were advantageous in one way and then retained because they became useful in another.

Structure of the teeth in ichthyostegids

In addition to the internal nares, the teeth of the ichthyostegids were a legacy from the rhipidistians (Fig. 6.10). The teeth which edged the jaws and studded the palate in both forms were conical and sharp, like those of most fish-eating vertebrates, but they bore distinctive striations from the base to the tip of the crown. When a tooth is cut in cross section, each striation can be seen to mark an infolding of the tooth wall. In the interior of the tooth, the enamel is further convoluted to form a labyrinthine pattern. Possibly because teeth constructed in this manner were especially strong, they persisted beyond the ichthyostegids in several lines of Paleozoic amphibians, Assuming that these curious teeth did not evolve more than once, and noting that the tetrapods which bore them resembled each other

Figure 6.10. Cross sections of teeth of **(a)** *Polyplocodus,* a rhipidistian fish, and **(b)** *Benthosuchus,* a labyrinthodont amphibian, showing labyrinthine infoldings of the enamel. (From Bystrow.)

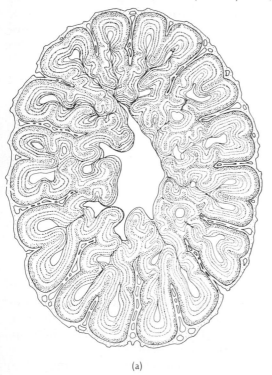
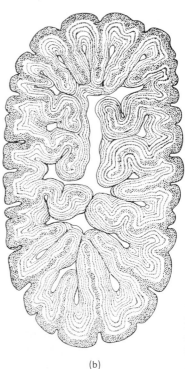

(a) (b)

in a number of ways, paleontologists assigned all these early amphibians to a single group, the Labyrinthodontia.

When the ichthyostegids were discovered, it was hoped that their postcranial remains would yield the missing link between the piscine skeleton of the rhipidistians and the land-adapted frame of the later labyrinthodonts. Although the vertebral column and the pectoral girdle were easily comparable to the rhipidistian structures, the fully formed five-toed legs of the ichthyostegids provided no new clue to the evolution of limbs from fins. Despite their failure to find animals with appendages of an intermediate type, investigators still believed that rhipidistian fins were the most likely precursors for the legs of land animals. In contrast to the fins of the actinopterygian fishes, in which several short basal elements articulate with the girdle, rhipidistian fins attached to the girdle through a single bone, as legs do. This most proximal bone was followed distally by a series of smaller ones, making a supportive axis that projected some distance beyond the trunk into the fin. Each axial element had braced against its anterior side a radial bone that broadened the surface available for the insertion of appendicular muscles. These elements, interposed between the girdle and the fin rays, seemed a better source for the bones of the tetrapod limb than the transverse rows of basal rods present in the ray-finned fishes. Taking into account anatomical and embryological evidence, Westoll was able to correlate each of the rhipidistian ossifications with a counterpart in the pentadactyl appendage (Fig. 6.11). The proximal element, which he called the first mesomere, he considered homologous to the humerus in the pectoral limb. It projected directly outward from the body in the early limb, just as it did in the ancestral fin. The second mesomere, or axial element, became the ulna and twisted forward to lie beside the first radial, now the radius. The change in the orientation of the second mesomere produced the sharp bend at the future elbow joint and also crowded the second radial element, which subsequently disappeared. The third mesomere with its radial formed the most proximal of the carpal bones, and the remaining mesomeres and radials became more distal carpal elements bracing digits I and II. The lateral carpal elements, Westoll thought, were derived from postaxial processes of the central mesomeres which eventually became independent bones. The femur, fibula, tibia, and tarsals in the pelvic appendage arose through the same kind of transformation. According to Westoll's theory, the bones of the digits, their

Transition from rhipidistian fin to tetrapod limb

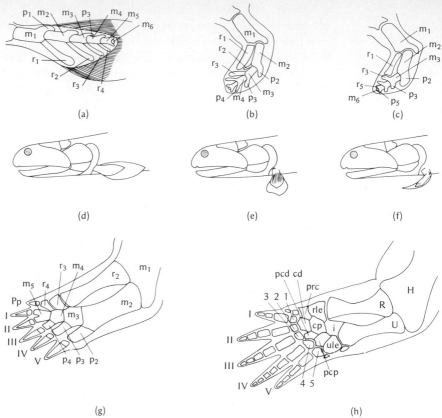

Figure 6.11. Westoll's conception of the evolution of the tetrapod pectoral appendage from the rhipidistian pectoral fin. **(a)** Skeleton of the rhipidistian fin. **(b,c)** The skeleton of the fin in protoamphibians, showing change in relative orientation of the bones. **(d,e,f)** Changes in orientation of the appendage to the body: **(d)** fin of fish in trailing position; **(e)** appendage of protoamphibian bent downward at elbow; **(f)** leg of amphibian with foot twisted forward. **(g)** Appendage of protoamphibian, showing relationship of newly evolved bones of digits and distal metacarpals to older elements of the manus. **(h)** Possible basic pattern for skeleton of primitive amphibian forelimb. Abbreviations: *cd, cp,* distal and proximal axial centralia; *H,* humerus; *i,* intermedium; m_{1-6}, meso-meres; p_{1-5}, postaxial processes of mesomeres; *pcd, pcp,* distal and proximal postcentralia; *pm,* postminimus; *Rp,* prepollex; *ppm,* post-postminimus; *prc,* precentrale; *R,* radius; r_{1-5}, preaxial rays of mesomeres; *rle,* radiale; *U,* ulna; *ule,* ulnare; I–V, digits of the pentadactyl limb. (From Westoll.)

supporting metacarpals (or metatarsals), and the most distal row of carpals (or tarsals) were not derived from any elements identifiable in the rhipidistian fin but evolved as wholly new structures as the fin rays were lost. Westoll defended this concept by pointing out that in the embryonic development of the limb in many terrestrial vertebrates the bones of the hand and foot originate separately from the more proximal elements of the appendage.

In examining embryos for clues to the evolution of the tetrapod limb, Westoll was using an approach that had been tried a decade earlier by N. Holmgren. Holmgren had also concluded that the skeleton of the leg was derived from appendicular elements present in lobe-finned fishes. He proposed, in addition, one hypothesis for which Westoll could find no support from his own research—that the legs of all tetrapods were not traceable to a single source. He pointed out that the embryonic rudiments of the limb bones in urodeles (amphibians known commonly as salamanders and newts) appeared in an arrangement slightly different from that of other forms. Holmgren interpreted his finding as evidence that the urodeles had arisen from lobe-finned fishes independently of the labyrinthodont amphibians. Because he found similarities between the embryonic appendicular skeleton in urodeles and lungfishes, he theorized that the urodeles were not descendants of any rhipidistian group but an offshoot of the dipnoan line.

Diphyletic origin of amphibians proposed and disputed

Since Holmgren stated his opinion in 1933, controversy over the origin of the amphibians has continued. Not only appendages but skulls and vertebrae of fossil forms have been studied minutely in an effort to determine whether tetrapods came from one stock among the fishes or from two or more by parallel evolution. Paleontologists who believe amphibians to be a natural group argue that the five-toed limb was unlikely to have evolved more than once; those supporting a dual origin voice doubt that the widely different amphibians which lived in the Carboniferous period and afterward could have descended from a common ancestor. Workers on both sides are forced to base their conclusions on studies of animals well beyond the first stages of transition to the tetrapod level. As primitive as the ichthyostegids of the Upper Devonian period were, they seem to have been at the end, or near the end, of a side branch of the labyrinthodont line. They had already diversified—at least three genera are discernible in the material from eastern Greenland —and had specialized to the extent of losing the intertemporal bones, elements of the rhipidistian skull that persisted among many labyrinthodonts of Carboniferous and Permian time. The fully formed legs of the ichthyostegids and their completely un-fish-like pelvic girdle indicate also that these animals lived long after the appendicular skeleton of the fishes began to change. To verify their theories of amphibian origin, paleontologists have to find the remains of earlier members of the tetrapod line. It is conceivable that some of these forms may yet come to light in rocks of Middle Devonian age. The lineages that were to produce terrestrial species may have been segregated as much as 30 million years before the ichthyostegids inhabited their swamps, at a time when lobe-finned fishes were the predominant vertebrates in fresh water. Small populations of individuals increasingly capable of amphibious behavior existed, no doubt, and underwent some degree of diversification throughout the second half of the Devonian period. Unfortunately, the possibility of finding fossils of these rare proto-amphibians seems small.

Variety of Amphibia during the Carboniferous

Aside from the ichthyostegids, the earliest amphibian remains in paleontologists' possession date from the Carboniferous period, especially the latter part of it, or Pennsylvanian, when favorable climatic conditions allowed the new tetrapods to extend their range and multiply dramatically. The weather warmed over much of the world at that time, and plentiful rainfall on tree-fern-shaded lowlands produced the swamps in which the amphibians thrived. Besides the labyrinthodonts, whose structural kinship to the rhipidistians was still apparent, other types of amphibians flourished whose origins are far less clear. Most of these animals were sharp-toothed, elongated little creatures with legs weakly developed or missing entirely. Fitted more for swimming than crawling, they must have stayed in the water almost continuously, hunting small invertebrates which lay concealed in the vegetation and the mud. Although these forms were diversely specialized, they were alike in lacking the separate intercentral and pleurocentral elements definable in the vertebrae of labyrinthodonts. They had, instead, vertebrae with centra that developed as indivisible bony cylinders, each enclosing the notochord and fusing with the neural arch above. On the basis of this common characteristic, Romer housed all these early amphibians in the suborder Lepospondyli.

The lepospondyls

The known lepospondyls belonged to three divergent groups (Fig. 6.12). One consisted of animals, called aistopods, that by the beginning of the Carboniferous period had already lost their legs and evolved a serpentine body. Another, the Nectridea, included forms distinguished by caudal vertebrae with peculiarly expanded neural and hemal spines. The members of this lineage developed a variety of specialized body shapes. Some

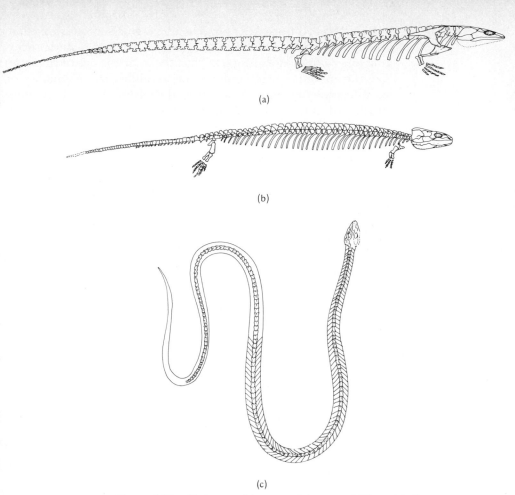

(a)

(b)

(c)

Figure 6.12. Skeletons of lepospondylous amphibians. **(a)** *Keraterpeton*, a nectridean. **(b)** *Microbrachis*, a microsaur. **(c)** *Dolichosoma*, an aistopod. (*a,b* from Steen, *c* from Piveteau.)

were long and limbless like the aistopods, but others broadened and evolved skulls with dermal bones prolonged into great prongs at the back. The third group of lepospondyls, the Microsauria, embraced a number of less extreme forms first thought to be little reptiles. Attempts to interrelate the lepospondyls or to trace their ancestry have failed for want of evidence. The presence of cylindrical centra in all these animals is not proof that they are a natural group. The fact that in aistopods the spinal nerves passed through the centra while in nectrideans they passed between them may indicate that the centra arose differently and that the animals possessing them evolved independently. The structure of the skull in nectrideans and microsaurs does suggest that these forms stemmed from labyrinthodont stock, and if liberal allowance is made for modification by fusion and loss, the aistopod skulls may be attributed to the same source; but nothing precludes the origin of these diverse lepospondyls from different groups of early labyrinthodonts or even from separate rhipidistian precursors.

If sufficient evidence is ever accumulated to show that the various lepospondyls descended separately from lobe-finned fishes or, as has been argued more strongly, that the lepospondyls as a group are derived from a different piscine stock than the labyrinthodonts, the proponents of a diphyletic origin

Arguments against diphyletic origin of lepospondyls and labyrinthodonts

for the amphibians will be proved correct. The possibility that the Carboniferous tetrapods arose from two sources has been judged unlikely, however, by several workers. H. Szarski maintained that two lines of animals could not have established themselves simultaneously in the same ecological niche. Furthermore, he refused to believe that labyrinthodonts and lepospondyls gained their virtually identical appendicular skeletons by convergent evolution. Adaptation for a similar function might have resulted in bent limbs and feet with spread-out toes, but the evolution in both lines of a pelvic girdle with three paired bones rather than two or four he regarded as an improbable coincidence. E. E. Williams interpreted the similar evolutionary tendencies shown by lepospondyls, labyrinthodonts, and modern amphibians as evidence that all of them emanated from a common ancestor. In undergoing loss of limbs, reduction of bone, and solidification of vertebrae before the middle of the Carboniferous period, lepospondyls exhibited precociously the kinds of change that took place later in the other lines. In Williams' opinion, vertebral structure was dissimilar in lepospondyls and labyrinthodonts, not because the two groups were derived from different sorts of fishes, but because the lepospondyls had given up the two-part centrum of the rhipidistians completely while the early labyrinthodonts had not. The cylindrical centrum of lepospondyls was evolved, Williams believed (like the similarly shaped centrum of salamanders long afterward), by expansion of the pleurocentrum at the expense of the intercentrum. Williams based his inference on a review of vertebral development in amphibian embryos: he found that in the living forms centra arise in a location equivalent to that of the pleurocentrum in the early tetrapods. The close correspondence between the appearance of the spool-shaped centra in lepospondyls and urodeles and the underlying homology between the structures of the latter and those of other amphibians increased the probability of a single stem for all the amphibians.

Jarvik's support of amphibian diphyly

Jarvik regards the suppositions of Szarski and Williams as misleading. From his own research on fossil fishes and modern amphibians, Jarvik has concluded that there is significant anatomical evidence in favor of the diphyletic or even polyphyletic origin of tetrapods. He agrees with Holmgren that the urodeles came from a different line than other tetrapods, but he traces them to a separate branch of the rhipidistians rather than to the Devonian lungfishes. Jarvik recognizes a cleavage of the early rhipidistians into two orders, Porolepiformes and Osteolepiformes (see page 148), and believes that each group had given rise independently to terrestrial forms. His studies show that modern urodeles still retain structural patterns inherited from the porolepiform fishes, and living anurans (the frogs and toads), comparable ones derived from the osteolepiforms. In examining the cavities and foramina in the cranium of the rhipidistian *Porolepis,* he found that the area for the anterior end of the brain, the passageway for the olfactory nerves, and the form of the nasal cavities were fundamentally like the corresponding parts of the urodelan head. His reconstruction of the cranial nerves and blood vessels furthered the resemblance of *Porolepis* to salamanders. In the porolepiform palate, Jarvik found a pair of deep recesses between the premaxillary bones and the vomerine elements which he believed held intermaxillary glands like those of modern urodeles. By contrast, *Eusthenopteron,* an osteolepiform rhipidistian, showed none of the peculiar specializations of *Porolepis* and the salamanders. Instead, the spaces for the brain, nerves, and vessels conformed to those existing in anurans and the tetrapods of more advanced classes.

Additional evidence for Jarvik's hypothesis came from the anatomy of the branchial skeleton. From the ventral parts of it preserved in *Eusthenopteron* and in the porolepiforms *Glyptolepis* and *Holoptychus,* Jarvik distinguished, again, two different structural designs. In the porolepiform fishes, a single median element, the basibranchial, had braced upon it all but the most posterior branchial arches as well as a long rod-like bone, the urohyal, which had a bifurcated end. In *Eusthenopteron* the basibranchial was divided into two segments, which together braced all the ventral arch elements. The urohyal lacked the bifurcated end and had in front of it an elongated sublingual rod, which did not appear in the porolepiforms. Jarvik pointed out the similarity of the branchial skeleton of salamanders to that of the porolepiforms and suggested that inheritance of the porolepiform design determined the peculiar mode of tongue development observable in modern urodeles. In these animals the tongue arises anterior to the branchial skeleton from a crescent-shaped glandular field present in the larva. According to Jarvik, such a glandular area existed in the porolepiform rhipidistians, flanked by denticles, of which vestiges appear in certain young salamanders. The region from which the tongue develops in anurans and higher tetrapods is that which was supported by the sublingual rod in the osteolepiforms. A homologue of that median element is present in the modern animals in the form of an anteriorly projecting skeletal process which underlies the tuberculum impar, the earliest part of the tongue to appear in the embryo. The embryonic derivation of the tongue is regarded by Jarvik as important because he believes that the developmental process does reflect accurately the separate history of the urodeles and the anurans. For this reason, he also included in his argument for the diphyletic origin of the amphibians the resemblance of the developing anuran limb to the osteolepiform appendage. He pointed out that the embryonic skeleton of the urodelan appendage exhibits a pattern that is different from the osteolepiform-anuran one. Since the fins of the porolepiforms are known only in outline, he had to rest his case without demonstrating that the endoskeleton of the urodele limb was in any way like that of those fishes.

Thomson's rebuttal of Jarvik's hypothesis

Few paleontologists take issue with Jarvik's meticulous anatomical descriptions, but many argue that his interpretation of the facts is open to question. Thomson does not accept Jarvik's conclusions and has attempted to show that none of his objections to monophyly is valid. If the development of the urodelan limb is thoroughly reviewed, Thomson maintains, it will be seen that the skeletal elements of the foot do not arise differently. They form and shift their position during growth, much as they do in other tetrapods. Certain rearrangements of the bones occur earlier in urodeles, but the mature skeleton of the foot resembles that of any unspecialized amphibian or reptile. Similarly, Thomson thinks that in the development of the tongue in urodeles, all trace of a sublingual element may be missing, not because the urodeles arose from rhipidistians which lacked it, but because the sublingual was reduced earlier in the urodelan line than in other amphibian stocks. Since anurans and higher tetrapods have only a rudimentary element under the tuberculum impar, Thomson supposed that the tendency toward reduction of the large osteolepiform sublingual rod was inherited in common by all the tetrapods. Rather than signifying the diphyletic origin of urodeles and anurans, the absence of the sublingual element in one group and its presence in the other may indicate simply that the rate at which the bone diminished differed in the immediate ancestors of the modern animals.

Thomson thinks that Jarvik has based his argument for diphyly on insufficient evidence and an exaggerated view of the dichotomy between the two groups of rhipidistian fishes. The porolepiforms especially are poorly known: it cannot be ascertained whether the structures Jarvik has described for *Porolepis* and *Glyptolepis* are distinguishing characteristics of all the porolepiforms or peculiarities of these forms. In Thomson's opinion, Jarvik has made the former assumption too readily and has overestimated the divergence these traits imply. Thomson insists that the differences now recognized between the porolepiforms and osteolepiforms are of minor significance. He minimizes the importance of the variations in the course of nerves and vessels through the snout and suggests that the urodele-like bulge in the anterior end of the cranial cavity of *Porolepis* may be an artifact. The anterior palatal recesses described by Jarvik do not seem to Thomson a good diagnostic character. His research showed that these depressions are basically similar in location in all rhipidistians and are shallow or deep depending upon the size of the mandibular teeth that fit into them when the mouth is closed. If Thomson is correct and the porolepiforms and osteolepiforms are members of the same order rather than two separate ones, Jarvik's hypothesis that urodeles and other tetrapods spring from a dual source is undermined.

Thomson finds still another reason for abandoning Jarvik's version of diphyly: it seems to him that in comparing modern amphibians to rhipidistians directly, Jarvik has used an approach of dubious validity. Jarvik has assumed for the purpose of his argument that urodeles and anurans have retained minute anatomical configurations present in Devonian fishes. To Thomson such conservatism on the part of the amphibians seems unlikely. The similarities that exist, he thinks, are more probably the result of convergence or parallelism than of phylogenetic relationship. Proof of the latter cannot be established without tracing the ancestry of modern amphibians back through the forms which preceded them in Paleozoic and early Mesozoic times. Jarvik made no attempt to find among these earlier animals, the labyrinthodonts and lepospondyls, any types which showed structures intermediate between osteolepiforms and anurans or between porolepiforms and urodeles. Without such evidence, Thomson believes, Jarvik's theory of amphibian origins cannot stand. He thinks that neither Jarvik nor any one else can tell how the rhipidistians gave rise to the tetrapods until many more of the transitional forms are known.

Fossil record and history of the early amphibians

Paleontologists hunting for material that would fill out the record of amphibian evolution have difficulty for two reasons. First, the small size and delicacy of many of the extinct forms make it certain that their preservation was a rare occurrence. Second, the continental deposits that contain amphibian fossils are accessible in only a few places in the world. Sediments laid down in swamps or in streams of the late Devonian and early Carboniferous years, when the amphibians underwent their initial radiation, have been explored in no more than a handful of isolated locations in the United States and Europe. Workers are aware that their picture of the ancient amphibian community is extrapolated from the small and perhaps not entirely representative sample of fossil forms they have gathered from these sites, but they feel, nevertheless, that they understand the broad trends in the history of these early tetrapods. However the group arose, its diversification was rapid: from what must have been a limited base in Devonian times, numerous tribes of labyrinthodonts and the highly differentiated lepospondyls evolved before the Carboniferous period was far advanced. Although certain types of labyrinthodonts exhibited the stout body and strong legs that characterized the ichthyostegids, many paralleled the lepospondyls in developing the elongated body and weak appendages that apparently facilitated swimming in swampland waterways. Other labyrinthodonts grew large, broad-headed, and flat, like the horned nectridean lepospondyls, and lay on the bottom using their wide jaws as deadly fish traps. Adaptation to continuous underwater living was accompanied by selection for neotenic forms—species retaining the external gills and partially cartilaginous skeletons that fitted them as larvae for their aquatic environment. The changes in body form that assured the early amphibians success in the swamps also assured their eventual failure to survive. As the marshy lowlands diminished in the Permian period, the animals that could live nowhere else began to die out. The extremely specialized little lepospondyls disappeared first. The labyrinthodonts, which were larger, staved off extinction longer, but by the end of the Triassic period they are also absent from the fossil record.

Watson's analysis of labyrinthodont evolution

The longer history and more numerous remains of the labyrinthodonts have made it possible to study their interrelationships in greater detail than those of the lepospondyls. Systematic analysis of these amphibians began as long ago as 1842, when Owen first called attention to the complex infolding of the dentin that distinguished their teeth from those of other tetrapods. Although he coined the term "labyrinthodont" to describe them, the animals with these teeth continued for another 75 years to be lumped with all pre-Jurassic amphibians as "stegocephalians." When, in the second decade of the twentieth century, Watson undertook to investigate the labyrinthodonts specifically, he found that fossils belonging to more than a hundred genera of these animals had accumulated in collectors' closets. After sorting those available to him according to age and reviewing their structure comparatively, Watson was able to outline the course of labyrinthodont evolution (Fig. 6.13). The earliest members of the group had had a relatively high, rounded skull with a well-ossified braincase. Characteristically, the ventral basioccipital and lateral exoccipital bones at the rear of the braincase contributed to a single condyle, through which the skull articulated with the vertebral column. The braincase was movably articulated also with the epipterygoid ossification of the upper jaw. Palate as well as jaw was suspended flexibly, because the pterygoid bones could move upon the medial parasphenoid, which adhered to the underside of the cranial floor. In these primitive labyrinthodonts, the pterygoid elements were broad and filled the central region of the palate, leaving minimal spaces, or interpterygoid vacuities, between them on either side of the parasphenoid. In more advanced labyrinthodonts Watson noted that the skull became depressed and less extensively ossified. The basioccipital bone withdrew slightly from the condyle, allowing enlarged exoccipitals to bear most of the stress of the cranio-vertebral joint. The palate lost its mobility through the development of a firm attachment between the pterygoid and parasphenoid bones, where a movable articulation had existed before, and wide interpterygoid vacuities appeared as the two pterygoid elements spread away from the

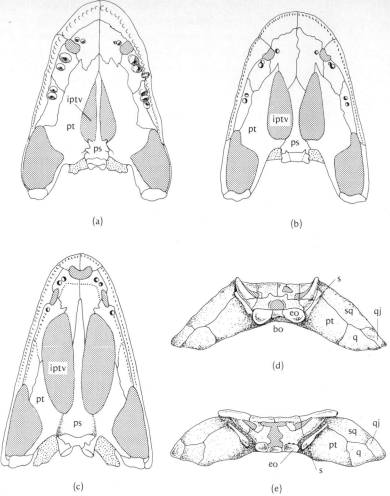

Figure 6.13. Labyrinthodont skulls in a morphological series illustrating the evolutionary trends observed by Watson. Palatal view of skull of **(a)** *Edops*, **(b)** *Eryops*, and **(c)** *Capitosaurus*, showing gradual enlargement of the interpterygoid vacuities (*iptv*), reduction of the palatal region of the pterygoid bone, and immobilization of the palate by union of the pterygoids with the parasphenoid. Occipital view of the skull of **(d)** *Eryops* and **(e)** *Capitosaurus*, showing reduction in the height of the cranium and division of the occipital condyle. Abbreviations as in Fig. 6.7. (*a* after Romer and Witter; *b,c,d,e* from Romer.)

knife-like process of the parasphenoid which lay in the midline. During the final phase of labyrinthodont evolution in the Triassic period, the trends established earlier persisted. Although superficial differences distinguished one family of late labyrinthodonts from another, Watson perceived that in all the forms he examined the skull was extremely flattened, the palate lightly built and broadly connected to the parasphenoid, and the condyle effectively paired because of the complete absence of the basioccipital from the articular surface.

Watson based his account of the evolution of the labyrinthodonts largely upon changes in the structure of the skull, because in amphibians, as in other aquatic vertebrates, the skull was the part of the skeleton most frequently preserved. The collections to which he had access included some postcranial remains, however, and he studied them also for evidence of progressive modification. His observations convinced him that labyrinthodonts could be categorized as primitive or advanced on the basis of their vertebrae as well as their skull (Fig. 6.14). The few vertebrae that were known from the Carboniferous period had centra composed of double rings: in each segment both the intercentrum and pleurocentrum encircled the notochord and constricted it. These vertebrae, described as embolomerous, Watson accepted as being primary for the labyrinthodonts since they were associated with the most ancient members of the group. In postcranial material of Permian age, Watson found that the typical vertebra was rhachitomous: neither part of the centrum formed a complete ring; the

Figure 6.14. Labyrinthodont vertebrae. **(a)** Rhachitomous type, from *Eryops*. **(b)** Embolomerous type, from *Cricotus*. **(c)** Stereospondylous type, from *Mastodonsaurus*. Abbreviations: *ic,* intercentrum; *na,* neural arch; *p,* pleurocentrum. *(a,c* from Romer; *b* from Goodrich.)

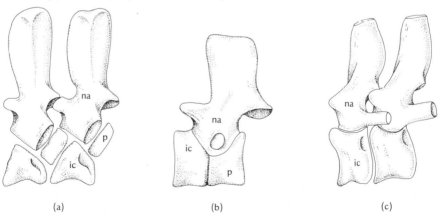

(a)   (b)   (c)

intercentrum, wedge-shaped, clasped the notochord from below, and the smaller pleurocentrum, consisting of separate left and right halves, braced the notochord dorsally. As the Permian labyrinthodonts gave way to the extreme forms of the Triassic period, the pleurocentra dwindled to cartilaginous elements and then disappeared entirely. The flat-headed amphibians at the end of the labyrinthodont line had stereospondylous vertebrae, in which the notochord was encased by expanded intercentra articulating firmly with one another. The sequence of vertebral types led Watson to conclude that the primitive labyrinthodonts, embolomeres, had given rise to the rhachitomes and that these forms had eventually produced the ungainly, doomed stereospondyls.

By fitting the hitherto uncorrelated labyrinthodont fossils into an evolutionary sequence, Watson reshaped paleontologists' understanding of a major group of the ancient amphibians. Subsequent research bore out his contention that the stereospondyls were a terminal branch of the family tree, but his idea that all primitive labyrinthodonts were embolomeres did not prove true. As Carboniferous labyrinthodonts became better known, it was evident that rhachitomous forms existed among them. Their characteristic vertebrae were not found at first, but skulls were unearthed that resembled those of Permian labyrinthodonts with rhachitomous centra. By 1947, Romer decided that he could not justify the maintenance of a separate order for the genera represented by these remains on the basis of Watson's supposition that, being of Carboniferous age, they could not belong to the Rhachitomi. Such a category, the order Phyllospondyli, had been invented earlier for tiny, gilled amphibians in some ways reminiscent of the rhachitomes and then used to house the increasing number of Carboniferous forms that looked more like rhachitomes than like proper embolomeres. It occurred to Romer that all the "phyllospondyls" were really rhachitomous labyrinthodonts; the gilled individuals, called branchiosaurs, were larvae, and the larger, early specimens were adults antecedent to the rhachitomes of the Permian period. Even before the discovery of their typical vertebrae in Carboniferous deposits confirmed the antiquity of the rhachitomes, Romer's hypothesis that they, not embolomeres, constituted the original labyrinthodont stock was strengthened by evidence from another source: Jarvik's studies showed that ichthyostegids, the oldest labyrinthodonts of all, had had rhachitomous vertebrae. Since the structure of their

Romer's modification of Watson's evolutionary scheme

centra could be traced to the ancestral rhipidistians, it seemed certain that the rhachitomous condition was the primitive one in labyrinthodonts and that embolomerous vertebrae were an early, but secondary, development.

Romer revised Watson's classification to reflect the division of the Carboniferous labyrinthodonts into two lines. The known rhachitomes of that period he assigned to the order Temno-spondyli, a group which also included the later rhachitomous and stereospondylous animals. The embolomeres he placed near the base of the order Anthracosauria, whose members split away from the ancestors of the temnospondyls and eventually gave rise to reptiles. If the fossil record presents an accurate picture, the temnospondyls were the more numerous and progressive of the two kinds of amphibians. In each period they produced populations highly adapted for the narrow niche which they occupied, and when environmental upheavals extinguished these overspecialized tribes, they managed until the end of the Triassic to replenish their kind. The hardiest temnospondyls exhibited the same evolutionary tendencies as the more vulnerable members of the group but did not express them to such an extreme degree. They seem not to have lost the robust body and supportive legs that enabled them to leave the water. *Edops* and its relatives, representatives of the early central temnospondyl stock, apparently remained sufficiently terrestrial to survive the droughts that stranded species whose legs had become too small to bear them. The descendants of the edopoid amphibians, whose skulls had changed in the way Watson described, were creatures like *Eryops*, strong and stocky animals that probably spent as much time out of the water as in it (Fig.

Figure 6.15. Restorations of two Lower Permian temnospondylous labyrinthodonts. **(a)** *Eryops*, an animal about 5 feet in length. (From Colbert.) **(b)** *Cacops*, a smaller form with a row of armor plates down its back. Posterior to the eye in *Cacops*, note the dorsally situated tympanic membrane. (Drawing by L. I. Price from Raymond.)

(a)   (b)

The Amphibians: Gaining the Land

6.15). Several lines of temnospondyls, derived either from eryopoids or their immediate predecessors, radiated landward early in the Permian period. They lost all trace of lateral lines in the skull roof, grew well-ossified appendages, and in one group, the dissorophids, even developed a row or two of protective armor plates down the back. By Permian time, however, the opportunity amphibians once had to live on land unchallenged had passed. The dissorophids and their allies surely had to compete with the emerging reptiles, and defeat in that encounter rather than grievous structural or physiological limitations probably brought about their extinction.

Despite the impermanence of the swamps, the temnospondyls were forced to take refuge in them. There they diversified, producing a number of families whose interrelationships paleontologists are still trying to determine. Tracing the evolution of the various groups is difficult because, in addition to the traits that distinguished them, temnospondyls of different kinds developed similar adaptive features. Since it is often impossible to discriminate between likenesses due to phylogenetic connection and those due to parallelism, hereditary lines are hard to define. The eryopoid descendants that survived beyond the Permian period showed gradual enlargement of the intercentra: paleontologists believe that they have identified correctly the sequence of forms leading from the rhinesuchoids, Permian temnospondyls with small pleurocentra, to the Triassic capitosaurs, great flat-headed animals in which intercentra alone enclose the notochord (Fig. 6.16). Several Triassic temnospondyls are known, however, which do not fit into this lineage. The trematosaurs, for instance, exhibited a parallel reduction of the pleurocentra, immobilization of the palate, and pairing of the occipital condyles, but they had high, often long-snouted skulls that were constructed somewhat differently from those of the rhinesuchoid animals. Romer suggested their derivation from the archegosaurs, early Permian rhachitomes that had evolved the long snout and jaws presumably as an adaptation for hunting fishes. Contemporary with the trematosaurs and rhinesuchoids were temnospondyls, with eyes far forward in the skull, whose origins are also not well understood. Watson sought to ally these forms, the brachyopids, with a much older stock, the trimerorhachoids, relatives of the eryopoids that had developed a short face as early as Carboniferous times. Romer objected to Watson's scheme because *Trimerorhachis* and its kin, despite their short faces, had skulls that were longer than those of the

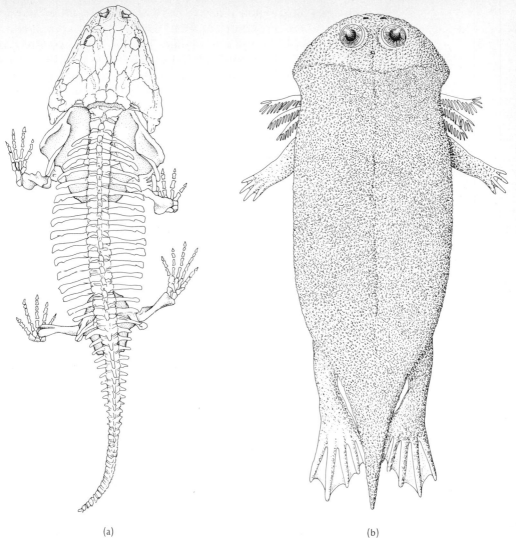

(a)   (b)

Figure 6.16. Triassic stereospondylous labyrinthodonts. **(a)** Skeleton of *Meto-posaurus* (=*Buettneria*), about 8 feet in length. (From Sawin.) **(b)** Restoration of *Gerrothorax*, a brachyopid, about 3 feet long. (From Nilsson.)

brachyopids and unlike them in the arrangement of the bones of the postorbital region. He pointed out, too, that millions of years elapsed between the time when the trimerorhachoids flourished and the brachyopids appeared. He explained the similarities between the two groups as another example of parallelism: in both lines the animals retained the underdeveloped snout of the immature branchiosaur stage as part of their adaptation for aquatic living. Two kinds of late Triassic stereospondyls, descended perhaps from the brachyopids, were also short-faced. That larval traits were maintained in these forms and in the earlier ones is borne out by the discovery of specimens with branchial arches and external gills. The short face is further evidence of the often repeated lapse into neoteny, it seems, rather than a clue that would help paleontologists untangle the family history of the swamp-bound temnospondyls.

Rarity of fossils of early temno-spondyls

Although relationships among the late temnospondyls are not always clear, enough fossils are known to elucidate the nature of the group as a whole. By contrast, the early Carboniferous animals are shadowy creatures whose remains have eluded collectors almost completely. The rare rhachitomes that have been discovered in Lower Carboniferous deposits are all of questionable relationship to the main stock. *Otocratia*, known from a single skull roof found in Scotland, resembled the ichthyostegids in lacking intertemporal bones and having nostrils very near the edge of the upper jaw. Certain temnospondyls of the first part of the Carboniferous period from Scotland, Nova Scotia, and Linton, Ohio, had oddly elongated orbits that mark them as divergent forms (Fig. 6.17). Called loxommids, the animals with the peculiarly shaped orbits have been described as rhachitomes which underwent at an early date the flattening of the head and the adaptation for bottom living that occurred again and again in temnospondyl history. For a while, paleontologists suspected that the remains of an amphibian taken from Lower Carboniferous strata at Greer, West Virginia, were those of a more generalized type. Romer, who studied the specimen, finally concluded that it did not represent the long-sought ancestor of the later temnospondyls. *Greererpeton*, as he named the form, had orbits set far forward, shallow otic notches, and no intertemporal bones, characteristics which rendered it unsuitable as a precursor for rhachitomes like *Edops*. The structure of the skull of *Greererpeton* led Romer to ally it to *Otocratia* and to two later Carboniferous amphibians, *Colosteus* and *Erpetosaurus*. The group stemmed, possibly, from the old ichthyostegid stock and represented a line evolving independently of the contemporary temnospondyls. The lengthening list of divergent families which existed during the first half of the Carboniferous makes it evident to paleontologists that a major diversification of rhachitomous amphibians

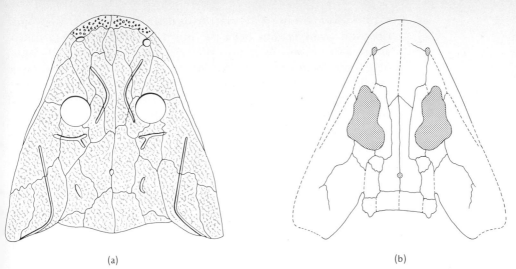

(a)   (b)

Figure 6.17. Skulls of early temnospondylous labyrinthodonts. **(a)** *Greererpeton,*
a colosteid. (From Romer.) **(b)** *Loxomma,* a loxommid. (From Watson.) Both
animals lived in mid-Carboniferous (Upper Mississippian) times. The restorations
represent their skulls as flatter and wider than they actually were.

had occurred near the beginning of the period or perhaps even
earlier. The still undiscovered ancestors of *Edops* and later forms
had apparently been part of a rich labyrinthodont fauna that is
only beginning to be known.

The anthracosaurs   In the Carboniferous and Permian periods, the labyrinthodont assemblage included the animals that Romer has classified as anthracosaurs. These amphibians resembled the temnospondyls superficially but showed a number of skeletal traits that
distinguished them as a separate group. Paleontologists recognize an anthracosaurian skull by its especially large tabular
bones: whereas in temnospondyls these elements lie beside the
postparietal bones, in anthracosaurs they extend forward to
border the more anterior parietals. In early anthracosaurs the
otic notch at the rear of the skull was slit-shaped instead of
scalloped or ovoid, as in other labyrinthodonts. In front of the
notch, the union of cheek and roof bones was so loose that the
two parts of the skull almost always separated after death.
Although the form of the otic notch changed in later anthracosaurs and the skull roof and cheek eventually became
firmly united, the skull remained generally more primitive in
the anthracosaurian amphibians than it did in the rhachitomes
and stereospondyls. With few exceptions, the anthracosaurian

skull never broadened and flattened. It retained the single occipital condyle and even the intertemporal bone that disappeared in numerous temnospondyls. The palate kept its movable articulation with the braincase and did not develop large medial vacuities at the expense of the pterygoid bones.

The anthracosaurs were less conservative in the structure of the vertebral column. They differed from their rhachitomous ancestors in evolving first enlarged, then fused pleurocentra that encircled the notochord. On the basis of centrum construction, two lines of anthracosaurs have been discerned: in one, the Carboniferous embolomeres, the intercentrum grew ring-shaped also, so that pleurocentrum and intercentrum contributed equally to the strength of the column; in the other, known from Permian representatives called seymouriamorphs, the intercentra gradually diminished in size until the pleurocentra were left as the principal supporting elements (Fig. 6.18). The two lines apparently diverged early in the Carboniferous period, for primitive members of both are already present in the deposits at Greer. *Proterogyrinus* seems well on its way toward the development of equal-sized, ring-form pleurocentra and intercentra, there being only a small area in the middorsal region of each element that remained unossified. *Mauchchunkia*, on the other hand, had wedge-shaped intercentra beneath the notochord and pleurocentra either in the form of rings around it or completely ossified disks.

It is impossible to tell from the fossil evidence why the anthracosaurs did not radiate as widely as the temnospondyls. Either the rhachitomes and stereospondyls were inherently more adaptable, or, by virtue of superior design, they curbed the spread of the anthracosaurs in every niche to which both had access. Whatever the reason, the anthracosaurs were a smaller and less enduring group than the other labyrinthodonts: the embolomeres became extinct at the end of the Carboniferous period, and the seymouriamorphs disappeared by the end of the Permian. The relative failure of the anthracosaurs as amphibians is paradoxical because the characteristics which distinguished the seymouriamorphs survived in the reptiles and played a part in their success. For a time, paleontologists speculated that the stout-legged seymouriamorphs might have been the first reptiles, but the discovery of a branchiosaur larva of a member of the group and of adult forms with typical amphibian modifications for aquatic life led them to keep the seymouriamorphs in the amphibian fold. Finding true reptiles

Anthracosaurs ancestral to the reptiles

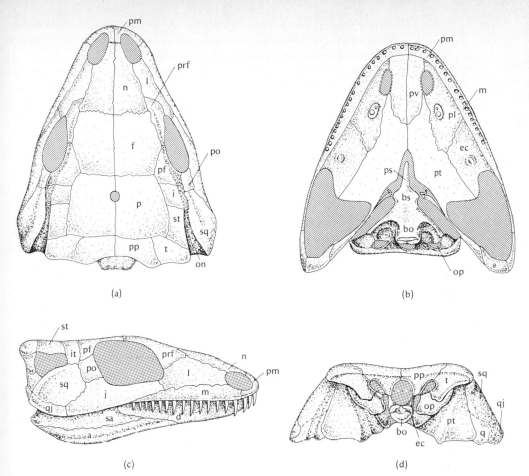

Figure 6.18. Skull of the Lower Permian anthracosaur *Seymouria*: **(a)** dorsal, **(b)** palatal, **(c)** lateral, and **(d)** posterior view. Compare structure with that of advanced temnospondylous forms shown in Fig. 6.13. Abbreviations as in Fig. 6.7. (From Romer.)

of Carboniferous age indicates that the first members of the new class must have sprung from an older part of the anthracosaur line, perhaps from relatives of the Upper Carboniferous *Gephyrostegus* (=*Diplovertebron*), a form with small intercentra and ring-shaped pleurocentra. If, as Hotton believes, the Lower Carboniferous *Mauchchunkia* is morphologically intermediate between *Gephyrostegus* and the Upper Devonian *Ichthyostega*, the branch of the anthracosaurs leading to reptiles may have separated very early from the rest of the assemblage. If that was the case, *Seymouria* and its Permian relatives were the very late and somewhat specialized remnants of a stock that had long ago given rise to more promising offspring.

Origin of modern amphibians is obscure

The origin of reptiles from archaic amphibians is better understood than the emergence of the frogs and toads, the salamanders and newts, and the caecilians (Fig. 6.19) that constitute the modern members of the class Amphibia. The caecilians, legless, worm-like animals which burrow in the soil of the tropics, are represented by only one specimen in the fossil record. The living forms have been set apart in the order Apoda (Gymnophiona) but are often thought of as having at least a connection with the salamanders and newts of the order Urodela (Caudata). Those animals and the frogs and toads which are members of the order Anura (Salientia) have been found as fossils, but known extinct forms are essentially modern in structure and give no hint of the older amphibians from which they have descended. Most of the urodelan and anuran fossils come from rocks of the Cenozoic era, which began only 65 million years ago. The specimens can usually be assigned to families that have living representatives or even to extant genera. Remains from the Mesozoic era are much more sparse: less than a dozen frogs and about the same number of urodeles have been described. Only one of these forms, the frog *Triadobatrachus* (=*Protobatrachus*) lived in the Triassic period (Fig. 6.20). The others are of Jurassic and Cretaceous age and already exhibit the skeletal modifications that distinguish the members of the modern orders from the labyrinthodonts and lepospondyls. Even *Triadobatrachus* had an essentially frog-like skull. Its iliac bones jutted forward, as they do in typical anurans, and the vertebral column had shortened, suggesting the reduction that was to characterize the later frogs. Although the two bones in the lower leg were not fused, as they are in modern forms, the hind limbs were larger than the front ones, an indication that the animal may have been able to jump. *Triadobatrachus* was a contemporary of the late temnospondyls but does not seem to have been closely related to any of them.

The lack of fossil specimens intermediate between anurans or urodeles and the older amphibians has forced paleontologists and students of the living animals to base their speculations

Figure 6.19. A caecilian, *Siphonops mexicana.*

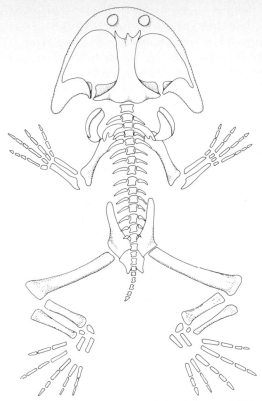

Figure 6.20. A restoration of the skeleton of *Triadobatrachus (= Protobatrachus)*, a Lower Triassic amphibian thought to be a primitive frog. (From Romer.)

about the evolution of the group upon evidence from the anatomy and embryology of modern species. This approach has presented difficulties that have so far proved insurmountable. The structure of the existing amphibians is so specialized that the more generalized condition from which it derived is almost completely obscured. Efforts to learn something about the early history of either urodeles or anurans by studying immature forms have been nearly fruitless because of the shortcuts which have appeared in the development of the embryo and, in the case of many anurans, the adaptation of the larvae for a special environmental niche. The frequency of neoteny among the salamanders and parallel evolution in different lines of both groups have added to the confusion. Nevertheless, there are certain questions which paleontologists and biologists still try to answer. They would like to determine whether the three modern orders arose independently from Paleozoic ancestors or whether they sprang from one stock and so constitute a natural group. They would also like to know from which of the archaic amphibians the living forms have come and how, within each order, the families are interrelated.

Diphyly or mono-phyly of the mod-ern amphibians

To ascertain whether or not anurans, urodeles, and apodans are monophyletic, investigators have scrutinized the differences and similarities in their organ systems and, where possible, have compared them to fossil forms. Workers, like Schmalhausen, who judge the modern animals to be diphyletic argue that members of the different orders show unmistakable affinities to separate groups of Paleozoic amphibians and that these affinities are more significant than mutual similarities which could be due to parallelism. The advocates of monophyly take the opposite view: they contend that the large number of peculiar characteristics which anurans, urodeles, and apodans share is strong evidence of their common ancestry. They question the conclusive resemblance of these forms to unlike Paleozoic types and minimize the importance of the structural differences which set the three groups apart. Parsons and Williams, who have declared themselves for monophyly, attribute the salient differences among the modern amphibians to their specialization for different modes of life. While the urodeles have maintained the long, muscular body and tail that characterize animals which undulate the body as they move, the anurans have turned to a peculiar form of locomotion dependent upon great enlargement of the hind legs. The caecilians evolved secondarily the solidly ossified skull and limbless trunk that are advantageous in burrowing. Parsons and Williams point out that despite the radical modification of the skeleton associated with their diverse habits, urodeles, anurans, and caecilians have in common certain traits that are not found in any other vertebrates. For instance, the crown and base of each tooth are separated by a narrow band of uncalcified dentin or fibrous tissue, a condition closely paralleled only in one isolated family of teleost fishes. In the middle ear, the end of the stapes implanted against the fenestra ovalis has, either next to it or fused with it, a bit of bone derived from the embryonic otic capsule. In the sensory portion of the ear, there is a patch of receptor cells so peculiar to the amphibians in its location that it is called the papilla amphibiorum. Other parts of the sensory and nervous systems of modern amphibians contain similar structures as well. Eyes, except in the caecilians, where they are extremely reduced, contain unusual light-sensitive elements known as green rods. The brain in all three types seems simplified in the same manner, the cerebellum being strikingly reduced even in the highly active frogs and toads. Anatomists acknowledge, also, that no other vertebrates have fat bodies anterior to the gonads for storage of reserve nutriment or glands in the skin like the mucous glands and poison glands of the amphibians.

The Lissamphibia and the "proto-lissamphibian"

Convinced that the unique structures shared by modern amphibians are too numerous to attribute to coincidental independent evolution, Parsons and Williams regard the living animals as a natural group, which they have called the Lissamphibia. In an effort to locate among the archaic amphibians the group that might have served as the common stem of this assemblage, they have listed the traits that the ancestral animals should have had. The "protolissamphibian" that emerges from their description, however, does not correspond closely to any known fossil type. No specimen presently in paleontologists' collections has the additional middle ear bone, the peculiar teeth, and the extra bones in the lower jaw that Parsons and Williams assumed would have appeared in the precursor of the lissamphibians. Only one form, a lower Permian rhachitome, which Bolt described in 1969 and named *Doleserpeton*, seems possibly a representative of the protolissamphibian stock. A small upland animal with strong legs and a compact body, *Doleserpeton* had teeth like those of modern amphibians and showed no peculiar characteristics in the skull or postcranial skeleton that would have debarred it from the ancestry of the extant animals. The vertebrae of *Doleserpeton* still possessed both intercentra and pleurocentra, but one of the elements (Bolt presumes it is the intercentrum) is much reduced. Its disappearance in the descendants of *Doleserpeton* would have left a single, cylindrical centrum in each vertebra, like that in the lissamphibian structure.

Before the discovery of *Doleserpeton*, the known temnospondyls, anthracosaurs, and lepospondyls had each been recognized as showing some similarity to the lissamphibian plan but also exhibiting traits or trends which excluded them from the lineage of the modern animals. When the skull was considered, the temnospondyls seemed the group most likely to have given rise to the three orders of modern amphibians. The advanced rhachitomes and stereospondyls evolved the broad, flat skull with two occipital condyles and the immobilized, open palate that are characteristic of the living animals. The roof of the

The Amphibians: Gaining the Land 243

temnospondyl skull was generalized enough in many forms to have served as a precursor for that of the later amphibians, but it showed no tendency toward the extensive loss of bones that was to leave vast orbital spaces and openings in the cheek region of most modern members of the class. The structure of their vertebrae made the temnospondyls less probable as ancestral stock. Whereas the temnospondyls exhibited separate intercentra and pleurocentra until the latter elements disappeared late in the history of the group, the modern amphibians appear lepospondylous upon their entrance into the fossil record. It is possible that the protolissamphibians underwent rapid solidification of the centra, as the lepospondyls may have done at an earlier time; but if Williams is correct in believing the modern amphibian centrum to be a pleurocentrum, the anthracosaurs with enlarged pleurocentra would make better ancestors than the temnospondyls, in which those elements tended to diminish. The skull of known anthracosaurs gave no hint, even in the Permian period, of evolving lissamphibian characters, however. It not only retained all the bones of the roof but also, unlike the temnospondyls and modern amphibians, kept the primitive labyrinthodont palate and single occipital condyle.

Schmalhausen's ideas concerning lissamphibian origins

No one has supported the monophyletic origin of the lissamphibians from the Carboniferous lepospondyls, although some paleontologists have suggested that the urodeles and apodans sprang from that source. Schmalhausen, who upheld the latter theory, agreed that aistopods and nectrideans were too specialized to have served as ancestral stock but believed that microsaurs may have done so. Except for their large eyes, the microsaurs, in Schmalhausen's opinion, were adapted for burrowing or at least concealing themselves under moist vegetation or mossy stones. He regarded the well-ossified skulls in microsaurs and apodans not as parallel developments for similar behavior but as evidence of common lineage, pointing out that other resemblances between the two groups existed in the structure of their vertebrae, rib heads, and scales. The reduced number of bones in the skull roof of apodans Schmalhausen viewed as no objection to relating apodans and microsaurs. The appearance in the embryonic apodan skull of numerous separate ossifications which are later lost through fusion with neighboring ones seemed to him additional proof that the modern animals evolved from the generalized microsaurs, the only lepospondyls to maintain a normal complement of dermal elements into the Permian period. The connection of the microsaurs and urodeles he postulated on the basis of their common possession of a first vertebra with a uniquely shaped surface for articulation with the back of the skull. He imagined the salamanders, with their long bodies and small legs, as retaining even more faithfully than the degenerate apodans the outward appearance and habits of the ancestral group.

Schmalhausen found no significant similarities between microsaurs or any other lepospondyls and the anuran amphibians. He concurred with Watson that the frogs and toads were much more likely to have come from labyrinthodont stock. He accepted Watson's suggestion that *Triadobatrachus*, the Lower Triassic frog, had descended from animals like the eryopoid temnospondyl *Amphibamus*, which, with their somewhat shortened trunk, relatively strong legs, and slightly flattened head, were beginning to assume anuran form. Although Carroll cited reasons for regarding *Amphibamus* as an incipient terrestrial dissorophid, Schmalhausen (who worked in the Soviet Union and may not have had access to Carroll's publication) believed that the Carboniferous form was becoming capable of the agile movement which was to save its protoanuran descendants from the predaceous reptiles that stalked them. Had he lived to see Bolt's *Doleserpeton*, Schmalhausen might have agreed that the dissorophoid Lower Permian rhachitome with its pedicellate teeth was an even closer relative of the first anurans.

Classifying the modern amphibians

In Schmalhausen's opinion the growing menace of the reptiles on land and the secondarily aquatic reptiles and stereospondyls in the water was important in shaping the course of the evolution of the modern amphibians. The early urodeles and anurans which did not go underground permanently, like the apodans, survived only because they were able to hide or live where the other animals could not go. Schmalhausen believed that the Permian and Triassic forebears of the modern forms migrated into streams too shallow for the ponderous temnospondyls and to uplands too cold for the sun-loving reptiles to tolerate. Their confinement to such areas throughout the Mesozoic era he regarded as borne out by their virtual absence from the fossil record, for highland animals are rarely preserved. When their tetrapod enemies disappeared from the waters downstream during the Cretaceous period, a number of anurans and urodeles returned to the habitat of their ancestors, but many species, including the most primitive frogs, remained in mountainous regions. As additional evidence for this theory, Schmalhausen pointed out that the tadpole form of the larva, universal among anurans, would have evolved only as an adaptation for existence in swift currents of upland brooks: the streamlined shape and sucker-like organs of attachment of the newly hatched animal enabled it to feed on algae without being battered by the water or swept away. The less radical modification of the urodele larvae Schmalhausen attributed to their always having remained on the bottom to feed, where currents were less strong.

Whether or not Schmalhausen's speculations are correct, it does seem that diversification within the anuran and urodelan groups took place where conditions were not conducive to fossilization. Since forms transitional between primitive types and more advanced ones are not available, investigators have had difficulty in constructing a phylogenetic scheme to underpin the classification of the modern animals. Like students of lissamphibian ancestry, biologists interested in the subsequent evolution of the anurans and urodeles have had to rely on *Determining rela-* analysis of the characters of the living species. In categorizing *tionships among* the frogs and toads, Noble took as his guide the shape of the *anurans* vertebral centra. He decided that the amphicoelous form, in which both ends were concave as in fishes, was more primitive than either opisthocoelous centra, with a concavity only at the back, or procoelous ones, in which the anterior surface was depressed. He regarded bell toads (ascaphids) of New Zealand and the northwestern United States as representing the lowest level of anuran differentiation, because besides having amphicoelous centra, they retained an extra trunk vertebra, free ribs, and certain tail muscles not present in other anurans. The opisthocoelous forms he viewed as being somewhat more advanced in that they had lost the trunk vertebra and tail muscles and, in the curious tongueless toads, the free ribs as well. The large number of frogs and toads with procoelous vertebrae seemed to be the most highly evolved of the anurans, and Noble set them at the top of the scale.

More recently, Orton pointed out that anuran larvae could be sorted into four types and suggested that the different kinds of tadpoles were a key to anuran diversification. Although many investigators agree that the tadpole is preferable to the shape of the centrum as a basis for classification, they have not been able to ascertain the evolutionary relationship among the anurans more surely than before (Fig. 6.21). Most workers assume that tadpoles with simple mouthparts and a pair of spiracles are more primitive than tadpoles with horny lips and a single,

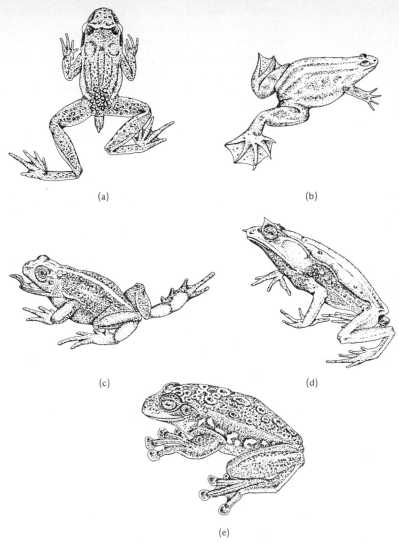

Figure 6.21. Recent anurans. **(a)** *Ascaphus*, an ascaphid. **(b)** *Xenopus*, an aglossid. **(c)** *Nectophrynoides*, a bufonid. **(d)** *Megophrys*, a pelobatid. **(e)** *Polypedates*, a tree frog. (From Noble.)

midventral spiracle, but Tihen argues that the reverse is possible. He does accept as an advanced condition the appearance of a single spiracle on the left side. Since the sinistral spiracle characterizes the tadpoles of Noble's procoelous forms, it does seem as though the procoelous anurans have been correctly identified as the most progressive members of the order. Hecht speculates that these animals became predominant during the last great radiation of the frogs and toads at the end of the Jurassic period and that anurans with other types of tadpoles are survivors of earlier populations the procoelous forms replaced. No one believes that any frog or toad now living can be likened to the ancestor of any other. Since a procoelous frog is known from the Jurassic period, paleontologists suppose that anurans of the major lines diverged from a common ancestral stock that existed at the beginning of Mesozoic time. Whether this group included allies or descendants of *Triadobatrachus* will be clear only when more fossil material becomes available.

Difficulty of discerning relationships among the urodeles

The history of the urodeles is even more obscure than that of the frogs (Fig. 6.22). Biologists recognize the living newts and salamanders as belonging to eight families but can say little or nothing about their interrelationship. Although it was suspected that animals with external fertilization were less advanced than those in which fertilization is internal, the supposition that families whose members fertilize their eggs externally were phylogenetically distant from the rest ran counter to the anatomical evidence. The tendency toward neoteny was also found unreliable as a guide to relationship. The sirenid salamanders, or "mud eels," of the United States retain larval gills and are permanently aquatic like the European *Proteus* and American "mud puppy" *Necturus*, but the unique specializations of the sirenids preclude their being allied closely to the latter animals. The "congo eel" *Amphiuma*, another salamander which retains gills and gill clefts, appears related to the gill-less salamandrids and the gill-less and lungless plethodontids on the basis of other characters. The idea that neoteny might be a primitive trait was put forth at first because the oldest known salamanders were elongated, aquatic forms. However, Estes has guessed that if Mesozoic deposits from drier terrain could be examined, remains of more terrestrial urodeles would be discovered. Neoteny was, in his opinion, an adaptation which occurred in several families of urodeles independently, not a primary characteristic of the group. Extinct neotenic salamanders may have been as distantly related as the living ones seem to be. Most of them belonged to modern families, and there is no evidence of how recently the families diverged or which ones shared a common lineage. All but two of the existing families have been traced to the Cretaceous period, when an abundance of streams and swampland fostered the proliferation of urodele amphibi-

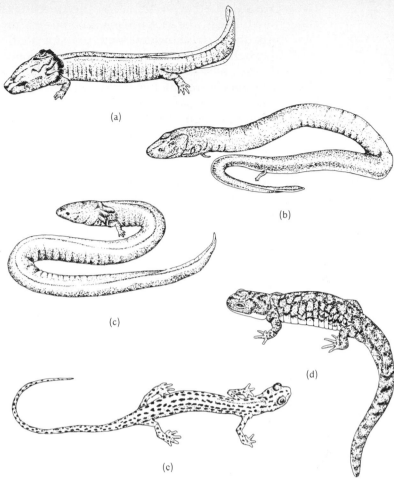

Figure 6.22. Recent urodeles. **(a)** *Necturus*, a proteid. **(b)** *Amphiuma*, an amphiumid. **(c)** *Siren*, a sirenid. **(d)** *Ambystoma*, an ambystomid. **(e)** *Eurycea*, a plethodontid. (From Noble.)

ans. Except for a late Jurassic specimen, the fossil record before the Cretaceous period is blank. With neither fossil evidence nor a consistent key from the anatomy of the living animals to help them, investigators cannot even speculate upon the radiation that produced the modern forms.

Fate of the Am-
phibia

Although it has been impossible to follow the evolution of the lissamphibians in detail, it is plain that the group as a whole has become highly adapted for life in the limited areas where amphibian physiological requirements are satisfied and competition with other vertebrates is not too intense. In abandoning fins for legs, the amphibians relinquished the ability to survive broadly in fresh waters and perhaps to find a wider range through migration to the sea. The lepospondyls and labyrinthodonts which kept the unprotected embryos of their piscine ancestors were unable to accompany the more advanced tetrapods to regions beyond the wet lands they originally inhabited. Caught, finally, between the vigorous fishes that held the waters and the reptiles and mammals superiorly fitted for the land environment, the amphibians were reduced to forms of small size and secretive behavior. Whether the lissamphibians represent animals on their way toward extinction or forms that will hold their place as long as any vertebrates is another question that no biologist would attempt to answer.

7

The Rise
and Fall
of the Reptiles

Progressive adaptation for terrestrial life in early tetrapods

Despite the sudden appearance of the four-footed ichthyostegids among late Devonian fossils, adaptation of the vertebrate body for life on land was a gradual and halting process. The protoamphibians, after crossing one of the most formidable boundaries that living things encounter, gave rise to animals that evolved new structures for terrestrial living at an unequal rate. In one tetrapod assemblage, fundamental modification of the body plan virtually ceased. To be sure, superficial variations continued to occur, and older genera were replaced by new ones, but the amphibian level of organization, with all its limitations, was never transcended. In another lineage, new characters appeared in combinations which raised the terrestrial adaptation of the vertebrates to a new height. The two innovations which were perhaps most crucial in effecting the transition from the amphibian to the reptilian state were the cornification of the skin and the protection of the embryo by membranes and a shell. The first, by preventing transcutaneous passage of water, made it possible for tetrapods to live their lives in exposed uplands, in arid regions, and even in salt water—surroundings in which amphibians would have lost their body fluid in a short period of time. The second, achieved by the evolution of what is known as the "land egg," enabled reptiles to reproduce under conditions of minimal moisture and thus to live permanently in the many areas to which their new skin gave them access.

Innovations in the organ systems of reptiles

The evolution of the reptiles involved far more than changes in the structure of the skin and the housing of the embryo, however. Dissection of extant forms and deductions from fossil evidence indicate that each of the organ systems underwent progressive modifications and eventually, in some lines at least, permitted activities far beyond the range of amphibian capabilities. Casts of fossil reptile skulls reveal an increase in the size of the cerebrum relative to other parts of the brain, and histological studies of brains of living reptiles demonstrate the presence of nerve pathways unknown in lower tetrapods. As it evolved, greater complexity of central nervous connections surely permitted the development of more intricate behavioral patterns which had survival value. Associated with new habits there were changes in the viscera. Zoologists assume that, in primitive reptiles, the ventricle of the heart became partially divided by a septum so that oxygenated and unoxygenated blood were less readily mixed than they had been in amphibians (Fig. 7.1). As a result, the brain, supplied by arteries that stemmed from the left side of the heart, received blood carrying a maximum supply of oxygen.

Aerated blood was not wastefully recirculated through the capillaries of the lung wall, because the pulmonary vessel drew an almost entirely venous stream from the right side of the pumping chamber. With gills and thin skin abandoned forever, the lungs showed a tendency to increase their respiratory surface. Although the lung walls may have developed a minimum of infolding in some reptilian orders, in others considerable subdivision into alveoli occurred and air sacs paralleling those of birds evolved. The kidneys, which had shared a pair of ducts with the reproductive system in male amphibians, acquired their own passageways to the exterior. In animals of both sexes, a pair of channels, the ureters, advanced inward from the cloacal wall and, invading the substance of the kidney, sprouted numerous collecting tubules which drained away the urine. Renal tissue homologous to the anterior part of the amphibian kidney, not reached by the newly evolved collecting tubules, degenerated after its characteristic appearance in the embryo.

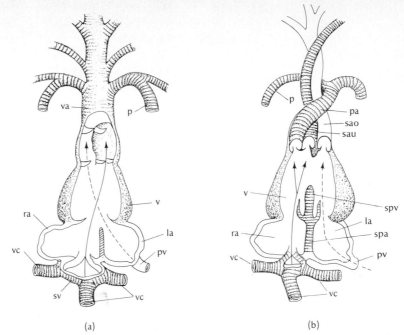

(a)   (b)

Figure 7.1. Diagram of the heart of **(a)** an amphibian and **(b)** a reptile, sectioned to show interior divisions. Arrows indicate the pathway of the blood through the organ. Abbreviations: *la*, left auricle; *p*, pulmonary artery; *pa*, pulmonary aorta; *pv*, pulmonary vein; *ra*, right auricle; *sao*, systemic aorta carrying oxygenated blood; *sau*, systemic aorta carrying unoxygenated blood; *spa*, interauricular septum; *spv*, interventricular septum; *sv*, sinus venosus; *v*, ventricle; *va*, ventral aorta; *vc*, systemic vein. (Modified from Goodrich.)

Like the heart and lungs, the kidneys may have been at first slightly more efficient than the amphibian organs and eventually, in some reptilian lines, much more so. The kidneys in living reptiles highly adapted for terrestrial life are able to conserve water by excreting nitrogenous wastes preponderantly in the form of urea and uric acid rather than as ammonia, a highly toxic compound which must be diluted in large amounts of fluid. To what extent the kidneys of ancient reptiles were capable of functioning in this way cannot be known: changes in the gross form of organs are not always strong clues to their physiological sophistication. It is safe only to say that the modifications of the soft organs that were part of the transition to the reptilian level endowed the new stock with greater potential for terrestrial living than the first tetrapods possessed.

Like contemporary amphibians, modern reptiles are mere remnants of their class and thus offer a limited amount of information about the physiology, structure, and behavior of the entire group. The traits shared by snakes, lizards, crocodiles, turtles, and the tuatara, *Sphenodon*, have been accepted as distinctive of reptiles in general, but the sum of the characters of these forms falls far short of demonstrating the adaptive range of the reptilian assemblage. In the 250 million years which elapsed between the appearance of the first reptiles and the establishment of the modern forms, species evolved with almost every imaginable kind of locomotor device and with teeth suitable for eating all sorts of organisms, plant and animal. It is probable, moreover, that the full extent of reptilian versatility is unknown, for the fossil record is incomplete for reptiles, as it is for other classes of vertebrates.

Versatility of the reptiles

Although paleontologists are aware that their understanding of extinct reptilian life is based for the most part upon forms which died in water or near it, they believe that they have been able to discern the broad outlines of reptilian evolution with reasonable accuracy. That the forerunners of the reptiles came of early rather than late amphibian stock they are sure, because primitive members of the new class have been found in deposits dating from mid-Carboniferous times. The first reptiles seem to have arisen as the labyrinthodont amphibians were entering their most progressive phase. Both kinds of animals inhabited the well-watered ground of the coal forests and produced descendants which vied for the land. Although the labyrinthodonts, especially the rhachitomes, had already developed some large, stout-legged forms, the reptiles secured a place for themselves. Like many newcomers in an established community, they were small and at first no threat to the older tetrapods. Lightly built and rather lizard-like, they differed from the amphibians in showing no tendency toward flattening of the skull or weakening of the limbs. They seem to have been alert little animals with big eyes and sharp teeth whose slender, long-footed legs enabled them to outrun their competitors. The primitive captorhinomorphs, as these earliest reptiles were called, soon gave evidence of their success: they produced some large forms, diversified, and spread into new territories.

Establishment of the reptiles

By early Permian time, the captorhinomorphs had become formidable animals that lived in deltaic regions. They had evolved special features of the skull and postcranial skeleton which fitted them more closely for their particular environment but which ultimately made them less adaptable than other advancing tetrapods. Their four-footed contemporaries included, besides the still flourishing labyrinthodonts, several varieties of newer reptiles, called pelycosaurs, which had originated from basal captorhinomorph stock. These reptiles were the dominant members of their class in the first part of the Permian period, having diversified into three or four lines each of which exploited a different food supply (Fig. 7.2). One group of pelycosaurs refined the ancient habit of catching fish. Two others, apparently closely related, evolved blunted teeth for crushing plant material. The most progressive forms became carnivores, preying in all probability upon their herbivorous cousins. The pelycosaurs are known primarily from fossils found in North America, as are the captorhinomorphs, but rare remains uncovered in western Europe and Russia prove that these early reptiles attained a wide distribution in continental areas.

Pelycosaurs and therapsids

The pelycosaurs were superseded by their own descendants, the therapsids. Paleontologists have identified these animals from deposits in South Africa: discoveries of numerous fossils in Upper Permian beds in the Karroo basin there have given investigators a peephole view, as it were, of a fauna that had become increasingly land-based. Although some forms still frequented marshes, the majority were no longer dependent upon aquatic plants and animals for food. They developed on a grander scale the pattern of feeding that had appeared among the most advanced pelycosaurs: herbivorous species evolved that fed upon terrestrial vegetation, and carnivores continued beside them, subsisting upon their flesh. Some of the therapsids were small animals, but many of them grew 10 feet long or more. The herbivores, especially, often became hulking beasts that roamed the open land in great numbers. Unchallenged by other tetrapods, the therapsids occupied regions of the Northern Hemisphere as well as the Southern until most of their kind disappeared during the first half of the Triassic period. Forms that escaped extinction at that time lost their dominance to another race of reptiles but lived to produce the ancestral stock of the mammals before succumbing in their turn.

The decline of the therapsids left the way open for the expansion of reptilian groups that had hitherto been of modest size. Many of the animals belonging to these groups had lived through the Permian period as conservative but not retrograde lizard-like forms. After resisting successfully the stresses that

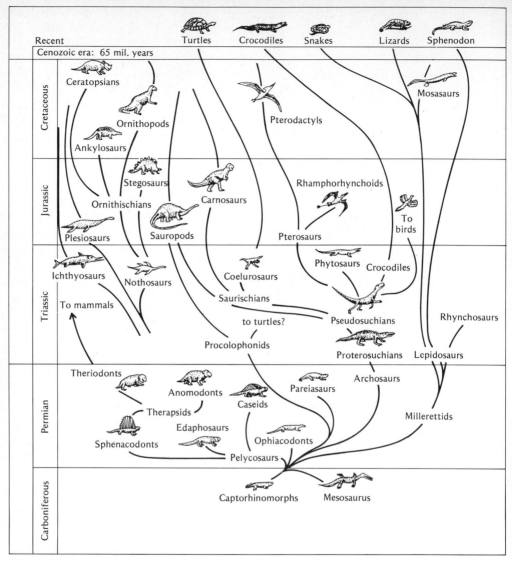

Figure 7.2. The radiation of the reptiles from their origin in the Carboniferous period to the extinction of their greatest forms at the end of the Cretaceous period. Reptiles belonging to lines which continued through the Cenozoic era are represented at the top of the scheme. Drawings of the various animals are not to scale.

destroyed more specialized reptiles, they diversified and multiplied rapidly during Triassic time. By far the most explosive radiation stemmed from a tribe known as the thecodonts, reptiles whose teeth were set in deep sockets rather than fused to the jaw. When the Mesozoic era opened, these animals were already distributed in wet lands and dry. The earliest of the thecodonts that took to the water had established themselves there in competition with the older labyrinthodonts and semi-aquatic therapsids. The proterosuchians, as these water-loving thecodonts were called, and their more terrestrial relatives, the pseudosuchians, together spawned the vast array of archosaurian reptiles which dominated the fauna of the Mesozoic years. A profusion of genera succeeded one another in every niche. The long-jawed, carnivorous proterosuchians gave way to similarly adapted forms, the phytosaurs, and these animals were eventually replaced by the ancestors of the modern crocodiles. The pseudosuchians, small- to medium-sized thecodonts capable of walking on four legs or running on two, claimed the land until they lost it to the dinosaurs. The latter reptiles, like the therapsids before them, became differentiated into herbivorous and carnivorous types. The vegetarians prospered as new and lush plants evolved, keeping the meat eaters well provided with food. As the Mesozoic era wore on, the dinosaurs produced the enormous and highly specialized forms for which the group is famous. Many of the larger herbivores wallowed offshore in lakes and rivers, where the water at once buoyed them up and protected them from their voracious enemies. Others developed armor, bony neck frills, or tail spikes, which made them difficult or dangerous to attack.

While the dinosaurs stayed on land or in freshwater shallows, some of the archosaurs became airborne and other kinds of reptiles entered the sea. One group of archosaurs that gained the air sired the birds. Less progressive types produced the Jurassic rhamphorhynchoids and Cretaceous pterodactyls, the pterosaurs which stayed within the reptilian fold and confined their flight to gliding. Although the flying reptiles diversified somewhat, finally evolving great crested forms like *Pteranodon*, they were a relatively homogeneous group. The reptiles that

lived in the sea, however, showed considerable variety. By the middle of the Triassic period, at least three kinds of marine forms had appeared. The placodonts, long-tailed but stocky animals with legs little different from those of terrestrial reptiles, paddled through the water hunting mollusks, which they cracked apart with massive teeth. In deeper but still coastal waters, perhaps, the nothosaurs swam, still paddling with large, five-fingered hands and feet. With a slenderer body than that of the placodonts, a longer neck, and sharper teeth, they were readapted for catching fish. Their descendants, the plesiosaurs, developed more blade-like appendages by increasing the number of the toe and finger bones, but neither the nothosaurs nor the plesiosaurs ever became as highly specialized for ocean life as the third group of marine reptiles which appeared in the Triassic, the ichthyosaurs. These animals were entirely fish-like in form. Their appendages had been transformed into flippers, the pectoral pair larger than the pelvic pair, and the body streamlined and reequipped with a dorsal and a caudal fin. Their thorough remodification for swimming benefited them for a relatively short time: their numbers dwindled during the Cretaceous period even as the plesiosaurs were becoming tremendous monsters and other reptiles like the mosasaurian lizards and the turtles were finding new living space in the sea.

The reptiles of the Mesozoic era, as widely dispersed and as well adapted as they were, did not survive the end of the Cretaceous period in large numbers. When the Cenozoic era dawned, the sea was clear of all the reptiles that had returned to it except the turtles. No gliding pterosaurs remained in the air, and on land the archosaurs had been almost entirely swept away. Only the crocodiles still existed, forms which had kept conservatively the amphibious habits of the most primitive tetrapods. Of the wholly terrestrial reptiles, none was left except the lizards, the newly evolved snakes, and *Sphenodon*, animals which lived the kind of secretive lives that had allowed their ancestors to survive while more powerful land animals held the field. No new radiation of the reptiles took place from the stocks that weathered the rigorous conditions prevalent at the close of Mesozoic time. Confined within narrow limits by the ascendant mammals, the reptiles became a relict group.

Recognizing Carboniferous reptiles One of the most difficult tasks faced by paleontologists who investigate reptilian history is distinguishing the first reptiles from the amphibians among which they lived. Since reptiles came originally of amphibian stock, the more generalized members of the two classes living in the Carboniferous period had many characteristics in common. Transitional forms, especially, constitute a problem in classification. Paleontologists would like to base their decision that an animal had reached reptilian status on its mode of reproduction: if a species produced eggs that were fertilized internally and gave rise to an embryo enclosed in an amniotic membrane, it would be judged a reptile regardless of how closely it resembled an amphibian in outward form. Unfortunately, it is impossible to use the attainment of internal

fertilization and a protected embryo as a criterion. The reproductive habits of extinct animals are unfathomable, even
through the use of indirect evidence, and the "land eggs"
themselves delicate structures which were not often fossilized.
Preserved eggs that date from Cretaceous time have been
discovered and associated with dinosaurs, but the few eggs that
are known from older deposits are harder to assign. The oldest
egg in paleontologists' possession was found in the Texas red
beds formed during the early part of the Permian period (Fig.
7.3). It might have been laid by any of the animals whose
remains were found in those beds and so proves nothing about
the reptilian status of doubtful forms.

To distinguish the early reptiles, paleontologists have had to
rely solely upon osteological criteria. They have defined skeletal
patterns that depart from the amphibian plan and verge on that
of the later reptiles, about whose identity there is less question.
When postcranial material of a specimen is available, they note

Structure of rep-
tilian vertebrae
the construction of the vertebrae: the reptilian structures were
stronger than those of the labyrinthodont amphibians because
of the increased ossification and fusion of their component parts
(Fig. 7.4). The pleurocentra were enlarged until they alone were

Figure 7.3. Oldest known reptilian egg, found in Texas red beds dating from
early Permian time. (From Romer.)

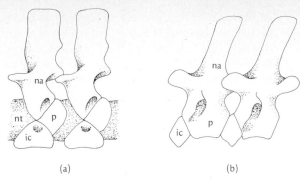

(a)   (b)

Figure 7.4. Diagram of trunk vertebrae, showing component elements. **(a)** *Eryops*, a labyrinthodont amphibian. **(b)** *Sphenodon*, a modern reptile of archaic form. Abbreviations: *ic*, intercentrum; *na*, neural arch; *nt*, notochord; *p*, pleurocentrum. (Modified from Jollie.)

the supportive elements of the column. Intercentra remained only as nubbins beneath the notochord, bracing the hemal arches which still protected the major blood vessels in the tail. In adult reptiles, the pleurocentra were routinely fused to the neural arches instead of articulated with them. Solidification of each vertebra reduced the flexibility of the vertebral column somewhat, but the increased rigidity was not disadvantageous for a terrestrial form. Modification of the first two vertebrae assured mobility of the head. The lowering of the neural spine of the first vertebra, or atlas, allowed the skull to rock up and down more freely than it had in amphibians. The second vertebra became an axis, articulating with the atlas in such a way that a twisting motion was possible and producing a large, compressed neural spine that anchored muscles which elevated the head.

Detecting primitive reptiles on the basis of these vertebral characters is not easy, because seymouriamorph amphibians, allied closely to protoreptilian stock, evolved similar traits. Adding evidence from observation of the appendicular skeleton frequently does not make a surer diagnosis possible. In both the amphibians and early reptiles the proportions of the limb bones varied with the weight of the animal. In a lightly built form, the bones tended to be thinner; when the body was bulkier, the bones of the appendages were more squat. Since the first reptiles continued to walk like the amphibians, with the body slung between sprawled legs, the design of the girdles initially was little changed. The pectoral and pelvic girdles kept the broad ventral

The appendicular skeleton in reptiles

surfaces to which were affixed the massive muscles required to keep the upper limbs horizontal and the body from sagging between them (Fig. 7.5). As time went on, however, the reptilian appendicular skeleton did become different from the amphibian one. In the pectoral girdle, the cleithral bones that amphibians had retained from the dermal shoulder shield of fishes disappeared entirely, and the endochondral elements to which the limb bones attached grew more elaborate. The scapula always ossified independently of the coracoid, as it had done in seymouriamorphs exceptionally among the amphibians, and a second coracoid bone formed behind the first in several early reptilian lines.

The pelvic girdle evolved distinctive hallmarks as the reptiles became increasingly terrestrial. The iliac bone gradually broadened its articulation with the vertebral column. Whereas the amphibians had one sacral vertebra or at most two, in reptiles the ilium established contact with at least two vertebrae and, in forms that walked on their hind legs, often with many more. In therapsid and archosaurian reptiles, both bipedal and quadrupedal, the hind limbs were drawn more directly beneath the body, and the musculature was adapted to hold them in that position. The underside of the pelvic girdle then changed markedly. The pubis and ischium became less plate-like, and a space appeared between them through which muscles could pass to gain attachment to their upper side. In Permocarboniferous specimens that still walked with limbs outspread, reptilian character was better ascertained from the lower leg

Figure 7.5. The pectoral girdle in **(a)** *Eryops*, a primitive amphibian; **(b)** *Dimetrodon*, a primitive synapsid reptile; and **(c)** *Iguana*, a modern lizard. Abbreviations: *ac*, anterior coracoid; *cl*, clavicle; *cor*, coracoid; *cth*, cleithrum; *glen*, glenoid fossa (for reception of head of humerus); *icl*, interclavicle; *sc*, scapula. (From Romer.)

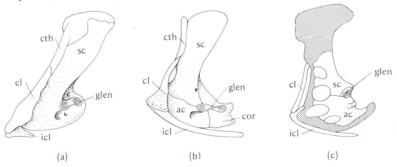

than the girdle. The digits in the reptiles of that time remained five in number, and the outer three, especially, grew longer than those of amphibians by increasing the number of phalangeal bones from three to four or five. The bones in the wrist and ankle showed a tendency to consolidate that was to continue in higher vertebrates (Fig. 7.6). In certain of the earliest reptiles, the three tarsal bones immediately below the tibia fused to form an astragulus. This element, with the more lateral calcaneus, became immovably interlocked with the tibia and fibula in several groups of early reptiles, so that the functional ankle joint was located between the astragulus and calcaneus and the more distal tarsal elements. Exceptionally, some thecodonts developed motion between the astragulus and calcaneus. In pelycosaurs and their descendants, however, the bones were arranged so that the midtarsal region was firm and the foot flexed at the articulation between the ankle and the lower leg.

Figure 7.6. Comparative structure of the foot and ankle in amphibian and reptilian forms. **(a)** *Trematops*, a primitive labyrinthodont amphibian with tarsals adjacent to the tibia existing as independent elements. **(b)** *Leptoceratops*, an ornithischian dinosaur, showing tibiale, intermedium, and central tarsal bones united to form astragulus. Greatest motion occurred between proximal and distal tarsal bones. **(c)** *Lycaenops*, a therapsid. In this form, astragulus and calcaneus interlocked with the distal tarsals; movement occurred between proximal tarsals and bones of the lower leg. Abbreviations: *a*, astragulus; *c*, central tarsal; *ca*, calcaneus; *F*, fibula; *f*, fibulare, a proximal tarsal; *i*, intermedium, a proximal tarsal; *T*, tibia; *t*, tibiale, a proximal tarsal; 1–5, distal tarsal bones. (From Romer.)

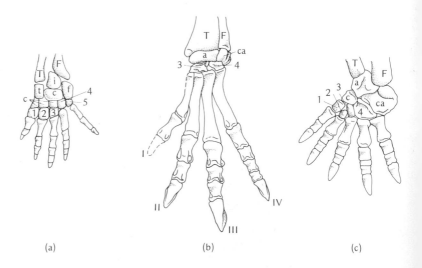

(a)   (b)   (c)

Like the postcranial skeleton, the skull in primitive reptiles carried with its amphibian heritage new traits that allow paleontologists to distinguish it (Fig. 7.7). Most noticeable is the absence of the otic notch. The deep cleft above the squamosal bone of each side, closed in labyrinthodonts of all kinds by the tympanic membrane, had disappeared completely in the captorhinomorphs. The cheek in those animals was solidly covered with bone and the eardrum relegated to a less exposed position. The skull remained high-domed and deep-jawed, never flattening like that of the progressive labyrinthodonts. It retained the dermal roofing pattern shown by the seymouriamorph amphibians but changed somewhat in its proportions. The snout narrowed, and the postorbital region shortened. The parietal bones became the most prominent elements at the rear of the roof. They replaced the lost intertemporal plates and spread outward to meet the squamosal bones as the supratemporals, tabulars, and postparietals diminished in area and migrated downward to the occipital surface. The supratemporal and tabular elements eventually vanished in more advanced reptiles, but in primitive forms one or the other often served to receive a process from a bone of the otic capsule which braced the back of the cheek over the jaw joint.

Although the rear of the skull roof and cheek initially became more solid, the palate of the first reptiles remained mobile, as it had been in the ancestral amphibians. The arrangement of the palatal bones changed little, but the pterygoid evolved, just in front of its articulation with the braincase, an outward-jutting shelf, or flange, that usually bore a row of small teeth. Some forms had teeth elsewhere on the palate, but the large fangs that had characterized the labyrinthodonts were gone. The regular, labyrinthine marginal teeth also disappeared as the reptiles emerged. In captorhinomorphs, the teeth that edged the jaw were frequently unequal in size. Either anteriorly or in the canine position, several teeth grew bigger than the others, an adaptation, apparently, for predation on land.

The innovations that appeared initially in the reptilian skull and those that differentiated it further from the amphibian skull as time went on can almost all be related to improvements in the jaw mechanism of the new animals. In contrast to the amphibian jaws, which were little more than a snapping device, the reptilian structures could be closed quickly and then pressed together with increasing force. Since the ability to generate pressure between the closed jaws made it possible to chew hard or coarse

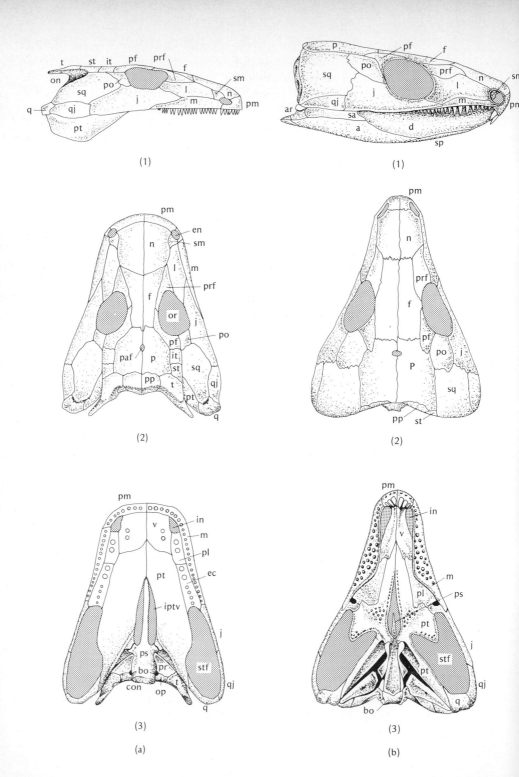

(1)

(1)

(2)

(2)

(3)

(3)

(a)

(b)

materials, the alterations in the skull associated with its development were rapidly selected for. Essentially, closing and compressing are independent actions that require separate sets of muscles. Very early in tetrapod history, two muscle masses did begin to differentiate from the ancient fan of fibers that originated under the skull roof and pulled the mandible upward. The fibers that originated farthest forward and were perpendicular to the lower jaw when it was depressed exerted the greatest force in raising it. When the mouth was closed, these fibers lost their mechanical advantage to the more posterior ones, which were then more nearly at right angles to the long axis of the mandible. In amphibians, the posterior fibers never became properly oriented to create strong tension on the lower jaw when it was raised and so to increase the pressure between the teeth. Since the amphibian jaw was long and the joint set far back beneath the rear of the skull, the posterior muscle fibers slanted forward from their insertion like the anterior ones. They reinforced the anterior muscle mass rather than performing a separate function. As the reptiles evolved, parallel changes took place in the various lineages, which resulted in greater differentiation of the jaw musculature and in realignment of the fibers for maximum efficiency.

The muscles of extinct reptiles, being soft tissue, are never preserved, of course, but paleontologists can reconstruct them by locating their sites of attachment on the fossil bones and by using as a guide the structure of the muscles in living reptiles, like lizards, whose cranial anatomy is similar in many respects to that of ancient forms. From restorations made in this manner it is apparent that the muscles in early reptiles became organized into increasingly discrete divisions which took origin from more narrowly limited areas of the skull than before (Fig. 7.8). The new flange on the pterygoid bone of the palate served to anchor the anterior fibers, now a well-segregated muscle called the pterygoideus, and the high postorbital region of the skull gave origin to the posterior part of the old fan which became the temporalis muscle. The pterygoideus performed the old function of lifting the lower jaw from its depressed position, while

Figure 7.7. Skulls of early amphibian and reptile forms in (1) lateral, (2) dorsal, and (3) palatal views. **(a)** *Palaeogyrinus*, a Carboniferous labyrinthodont amphibian. **(b)** *Captorhinus*, a stem reptile (cotylosaur) of Permian time. Abbreviations: *con*, occipital condyle; *iptv*, interpterygoid vacuity; *or*, orbit; *paf*, parietal foramen; *stf*, subtemporal fossa; other abbreviations as in Fig. 6.7. (From Romer.)

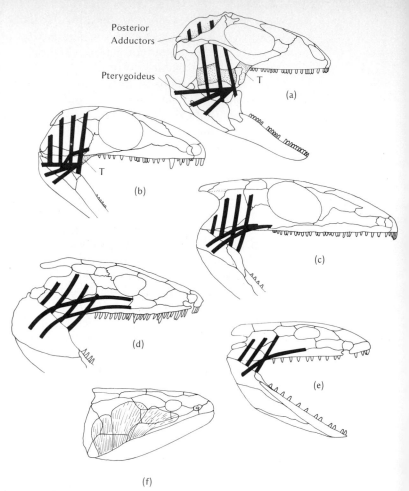

Figure 7.8. Gradual differentiation of adductor muscles which raise the lower jaw. **(a)** *Iguana*, a modern lizard. In *Iguana*, a temporal fossa has appeared, allowing the posterior adductor muscles (including the temporalis) to fasten upon the outer surface of the cranial roof. **(b)** *Paleothyris*, an early reptile. In *Paleothyris*, the pterygoideus has separated from the rest of the adductor mass and obtained an origin upon the transverse flange (*T*) of the pterygoid. (*c,d,e*) Amphibians with adductor muscle originating wholly upon the underside of the skull roof: **(c)** *Gephyrostegus*. **(d)** *Palaeogyrinus*. **(e)** *Ichthyostega*. **(f)** *Ectosteorhachis*, a rhipidistian fish with fan-like, virtually undifferentiated muscle mass. (From Carroll.)

changes in the configuration of the bones at the back of the cheek and mandible transformed the temporalis into a powerful compressor of the jaws. The first modifications to improve the efficiency of the temporalis muscle were those which brought the contractile fibers perpendicular to the part of the lower jaw on which they pulled. This reorientation was effected largely by movement forward of the jaw articulation. In its new position, the temporalis slanted in a direction opposite to that of the pterygoideus, backward from mandible to cheek. Commonly, the lower jaw developed a high-rising coronoid process behind the tooth row for attachment of the temporalis, allowing the muscle to spread posteriorly and still exert its force at right angles to the bone. The leverage of the muscle was increased further in some reptiles by lowering the mandibular joint below the tooth-bearing margin of the bone.

Further changes in the skull allowed the temporalis muscles to become stronger. Their size, and thus the force they could generate, had been limited in amphibians by the close quarters within which they had been confined. In those animals only as many fibers could exist as would fit under the skull roof between the cheek region and the braincase wall. In reptiles, when the temporalis fibers developed specific, restricted areas of origin, parts of bones not anchoring muscle tissue disappeared, leaving window-like spaces, or fenestrae, behind the eye. The temporalis, with room to bulge, increased its bulk and eventually expanded through the apertures to the outer surface of the skull roof.

Temporal fenestrae as basis of reptilian classification

The temporal fenestrae, as the openings behind the orbits were called, became such salient and stable features of the skull that Osborn and later Williston based the classification of the reptiles upon their location (Fig. 7.9). According to Williston's scheme, most of the reptiles fell into one of two groups, the Synapsida or the Diapsida. The former embraced the pelycosaurs and the therapsids, which had a single opening arched over by the postorbital and squamosal bones. The latter included the two-arched forms, the archosaurs and other reptiles with a second space above the postorbital and squamosal elements. The ichthyosaurs and a few isolated genera Williston lumped as parapsids, reptiles whose skulls showed only the upper opening. Captorhinomorphs and the turtles which lacked temporal fenestrae were categorized as anapsids, animals without spaces, hence without bony arches in the cheek region. In later years, as research made reptilian relationships clearer, Williston's classification was refined. The Diapsida was divided, when it became apparent that the group was not a natural one, into the subclasses Archosauria and Lepidosauria, the latter containing the lizards, snakes, and extinct allies of *Sphenodon*. The Parapsida was dismantled as well, although its members,

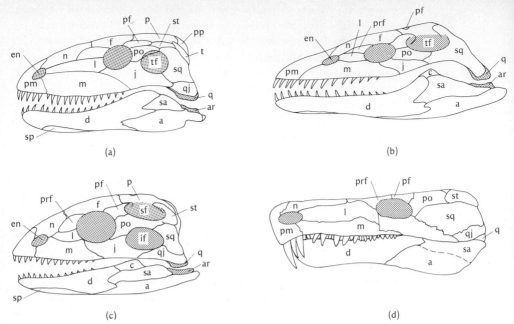

Figure 7.9. Reptilian skulls, showing locations of the temporal fenestrae. **(a)** A synapsid with temporal opening below the postorbital and squamosal bones. **(b)** A plesiosaur with temporal opening above the postorbital and squamosal bones, an arrangement described as parapsid. **(c)** A diapsid, having temporal openings above and below the postorbital and squamosal elements. **(d)** A cotylosaur, described as anapsid because cheek behind orbit is without an opening and thus without an arch-like configuration of the bones. Abbreviations: *if*, inferior temporal fenestra; *sf*, superior temporal fenestra; *tf*, temporal fenestra; other abbreviations as in Fig. 6.7. (*a,b,c* from Goodrich; *d* from Colbert.)

Evolution of the "land egg"

many of them marine forms of uncertain origin, have not yet been rehoused to everyone's satisfaction.

It is easier to understand the stages by which the reptiles evolved temporal fenestrae and other distinguishing skeletal characters than to imagine the steps that led to the development of the "land egg." Paleontologists continue to speculate upon the way in which the enclosure of the embryo came about, however, because the matter is central to the broad question of reptilian origins. Study of the eggs laid by living reptiles has provided little insight into the evolution of the extraembryonic structures which gave protoreptiles their first advantage over other tetrapods. Rather than recapitulating the process of its evolution, the "land egg" develops in a specialized manner derived, no doubt, by abbreviation and reordering of an earlier procedure.

In the present series of events, gelatinous albumen and the material of the shell are deposited around the fertilized ovum before the embryo or its membranes have proceeded far in their formation. The yolk sac, which is probably the oldest of the accessory structures phylogenetically, develops simultaneously with the other extraembryonic membranes (Fig. 7.10). An extension of the gut wall of the embryo, it spreads downward to cover the large mass of yolk upon which the growing embryo is perched. As it does so, a fold at the periphery of the animal's expanding body wall rises up and eventually covers the whole embryo. The inner side of the fold becomes the amnion, and the outer part, split away, the chorion, a membrane which envelops all the other living tissues within the shell. From the hindgut beyond the stalk of the yolk sac, the allantois begins as a pocket and then swells outward to fill the space within the chorion not already occupied by yolk and the embryo in its amniotic capsule. The capillary network of the embryo extends throughout the extraembryonic membranes as they develop and maintains them as adjuncts to the body until the time of hatching.

All the extraembryonic membranes in the "land egg" of a modern reptile must complete their formation normally if the embryo is to sustain itself. The yolk sac is of crucial importance, because nutritive materials from the yolk mass can enter the body only by passing through the vessels in its surface. The

Figure 7.10. Diagram showing amniote embryo (bird) with its extraembryonic membranes developing within the shell. (From Balinsky.)

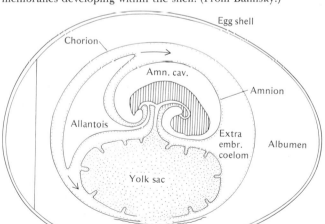

allantois also cannot fail: it serves as the respiratory organ for the embryo, since blood coursing through it loses carbon dioxide and receives oxygen by diffusion through the adjacent chorion and porous shell. In addition, its central cavity stores nitrogenous wastes produced by the actively metabolizing embryonic cells. Blood reentering the embryo from the allantoic vessels restores to the body water that has been resorbed from the excreted waste and also adds some that passes into the egg from the environmental air. The exterior of the embryo is kept wet by a liquid that accumulates within the amnion. Unlike pond water, to which it is often compared, the amniotic fluid does not act as an oxygen-bearing medium for the embryo. It is an adaptation for protecting the developing animal against shock and for preventing it from resting against the membranes in the shell and sticking to them.

Despite the difficulty of explaining how the embryo might have been served while the "land egg" was evolving to its present state, Szarski has suggested a series of steps by which the reptilian structure may have arisen. He observed first that a variety of embryonic membranes have appeared among the lower vertebrates as accessory nutritive or respiratory devices. Some of them, like those in cyprinodont fishes, make possible diffusion of substances from maternal to embryonic tissues. Others, like the ones in specialized amphibian larvae, provide increased surface area for respiratory exchange between the developing animal and its environment. Among the ancient amphibians, auxiliary respiratory membranes may have developed a number of times. Like some modern forms, the ancient animals may have then been able to produce embryos whose respiratory capacity enabled them to survive in moist places beyond the water's edge. If the ancient foetal membranes were similar to those of living amphibians, however, the adaptation of the embryo to terrestrial conditions would have stopped short of the evolution of a true "land egg." Since amphibian membranes are expansions of the body proper (usually of the gills or tail), they provide no reservoir for wastes. The embryo must excrete its waste material directly into the environment, from which therefore it can never be isolated by a shell. According to Szarski, the amphibians ancestral to the reptiles probably laid a relatively small number of large eggs from which embryos developed whose improved respiration depended not on expansion of any part of the body but upon the great surface area of the yolk sac and upon a rudimentary allantois. Once a pocket of the hindgut had arisen as a respiratory device, the embryo could contain its wastes and restrict its loss of fluid. Szarski thinks that the ability to retain fluid was eventually enhanced in these animals by producing urea as a waste substance rather than ammonia. The presence of urea in the system raises the osmotic tension of the tissues and thereby causes them to hold and attract water more strongly. Since embryos containing urea would swell to bursting in fresh water, ureotelism, as excreting urea is called, would not have evolved until the amphibians had begun to deposit their eggs on land. There, ureotelic embryos would have absorbed water passively as they utilized their yolk material and grew larger. The evolution of an amniotic fold increased the embryos' resistance to drying, and completion of the sac allowed the cushioning reservoir of fluid to form. It is probable that the first amphibian eggs laid out of water were hidden under damp vegetation and protected by a gelatinous substance manufactured by the maternal oviduct wall. A shell for each embryo and its membranes would have resulted from a change in the chemistry of the oviducal secretion. A substance which became fibrous rather than jelly-like would have given the embryos a leathery cover that was tough and in need of little water to maintain its own structure. The mineralization of the shell was evidently the last step in the evolution of the "land egg" and one that occurred to a variable degree in reptiles. Dependent on advanced and specialized calcium metabolism in the adult animal, mineralization reduced the water content of the shell to nil and rendered it a more efficient shield against loss of moisture from the embryo within.

Origin of reptiles: monophyletic or polyphyletic?

Szarski, Romer, and many other paleontologists believe that common possession of the "land egg" identifies reptiles as a monophyletic group. Since the development of the "land egg" involved such a long and complicated series of interdependent genetic mutations, it is highly improbable, they argue, that it evolved more than once. They find it reasonable to assume that all reptiles, despite their diversity, descended from the single amphibian line in which the "land egg" appeared. There are investigators, however, who question this view. They point out that ancient reptiles of different groups show some skeletal characters which are hardly likely to have had a common origin. They have found for these traits antecedents in different amphibian stocks and therefore think that the reptiles may be polyphyletic. If it could be demonstrated convincingly that reptiles stemmed from more than one source, they could no longer be regarded as constituting a class in the pure sense but simply a miscellaneous collection of tetrapods which had evolved independently to the reptilian grade.

Evidence from studies of Diadectes

One proposal of polyphyletic origin was occasioned by studies of the animal called *Diadectes* (Fig. 7.11). This form was a heavy-bodied, stocky-legged tetrapod of the Lower Permian, a contemporary of the captorhinomorphs and pelycosaurs as well as the flourishing labyrinthodonts. To E. D. Cope, who studied it in the 1870s, *Diadectes* seemed the prototype of the primitive reptile. He made it the basis of his order Cotylosauria, a category in which he placed all the early reptiles that had a solid skull roof behind the eye—the captorhinomorphs and forms known as procolophonids and pareiasaurs. In the structure of the post-cranial skeleton, *Diadectes* exhibited several of the reptilian modifications associated with terrestrial living. Its vertebral column had been strengthened by expansion of the pleurocentra at the expense of the intercentra and fusion of the former with the neural and hemal arches. The head was mounted on a well-developed atlas, which rotated upon a high-spined axis like that of the captorhinomorphs. The limbs were built for sustaining the animal's weight. The feet were strong and five-toed in both pairs of appendages, and the proximal bones of the tarsal region had coalesced to form an astragulus and calcaneus. Both front and hind limbs articulated with girdle elements of generous proportions, and the ilium of the pelvis pressed broadly against the vertebral column through three sacral ribs. From the characteristics of the skull, it was apparent that *Diadectes* enjoyed the advantages of an improved jaw mechanism. The quadrate bone stood upright, so that it articulated with the articular of the

Figure 7.11. Skeleton of *Diadectes*, an early Permian tetrapod apparently of seymouriamorph amphibian stock but exhibiting numerous reptilian characters. (From Gregory.)

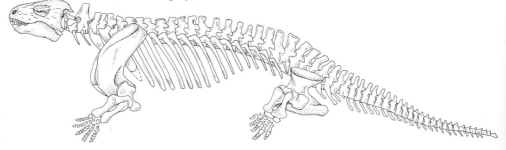

lower jaw in front of the occiput and below the tooth row. Anterior to the jaw joint, the mandible deepened and sent a process upward upon which the temporalis fibers could insert. The gape was short and the closure no doubt powerful. Like a number of reptiles, *Diadectes* had probably adapted to a diet of terrestrial vegetation. Its anterior teeth, conical pegs which met nearly edge to edge, seemed specialized for cutting plant material, and the back teeth, broadened and flattened, were certainly capable of crushing it. In connection, perhaps, with its habit of grinding coarse food, the palate of *Diadectes* had become fused to the underside of the braincase, and the dermal bones were thickened.

Paleontologists after Cope who studied the skull of *Diadectes* further began to find traits that separated *Diadectes* from the captorhinomorphs and allied it to *Seymouria*, a contemporary Permian amphibian with many reptilian characters. Instead of a closed cheek like that of the captorhinomorphs, *Diadectes*, like *Seymouria*, had a huge otic notch beneath the supratemporal and tabular bones. The tympanic membrane in *Diadectes* was stretched across the notch, just as it was in *Seymouria* and other labyrinthodonts, and so was located more dorsally than the captorhinomorph eardrum is thought to have been. In the region of the otic capsules, *Diadectes* showed specializations similar to those which set *Seymouria* apart from other tetrapods: in both animals, the bones surrounding the fenestra ovalis flared laterally, and the rear of the braincase was, as a result, unusually broad. Its resemblance to *Seymouria* suggests that *Diadectes* evolved from amphibians already advanced toward the seymourian state and hence was not closely related to the more primitive anthracosaur labyrinthodonts, which supposedly sired the captorhinomorphs. If this inference is correct, the transition from the amphibian to the reptilian condition could be interpreted as having occurred at least twice. *Diadectes* and the captorhinomorphs each arrived at the reptilian level, at least in osteological characters and, if the transformation was complete, developed the "land egg" independently through parallel evolution.

Those who accepted the idea that *Diadectes* and the captorhinomorphs had crossed the amphibian-reptilian line separately acquiesced in Watson's division of the reptiles into two groups, the Diadectomorpha and the Captorhinomorpha. Included with *Diadectes* in the former category were the procolophonids and the pareiasaurs, reptiles which were thought by Watson and

Evolution of the procolophonids and pareiasaurs

other workers to have sprung from a diadectid root. Although *Diadectes* was found in the Texas red beds and the procolophonids and pareiasaurs in the Old World, a relationship between the animals was supposed on the basis of similar features in the skull. The pareiasaurs and the procolophonids, which appear first in Middle and Upper Permian deposits in South Africa, were vegetarians like *Diadectes* and showed comparable modifications of the jaw. *Procolophon* paralleled *Diadectes* also in specialization of the teeth and in the widening of the rear of the skull. The pareiasaurs had leaf-shaped cutting teeth but exhibited a diadectid-like thickening and overgrowth of the dermal bones of the head. The procolophonids outlasted the pareiasaurs, perhaps because they were small forms that did not compete directly with the reigning therapsids. Pareiasaurs became large, grazing animals comparable to the therapsid herbivores of the same time and, like them, were virtually extinguished at the end of the Permian period. Continued study of the African members of the Diadectomorpha eventually raised doubt that they had descended from *Diadectes* or animals like it. Olson and other paleontologists who had worked on the problem of reptilian origins finally came to the conclusion that the ancestors of the procolophonids and pareiasaurs evolved to reptilian status independently of the diadectid line. Olson suggested that progress to the reptilian grade was made from several amphibian stocks and that the Cotylosauria, or stem reptiles, included all those tetrapods that actually achieved reptilian standing—captorhinomorphs, procolophonids, diadectids, and isolated forms, like *Bolosaurus,* which exhibited reptilian traits in combination with special features of their own.

Romer's argument for monophyly of the reptiles

To this rampant polyphyly, Romer took vehement exception. Reemphasizing the importance of the "land egg" as a criterion of reptilian status and the improbability of the repeated evolution of so complex a structure, he held to the idea that the reptiles are traceable to a single ancestral group and so constitute a valid class. He reduced the heterogeneity of the stem reptiles first by pruning away the diadectids. *Diadectes,* he declared, was closely related to *Seymouria* and should be classified with that form as an amphibian. Although the reproductive habits of *Diadectes* are unknown, Romer believed that an animal allied to seymouriamorphs with gilled larvae must have deposited its eggs unprotected in the water. The skeleton gave more tangible evidence of the amphibian character of *Diadectes.* The palate was missing the typical reptilian pterygoid flanges, and the skull roof had been interpreted as having a pair of intertemporal bones rather than the laterally expanded parietal lappets which replace them in reptiles. That the skeleton did have reptilian features Romer acknowledged, but these traits he regarded as the result of parallel evolution in animals becoming adapted for living on land.

After removing diadectids from the reptilian ranks, Romer further reduced the number of independently derived reptilian forms by cutting the direct ties of the procolophonid-pareiasaur group to amphibian ancestors. He saw no necessity of postulating a polyphyletic origin for the stem reptiles when the procolophonids and pareiasaurs could be plausibly interpreted as having descended from the captorhinomorphs. He pointed out that primitively constructed procolophonids resembled captorhinomorphs rather than amphibians and that the otic notch of advanced forms arose more likely by emargination at the back of the captorhinomorph cheek than by derivation from the more dorsally placed amphibian notch. In Romer's opinion, the captorhinomorphs alone are traceable to amphibian forebears and to a narrow enough base among the anthracosaur labyrinthodonts to leave the monophyletic origin of the reptiles uncompromised.

Microsaurs proposed as ancestors of the reptiles

While Romer and others were dismantling the Diadectomorpha and arguing against the multiple origin of reptiles from labyrinthodonts, Westoll was proposing still another group as ancestors of the captorhinomorphs. Because captorhinomorphs lacked an otic notch, he sought to ally them to microsaurs, lepospondylous amphibians which, in addition to having no otic notch, displayed some other similarities to early cotylosaurs. Critics of Westoll's theory were quick to point out that the microsaurs were too late and too specialized to have given rise to the reptiles. Some of the most ancient of the typical microsaurs were found in Pennsylvanian deposits side by side with captorhinomorphs. Their vertebrae already lacked completely the intercentra of which vestiges remained for a long time in reptiles, and their skulls showed a tendency to elongate in the postorbital region and to develop enlarged supratemporal bones which was quite the reverse of that observed in captorhinomorphs. Although there were microsaurs which persisted in their terrestrial adaptation, many of them became committed to aquatic life, undergoing weakening of the limbs and reduction in the number of digits. P. P. Vaughn, who seconded Westoll's suggestion that a microsaurian ancestry for reptiles be seriously considered, admitted that the late Carboniferous microsaurs were specializing divergently from the captorhinomorphs but maintained that the most primitive members of the two groups were close enough in structure to have had a common ancestor. Any linkage between microsaur and captorhinomorph precursors immediate enough to be meaningful for reptilian origins has been rejected by Carroll. He lists several characteristics of microsaurs—among them the structure of the temporal region of the skull, the design of the atlas-axis complex, and the nature of the dermal scales—which differ so radically from the reptilian pattern that they practically exclude close relationship between the two types of animals at any time in the Carboniferous period. The structural similarities which exist in microsaurs and captorhinomorphs are better explained, he thinks, by convergence and by the retention of traits widespread in primitive tetrapods than by supposing the two groups to have been connected phylogenetically.

Similarities between an- thracosaurs and primitive reptiles

A microsaurian origin for the reptiles is more easily set aside because the earliest captorhinomorphs resembled anthracosaurian labyrinthodonts so nearly. Unlike the later captorhinomorphs, which grew large and developed distinctive multiple rows of marginal teeth, the most ancient forms differed from anthracosaurs in little more than details of palatal structure, the absence of intertemporal bones from the skull roof, and the more solid construction of the vertebrae. Upper Carboniferous specimens of both kinds, found in the same location, can be sorted only after careful study. Two sites in Nova Scotia have given up a mixed fauna which includes some animals that seem transitional between one group and the other. All the fossils at these places were found in hollow trunks of the tree-like plants that constituted the coal forests. The small terrestrial and semiterrestrial tetrapods of the time apparently took refuge in stumps and then were buried there by sudden washes of mud. Since the tree-stump labyrinthodonts and reptiles were contemporaneous, paleontologists are still unable to document the emergence of the captorhinomorphs from anthracosaur stock. They have scrutinized reptiles which retain vestiges of amphibian structures, however, in an effort to determine as closely as possible the kind of anthracosaurs which served as the source of the higher forms.

Unraveling the tangled history of captorhinomorph origins has been a slow process because of the limited amount of useful fossil material and the difficulty of studying it. The kind of obstacles paleontologists have encountered can be made clear, perhaps, by an account of the research on one primitive form, *Solenodonsaurus* (Fig. 7.12). The type of this genus came from

Figure 7.12. The type of *Solenodonsaurus*, drawn as it appears exposed on the block in the collection of the Humboldt Museum, Berlin. Abbreviations: *C*, clavicle; *H*, humerus; *I*, intercentrum; *IC*, interclavicle; *P*, pleurocentrum; *R*, radius; *S*, scapulocoracoid; *U*, ulna; abbreviations for bones of the skull as in Fig. 6.7. (From Carroll.)

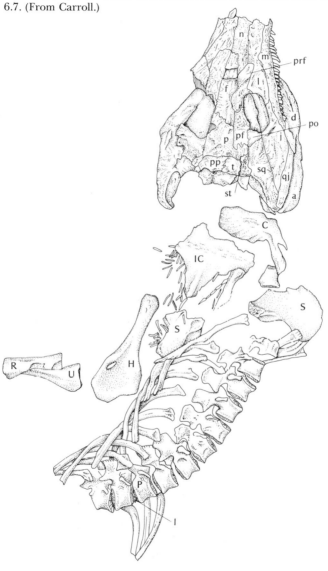

Upper Carboniferous rocks in Czechoslovakia. The block containing it was split, and the halves, each with a part of the fossil, were sent to different museums, one in Berlin and the other in Munich. Broili examined both pieces and published descriptions of the specimen in 1905 and 1924, but further examination of the material in Munich was made impossible by its loss during World War II. The part of the specimen still available in Berlin shows the skull roof, some teeth, a curving row of eleven vertebrae, one disarticulated front limb without the manus, and scattered elements of the pectoral girdle. Certain elements important in establishing the progress of *Solenodonsaurus* toward reptilian status are not in evidence. The palate is not visible, the first two vertebrae are missing, and the suture marks on the posterior part of the skull are so poorly preserved that the presence or absence of the intertemporal bones cannot be determined. Pearson, working in 1924 from a presumed second specimen in a halfblock which had found its way to England, was of the opinion that *Solenodonsaurus* should be considered a seymouriamorph. Twenty-five years later, when the other half of Pearson's specimen was found in a Berlin museum and the whole fossil restudied, it became apparent that the animal lacked definitive seymouriamorph characters and had, instead, traits which allied it more closely to the captorhinomorphs. Besides exhibiting reptilian pterygoid flanges in the palate and parietal lappets in the place of intertemporal bones, however, this specimen of *Solenodonsaurus,* like the type, showed anthracosaurian ties in the form of labyrinthodont teeth and a prominent otic notch above each cheek.

Animals near the amphibian-reptilian boundary

The anthracosaurs with which *Solenodonsaurus* had most in common were not seymouriamorphs or embolomeres but less specialized forms. A near contemporary of *Solenodonsaurus,* an amphibian called *Gephyrostegus* or *Diplovertebron,* may be a survivor of the true ancestral stock. Also found in Czechoslovakia, *Gephyrostegus* had no traits that would have barred its kind from the protoreptilian line. The extent of the resemblance between *Solenodonsaurus* and *Gephyrostegus* is such that the Broughs, working before the rediscovery of the second half of Pearson's specimen, discarded the name *Solenodonsaurus* and attributed the fossils studied by Broili and Pearson to the amphibian genus. Actually, *Gephyrostegus* is distinguished from early reptiles by its otic notches, its regular teeth, its intertemporal bones, its stout cleithra and two-pronged ilia in the appendicular skeleton, and the lack of fusion between the centra and arch elements in the vertebral column. The relatively minor changes necessary for conversion of the skeleton to the reptilian state occurred to a varying degree, apparently, in several lines that sprang from gephyrostegid amphibians. In the group to which *Solenodonsaurus* belonged, the uniform labyrinthodont teeth and the otic notches remained until the animals became extinct. In another descendant family, called the Limnoscelidae after the Permian form *Limnoscelis,* the transformation was more complete: the otic notches closed, and enlarged anterior teeth evolved, but the amphibian labyrinthine enamel persisted.

Because they exhibit most of the distinctive reptilian characteristics, *Solenodonsaurus* and the limnoscelids are now placed "over the line" in the class Reptilia, despite the primitive, amphibian traits which they retain. Although he regards them as relics of earlier protoreptilian groups, Romer lists *Solenodonsaurus* and the limnoscelids as primitive captorhinomorphs. Carroll, who agrees essentially with Romer concerning their history, prefers to exclude *Solenodonsaurus* and the limnoscelids from the Captorhinomorpha and count contemporaries of the limnoscelids, termed romeriids, as the base of the captorhinomorph line.

Romeriids and their descendants

The romeriids were small, lithe, long-tailed animals, from which, if the proponents of monophyly are right, all the later reptiles arose. Carroll believed them to be advanced in their adaptation to life on land. He imagined their filling the niche that the modern lizards occupy, eating insects, perhaps, and using the low vegetation of the time for cover. In Romer's opinion, the earliest reptiles were less terrestrial than Carroll assumed. Romer thought that they shared the swamp and pond water with aquatic amphibians, emerging primarily to lay their eggs out of reach of swimming predators.

Whatever their habitat, the romeriids soon diversified. In the Upper Carboniferous tree stumps, they already appear in the company of pelycosaurs surely descended from earlier members of the romeriid line. Although the pelycosaurs were at once successful and became numerous on stream banks and on drier terrain, the romeriids held their ground and produced still other new forms. Some, like the advanced captorhinomorphs, appeared before the Carboniferous period ended and left a fossil record that enabled paleontologists to follow their career. Others, though they may have branched away as early, are known from a single genus and not so well understood. *Mesosaurus,* for instance, occurs in freshwater deposits from the

Permocarboniferous boundary in South America and South Africa, far from the areas where romeriids have been found. *Mesosaurus* (Fig. 7.13) was a reptile almost 2 feet in length that paddled its way slowly through the water with broad-handed limbs or swam speedily by undulating its long, fish-like tail. It had become specialized for catching small aquatic animals by developing long jaws and needle-like teeth and also had evolved a low-set temporal fenestra behind the eye. Despite its geographical isolation and its skeletal peculiarities, paleontologists believe that *Mesosaurus* should probably be counted among the descendants of the romeriid captorhinomorphs. They reason that the generalized ancestor of so specialized a form must have existed far back in the Carboniferous period at a time when only the romeriids had completed the transition to the reptilian level.

Other, later reptiles are presumed to have sprung from romeriid (or at least early captorhinomorph) stock on evidence that is equally indirect. Although romeriids are not known later than the Lower Permian, many paleontologists regard them as possible ultimate ancestors for Upper Permian procolophonids and for two kinds of Upper Permian reptiles, the millerettids and the proterosuchians, which seem close to the base of the diapsid lepidosaurian and archosaurian lines respectively. The scarcity of all but therapsid fossil material from Middle Permian times has so far frustrated investigators' efforts to find forms which might be intermediate between primitive captorhinomorphs and these later animals.

Early dichotomy of the reptiles

Even though support has been growing for the idea that the nondiadectid reptiles originated monophyletically from some group of early, probably gephyrostegid, amphibians, workers still recognize and discuss a dichotomy that appears in the reptilian assemblage. Osborn was aware of it when, in 1903, he divided the reptiles into the Synapsida and the Diapsida. He had in mind not only the difference in the position of the temporal fenestrae in the two kinds of animals but also the fact that

Figure 7.13. Reconstruction of *Mesosaurus*, an aquatic reptile of late Carboniferous and earliest Permian times. (From Colbert.)

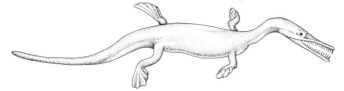

synapsids gave rise eventually to mammals and diapsids to surviving reptiles and the birds. Goodrich believed that the anatomy of living tetrapods yielded evidence that the reptiles had split into two groups very early in their history. He pointed out that the ventral aorta carrying blood from the heart is subdivided differently in reptiles and in mammals. During ontogeny, the reptilian vessel separates into three channels, a pulmonary aorta and two systemic ones. Birds follow the same developmental pattern but lose the systemic branch leading from the right side of the heart. Like the birds, mammals have a pulmonary aorta and a single systemic vessel, but the two vessels form by simple division of the embryonic ventral aorta. Since the mammalian arrangement is not derivable from the reptilian one, Goodrich supposed that the forebears of the mammals must have separated from other reptiles before the typical reptilian modification of the old, undivided amphibian aorta had taken place.

Because it is impossible to follow the evolution of different structural patterns in soft organs, Goodrich sought skeletal characteristics which would support his theory that ancient reptiles were split into theropsids and sauropsids, the former precursors of the mammals and the latter of the reptiles and the birds. Finding traits which differentiated the two groups was difficult, however, since the oldest reptiles shared many primitive features and later forms evolved special structures that were adaptive rather than indicative of phylogenetic relationship. He proposed, finally, the shape of the fifth metatarsal bone as a diagnostic character. In sauropsids it was hook-shaped; in theropsids, straight. For a time, Goodrich's way of distinguishing sauropsids from theropsids seemed promising, but when closely related animals were found with and without the hook-shaped element, paleontologists began to look for a more reliable criterion.

Structure of the middle ear in fossil reptiles

Watson thought he found one in the structure of the middle ear. In his opinion, sauropsids and theropsids differed so greatly in the position and attachment of the stapes that its divergent development must have begun in animals at the amphibian level. According to Watson's theory, sauropsids kept the dorsally placed stapes of their amphibian ancestors. It remained in contact with the tympanic membrane, which was visible behind the concave posterior edge of the quadrate bone. The lower end of the quadrate had swung away from the process of the stapes which once braced it. The forward rotation of the quadrate, which came about as the jaw shortened, widened the space for the eardrum behind it and left the bone standing vertically instead of resting horizontally, as in amphibians. Theropsids evolved a vertically aligned quadrate element, also, but in a way that caused a reorientation of the stapes. In theropsids, the upper end of the quadrate pivoted backward over the lower one, bringing with it an extension of the squamosal bone which closed the old otic notch. Watson believed that the tympanic membrane was lost in the process and that the stapes turned downward and reestablished its connection with the quadrate bone. In contrast to the sauropsids, whose hearing, like that of the amphibians, depended upon vibration of the eardrum, the theropsid reptiles could hear only by transmission of sound waves from bone to bone.

Although not everyone agreed with Watson that the theropsids had lost the tympanic membrane entirely, several investigators used the structure of the middle ear as a guide in determining the affinities of specimens which had been difficult to classify. If the stapes was present, its position high on the braincase behind the quadrate denoted the animal's relationship to the diadectids or to the reptiles of the main stream. A downward-directed stapes which rested against the quadrate signified that the animal had belonged specifically to the captorhinomorph or pelycosaur-therapsid lineages, from which the mammals arose. If the stapes was not found, its orientation could be discovered indirectly by examining the quadrate element. In theropsids, the quadrate had a recess on its medial side for reception of the stapes, but in sauropsids the quadrate showed no such mark.

Araeoscelis

Vaughn relied on this key in his reassessment of the relationships of *Araeoscelis* (Fig. 7.14), a somewhat specialized little reptile from the Lower Permian beds of Texas. *Araeoscelis* had been hard to place because it was like the primitive captorhinomorphs in many features yet similar in others to forms which existed only much later, in Mesozoic times. It had in the temporal region a fenestra above the squamosal and postorbital bones like that of the plesiosaurs and their marine associates. The suggestion was made that *Araeoscelis* was an early precursor of these animals because, in addition to the same type of temporal fenestra, *Araeoscelis* displayed elongated cervical vertebrae, a trait which became pronounced in the Mesozoic marine forms. Those who considered the long neck and high-set temporal fenestra of *Araeoscelis* as insufficient grounds for

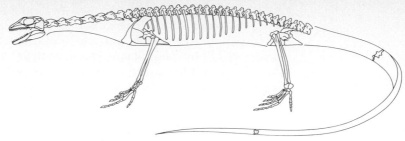

Figure 7.14. Skeleton of *Araeoscelis*, a Lower Permian reptile of uncertain relationships. (From Vaughn.)

linking it to reptiles that lived so much later in time argued that *Araeoscelis* could as reasonably be placed in the lineage of true lizards. Its legs were lizard-like in their general proportions, and the fifth metatarsal, though not hooked, diverged from the others in lepidosaurian fashion. The ankle joint was located between the tarsal bones, as it was in lizards, rather than above them, as in captorhinomorphs and pelycosaurs. Vaughn re-analyzed the characteristics which *Araeoscelis* shared with lizards, with plesiosaurs, and with more primitive reptiles and, finding them ambiguous, turned to the structure of the middle ear. Since it conformed to Watson's theropsid type, Vaughn decided that the captorhinomorph traits of *Araeoscelis* were indicative of the animal's phylogenetic position. Since Vaughn adopted Watson's thesis that the theropsid and sauropsid stems were separate and divergent, he held that *Araeoscelis* could not have been intermediate between the captorhinomorphs and either lizards or plesiosaurs, as several paleontologists had supposed. The special features it shared with those sauropsid forms were, in his opinion, only parallel or convergent traits selected for because of their adaptive value.

Another view of early reptilian phylogeny

Romer is not willing to rule out the possibility of *Araeoscelis'* relationship to the plesiosaurs and their kin. He does not agree with Watson that theropsids and sauropsids are traceable separately to the amphibian level and therefore with Vaughn's corollary that animals with captorhinomorph affinities cannot stand in the lineage of any member of the sauropsid tribe. Romer doubts that the middle ear of the captorhinomorphs was as different from that of sauropsid reptiles as Watson believed and also that reptiles with a broadly exposed tympanic membrane inherited it directly from the anthracosaurs. He believes, in short, that Watson's theory is untenable, and that the sauropsid-theropsid dichotomy, insofar as it exists, has little phylogenetic significance. According to his reconstruction of the evolution of the middle ear, the dorsally placed, superficial eardrum of the amphibians diminished in size and migrated downward and backward as the otic notch closed in the emerging captorhinomorphs. The tympanic membrane never disappeared completely in captorhinomorphs, as Watson postulated, but remained concealed a little below the surface, just behind the quadrate bone, to which it became attached. Romer is convinced that the structure of the middle ear described by Watson as sauropsid could easily have been derived from that of the captorhinomorph reptiles. The tympanic membrane in Watson's sauropsids was set low and bounded anteriorly by the quadrate, as Romer supposed it to be in captorhinomorphs, not mounted above the squamosal bone, as in amphibians. It was visible through an otic notch which Romer interpreted as a new excavation in the back of the solid captorhinomorph cheek rather than a space equivalent to the more dorsal one in the ancestral anthracosaurs. The changes in the stapes he found more difficult to explain, but he believed that its articulation with the quadrate persisted in cartilage instead of bone. Romer's analysis of the changes in the middle ear reinforced conclusions he had drawn from his study of other parts of the skeleton: the captorhinomorphs had not, as Watson thought, produced only the pelycosaurs but had radiated widely. Since an otic notch and an enlarged tympanic membrane apparently evolved independently in at least three groups of captorhinomorph descendants, the term "sauropsid" had no phylogenetic connotation. There was no simple way of dividing the diverse assemblage of reptiles. No single structure could be used as a key to common ancestry. Those who sought the ancestry of special groups of reptiles would have to follow the old method of establishing a morphological series of forms and then trying to prove, by study of the temporal and geographical distribution of its members, that it represented a phylogenetic series as well.

Evolution of the turtles

If sufficient fossil material were available, paleontologists think that the turtles* would be traceable in this way almost directly to the early cotylosaurs. Because turtles enter the record in a nearly modern state in the Triassic period, the supposition

*The term "turtle" is used here to refer to the chelonians generally. Many authorities prefer to restrict "turtle" to the marine forms, calling terrestrial chelonians tortoises and amphibious types inhabiting freshwater or brackish marshes terrapins.

rests on inferences from the anatomy of their skull rather than on evidence of known transitional forms. At first it might seem curious to link turtles with the most primitive reptiles. Their shell is a highly complex structure of dermal plates overlain by horny scutes, and their skull is far more specialized than that of many other members of their class. Even the first known turtle, *Proganochelys (= Triassochelys)* (Fig. 7.15), showed the short face, toothless jaws, and immovable palate that have remained the outstanding characteristics of chelonians down to the present day. The shell of *Proganochelys* was equally modern. The arrangement of the dermal plates in the carapace was the same as in living turtles: the median row of plates fused to the underlying trunk vertebrae was flanked on either side by a series attached to the ribs and the whole bordered by a rim of smaller, marginal elements. The carapace was connected to the plastron, which protected the underside of the body. That part of the shell had fused among its plates the dermal clavicles and interclavicle of the pectoral girdle, as it does in the Recent forms. Besides its specialized chelonian characteristics, however, *Proganochelys* had one trait which, if paleontologists have assessed it correctly, indicates a cotylosaurian origin for the turtle clan. The temporal region in the Triassic turtle was covered solidly with dermal bone. Although several bones which modern turtles lack in the rear of the skull roof were already missing in *Proganochelys,* there was no sign that fenestrae had ever existed in the cheek. Before the discovery of *Proganochelys,* there had been

Figure 7.15. *Proganochelys (= Triassochelys),* a Triassic turtle. **(a)** The skull and carapace in dorsal view. **(b)** The skull in lateral view. Abbreviations as in Fig. 6.7. (From Romer.)

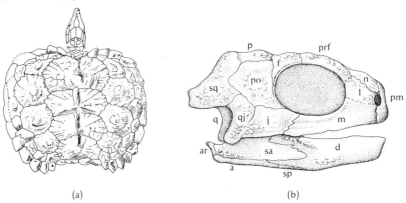

(a)   (b)

speculation that chelonians might be descended from Triassic diapsids, for Recent turtles generally have the upper and lower part of the temporal region widely open for the attachment of jaw muscles. The presence of a completely closed cheek in *Proganochelys* and several other primitive turtles caused investigators to abandon this idea and to relate the turtles instead to the anapsid cotylosaurs. In the line leading from cotylosaurs to turtles, paleontologists now thought, the dermal bone covering the cheek remained intact until the quadrate element became solidly bound to the braincase and the palate. When the cheekbones no longer served as a necessary brace for the jaw joint, they became deeply excavated and, as in other advanced reptiles, the jaw musculature spread to the outer surface of the skull roof.

Linking the turtles directly to anapsid precursors implied the rejection of older theories connecting turtles with more progressive forms that showed similar specializations in the skeleton. Jaekel and later Broom had suggested, for instance, that the turtles might have come from the same stock as the marine tetrapods which were also making their first appearance in the Triassic period. Like the turtles, the nothosaurs and the early plesiosaurs were evolving a long neck and a shortened trunk as they acclimated to life in the sea, and placodonts were producing species which were toothless and protected by a shell. Comparative studies of the skull in these animals and in turtles persuaded paleontologists, finally, that similarities between the two groups must have been due to convergence. The most aberrant of the marine forms were the most turtle-like; the more primitive genera resembled neither chelonians nor any other known reptiles closely enough to make relationship probable. Because they had one dorsally located temporal fenestra, Romer thought that the early marine tetrapods might have evolved from animals like *Araeoscelis,* a form which no one would place among the forerunners of the turtles.

In 1914, Watson described a reptile, *Eunotosaurus africanus* (Fig. 7.16), which he believed to be a turtle from the Permian period. Since the remains of the animal were exposed in ventral view, he could not tell whether its temporal region was anapsid, but the palate, ribs, trunk vertebrae, and appendicular elements which could be seen were like those Watson expected to find in a very early chelonian. The neck was already long and flexible in *Eunotosaurus* and the trunk vertebrae few in number. Nine or ten broad, flat ribs spread outward to either side, together

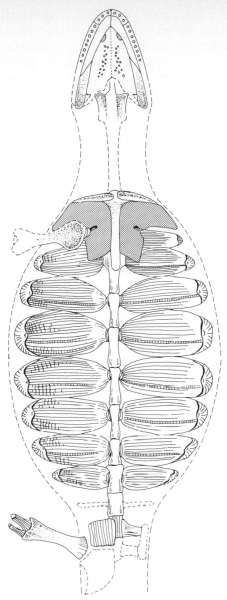

Figure 7.16. *Eunotosaurus*, the Permian form believed by Watson to be ancestral to the turtles. Ventral view. (From Watson.)

constituting a kind of shell which Watson suspected was reinforced by dermal ossifications. Both girdles were narrow, and the pectoral one was already well within the arch of the ribs as it is in the later turtles. Marginal and palatal teeth were still present, but the pterygoid bones were broad, promising the palatal solidity characteristic of chelonians. Although many paleontologists accepted *Eunotosaurus* as an ally of the later turtles, there were some who thought it might be another unrelated, convergent form. Parsons and Williams pointed out that *Eunotosaurus* was similar to Recent turtles rather than to Mesozoic ones. They believed that *Proganochelys* was actually more primitive and that the Permian ancestors of the turtles would have been narrow-ribbed animals closer to the root of the reptilian line.

Cope had guessed that terrestrial tetrapods of the early Permian like the dissorophids or the diadectids might have served as the source of chelonian stock. The dissorophids were rejected as antecedents of the turtles because, despite their dermal armor and terrestrial habits, they were discovered to be rhachitomous labyrinthodonts, not amphibians approaching the reptilian level. The diadectids were considered more seriously as possible ancestors. Although *Diadectes* was not armored, its postcranial skeleton could have undergone chelonian modifications in association with the evolution of a shell. The structure of the skull also seemed susceptible of change in a chelonian direction. The lateral extension of the otic capsules might have augured the bracing of the quadrate bones against the ear region of the braincase in turtles, and the tendency toward fusion of the bones in the skull roof could have foreshadowed the disappearance in chelonians of certain dermal roofing elements. Despite these possibilities, however, the idea that diadectids gave rise to the turtles was eventually set aside. At first, the minor specializations of *Diadectes* itself led paleontologists to exclude that form from the chelonian lineage. Then, on the basis of his studies of the middle ear and the posterior part of the braincase, E. C. Olson stated that the diadectids were closer to *Seymouria* than to the turtles. He concurred with investigators, like Romer, who supposed the diadectids to have left no survivors and proposed the derivation of turtles from the procolophon-pareiasaur branch of the stem reptiles. Since serial sections of the skulls of procolophonids and pareiasaurs have not been made, Olson was unable to compare the deep structure of the middle ear and braincase of these animals with that of the turtles he had examined. Like W. K. Gregory, who first associated turtles with pareiasaurs, Olson saw in these Permian forms anapsid reptiles whose general skeletal pattern would have provided a good starting point for the evolution of the Chelonia. Neither these forms nor any others gave a clue to the origin of the peculiar postcranial skeleton of the turtles, but the skull in early procolophonids and pareiasaurs had developed the low, broad shape and immobile palate which became characteristic in chelonians. Few more specific similarities were evident, but nothing in the arrangement of the bones in the skull appeared to bar the derivation of chelonian traits. Although the evidence was tenuous, it seemed to Olson that the pareiasaurs may have arisen from primitive procolophonid stock and then, before reaching great size, produced the armored ancestors of the Triassic turtles.

Until remains of animals closer in age to *Proganochelys* are found, paleontologists can say nothing about the establishment of the body form which has sustained the turtles from the Triassic period to modern times. From the post-Triassic fossil record it is clear that extreme specialization of the armor and the endoskeleton led not to a decline of the race but to an adaptive

Adaptive radiation of the turtles

radiation which carried the chelonians into almost every environment accessible to reptiles. Even though the box-like shell was a cumbersome burden and the legs of the animal were fixed in their primitive sprawling position, the turtle competed successfully against other kinds of vertebrates first in the marshes and later on drier land and at sea. Most paleontologists agree that early turtles were, like many Permian and Triassic tetrapods, land animals which frequented freshwater shallows, where fishes and invertebrates provided an abundant supply of food. As slow-moving, perhaps, as the late, heavy-headed labyrinthodonts, the turtles survived because their shells made them unappetizing even to the largest predators. *Proganochelys* and forms like it could not protect themselves by drawing the head, tail, and legs within the shell, but in their descendants changes occurred in the skeleton which made retraction of these vulnerable body parts possible. Two mechanisms of folding back the long neck were selected for. In one line which diverged from the old chelonian stem, the neck was pulled in by zigzagging it sideways. The turtles of this group, called pleurodires, spread throughout the world during late Mesozoic time and then, for reasons that are poorly understood, dwindled like the lungfishes to a few genera distributed discontinuously through the Southern Hemisphere. Another group of turtles evolved the ability to fold the neck upon itself vertically. These animals, the cryptodires, fared better than the side-neck forms. During the Upper

Cretaceous period, instead of declining, as many of the Mesozoic reptiles did, they diversified. While some species continued their amphibious habits, others became more terrestrial, and whole families appeared adapted for aquatic life or for swimming in the sea. Land turtles, or tortoises, retained a heavily ossified shell, sometimes evolving a hinged attachment between the carapace and plastron which allowed the two parts to be pulled together for better protection of the body. Being unable to pursue terrestrial animals, tortoises survived by adapting to a largely herbivorous diet. Wherever there was suitable vegetation and warmth, they multiplied—in forests, in the fields, and even in the desert. Some of them remained small and could, if necessary, conceal themselves, but others came to rely on great size as well as the shell for protection. Gigantism seems to have boded ill for the tortoises, however, as it has for many vertebrates. Many of the largest forms are already extinct, and those that remain are seriously threatened now by man.

Although the tortoises and their more primitive amphibious relatives like the mud turtles and the snappers kept the ancient heavy shell, chelonians which lived continuously in the water evolved lighter armor. The freshwater cryptodires of the family Trionychidae lost the horny scutes and bore only thin dermal plates beneath the skin. The advantage of increased buoyancy apparently outweighed the loss of the hard shield, for the trionychids became widespread in the Cretaceous period and remain so today. Study of the living forms suggests that the longevity of the group may be due in some part to their thorough adaptation for underwater existence (Fig. 7.17). Since their mucous membranes in the pharynx or cloaca serve a respiratory function, the trionychids can stay submerged or buried in the mud. Their feet are webbed for swimming, the head streamlined and the body flattened for smooth passage through the water. Not placid animals, they attack fish or crush mollusks and defend themselves effectively with their sharp, strong beaks.

The marine turtles committed themselves even more completely to water than the trionychids by evolving limbs excellent for swimming but almost useless on dry ground. When their

Figure 7.17. Recent turtles (order Chelonia). **(a)** *Trionyx*, freshwater soft-shelled turtle. **(b)** *Emys*, pond terrapin. **(c)** *Chrysemys*, painted terrapin. **(d)** *Macroclemmys*, snapping turtle. **(e)** *Testudo*, tortoise. **(f)** *Chelone*, hawksbill turtle. **(g)** *Dermochelys*, leathery turtle.

(a)

(b)

(c)

(d)

(e)

(f)

(g)

fossil history begins, late in the Cretaceous period, marine turtles existed showing appendages in various stages of transition from the normal tetrapod condition to the flipper shape which was most advanced. The most primitive genera, now all extinct, kept the large hind legs that amphibious turtles had used to propel themselves forward but had already exchanged the forefeet for flexible flippers more advantageous for paddling. In more progressive forms, the rear appendages had diminished in size and were acting chiefly as rudders, leaving the enlarged pectoral flippers as the principal swimming organs. The most advanced of the Cretaceous sea turtles, the ancestors of those living today, had increased the rigidity of the flippers by restricting mobility at the joints in the wrist and hand. In the extant loggerhead, hawksbill, and green turtles, heavy ligaments binding the bones and an especially tough skin also contribute to stiffening the flippers. These turtles, members of the family Cheloniidae, no longer move the flippers back and forth but beat them vertically like wings. Unlike their terrestrial cousins, the tortoises, the chelonids are capable of graceful motion and great speed. Since Cretaceous time they have grown large, but their buoyancy has been maintained by reduction in the bone of the shell.

Other marine chelonians, the leathery turtles, have paralleled the chelonids in the evolution of flippers and the loss of much of their armor. In the living *Dermochelys* and its ancestors, the ossification of the carapace is so weak that the ribs and backbone are freed of their attachment to the exoskeleton. For a time, paleontologists thought they might have discovered in the leathery turtles relicts of primitive forms which had not yet evolved a heavy shell, but the apparent terrestrial origin of the Chelonia and the late appearance of the group to which *Dermochelys* belongs rendered that possibility unlikely.

The consensus is that all the marine turtles are highly specialized and exceedingly well adapted for an environment generally hostile to tetrapods. They failed in only one way to suit themselves for the sea: not having become viviparous, the females must swim to shore and climb the beach to bury their eggs in the sand. The eggs and the newly hatched animals are at the mercy of all the terrestrial enemies from which the adults have escaped. The eggs are sought and eaten both by four-footed beach dwellers and by man, and the hatchlings are seized by sea birds as they scramble from the nest toward the surf and the deep water beyond. Against these depredations the turtles have defended themselves by laying more eggs in a clutch than any other reptiles and by evolving an instinctual sense for navigation that brings them unerringly to their breeding areas from far out at sea.

If marine turtles had become extinct after their Cretaceous radiation, paleontologists doubtless would have attributed their disappearance to the inadequacy of their reproductive habits or to the failure of their specialized body form. Neither of these explanations seems logical as a way of accounting for the extinction of the ichthyosaurs. These reptiles had become completely fish-like in shape by the middle of the Triassic period and are known, from a specimen preserved with embryos inside, to have evolved the ability to release their young directly into the water. The beautiful fossils compressed in shale from Europe, North America, and continents in the Southern Hemisphere reveal animals which were structurally as well adapted for survival in the open ocean as any vertebrates could be. It is difficult to believe that they would have fallen prey to the more awkward-looking nothosaurs and plesiosaurs or that they would have been less able than these reptiles to escape from the hybodont sharks or the large and carnivorous bony fishes with which they shared the sea. It is possible that their downfall was due to some physiological shortcoming, to the loss of their food supply, or to behavior which led to their selective destruction, but evidence for these speculations is not forthcoming from fossil remains.

There is no better example of convergent evolution than that provided by the ichthyosaurs' resumption of fish-like form (Fig. 7.18). The flexible neck characteristic of tetrapods disappeared, and the head merged once again with the trunk, which broadened around the viscera and then tapered toward the tail. The fusiform body was stabilized in the water by a large, triangular

Figure 7.18. An ichthyosaur of Jurassic time. (From Romer.)

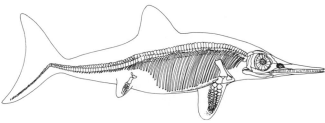

dorsal fin and by the paired appendages which had become broad-based projections similar in shape to the fins of archaic sharks. Like the fishes, the ichthyosaurs depended for locomotion upon undulations of the body, and by Jurassic time they had evolved a great crescent-shaped caudal fin stiffened in the lower lobe by the end of the vertebral column. The column itself regained flexibility by failing to develop the reinforced structure which characterized it in reptiles that walked on land. The centra remained amphicoelous and thus met edge to edge without interlocking, and the zygapophyseal joints between adjacent neural arches grew weaker as the ichthyosaurs evolved. As in many more ancient swimmers, the neural arch in each vertebra was not fused to the centrum but separated from it by a pad of cartilage. Built in this way, the vertebrae were no longer adapted to withstand stress projected from the appendicular skeleton. Indeed, there was none, since the pelvic girdle lost its connection to the column and the body became entirely waterborne.

In redeveloping the form and swimming habits characteristic of fishes, the ichthyosaurs did not controvert the rule, known as Dollo's law, that evolution is irreversible. The structure and physiological attributes of the fishes ancestral to tetrapods depended upon a particular genetic complex which was lost little by little as mutations altered the hereditary material over millions of years. The reacquisition of the old genotype of fishes was impossible, as reconstitution of any former genotype is, because environmental conditions and internal factors do not repeat themselves in such a way that a series of genetic changes occurs in reverse order. Animals like ichthyosaurs do not appear as a result of reversion or even partial reversion of an evolutionary progression that has taken place. They retain the underlying characteristics of the group from which they sprang and add new traits, imitative of ancient ones perhaps but based upon a different and unique configuration of the genes. Even a cursory examination of the ichthyosaur skeleton reveals that its possessor was a perfectly good reptile. The shoulder girdle, for example, though not extensive dorsally, was clearly built according to the reptilian pattern, with separate scapular and coracoid ossifications. Its dermal bones consisted of a pair of clavicles and an interclavicle, there being no trace of a cleithrum or any of the elements which in fishes connected the girdle with the skull. Although the pelvic girdle did not articulate with the axial skeleton, it retained the ilium, ischium, and pubis unfused and, in Jurassic ichthyosaurs, vestigial, but recognizable, neverthe-

Reptilian nature of ichthyosaur skeleton

less. The bones of the extremities are more difficult to identify because they multiplied and lost their distinctive shape in forming a support for the fin-like appendages. It is obvious, however, that the appendicular elements are derivatives of the bones in the regular pentadactyl leg and are not at all similar to the basal and radial pterygiophores of fishes. Since fin rays lost in the transition from rhipidistians to tetrapods could not be re-evolved and no substitute for them appeared, the part of the appendage beyond the bones was supported in ichthyosaurs only by tough connective tissue. No skeletal material of any kind formed in the dorsal fin or in the caudal fin above the vertebrae, so that these structures, too, while resembling those of fishes, were in fact ichthyosaurian innovations rather than reconstituted piscine traits.

The axial skeleton shows clearly the reptilian character of the ichthyosaurs. Despite the absence of sacral vertebrae and the similarity of those in the neck and trunk, the vertebral column could not have been mistaken for that of a fish. The centra were pleurocentra, bearing two-headed ribs in the trunk region and articulating with reduced chevron bones rather than well-developed hemal arches in the tail. In most ichthyosaurs the neural spines were low, not tall and slender as they are in fishes, and the neural arches articulated with one another substantially as they did in other reptiles. The skull was also thoroughly reptilian (Fig. 7.19). In a Triassic ichthyosaur like *Grippia,* the reptilian arrangement of the bones was quite evident. In more specialized forms, the shape of the skull changed radically, but its reptilian structure remained discernible. As the ichthyosaurs became adapted for marine life, the snout became an elongated, toothed beak and the eyes enlarged at the expense of the cheek. Even though the elements covering the cheek were crowded and reduced in size, a temporal fenestra developed high behind the eye. At first it seemed unique among reptilian fenestrae in being bounded posteriorly by a much expanded supratemporal bone, which extended downward and backward to intervene between the quadrate and the braincase as the squamosal usually did. After restudying several ichthyosaurian skulls, however, Romer suggested that the bone originally termed the squamosal was, in fact, the upper part of the quadratojugal element, and that the "supratemporal" bordering the fenestra was the true squamosal. If Romer's interpretation is correct, the temporal fenestra of the ichthyosaurs would be comparable to that in the nothosaurs, plesiosaurs, and placodonts.

Uncertain ancestry of the ichthyosaurs

The similarity in location of the temporal fenestrae has not

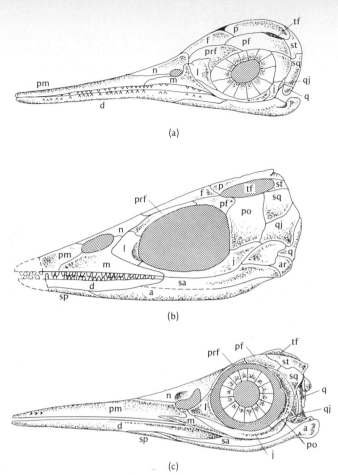

Figure 7.19. Skulls of ichthyosaurs. Triassic genera showing primitive traits: **(a)** *Mixosaurus* with small temporal opening (*tf*); **(b)** *Grippia* with relatively short snout; **(c)** *Ophthalmosaurus*, an advanced Jurassic ichthyosaur with long, toothless jaws and large, dorsally situated temporal opening. Abbreviations as in Fig. 6.7. (From Romer.)

convinced paleontologists of a relationship between the paddle-legged marine reptiles and the ichthyosaurs. Aside from their common reptilian features, they were little like one another. The adaptation of the ichthyosaurs for the sea was entirely different and more extreme. The magnitude of the changes involved in the evolution of the ichthyosaurs suggested to investigators that rather than diverging from the ancestral stock of the nothosaurs and their kin, the forerunners of the ichthyosaurs stemmed independently from animals near the base of the reptilian line. This idea was seconded by the apparently primitive lack of an otic notch in the ichthyosaur skull and the large stapes which fitted against the quadrate bone, as it did in the earliest reptiles. Assuming that the ichthyosaurian disposition toward aquatic living was of ancient origin, several paleontologists have proposed their descent from forms which never became wholly terrestrial. Since the teeth of ichthyosaurs show infolding of the enamel, the possibility of their evolution directly from labyrinthodont amphibians has been raised. Romer believes it more likely that the precursors of the ichthyosaurs were allied in some way to the ophiacodonts, primitive pelycosaurian reptiles of amphibious habits. Although ophiacodonts exhibited none of the specialized characteristics of ichthyosaurs, their type of skull seemed more suitable than any other as an antecedent for that of the Triassic seafarers. Only one trait set them clearly apart from the ichthyosaur line: whereas the temporal fenestra of ichthyosaurs was high on the head, in ophiacodonts it was established in the synapsid position, below the postorbital and squamosal bones. There was, among the reptiles surely descended from the pelycosaurs, a Triassic form, *Cynognathus,* in which the temporal fenestra had moved dorsally, but Romer found no grounds for speculating that a similar change in the position of the opening had taken place in the evolution of the ichthyosaurs. In his opinion, the ancestors of the ichthyosaurs might have branched away from the primitive anapsid animals which were the direct forebears of the ophiacodonts. Since these "preophiacodont" reptiles are not recognizable in the fossil record, however, the origin of the ichthyosaurs is a problem which remains wholly unresolved.

Paleontologists have made better progress in tracing the history of the reptiles which remained on land. Although the first known members of a terrestrial group may appear with their characteristic traits fully developed, their primitive features are often less completely masked than those of marine *Synapsid reptiles* reptiles. The synapsid pelycosaurs, for instance, when they entered the fossil record in the Carboniferous period, already showed the low-set temporal fenestra but retained so close a resemblance to the romeriid captorhinomorphs that their descent from them is indisputable. In the case of the pelycosaurs, diversification produced a variety of adaptive types whose interrelationship can be analyzed in some detail (Fig. 7.20). Because they lacked the specialization of other pelycosaurs and

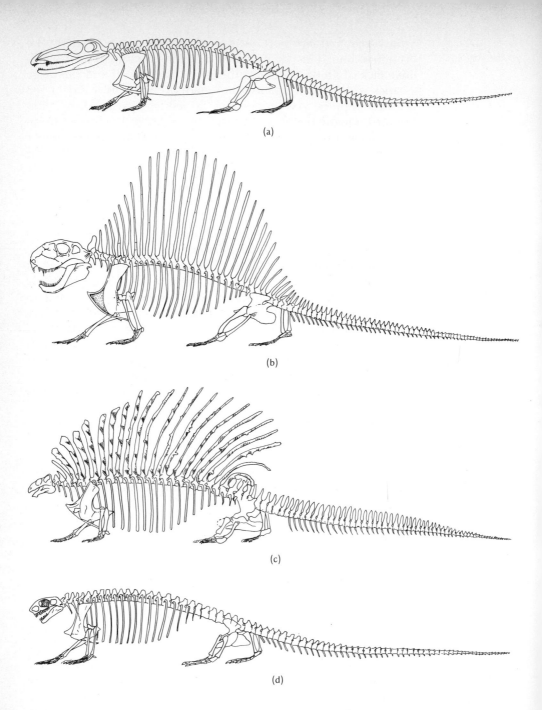

(a)

(b)

(c)

(d)

differed least from the romeriid captorhinomorphs, the ophiac- odonts are considered to represent the root of the synapsid line. Although they were still amphibious like the oldest tet- rapods, their skull had become higher, narrower, and less ponderous-looking than that of their predecessors. An animal like *Varanosaurus,* with its long jaws and many teeth, probably still chased fish, but its two maxillary fangs and large, elevated eyes suggested its relationship to the Permian pelycosaurs which hunted four-legged prey. Having made comparative studies of the postcranial skeleton as well as the skull, paleontologists have concluded that one or possibly two lines of terrestrial carnivores descended from ophiacodont reptiles.

The group of paramount importance was that of the sphenac- odonts. In these animals, the teeth were large and knife-like. The most advanced sphenacodonts evolved great tusks in the front of the mouth and in the canine position that could have been used only for biting and tearing flesh. Since the back of the skull had deepened considerably and the jaw joint had dropped below the tooth row, the gape was wide and the closure powerful. The structure of their appendages implies that the sphenacodonts could overtake their victims easily. They still walked with limbs outspread, but their legs were less stocky than those of the ophiacodonts and presumably allowed them to cover the ground faster. Besides becoming structurally adapted for a more active life, the sphenacodonts may have undergone physiological changes as well. Many of them had evolved a huge, dorsal "sail" in the midline—a vertical fold of skin supported by mast-like extensions of the neural spines—which seems to have been a device for regulating the temperature of the body. If the animals did gain or lose significant amounts of heat through the vascular sail, they would have recovered more quickly from the sluggish state that nocturnal cold induced and been able to stay abroad through a larger part of the sunny day. The sphenacodonts were the most progressive but perhaps not the only carnivores to develop from ophiacodont stock. Some forms with enlarged teeth, like *Eothyris,* lacked sphenacodont cranial characters and apparently belonged to another line, but the fragmentary nature of their fossils has led to disagreement concerning the group with which they should be allied.

Figure 7.20. Skeletons of Permian pelycosaurs (subclass Synapsida). **(a)** *Varan- osaurus,* an ophiacodont. **(b)** *Dimetrodon,* a sphenacodont. **(c)** *Edaphosaurus* and **(d)** *Casea,* two herbivorous forms. (From Romer.)

Although the carnivorous pelycosaurs might have sustained themselves on a diet of cotylosaurs and terrestrial amphibians, they would also have found other synapsids available. The latter were herbivores that evolved during late Carboniferous times and maintained themselves through the early Permian years in a highly specialized state. Like the sphenacodonts, these pelycosaurs kept traces of their ophiacodont ancestry but acquired traits that suited them for a different mode of life. Their teeth were strong, even-sized, and not so sharp as those of meat eaters. They were implanted at the edge of the jaws and also on the palate. The palatal teeth were usually aligned in rows, but *Edaphosaurus,* one of the earliest herbivores, bore them on a pair of plates which met counterparts mounted on the inner side of the mandible below. Except for its dorsal sail supported by peculiar, knobby spines, *Edaphosaurus* resembled other pelycosaurian herbivores. They were all small-headed forms with bulbous bodies slung low on strong, squat legs. There seem to have been two separate radiations of these animals. The associates of *Edaphosaurus* flourished at the beginning of the Permian period, grew large, as all successful pelycosaurs did, and then disappeared. Their niche was filled by the caseids, herbivores related to the edaphosaurs but not derived directly from them. Although the body form of *Casea* and its allies was like that of the earlier plant eaters, the skull was oddly shaped, the short facial region being entirely given over to enormous nasal openings. The ancestral source of the caseids is not known exactly, but paleontologists have theorized that they may have branched away at a later time from the same stock that gave rise to the edaphosaurs. Since their remains have been found in the southwestern United States and in Russia, the caseids must have become an important element of the continental fauna before their extinction in mid-Permian time. The habits of life on which their success was based are not quite clear. Their feet were clawed and the legs equipped with strong extensor muscles as if adapted for digging, but the purpose such activity would have served is difficult to imagine. Permian plants had evolved no taproots or tubers to uncover, and the large size of some of the caseids makes it unlikely that they were burrowing animals. Olson suggested that they might have scratched the edible exterior from the conifers and seed-ferns which grew in the forests, a method of food gathering not used by any reptile today.

Therapsids

Long before the caseids and other pelycosaurs were described, another group of synapsids had been discovered in late Middle and Upper Permian strata in South Africa. Owen published studies of some of the South African forms, called therapsids, in 1876, when modern paleontology was in its infancy, and Broom, Watson, Boonstra and other investigators continued to collect and examine new specimens afterward until hundreds of genera were defined. What fascinated paleontologists at first was the resemblance of these animals to mammals. Eventually, they found the relationship of the therapsids to the pelycosaurs equally intriguing. Like the pelycosaurs of the American southwest, the African therapsids had diversified into carnivores and herbivores. Both types showed evidence of affinity to the pelycosaurs in their skeletal structure but had advanced beyond the primitive reptilian state. Their legs were pulled under them instead of sprawled on either side of the trunk, the vertebral column was strengthened by increased ossification of the centra, and the tail was no longer long and lizard-like. The skull had a much enlarged temporal fenestra, a shorter jaw, and a palate no longer movable upon the cranium. There was a tendency toward loss of some of the dermal roofing elements which the reptiles had inherited from early tetrapods, and the ancient intercentra were also missing from the column.

The therapsids dominated the fauna of their time and diversified so widely that establishing interrelationships among the known forms has been a herculean task. Although a number of different assemblages can be identified, paleontologists do not agree exactly upon their arrangement in the phylogenetic scheme or upon the names to be used for the various groups. The term "Anomodontia," for instance, was applied by Watson and Romer to the suborder inclusive of all the herbivores, whereas other investigators had used it for one particular group of plant-eating therapsids also called dicynodonts. The dicynodonts were by far the most successful of the herbivores and the only ones to survive throughout the Upper Permian and Triassic periods. They cropped vegetation with horn-covered jaws, usually toothless except for two canine spikes, and eventually became large, lumbering, barrel-bodied animals similar in general appearance to some of the modern herbivorous mammals. Having few rivals in Upper Permian times, they distributed themselves over lands which now constitute parts of Africa, South America, North America, and Asia. One of their number, *Lystrosaurus,* was found in Lower Triassic rocks of Antarctica, an area which, near the beginning of the Mesozoic era, may have been contiguous with southern Africa and was in any case much warmer than it is now. The awesome multiplication of the dicynodonts was perhaps responsible for the extinction of other kinds of herbivorous therapsids. There had been, in mid-Permian times, forms called dromosaurs and dinocephalians, both of which developed into plant eaters without losing their teeth. The dromosaurs evolved as rat-sized animals and the dinocephalians as ever enlarging, heavy-headed beasts, but neither group approached the dicynodonts in numbers or longevity.

Besides dicynodonts, Permian and Triassic rocks in South Africa yielded a variety of carnivores collectively assigned to the suborder Theriodontia (Fig. 7.21). The earlier ones were reminiscent of the sphenacodont pelycosaurs but were already more advanced in the differentiation of their teeth and the design of

Figure 7.21. Two therapsids from the Lower Triassic of South Africa (subclass Synapsida). **(a)** *Kannemeyeria*, a dicynodont herbivore. (From Pearson.) **(b)** *Cynognathus*, a cynodont carnivore. (From Gregory and Camp.)

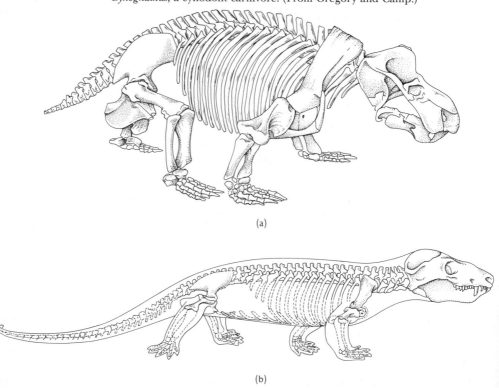

(a)

(b)

their jaw mechanism. Rarely was there more than a single canine tooth on each maxilla, and the distinction between the shape of the biting teeth at the front of the mouth and those in the cheek region was quite clear. The dentary bone in the mandible had enlarged at the expense of the other elements and sent a coronoid process upward under the edge of the skull roof behind the eye, providing for the insertion of a powerful temporalis muscle which could bring about compression of the jaws. From the structure of the available skulls, paleontologists knew that there must have been, even in the lowest fossiliferous stratum in the South African beds, at least two lines of carnivorous therapsids represented. In the *Tapinocephalus* zone (a layer named after a dinocephalian whose remains are commonly found in it), most of the carnivorous genera discovered could be classified as therocephalians. These animals possessed the most primitive-looking teeth of all the theriodonts but had developed large, elongate temporal fenestrae and a middorsal bony crest that allowed ample space for the attachment and expansion of the jaw muscles. The other carnivorous forms of the time, known as gorgonopsians, were less advanced in the structure of the temporal region and might have been considered as representative of the ancestral stock of the theriodonts had it not been for the specializations they showed—reduced cheek teeth and an odd, median preparietal bone in the skull roof. The gorgonopsians outnumbered the therocephalians by the end of the Permian period, but neither group survived into the Triassic. Their place was taken by more advanced carnivores, the bauriamorphs and the cynodonts, animals which paralleled one another in evolving traits approaching those of mammals. The bauriamorphs, apparently descended from therocephalians, fade from the fossil record by the middle of the Triassic period. The cynodonts survived somewhat longer, preying upon the few remaining herbivorous dicynodonts and competing with the diapsid reptiles which were becoming predominant. Near the end of their existence, the cynodonts gave rise to some aberrant forms, called gomphodonts, which may have been able to add plant material to their diet. With their cheek teeth molarized, these animals prospered briefly in Africa and South America but then vanished with the rest of the therapsids. No synapsids resisted the onslaught of the diapsid reptiles, it seems, except those which neared or attained the mammalian level.

Origin of therapsids from pelycosaurs

Paleontologists studying the beginning of therapsid history rather than the end of it suggested at once that the South

African therapsids might be allied to sphenacodont pelycosaurs which had not yet evolved the specialized middorsal sail. At the time it was proposed, this theory was based entirely upon comparative anatomical studies, because no intermediate forms were known. There was, in fact, a long gap in the fossil record that could not be filled. The sphenacodonts were collected from Lower Permian deposits in Texas widely separated in time from the late Middle and Upper Permian rocks of South Africa in which the therapsids had been found. Fossiliferous strata dating from the years between were unavailable. After World War II, when international communication was again possible, British and American paleontologists turned their attention to sites their colleagues were exploring in Russia. In several localities between the Caspian Sea and the Mesen and Dvina Rivers in the north, Russian investigators had discovered a wealth of Permian tetrapods. Because there were labyrinthodont amphibians, cotylosaurs, pelycosaurs, and several kinds of therapsids, it was suspected that the Russian deposits might include the Lower Middle Permian strata missing elsewhere. Correlating the Russian beds with those in the American southwest and in Africa was difficult, however. The sediments in Russia had settled in lowlands and deltaic regions instead of on higher, drier terrain, as they had in South Africa, so that the difference in the vertebrate fauna in the two areas could not be ascribed solely to a discrepancy in their age. The invertebrate genera, useful as a rule in correlating deposits, were not very helpful in this case. Many of them were provincial, and those that were represented in more than one region endured too long to serve as markers of specific time zones within the Permian period.

As nearly as paleontologists could tell, the youngest fossiliferous rock formations in Texas and Oklahoma explored by Olson were approximately contemporary with certain of the lower Russian ones, and the latter were somewhat older than those at the base of the African series. Both the American and the Russian deposits gave evidence of the gradual triumph of the early reptiles over the most terrestrial of the amphibians, but the Russian beds yielded therapsid animals unknown in the Permian sediments of the Western Hemisphere. Some of them were, as paleontologists had hoped, forms more primitive than the African species. A half dozen genera assigned to the family Phthinosuchidae were almost surely the long-sought intermediates between the carnivorous therapsid gorgonopsians and the sphenacodont pelycosaurs. Their relationship to the sphenacodonts was indicated not only by a general similarity in the structure of the skull but also by the possession of peculiarly sphenacodontid traits like the protuberance on the angular bone at the rear of the lower jaw. The therapsid character of the phthinosuchids was also perfectly clear. They had evolved the modifications of the temporal region and the mandible that allowed increased compression of the jaws and had developed the huge canines that distinguished all the later theriodonts.

The Russian deposits have given up no specimens which link herbivorous therapsids to animals at the pelycosaurian level. Romer believes that there were no such forms and that the earliest dinocephalians evolved from primitive carnivorous therapsids of phthinosuchid ancestry. He regards the South African titanosuchids and the Russian brithopodids, animals bearing carnivorous teeth but otherwise similar to dinocephalians, as survivors of this ancient stock. The dicynodonts split away, he thinks, from the base of the dinocephalian line, since *Venjukovia* of the Russian Middle Permian appears transitional between the early toothed herbivores and the later toothless forms. Olson, who studied the Russian and African therapsid herbivores intensively, has proposed an alternate version of their history. In his opinion, the group may not have been a natural one but an assemblage of animals which reached the same grade. He is not convinced, either, that any of the herbivores were derived from the phthinosuchid carnivores. He acknowledges the relationship of the brithopodids to the phthinosuchids but does not agree with Romer that there was a progression from those animals to the titanosuchids and thence to herbivorous dinocephalians. He separates the herbivores from the carnivores completely, deriving them independently from the pelycosaurs. Although the connection is less sure than that of the carnivores and the sphenacodonts, he suggests an alliance between the herbivorous therapsids and the caseids. Rather than linking the dicynodonts to dinocephalian ancestors through *Venjukovia,* as Romer did, he traces both lines to a source among the caseoid pelycosaurs. He emphasizes the tentative nature of these hypotheses, however, and supposes that new evidence will show them to be oversimplified.

Decline of the therapsids

Since the large and well-adapted therapsid reptiles had virtually no enemies except each other, the extinction of most of them at the beginning of the Triassic period seems to indicate some change in the environment which they could not withstand. Paleobotanists know that the flora in the Southern

Hemisphere changed markedly as the Mesozoic era opened, and Watson speculated that the disappearance of the therapsids may have followed the loss of the particular plants on which the herbivorous forms depended. The reptiles best fitted to survive under such circumstances were those that fed upon invertebrates, especially upon the insects which by the Triassic period were everywhere on land. Insectivores, small animals with numerous sharp conical teeth, were not found among the therapsids or specialized cotylosaurs like the procolophonids and pareiasaurs but among the diapsid forms newly evolved in Upper Permian time. These reptiles prospered as the others dwindled and, before the Triassic period closed, produced aquatic forms and, on land, carnivorous and herbivorous variants of their kind. They radiated so profusely, in fact, that paleontologists are still struggling to trace the intricate branching of their family tree.

Ancestry of the diapsid reptiles

Initially, all the reptiles with two temporal fenestrae were assumed to be closely related. Watson identified them as sauropsids on the basis of the structure of their middle ear and traced them back, separately from the cotylosaurs and their synapsid kin, to the anthracosaur amphibians. As intermediates between the ancient anthracosaurs and the Upper Permian diapsids, Watson proposed the millerettids, primitive Permian reptiles similar in some ways to captorhinomorphs but having a sauropsid middle ear. *Millerosaurus* had first been described by Broom as a forerunner of the mammal-like reptiles because it displayed a temporal fenestra below the postorbital and squamosal bones. When he restudied the specimen, Watson noticed that the posterior border of the cheek was concave and the quadrate bone was exposed in lateral view, as in diapsids. The quadrate fitted into a notch in the squamosal above it, just as it did in early lizards and primitive archosaurs. Since the remaining part of the skull was generalized in its structure, the appearance of a second temporal opening above the first one would have transformed the millerettid into an animal perfectly acceptable as a diapsid.

Although Romer was willing to consider the millerettids as possible ancestors of diapsid reptiles, he disagreed with Watson's theory that they fathered the whole assemblage of two-arched forms and could themselves have been traced back to anthracosaurs independently of the captorhinomorphs. The resemblance between millerettids and captorhinomorphs he interpreted as evidence of their belonging to the same line, not as the result of parallel inheritance of anthracosaur traits. Since he believed that the difference between the middle ear in captorhinomorphs and other reptiles was minimal, he found little difficulty in deriving the latter from the former. The exposed quadrate could have evolved, as Parrington suggested, by the reduction of the squamosal bone beside it. The shallow otic notch that appeared in the skull roof of the millerettids as the shape of the squamosal changed and the jaw shortened did ally those animals to the lizards and their relatives, but Romer pointed out that the same feature made it unlikely that the millerettids were also the ancestors of the archosaurian reptiles. Although the crocodiles, the pterosaurs, the dinosaurs, and their immediate predecessors had an otic notch, the earliest archosaurs, the proterosuchians, did not. Romer viewed the otic notch of the advanced archosaurs as an independent development and asserted the necessity of looking for ancestors of the primitive proterosuchians among the captorhinomorphs that

still had a long jaw and an unindented skull roof. His belief that the ancestors of the archosaurs differed from those of other diapsids led him to propose a new classification of the two-arched reptiles. He recognized the archosaurs as a natural group whose characteristics (except for the diapsid condition of the cheek) were not especially similar to those of the contemporary Mesozoic lizards, rhynchosaurs, and rhynchocephalians. The latter forms, which bore an underlying resemblance to one another, he also accepted as a monophyletic assemblage, categorizing them as lepidosaurs. Since Romer's view prevailed, Osborn's old subclass Diapsida was replaced in the phylogenetic scheme by the subclasses Archosauria and Lepidosauria.

If the lepidosaurs were, in fact, descended from the millerettids, they must have come from members of that group anterior to *Millerosaurus* itself (Fig. 7.22). The oldest known diapsid reptiles were contemporaries of *Millerosaurus* in Upper Permian South Africa, living on a diet of invertebrates and keeping out of the way of the large therapsid carnivores. Although they were as primitive in the structure of their skull as the millerettids, by evolving upper and lower temporal fenestrae early diapsids like *Youngina* had abandoned the solid skull roof that was a characteristic feature of their ancestors. Rather than gaining its strength from the thickness and continuity of its dermal elements, the posterior part of the roof was supported in these animals as the wall of a gothic cathedral is by piers properly located to receive and transfer stress. The lighter construction of the skull and the increased bulk of the jaw musculature it

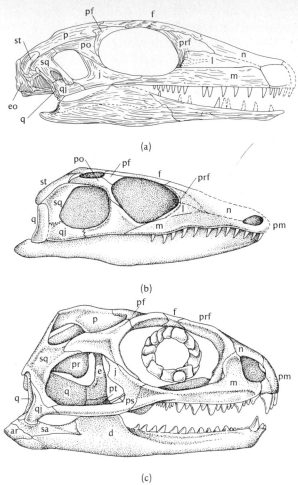

Figure 7.22. Skulls of **(a)** *Millerosaurus*, an Upper Permian form, of stock perhaps ancestral to diapsid reptiles; **(b)** *Youngina*, an early diapsid contemporary of *Millerosaurus*; **(c)** *Sphenodon*, a lepidosaurian diapsid and the sole surviving rhynchocephalian. Abbreviations as in Fig. 6.7. (*a* from Watson; *b,c* from Romer.)

allowed were evidently highly advantageous to reptiles otherwise not very progressive, for descendants of *Youngina* and its relatives have survived to the present day.

Sphenodon and other rhynchocephalians

The living animal most like the Upper Permian diapsids is *Sphenodon,* the 2- to 3-foot-long tuatara which inhabits islands off the coast of New Zealand. It seems to be the last of the rhynchocephalians, reptiles whose skeleton, at least, differed very little from that of *Youngina. Sphenodon,* like the Mesozoic rhynchocephalians that preceded it, lost the supratemporal and tabular bones from the back of the skull roof and developed a beak-like projection of the upper jaw, for which the group is named. The teeth under the beak, as well as the more posterior ones, have become fused to the rim of the jaw instead of being sunk into sockets, as they were in *Youngina* and its fellows. *Sphenodon* has been called a living fossil because of its archaic skeletal structure, but it is hazardous to assume that the animal is closely representative of the primitive diapsids. During its long confinement to New Zealand, *Sphenodon* has undoubtedly evolved specialized traits. It is hard to believe that its sluggish behavior could have characterized the Mesozoic rhynchocephalians which for millions of years maintained themselves in the presence of the agile, carnivorous archosaurs. Also, *Sphenodon* has developed an extraordinary tolerance for cold, remaining active at temperatures low enough to prevent other reptiles from moving about. In its outward appearance, however, *Sphenodon* is surely similar to its forebears. It has kept the relatively small, long-tailed body supported between outspread legs that was characteristic of Permian and Mesozoic forms and, like them, has retained the sharp teeth of an insectivore.

Rhynchocephalians which departed further than *Sphenodon* from the ancestral condition came to grief before the end of the Triassic period. Such forms were the rhynchosaurs, animals that enlarged, pulled their legs under them, and evolved teeth designed for a specialized diet (Fig. 7.23). Their beak was

Figure 7.23. Skull of *Scaphonyx,* a rhynchosaur from the Middle Triassic of South America. Abbreviations as in Fig. 6.7. (From Romer.)

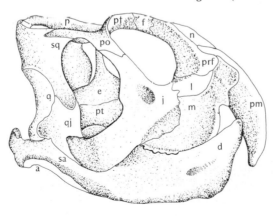

toothless, consisting of grotesque extensions of the premaxillary bones which overhung the upward-curving anterior end of the mandible. Large toothplates on the upper jaw bore numerous rounded teeth and received the cutting edge of the lower jaw in a long groove. The muscles which operated this strange dental battery were massive. The lower jaw deepened to allow their insertion, and the jugal bone beneath the orbit flared widely, affording them adequate passage upward to the skull roof. Whatever these reptiles ate—suggestions range from mollusks to heavy-husked fruits—must have been available almost everywhere, because, in the Middle Triassic, rhynchosaurs ranged throughout the Southern Hemisphere and (scanty remains reveal) into the Northern Hemisphere as well. Since their extinction coincided with the rise of the dinosaurs, it is probable that the rhynchosaurs, as well adapted as they were for their special mode of life, had no defense against these fiercer animals.

Appearance of the lizards

The only lepidosaurs which held their ground and diversified in the presence of the dinosaurs and other archosaurs were the true lizards. Almost as conservative in their postcranial skeleton as *Sphenodon,* the lizards seem to owe their success to the evolution of a highly mobile and versatile jaw apparatus. They appeared in the Triassic period as small reptiles in which the quadrate bone was no longer held in place by being sutured to the squamosal and quadratojugal bones at the back of the cheek. Since the quadratojugal bone, which had formed the ventral boundary of the lower temporal fenestra, was gone, the quadrate stood free beside the scooped-out cheek and could swing back and forth at its contact with the squamosal. As a consequence, the mandible could be lowered by pivoting at its ancient hinge and then dropped still farther by retraction of the quadrate to which it was attached. In time, the design of the jaws was further refined, and mobility developed between parts of the skull, aiding the ingestion of large or struggling prey. Late in the Cretaceous period, the lizards produced, as an offshoot from their line, the snakes, animals in which the jaw and skull bones became even more loosely connected and extreme specialization of the body arose. Lizards and snakes, together constituting the Squamata, are the most numerous of the modern reptiles, a fact attesting to the advantage of the flexible mouthparts whether they are borne by reptiles conservative or specialized in the rest of their anatomy.

Diapsid precursor of the lizards

The lizards were once thought to have been derived from

parapsid forms by excavation of the cheek below the dorsally located temporal opening, but the discovery in Lower Triassic rocks of South Africa of a near lizard with an almost completely enclosed lower temporal fenestra proved that the lizards were modified diapsid reptiles. *Prolacerta,* as the specimen was called, had a skull which differed little from that of *Youngina* except for its much reduced quadratojugal element (Fig. 7.24). Since that bone failed to reach the process of the jugal bone in front of it, there was a small gap in the rim of the temporal fenestra at its posteroventral corner. *Prolacerta* and the two or three forms apparently allied to it could not be included in the order Squamata because the squamosal bone was still too extensively attached to the quadrate to allow that bone to move. By late Triassic time, however, the part of the squamosal that had descended along the anterior edge of the quadrate had disappeared, so that the contact between the two bones was reduced substantially. *Kuehneosaurus,* found in Upper Triassic sediments in Britain, showed a quadrate which fitted into a rounded depression on the underside of the squamosal bone. Since the arrangement was one that would have permitted the quadrate to

Figure 7.24. Skulls of **(a)** *Prolacerta,* a Lower Triassic lepidosaur possibly ancestral to the lizards; **(b)** *Kuehneosaurus,* an Upper Triassic lizard specialized for gliding flight; **(c)** *Clidastes,* an Upper Cretaceous marine lizard (a mosasaur); **(d)** *Iguana,* a modern lizard. Abbreviations as in Fig. 6.7. (From Romer.)

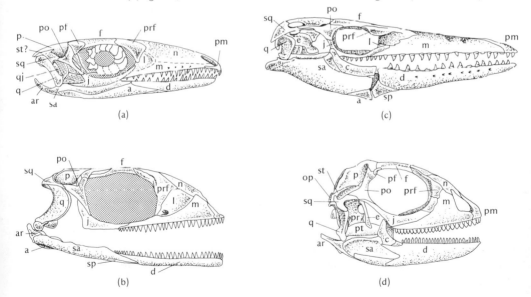

swing, *Kuehneosaurus* is considered a very early true lizard.

Early specialization in lizards

Despite its retention of palatal teeth and paired dermal elements in the skull roof, traits which most modern lizards do not display, *Kuehneosaurus* can hardly be regarded as a primitive member of the Squamata. It was already specialized in having elongated ribs, doubtless supports for an extension of the body wall which acted as a flight membrane. That gliding lizards were well established by the end of the Triassic period is indicated by the finding of another such form, *Icarosaurus,* in Upper Triassic strata in New Jersey on the North American east coast. The evolution of the lacertilian jaw mechanism evidently led to the immediate diversification of the group in which it appeared. Robinson reports the discovery of lizards contemporary with *Kuehneosaurus* and *Icarosaurus* that lack the specialized ribs but show a number of more advanced features in the skull. As the lizards radiated, there were apparently parallel developments in the different lines that emerged, because the known Triassic forms exhibit various assortments of old and new characters.

Although the fossil record of lizards of Jurassic and Lower Cretaceous age is poor, paleontologists feel certain that the modern lacertilian fauna evolved during that interval and replaced the older forms. The new lizards radiated, as the earlier ones presumably had, into a variety of ecological niches.

Mosasaurs

Some small semiaquatic species gave rise to the marine mosasaurs, animals which may have divided their time between basking on the beach and hunting large fishes and other dwellers in the sea. These animals, which grew monstrous, had a lower jaw with a joint midway along its length on each side so that the front part of it could be depressed farther than usual. This construction and the replacement of the tight connection between the left and right halves of the mandible by a ligamentous one fitted the mosasaurs for ingesting huge animals without first tearing them apart. Although terrestrial lizards related to the mosasaurs have survived to the present day, the mosasaurs themselves disappeared at the end of the Cretaceous period. Besides the Mosasauridae, no other family of lizards became adapted for life in the ocean. Only one modern form, *Amblyrhynchus* of the Galapagos Islands, swims in salt water, and that lizard is a herbivore, subsisting on the seaweed which grows off the rocky shore.

Modern lizards

The members of the family Iguanidae, to which *Amblyrhynchus* belongs, spread into warm areas worldwide and habituated themselves to living in the trees as well as on the ground. Among the existing iguanids there are many that can navigate in fresh water, some that live in the desert, and at least one genus that survives in the cold of the Andes Mountains of South America. Despite their adaptability, it seems that the iguanids were replaced in the Old World by the Agamidae, a group perhaps descended from them. The agamids, the oldest of which are known from Upper Cretaceous times, also acclimated themselves to several different habitats and produced in *Draco* a gliding form which, like *Kuehneosaurus* and *Icarosaurus,* has a flight membrane supported by ribs. Other kinds of lizards, traceable to the Cretaceous period or through rare fossils even to the Jurassic, collectively exhibit as great a range as the iguanids and agamids but have become more specialized in their structure (Fig. 7.25). Although the diminutive, nocturnal geckos and the much larger monitor lizards retain their legs, in a number of families the paired appendages have been reduced or lost. Lizards, like the skinks, that have an elongated body and small legs can slither among tangled roots or slip easily through loose sand. The amphisbaenids, limbless entirely (except for the genus *Chirotes*), are committed to burrowing and, having extremely reduced eyes, are worm-like in appearance.

In the more specialized lizards, the temporal fenestra is often decreased in size or obliterated altogether as the skull is reinforced by the expansion of neighboring bones. The amphisbaenids in becoming modified for burrowing have not only lost their temporal fenestrae but have given up the mobility of the skull roof characteristic of modern lizards and, by extension of the parietal bones and ossification anteriorly, have enclosed the brain completely in bone. The squamosal has disappeared from the skull of these animals, apparently not because of their specialization for burrowing but in accordance with a general tendency toward diminution of that element among the lacertilian reptiles. In contrast to the squamosal of the Triassic lizards, which bore the cup-shaped depression for the head of the quadrate, that of the most conservative forms stemming from the Cretaceous radiation is a delicate bone characterized by a peg which fits into a notch in the quadrate and by a dorsal process that abuts the skull roof. In skinks and other advanced lizards, the dorsal process is lost, and the squamosal becomes a narrow bar with its posterior end turned down to meet the quadrate. The amphisbaenids, the geckos, and the poisonous heloderms, in losing the squamosal have apparently carried the trend toward reduction of that element to its end point.

(a)

(b)

(c)

(d)

(e)

(f)

Snakes: modifica-
tion of the jaw ap-
paratus
The squamosal may also be reduced or absent in snakes. In these animals, increased mobility of the jaw was attained in part by the loss of the last vestiges of the bony arcade which bounded the temporal fenestra and braced the quadrate (Fig. 7.26). That element is held in snakes only by a loose attachment to the squamosal or another bone at the back of the skull roof and usually by a flexible connection to the pterygoid bone of the palate. When the mouth is opened, movement of the quadrate contributes to the widening of the gape, as does the spreading apart of the left and right halves of both the lower and the upper jaws. Since the bones of the snout, the upper jaw, and the palate are not tightly bound to one another, the skull in the vicinity of the mouth cavity can be temporarily deformed to accommodate the passage of large prey. Because of their elastic interconnections, the bones exert pressure upon the victim within the mouth, and the recurved teeth on the jaws and palate prevent it from wriggling free. The independence of the two sides of the jaws allows the snake to advance them alternately over the body of the trapped animal and so to hitch the animal backward toward the esophagus without ever relaxing its grip. The brain, lying above the mouth cavity, is protected from compression during the process of ingestion by a solid casing of bone. The floor and back of the shield are composed as usual of elements of the otic capsule and occipital complex, but the anterior region, normally cartilaginous in reptiles, is covered by downgrowths of the parietal and frontal bones, somewhat as in burrowing lizards.

Specialization of
the body in snakes
Besides the loosening of the jaws, the elongation of the body and reduction of the legs first evident in the lizards have been carried to an extreme in snakes. Loss of the entire appendicular skeleton (except for a pelvic bone or two in certain genera) was accompanied by a return to locomotion by undulation of the trunk and tail. This kind of movement was not effected as simply on land as it was in water, however. Pressure against the surrounding air contributed nothing to the forward motion of the animal. Thrust was obtained only by pushing against a solid surface. Purchase had to be gained on the substratum, or the animal would bend back and forth helplessly in place. As substitutes for the legs that kept the long-bodied lizards from

Figure 7.25. Recent lizards. **(a)** *Draco*, the flying lizard. **(b)** *Ptyodactylus*, a gecko. **(c)** *Varanus*, a monitor. **(d)** *Scincus*, a skink. **(e)** *Anguis*, a blindworm. **(f)** *Chirotes*, an amphisbaenid.

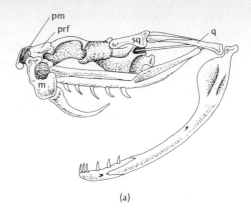

(a)

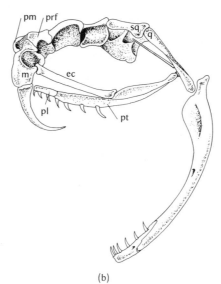

(b)

Figure 7.26. Skull of the rattlesnake, *Crotalus*: **(a)** mouth nearly closed; **(b)** mouth opened fully. Abbreviations as in Fig. 6.7.

slipping backward, snakes evolved eversible scales controlled by an incredibly complex arrangement of muscles. Since the stress incurred in legless locomotion was great, the vertebral column in snakes was reinforced in a way that made it more resistant to dislocation of its elements than the column of a waterborne animal. The centra of adjacent vertebrae interlocked, the convex end of one fitting into the concave face of the next, and another pair of articulating processes appeared on the neural arches above the old zygapophyses. The number of vertebrae increased far beyond that in any lizard, enabling the snakes to bend the body into many curves and thus enhance their ability to cover ground. The elongation and narrowing of the trunk that resulted from the lengthening of the column entrained modification in the structure of the viscera, asymmetry occurring in the development of paired organs, for instance, and the intestine looping less than usual.

Versatility and radiation of the snakes

The extreme specialization of the snakes did not confine them to a restricted habitat. Amazingly, they diversified and became adapted for as wide a range of environments as the lizards. The surviving snakes are divisible into at least eight families, most of which are distributed broadly over the globe. The members of several of these families are burrowers, snakes with reduced eyes and teeth and an upper jaw bound quite firmly to the skull. The remaining lineages contain some burrowers but consist largely of surface-dwelling species that may be found traveling on the ground, in the trees, or in the water in swamps, jungles, deserts, grasslands, or temperate forests.

Although the relationship of the families of burrowing snakes to the others is not clear, it seems that the snakes that live above ground represent a series of radiations which took place after the end of the Cretaceous period. From the structure of their skull and their lack of venom-forming organs, it is assumed that the boid snakes are the most archaic of the surface-dwelling types. Differentiated into the pythons of the Old World and the boas of the New, they have evolved, in constriction, a method of killing their prey which has rendered them competitive with more progressive snakes. Poisonous snakes seem to have arisen first in the colubrid group. Related to the many nonpoisonous members of this tribe like the garter snakes, racers, and grass snakes are forms which synthesize venom and deliver it through their posterior teeth. Though they have been successful worldwide, most of the poisonous colubrids are not nearly so deadly as the more advanced snakes, which have developed fangs at the front of the mouth. The terrestrial elapids, like the cobras and the mambas and their allies, the oceangoing hydrophiids, have a pair of enlarged teeth at the anterior end of the maxilla more deeply grooved for the flow of venom than the fangs of the poisonous colubrids. These teeth are immovable, unlike the fangs of the most advanced and presumably the most recently evolved of the venomous snakes, the viperids.

The vipers, including the rattlesnakes, the asps, and the

adders, have hollow rather than grooved fangs, which, because they can be folded back against the palate when the mouth is closed, can grow enormously long. These snakes secrete such a powerful venom that they need strike but once and withdraw. Their prey soon dies, and they take it without a struggle. Like the boids, colubrids, and elapids, the viperids have developed terrestrial, arboreal, and aquatic representatives as well as a few burrowing forms. Their adaptation for movement over every sort of terrain as well as their ability to eat almost any kind of animal they can catch or kill doubtless accounts for the cosmopolitan distribution of these most specialized of reptiles.

Origin of snakes

The origin of the snakes is still an unsolved problem. From the anatomy of living species it is obvious that the group is derived from the lizards, but the rarity of fossil snakes from the Mesozoic era has prevented paleontologists from ascertaining the particular lacertilian stock from which the snakes arose. There are almost no well-preserved skulls available, partly because the loosely associated dermal elements of the roof, palate, and jaws did not stay together after death. The preserved material in hand consists mostly of vertebrae found disarticulated in Cretaceous marine deposits. Study of these bones has convinced some workers that the first snakes were marine forms. Investigators who disagree point out that the most primitive snakes, those with remnants of pelvic bones and more rigid upper jaws, are almost exclusively burrowers and suggest that snakes were originally subterranean animals closely related to elongated, nearly legless lizards. The extant lizards whose anatomy is most like that of snakes, however, are the monitors, forms which live on the surface and show no sign of reduction of the limbs. A survey of the lizards of the superfamily Varanoidea (Platynota), to which the monitors belong, reveals that many traits characteristic of snakes are present in these animals as well. The extra articulating processes on the neural arches were common in the ancestors of the monitors, and the lower jaw in all the varanoids is flexibly joined in the midline, as it is in snakes. The entire skull of the primitive varanoid *Lanthanotus* strongly resembles that of a snake, even to the extent of lacking eardrums and retaining teeth on the palate. McDowell and Bogert, who studied this short-legged, long-bodied form from Borneo, believe that the ancestors of snakes might have been similar to it.

There is one obstacle to the derivation of snakes from varanoids or any other surface-dwelling lizards. Walls has called attention to peculiarities in the eyes of snakes and maintained that these organs, with their tinted lens, unique retinal cells, and unusual mechanism for accommodation, could never have come from the eyes of lizards that live above ground. He favors the origin of snakes from burrowing lizards, saying that the strange construction of the eye in snakes could be the result of its independent redevelopment from the reduced eye of a subterranean ancestor. Having reviewed all the arguments, Bellairs believes that the forms transitional between lizards and snakes, when they are discovered, will prove to be Jurassic animals of varanoid stock modified for burrowing. According to this theory, some of the early snakes remained subterranean in their habits, giving rise to the several families of burrowers still extant, and others produced descendants that moved above ground and sired the boids and more progressive members of the group.

The first archo-saurs

The origin of the snakes and other lepidosaurian reptiles is understood more thoroughly than that of the archosaurs. The latter diapsids appeared at the end of the Permian period, animals with teeth sunk into deep sockets and a tendency to form fenestrae in front of the eye and in the lower jaw. Although *Euparkeria*, a primitive member of this assemblage, was once thought traceable to *Youngina* or one of its allies, discovery of still earlier archosaurian reptiles, the proterosuchians *Chasmatosaurus* and *Erythrosuchus*, made the origin of the archosaurs from reptiles like *Youngina* highly unlikely. Whereas *Youngina* already had a shortened jaw and the beginnings of an otic notch, the proterosuchians possessed a long jaw and no indication of an otic notch at all. Because the more aquatic and, he believes, the more primitive proterosuchians have certain traits in common with the ophiacodont pelycosaurs, Reig has postulated the descent of the archosaurs from those early Permian synapsids. Romer, recognizing the proterosuchians as primitive but regarding them as somewhat specialized, hesitates to link the archosaurs to the ophiacodonts through forms like *Chasmatosaurus* (Fig. 7.27). He adheres to the hypothesis, which Reig thinks too general, that the archosaurs may have stemmed directly from cotylosaurian stock.

Although only three or four genera besides *Chasmatosaurus* and *Erythrosuchus* are well known, the proterosuchians seem to have been an important transitional group at the beginning of the Triassic period. Their postcranial skeleton was still like that of archaic reptiles. They walked with the upper part of the leg extended horizontally and retained in the vertebral column

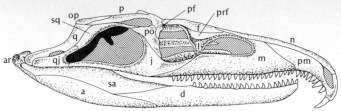

Figure 7.27. Skull of *Chasmatosaurus*, a proterosuchian thecodont of early Triassic time. Abbreviations as in Fig. 6.7. (From Romer.)

vestiges of the intercentra inherited from the most ancient tetrapods. That they were archosaurs is evident from the structure of the skull. Even though the otic notch was lacking and the teeth were not quite characteristically thecodont, the cheek and face were fenestrated in the archosaurian pattern and the arrangement of the bones was nearly the same. Apparently, the conservative design of the postcranial skeleton was associated with the animals' continued amphibious habits. The evolution of the advanced diapsid skull, besides fostering the proterosuchians' survival in watery regions, allowed the emergence of more terrestrial forms that eventually became the ruling reptiles of the Mesozoic era.

Euparkeria, a Lower Triassic pseudosuchian

The earliest upland forms derived from the proterosuchians were the primitive pseudosuchians like *Euparkeria* (Fig. 7.28). This little animal, one of the reptiles from the African Karroo described by Broom, had an otic notch and fully thecodont teeth but was still primitive in displaying teeth on the palate and intercentra in the vertebral column. Its stance was intermediate between that of the proterosuchians and the more advanced archosaurs. Ewer, who restudied Broom's specimens, believed that *Euparkeria* walked with its legs out to the side, like a proterosuchian, when it was moving slowly but when running rose up on its hind legs, pulling them more directly beneath the body. Other paleontologists agree with Ewer that *Euparkeria* was capable of the bipedal gait used by many of the later archosaurs. As in those forms, the hind legs of *Euparkeria* were longer and stronger than the front ones. The trunk was shortened, and its weight and that of the head were counterbalanced on the other side of the pelvic limbs by a heavy, muscular tail.

Locomotor posture in early archosaurs

Since the anatomy of *Euparkeria* indicated that bipedalism was established among the most primitive of the terrestrial archosaurs, it was supposed for a long time that the later members of the group, which walked on four legs, had reverted to that

Vertebrate History: Problems in Evolution

position secondarily. It was pointed out that quadrupeds like the armored *Stagonolepis*, the swamp-dwelling dinosaur *Brontosaurus*, and even the swimming forms like the phytosaurs and crocodiles had larger hind limbs than front ones as a result of their having descended from bipedal ancestors. As investigation of the early archosaurs continued, the idea that bipedalism was fundamental among the pseudosuchians and the other ruling reptiles began to be questioned. Reig asserted that though *Euparkeria* was in some measure bipedal, most of its near relatives were not. He believed that the several families of quadrupedal pseudosuchians were derived from euparkerids which had never departed from the primitive four-footed stance. He also regarded the quadrupedal posture of the phytosaurs, crocodiles, and sauropod dinosaurs as primary, deriving those animals not from the pseudosuchians but from the more primitive tetrapodal proterosuchians. The new theory made it unnecessary to suppose that groups of archosaurs had become highly modified for terrestrial life and then readapted for amphibious behavior within the confines of the Triassic period. The proterosuchians had retained the ancient reptilian habit of living in and beside the water; according to Reig's hypothesis, they passed this habit directly to many of their descendants.

Figure 7.28. Restoration of the skeleton of *Euparkeria*, a Lower Triassic pseudosuchian thecodont from South Africa. (From Ewer.)

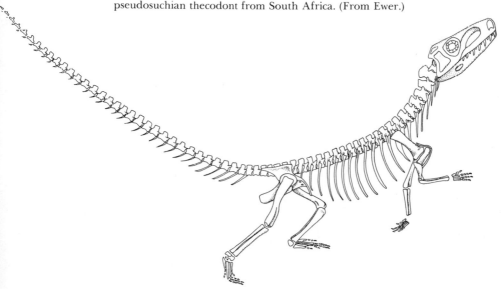

In Reig's opinion, the discovery of an animal thought to be a primitive crocodile in Middle Triassic rocks of South America added to the evidence in favor of his theory. Before Sill described *Proterochampsa* (Fig. 7.29), the oldest known crocodile was *Sphenosuchus* of the Upper Triassic, a form which resembled the pseudosuchians in being somewhat adapted for living on dry ground. The existence of a thecodont intermediate in structure between pseudosuchians and crocodiles was interpreted as signifying that the latter arose from the former near the end of the Triassic period, abandoning the bipedal gait and returning to the lakeshore and riverbank. The introduction of *Proterochampsa* cast doubt upon this version of crocodilian evolution. Although it lived long before the supposed protocrocodilian *Sphenosuchus*, *Proterochampsa* seemed to Sill to be definitely a crocodile. Besides the characteristic long snout and flattened head covered by roughened dermal bones, the animal had a secondary palate dividing the front of the mouth from the air passage above it and, in the postcranial skeleton, other distinctively crocodilian modifications. Sill observed that since *Proterochampsa* dated from mid-Triassic times, crocodiles must have split away from an early thecodont stock. Reig agreed and pointed out that the quadrupedal proterosuchians were structurally and temporally more probable ancestors for the group than even the most primitive of the pseudosuchians. Like the proterosuchians,

Figure 7.29. Skull of *Proterochampsa*, a thecodont from the Middle Triassic of South America. Dorsal view. Abbreviations as in Fig. 6.7. (From Romer, after Sill.)

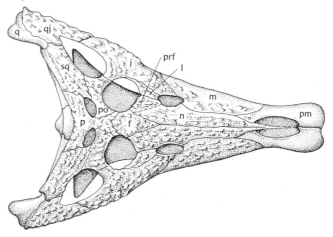

Proterochampsa lacked an otic notch and had a long jaw that articulated with the quadrate well behind the occiput. Reig maintained that these traits were indicative of phylogenetic relationship rather than parallelism, because *Proterochampsa* matched the proterosuchians in the shape and relative size of their cranial fenestrae and in other characters as well. The proterosuchians fell behind the bipedal pseudosuchians as likely ancestors for the crocodiles in only one respect: advanced pseudosuchians and crocodiles show the same kind of ankle joint, a peculiar hinge between the astragulus and calcaneus. Those who regard the connection between proterosuchians and crocodiles as probable must assume that the unusual arrangement of the tarsal bones evolved separately in the pseudosuchians and the crocodiles. No evidence on this point can be gleaned from the remains of *Proterochampsa*, as the lower part of the hind limb has not been found. Moreover, Romer has questioned the relevance of the structure of the ankle in *Proterochampsa* to an understanding of crocodilian origins because he does not agree with Sill that *Proterochampsa* was a crocodile and doubts even that the Middle Triassic form was a precursor of the crocodilian line.*

In Reig's account of crocodilian evolution, *Sphenosuchus* and its few known allies are regarded as aberrant forms which became adapted for terrestrial living but ultimately failed in competition with the pseudosuchians and the early dinosaurs. The crocodiles that survived into the Jurassic period were those that stayed in their original habitat. As predaceous swimmers, they radiated into fresh and salt waters worldwide. Since continental deposits from the Jurassic period are rare, the marine crocodiles of the time are better known than those which were confined to interior lakes and streams. The crocodiles that invaded the sea, like other great marine reptiles, became more and more specialized for their new environment and then died out completely in the Cretaceous period. Several families of freshwater forms evolved and disappeared during the same time, but one group emerged, from whose ranks the modern crocodiles came. These animals, with an extended secondary palate and other traits

Radiation of crocodiles and distribution of modern forms

*Romer (1971a) has opposed Reig's theory that the proterosuchians were directly ancestral to the crocodiles. Taking note of the primitive traits of *Proterochampsa* and other thecodonts closely related to it, he has housed the proterochampsids in the suborder Proterosuchia and described them as having paralleled the crocodiles. The latter animals, he assumes, have descended from pseudosuchians, thecodonts derived from the proterosuchians but more advanced.

characteristic of existing species, produced a new radiation of crocodilian reptiles, one that continued into Tertiary times until the climate became harsh over much of the globe. Thereafter, the crocodiles became restricted to the parts of the world where tropical and subtropical conditions prevailed. The extant members of the group, the last surviving archosaurs, represent the major varieties that evolved in the Cretaceous period. The gavials in India are the remnants of a family of long-snouted animals which once included species growing to a length of 60 feet. The remaining forms belong to the Crocodylidae and have been categorized on the basis of minor differences as alligators and crocodiles (Fig. 7.30). The alligators, most of which are relatively broad-snouted, are now confined to the Western Hemisphere except for a single species which has survived as a relict in the Yangtze River of China. The crocodiles, commonly considered narrow-faced but in reality exhibiting a broad range in the shape of the snout, are more widely distributed, being found in tropical America, Africa, and southern Asia. All these animals are formidable carnivores, vulnerable, it seems, only to cold, drought, and bullets.

Phytosaurs

Few other reptiles could compete with the crocodilians. The phytosaurs, which arose as crocodiles did in the Middle Triassic, multiplied briefly but failed everywhere before the end of the period. The reason for the superiority of the crocodiles is hard to tell from bare bones. The phytosaurs were modified in much the same way for life in the swamps and lakes. They were strong-legged quadrupedal forms that walked with their limbs well under them and swam with the aid of a long, crocodile-like tail. Protected by dorsal armor, they pursued their prey in the water, catching fish and unwary tetrapods, as crocodiles do, with a snap of their long, tooth-studded jaws. Because the phytosaurs paralleled the crocodiles in many characters, it has been suggested that the two groups emanated from a common source. A. D. Walker thought that the primitive members of each group diverged from proterosuchians resembling *Chasmatosaurus*. Reig also traced the phytosaurs to proterosuchian ancestors, being convinced that phytosaurs like crocodiles were persistently amphibious rather than secondarily adapted for semiaquatic living. Paleontologists who disagree with Reig still prefer to derive these animals from pseudosuchians. The similarities between phytosaurs and pseudosuchians that Reig dismissed as parallelisms they believe indicate relationship. There is no argument about the history of the phytosaurs after their first

(a)

(b)

(c)

Figure 7.30. Recent members of the order Crocodilia: **(a)** *Alligator mississippiensis*; **(b)** *Crocodilus vulgaris*; **(c)** *Gavialis gangetica*.

appearance. They became increasingly specialized, their nostrils migrating posteriorly to an eminence in front of the eyes so that they swam with only a small part of the head visible above the surface of the water. Although they spread throughout the Northern Hemisphere, they seem not to have crossed the equator in great numbers. If, as Sill maintains, the crocodiles originated in the area that is now South America and Africa, the phytosaurs may have been prevented from entering those

regions by the reptiles which later overwhelmed them on their
own ground.

Just as the phytosaurs became highly modified for a particular
mode of living and then were replaced by animals very similar to
them, the first reptiles to take to the air, the pterosaurs, were
driven from it by forms which shared their archosaurian heritage. Although the precise ancestry of the pterosaurs and their
successors, the birds, is not known, they are thought to have
stemmed from small, slim-boned, bipedal pseudosuchians which
were probably arboreal in habit. The pterosaurs, which appeared near salt water early in the Jurassic period, were less well
adapted for flying and even for walking than the birds. They
stayed aloft by catching the wind under flight membranes
supported by the elongated fourth finger of each hand. The
structure of the humerus indicates that strong muscles inserted
upon the upper arm, but the vigor with which the large
membranes could have been flapped is questionable. It is
possible to imagine the pterosaurs gliding on air currents,
swooping low over the sea to scoop fishes from the water, and
then twisting or beating their wings slowly to rise again. The
position these animals assumed at rest is not known. The hind
limbs were not arranged in such a way that they could have held
the body upright. Relatively slender and weak, they terminated
in an elongated foot, which, in some genera, bore digits that
diverged as if they were webbed. If the pterosaurs lighted upon
the water, they might have paddled forcefully with the hind feet
to gain the speed necessary to retake the air. On land, however,
it seems that the animals would have been terribly awkward.
Romer has suggested that they never descended to the ground
at all but hung like bats by the clawed digits in front of the wing
from cliffs beside the sea.

The pterosaurs did apparently stay by the ocean. Their
remains have been recovered almost entirely from marine beds,
and their elongated jaws, usually set with sharp teeth, were
obviously useful only for catching fish. *Dimorphodon*, the most
primitive pterosaur known, did not yet have the low, long beak
of the later forms, but it had already evolved the other skeletal
traits peculiar to the group (Fig. 7.31). Its large, fenestrated
skull was supported by a neck thick and long compared to the
trunk region behind. The pectoral appendages were fully
modified for the extension of the flight membranes, and, as in
all its closest relatives, the Jurassic rhamphorhynchoids, a long
tail was present. Its bones were hollow and light, as those of true

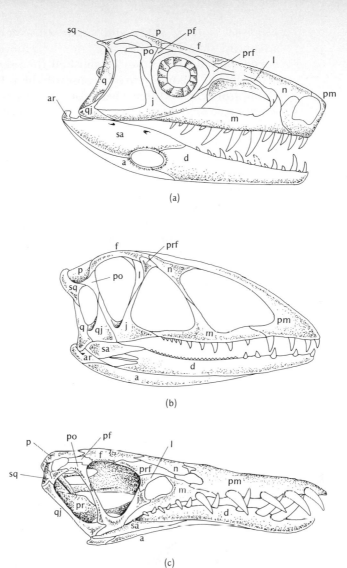

Figure 7.31. Skull of **(a)** the pseudosuchian *Euparkeria*, showing its general similarity to that of **(b)** the Lower Jurassic pterosaur *Dimorphodon*. **(c)** *Rhamphorhynchus*, an Upper Jurassic long-tailed pterosaur, showing the long jaws typical of the later forms. Abbreviations as in Fig. 6.7. (From Romer.)

avians are, and paleontologists suspect strongly that its soft organs may have been adapted in similar avian fashion to allow the sustained expenditure of energy necessary for flight. *Rham-*

phorhynchus (Fig. 7.32) and other Jurassic pterosaurs more advanced than *Dimorphodon* evidently achieved security as airborne animals, since they established themselves on shorelines everywhere. Though rare because of their fragility, fossils of rhamphorhynchoids have been found in Europe, western North America, Cuba, Argentina, East Africa, India, and possibly in Soviet Asia.

These first pterosaurs were superseded by more specialized members of their own group. The pterodactyloids, which appeared in the Middle Jurassic, were the only pterosaurs left by Cretaceous time. These forms had no tail and showed a tendency toward loss of teeth and development of curious skeletal modifications. Several vertebrae in the shoulder region of the most progressive pterodactyloids fused into an abutment for the upper ends of the scapulae and so helped to anchor wings that in some species extended outward 5 to 10 feet on either side of the body. *Pteranodon* (Fig. 7.33), one of the largest pterodactyloids, not only held out great wings as it soared over the sea but supported an enormous head. Although the weight of its long, toothless and perhaps horn-covered jaws was partly balanced by the bony crest that jutted backward from the rear of its skull, *Pteranodon* could not have been as agile or as tireless a flier as the primitive birds with which it was forced to share its habitat. Possibly less aggressive than the birds as well as less efficient

Figure 7.32. Remains of *Rhamphorhynchus* preserved in limestone from Bavaria, showing the wing membranes intact. (From Seeley.)

Vertebrate History: Problems in Evolution

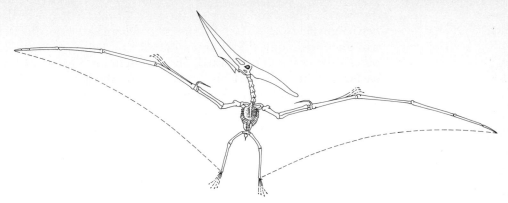

Figure 7.33. Skeleton of *Pteranodon*, a Cretaceous pterosaur. This form, the largest of the pterodactyloids, had a wingspread as broad as 25 feet. (From Eaton.)

Dinosaurs: their discovery and classification

mechanically, *Pteranodon* and the other pterodactyloids became extinct before the Tertiary period began.

While the pterosaurs fished in the sea, the dominant animals on land were the dinosaurs. Like the therapsids which they replaced, the dinosaurs distributed themselves wherever the herbivores among them could find the land plants and water-weeds upon which they depended. Since no high mountain ranges had yet risen to divert the flow of warm, moist air from inland regions, the dinosaurs found hospitable territory almost everywhere and for 100 million years continued to prosper and diversify. After Buckland unearthed remains of the fierce *Megalosaurus* from European Jurassic beds in 1824 and Mantell announced the discovery of *Iguanodon* 25 years later, pale-ontologists turned up scores of different genera on every continent. The large size of many of these animals intrigued investigators from the first, and Owen celebrated it by calling them dinosaurs, or "terrible lizards." The number of forms known increased rapidly in the 1870s as a result of the bitter competition between O. C. Marsh and E. D. Cope for the dinosaur bones that lay buried in Jurassic and Cretaceous rocks in western North America. The fact that the dinosaurs belonged to several different stocks was soon apparent. By 1890, Zittel had divided them into sauropods, theropods, and orthopods, a classification which Seeley modernized at the turn of the century by bracketing the first two groups in the order Saurischia and setting the last apart as the order Ornithischia. He differentiated the two orders of dinosaurs on the basis of the structure of the pelvic girdle (Fig. 7.34). In the saurischians, the girdle was triradiate in shape: the ilium extended upward, the elongated pubis downward and forward, and the ischium downward and backward. In ornithischians, the body of the pubis was bent backward beside the ischium and a new process developed that extended forward, making the girdle tetraradiate.

Saurischian di-nosaurs

The earliest dinosaurs were almost all saurischians. Mid-Triassic strata in South America have given up bones belonging to theropod members of the order and also to relatives of the heavy quadrupedal sauropods. The theropods were small, carnivorous coelurosaurs, animals which rather resembled *Euparkeria* in their general structure. They differed from that form chiefly in having a proportionately smaller head and more flexible neck, longer hind limbs, and shorter front ones. They also displayed two special dinosaurian traits: a perforated acetabulum in the pelvic girdle and, in the pectoral girdle, an absence of dermal bone. The coelurosaurs became increasingly bipedal in the Jurassic period, evolving three-fingered forelimbs better suited for handling food than aiding in locomotion (Fig. 7.35). Always light and hollow-boned, they produced late in their history a family of especially bird-like forms, the ornithomimids. Except for its saurian tail, *Ornithomimus* itself looked quite like an ostrich with its long bird-footed legs, delicate upper

Figure 7.34. The pelvic girdle in dinosaurs. Lateral view of the right half of the girdle of **(a)** *Allosaurus*, a carnosaur, showing the triradiate structure typical in saurischian dinosaurs. **(b)** *Thescelosaurus*, an ornithopod with pubis directed posteriorly and pubic process extending forward in the manner typical of ornithischian dinosaurs. Abbreviations: *a*, acetabulum; *ap*, anterior process of pubis; *il*, ilium; *is*, ischium; *p*, pubis. (From Romer.)

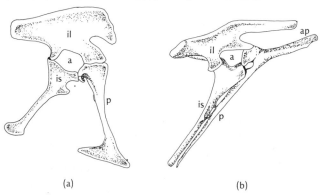

(a)   (b)

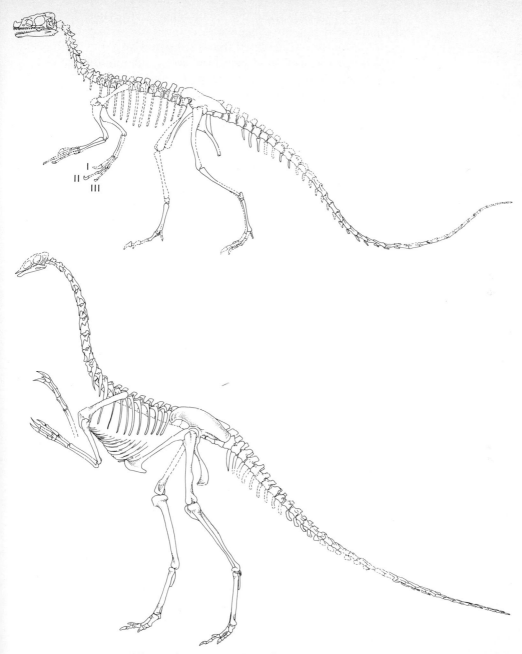

Figure 7.35. Skeletons of coelurosaurs. **(a)** *Coelurus* (=*Ornitholestes*) from the Upper Jurassic. **(b)** *Ornithomimus* (=*Struthiomimus*), a Cretaceous form. (From Osborn.)

limbs, narrow neck, and tiny head. Since it lacked teeth, *Ornithomimus* could not have been an ordinary carnivore. Some paleontologists have speculated that it subsisted on a diet of eggs, but it could just as well have eaten the frogs, salamanders, and little lizards which were in existence by Cretaceous time.

The largest carni-vores

The most ferocious carnivores of the Jurassic and Cretaceous years were not the coelurosaurs but their fellow theropods, the carnosaurs (Fig. 7.36). Gigantic bipeds which stood 15 to 20 feet high, these animals had a large and heavy head armed with great, recurved teeth. The weight of the skull was supported by interlocking cervical vertebrae and trunk vertebrae cantilevered from the sacrum over broad iliac bones and strong hind legs. Despite their bulk, the carnosaurs walked upon the three middle toes, leaving bird-like tracks when they ran over soft ground. *Allosaurus* and other Jurassic carnosaurs had three digits upon the hands, also, but in *Tyrannosaurus* and other late forms the pectoral appendages became very small and the number of digits fell to two. The bird-like feet of the carnosaurs as well as their hollow bones and bipedal habit have convinced many paleontologists that these dinosaurs were derived from coelurosaurs.

In the progression from coelurosaurs to carnosaurs, Reig finds another example of the tendency of dinosaurs to increase in size and to become more specialized for a particular way of life. Walker has dissented from this view. He believes that carnosaurs with large heads and short necks could not have descended from coelurosaurs, in which the head was small and the cervical vertebrae elongated. The carnosaurs were, in his opinion, derived independently from the pseudosuchians through intermediate forms like the Upper Triassic *Ornithosuchus.* That animal retained some pseudosuchian traits in the postcranial skeleton but had a skull of carnosaurian proportions and structure. Walker's theory is untenable, according to Reig, because *Ornithosuchus* possessed the typical pseudosuchian ankle joint, one that could not have been transformed into the kind of articulation present in carnosaurs. Reig regards *Ornithosuchus* as a true pseudosuchian which paralleled the carnosaurs in becoming strongly bipedal and highly predaceous, not as a form which gave rise to the advanced and monstrous theropods.

Origin of the sauropodomorphs

There has also been controversy concerning the origin of the other division of saurischians, the Sauropodomorpha (Fig. 7.37). The earliest members of this group were present, like the coelurosaurs, in Middle Triassic times. Most of them were fairly

Figure 7.36. Skeletons of carnosaurs. **(a)** *Tyrannosaurus*, a Cretaceous form that attained a height of 30 feet. (From Osborn.) **(b)** *Ornithosuchus*, a smaller Triassic dinosaur with carnosaurian traits. (From Walker.)

large, rather heavily built bipedal forms with an exceedingly small head. Their flattened teeth and great girth identified them as part of the vegetarian population upon which the carnivorous saurischians fed. These Triassic herbivores seemed to paleontologists to be ideal ancestors for the enormous quadrupedal plant eaters which succeeded them in the Jurassic period. As the descendants of animals like *Plateosaurus* grew bigger, it was

(a)

(b)

Figure 7.37. Skeletons of sauropodomorph saurischians. **(a)** *Thecodontosaurus* (=*Anchisaurus*), an Upper Triassic prosauropod. (From Huene.) **(b)** *Brontosaurus*, an Upper Jurassic sauropod. (From Marsh.)

thought, they returned to the four-legged stance and finally took refuge in lakes and deep swamps, where the water supported their great bulk.*

Since the trend in sauropod evolution appeared to be toward increased size and reversion to the quadrupedal position, paleontologists assumed that the precursors of the sauropodomorph dinosaurs must have been small or medium-sized bipedal forms. Both the coelurosaurs and the pseudosuchians answered that

*Bakker (1971) has challenged this long-held theory of sauropodomorph habits. From his study of brontosaur remains, he has concluded that the huge, straight-legged animals were terrestrial, traveling through forests like modern elephants.

description, and, in drawing up their phylogenetic schemes, workers placed one group or the other at the base of the sauropodomorph line. Reig, however, following Charig and his colleagues, came to believe that the sauropodomorphs, like the crocodiles, were not derived from bipedal ancestors. He pointed out that since these dinosaurs were already diversified by the middle of the Triassic period, the group must have originated somewhat earlier, at a time when the coelurosaurs were just emerging and the pseudosuchians were not far advanced. The well-established thecodonts from which the sauropodomorphs were most likely to have sprung, Reig thought, were the primitively quadrupedal proterosuchians. Just as the phytosaurs and crocodiles may have arisen from the proterosuchians which remained amphibious, so the sauropodomorph dinosaurs may have evolved from members of the group, like *Erythrosuchus,* which were somewhat modified for terrestrial living. In support of his theory, Reig noted that the Triassic sauropodomorphs were not exclusively bipedal, as paleontologists had once supposed. Remains have been found in South Africa of quadrupedal forms, called melanorosaurids, which were almost surely the precursors of the huge Jurassic and Cretaceous four-footed sauropod waders like *Apatosaurus* and *Brontosaurus.* Reig supposed that the melanorosaurids retained the posture primitive for the group and that members of the other Triassic sauropodomorph families abandoned it.

As a result of studies made in the 1960s and afterward, new theories concerning the history of the sauropodomorph dinosaurs have emerged. The Triassic bipedal forms, long called prosauropods in the belief that they sired the Jurassic sauropods, have been regarded, since the discovery of the melanorosaurids, as types that became extinct without leaving descendants. They have been accorded a larger place in the fauna of their day, however, since Charig and his colleagues recognized that their group included carnivores as well as herbivores. The carnivorous members had been categorized as theropod carnosaurs on the basis of tooth structure by earlier workers who had ignored the sauropodomorph traits in the skull and postcranial skeleton. The sauropodomorphs had evidently radiated to fill all the ecological niches that large terrestrial reptiles had occupied before them. They had been, it seems, hunters as well as the hunted. Their extinction at the end of the Triassic period was followed by a second wave of dinosaurian evolution. Again the predators were saurischians: the coelurosaurs multiplied, and the carnosaurs appeared for the first time. The sauropods were the only remaining saurischian herbivores. Most of the plant eaters belonged to the newly evolved ornithischian tribe.

Paleontologists have virtually no clues to the origin of the ornithischian dinosaurs. They appear in the fossil record in increasing numbers after the middle of the Jurassic period, already structurally distinct from the saurischians, without any telltale traits which might link them to a particular group of earlier archosaurs. Three or four genera have been identified from fragmentary remains found in Middle Triassic deposits in South America and Upper Triassic sediments in South Africa, but study of these forms has proved little except that the ornithischians were evolving while the prosauropods and coelurosaurs held sway. The Jurassic ornithischians included both bipedal and quadrupedal varieties. In this case, it seems clear that the bipedal position was the primitive one for the group. The least specialized ornithischians all held themselves erect; the more advanced forms, which walked on four feet, had such short forelimbs compared to the hind ones that their descent from forebears with upright posture is almost certain.

The first ornithischians may have been, like *Hypsilophodon,* small lightly armored herbivores that ran on two legs most of the time and used their clawed fingers for grasping or even as an aid in climbing (Fig. 7.38). By the Jurassic period, most of the ornithischians had become larger, more slow-moving animals that were in the habit of dropping down on their forelegs to graze. The majority of them, like many herbivores, lacked anterior teeth. The mandible bore a median predentary bone, probably horn-covered in life, which cut vegetation against the broad premaxillary elements of the upper jaw. The cheek teeth were adapted for cutting and grinding. Since the jaw joint was set low, all the teeth met simultaneously like a press when the mouth closed. The presence of a large coronoid process on the mandible attested to the existence of strong temporalis muscles and thus to powerful compression of the jaws as the animals chewed. The earlier and more generalized ornithopods, as the bipedal ornithischians are called, may have stayed on dry ground much of the time, but by the last half of the Cretaceous period, most members of the group had moved into shallow lake waters, perhaps the only place where they were safe from attack by the ferocious carnosaurs. The most successful of these late, amphibious ornithopods were the hadrosaurs, or duck-billed

dinosaurs. Paleontologists have had the extraordinary good luck to find mummies of these beasts, and so they know that the animals' feet were webbed as an adaptation for swimming or walking in the mud. Even if they had had only the skulls, however, workers could have guessed that these dinosaurs spent their time in the water. Their nostrils either opened some distance behind the flat-beaked snout or led to internal passages that looped through bony crests high above the skull roof. Both arrangements would have prevented the inrush of water when the animal dipped its bill beneath the surface.

Stegosaurs and ankylosaurs

The quadrupedal ornithischians, forms which did not take refuge in the water, were protected from the attack of carnivores by some sort of dermal armor. In the stegosaurs, large pentagonal plates stood upright on either side of the dorsal midline, edging the body that arched from a small head upward over high hips and downward at the tail. Two pairs of heavy spikes mounted on the end of the tail were the animals' only defensive weapons. The spikes and plates did not save the stegosaurs for long. They are the only group of ornithischians which disappear from the fossil record before the end of the Cretaceous period. The stegosaurs were succeeded by the ankylosaurs, quadrupeds with more extensive body armor. These dinosaurs were completely covered by closely fitted dermal plates. Even the tail was sheathed in bony rings. Broader and squatter than the stegosaurs and not similar to them in the structure of the skull, the ankylosaurs apparently evolved from a different ornithopod stock than the earlier quadrupeds.

Ceratopsians

Although no ornithopod type which might have given rise either to the stegosaurs or the ankylosaurs is known, one bipedal form, *Psittacosaurus,* provides a tenuous link to the ceratopsians, the last of the quadrupedal ornithischians to emerge. *Psittacosaurus* had developed the parrot-like beak underlain by a special rostral bone which was a prominent feature of the ceratopsian skull. The beak was already present in *Protoceratops,* the small Upper Cretaceous dinosaur from Mongolia which was doubtless close to the base of the ceratopsian line. *Protoceratops* also displayed the fan-shaped bony frill at the back of the skull roof that made the head of the ceratopsians seem enormous. Largely a prolongation of the parietal bones, it served as a place of attachment for the axial muscles which held up the head and, anteriorly, for those that elevated the lower jaw. Although the frill in all but one of the ceratopsian dinosaurs was pierced by a pair of wide fenestrae, it was solid enough to protect the neck of the animals from attack. The ceratopsians larger than *Protoceratops* had offensive weapons, also, in the form of horns that grew on the face (Fig. 7.39). In some genera there was a single horn above the snout; in others, a pair above the eyes. *Triceratops,* the ceratopsian with the unfenestrated frill, had all three horns to use against its carnivorous enemies and possibly,

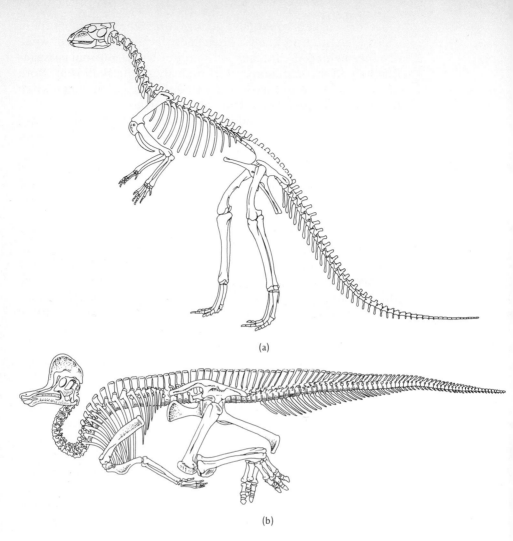

(a)

(b)

in the mating season, against others of its own kind. The ceratopsians spread eastward over the land bridge which connected Asia and North America in late Cretaceous time and multiplied in the region to the west of the sea that covered the area that is now the American Great Plains. Some forms may have reached the South American continent, but their wider distribution was prevented by the extinction of all the dinosaurs during the upheavals which brought the Cretaceous period and the Mesozoic era to an end.

Extinction of the dinosaurs and other Mesozoic forms

The extinction of the dinosaurs, as dramatic as it seems in retrospect, cannot be discussed as an isolated phenomenon. It was part of a change in the world's fauna more profound even

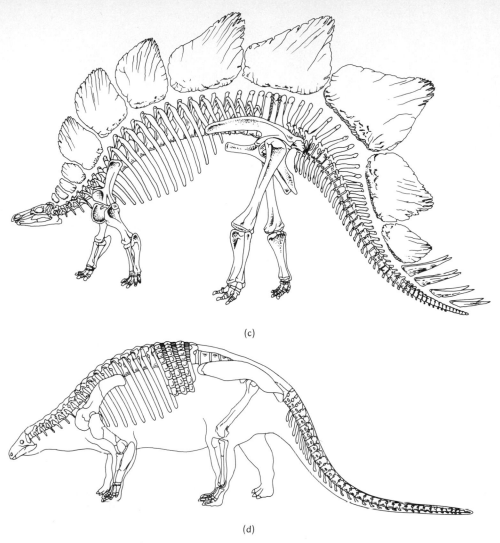

(c)

(d)

Figure 7.38. Skeletons of ornithischian dinosaurs. **(a)** *Hypsilophodon*, a primitive ornithopod from the Lower Cretaceous. **(b)** *Corythosaurus*, a hadrosaur, an advanced ornithopod from the Upper Cretaceous; skeleton reconstructed in the position in which it was found. **(c)** *Stegosaurus*, from the Upper Jurassic. **(d)** *Nodosaurus*, an ankylosaur from the Upper Cretaceous. (*a,c* from Marsh; *b* from Brown; *d* from Lull.)

than that which took place at the end of the Permian period, when the Paleozoic era drew to a close. As the dinosaurs disappeared, other reptiles did also. The pterosaurs, plesiosaurs, and mosasaurs vanished completely, and several families of crocodiles, lizards, and turtles followed them into oblivion.

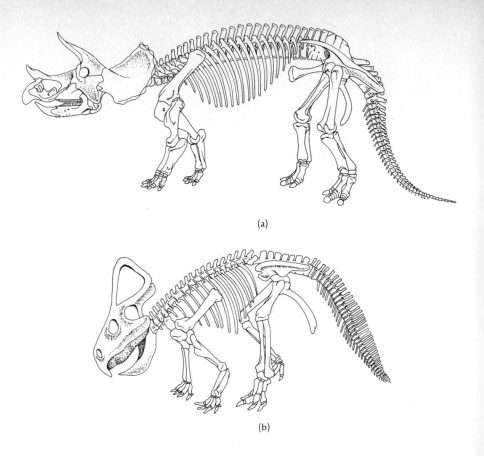

(a)

(b)

Archaic vertebrates of other kinds were displaced, and even among invertebrates of long standing important species were wiped away. The disappearance of some forms, like the majority of the holostean fishes, can be explained as the result of natural selection within continuing populations, for the teleosts were descendants of the holosteans that had evolved superior features. The dinosaurs and other prominent Mesozoic reptiles which became extinct, however, were the last members of their line. The niches they left empty were, in some cases, not filled for many millions of years.

Whereas causes can often be proposed for the disappearance of a single genus or family of organisms, it is more difficult to explain satisfactorily the reasons for the nearly simultaneous extinction of a large number of forms living in different environments. In the past, investigators unable to ascertain the external conditions responsible for such extinctions have alleged that groups carry within themselves the seeds of their own

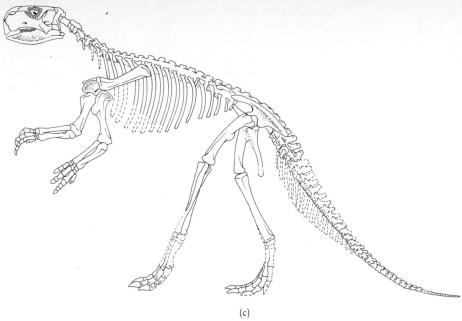

(c)

Figure 7.39. Skeletons of **(a)** *Triceratops*, an advanced ceratopsian dinosaur; **(b)** *Protoceratops*, a smaller, primitive ceratopsian; **(c)** *Psittacosaurus*, an ornithopod from eastern Asia, showing beak similar to that of the ceratopsians. (*a* from Marsh; *c* from Osborn.)

destruction. They argued that a race, like an individual, after its birth passes through a period of vigor and then eventually weakens and dies. A sure sign of racial senescence, according to this theory, is the appearance of nonadaptive characters which make their possessors incapable of coping with competitors or with changes in the environment. Scrutiny of the reptilian forms which were lost during the Cretaceous period seems not to bear out the idea of racial senescence, however. The ceratopsian dinosaurs, a relatively young group when they became extinct, were apparently diversifying and enlarging their range when they were cut off. Neither they nor, for example, the more ancient saurischian coelurosaurs were evolving structures that were obviously inadaptive for their way of life. Evidence for racial senescence has not come forth from the study of other groups either. Among vertebrates and invertebrates alike, aberrant forms and those with bizarre traits often survived beside more normal members of their line, all of them becoming extinct at the same time.

Racial senescence improbable

Although natural senescence does not occur, organisms may fail in competition if they become too narrowly specialized.

Specialization as a factor in extinction

Paleontologists have argued that the very characters which rendered Mesozoic reptiles advanced, progressive, and well adapted for their environment became insuperable liabilities as conditions changed at the end of the Cretaceous period. A dinosaur like *Brontosaurus,* for instance, which weighed 35 tons and consumed vast quantities of lakeweed while standing in the water out of reach of its enemies, would have been unable to survive as the warm, moist climate of the Cretaceous gave way to the more rugged, drier one of early Tertiary times. Even though the change in climate and the nature of the terrain was gradual, there was no possibility of the reappearance in *Brontosaurus* of teeth strong enough to chew tough terrestrial plants or of the slimming of the body necessary to allow the animals to escape predators on land. In short, in evolving its peculiarly large size and weak teeth, *Brontosaurus* had gained an immediate advantage but lost the versatility which would have allowed it to change its way of life as conditions changed. Since the dinosaurs and other reptiles which died out exhibited many structures that had undergone extreme modification in association with specialized behavior, many workers assumed that overspecialization was an important factor in the fall of the dominant animals of Mesozoic time.

Overspecialization, although it surely contributed to the disappearance of many forms, does not explain the pattern of Cretaceous extinctions entirely. If the degree of specialization were the only yardstick, it would be difficult to account for the survival of the legless snakes and the heavily armored turtles. The rather generalized lizards would be expected to continue, but why not also some of the smaller coelurosaurs? And since extremely specialized, well-adapted animals prosper so long as their environment does not change, why should not the mosasaurs, ichthyosaurs, and plesiosaurs be living still? Although the terrestrial environment changed markedly late in the Cretaceous period, conditions in the ocean altered less radically. If temperature and the organisms available for food became unsuitable in one region, the marine reptiles might well have migrated to another. It has often been said that the rise of the modern sharks was responsible for the extinction of the seagoing reptiles, but this explanation is not wholly convincing. The survival of the sea turtles argues against the ability of Recent sharks to clear the waters of all reptilian life, and the initial success of the Mesozoic marine tetrapods in the presence of the predaceous hybodont sharks proves that the reptiles were able to cope with aggressive elasmobranch fishes. Still, it is possible that the habits of the mosasaurs, ichthyosaurs, plesiosaurs, and emerging modern sharks brought them continuously into direct conflict and that this competition did result in the extinction of the reptilian participants.

Possible role of environmental change

On land, the decimation of the reptiles was so extensive that it was once suspected that some sort of geological or climatological catastrophe might have taken place. Study of the sediments laid down at the end of the Cretaceous period and the fossil plants found therein failed to bear out this theory. Although there were some spectacular changes in the topography of the earth, the means by which they were achieved and the rate at which they took place were not at all extraordinary. The rising of the Rocky and Andes Mountains along the western edge of the American continents and the spilling off of the continental seas that had covered North America and much of Europe were slow processes rather than disasters that would have wreaked havoc among the extant animals. The retreat of the seas and the walling off of interior regions by new mountains in several areas of the world did bring climatic changes, which, though not sudden, could have had a harmful effect upon the reptilian fauna. If the internal temperature of the dinosaurs changed with that of the environment as it does in modern reptiles, the Cretaceous animals could have continued to live only where heat and cold did not become extreme for any length of time. The ancient reptiles may have behaved like existing ones in remaining quiescent during parts of the day when their tissues would overheat or chill, but there would have been a limit to the range within which behavioral adaptations could protect them against death.

The largest reptiles may have faltered first in areas where the climate became harsh. The great size that made the animals better predators or rendered them less susceptible to attack also made them particularly vulnerable to seasonal fluctuations in temperature. Unlike small forms, whose large surface-to-volume ratio allowed rapid gain or loss of heat, the big animals would have undergone little change in body temperature during a single day. Presumably, more than 12 hours of nocturnal cold would have been necessary to sap the body of a large dinosaur of the heat required to keep its cells active. Similarly, 4 or 5 hours of midday sun would not have provided enough energy to raise the temperature of the animal significantly. Over a long period of time, however, an imbalance of heat and cold would have been fatal. In areas where the daytime heat was intense and the nights were short, the body temperature of a large reptile would have climbed gradually until it became intolerable. Protracted cold interrupted by brief periods of sunshine would have had the reverse effect. Prevented by their bulk from burrowing or otherwise shielding themselves from the heat or the cold, the large animals could have survived only by moving away from regions where the temperature became inequable. Although they might have prolonged their stay in their old habitats by changes in their behavior, smaller forms would have had to follow the giants when short-term cold or heat grew too intense. Reptiles of any size which could not reach territories with a suitable climate would have perished. Increase in the rigor of the climate would probably have affected the dinosaurs adversely even if they had evolved some degree of internal temperature regulation, as Bakker has suggested. Adult dinosaurs might have been able to raise their body temperature above that of the air around them and dissipate excess heat through air sacs in hollow bones or visceral spaces, as Bakker believes, but newly hatched animals would have survived cold hours or days with difficulty. With neither subcutaneous fat nor an outer covering of any sort to serve as insulation, the relatively small hatchlings would have succumbed if temperatures fell below a certain minimum during hatching season.

If the downfall of the Mesozoic ruling reptiles was due to the disappearance of the gentle climate they required, it seems that some forms besides the crocodiles should have survived as relicts in tropical or subtropical lands. To explain the extinction of the terrestrial archosaurs everywhere, investigators have suggested that in addition to the spread of insufferable temperatures at the end of the Cretaceous period, there was an unfavorable change in the food supply. Displacement of the plants customarily eaten by the herbivorous reptiles would have brought about first a decline in the number of these animals and then shrinkage in the population of the carnivores which subsisted upon them. There is some evidence for such a turn of events in the decreased number of herbivorous species found in the uppermost Cretaceous strata of North America, but paleobotanists insist that if a transformation of the flora were responsible for the extinction of the archosaurian reptiles, they should have departed from the scene much earlier in Mesozoic time. The significant change in the plant world took place not at the end of the Cretaceous period but at the beginning of it, when the flowering plants replaced the seed-ferns, cycads, and conifers as the dominant forms. Rather than diminishing then, when the older plants gave way, the archosaurs flourished. The ornithischian dinosaurs, in fact, did not multiply rapidly until the higher plants had become firmly established in Cretaceous time.

The Jurassic and Cretaceous periods had seen the radiation of several groups of early mammals, and some workers have thought that the advent of these animals, rather than the higher plants, doomed the reptilian fauna. It is true that the ruling reptiles were replaced by the fur-covered, warm-blooded quadrupeds, but the manner in which the two kinds of animals competed is not easily discernible from the fossil record. Remains of Mesozoic reptiles and mammals are not commonly found together, and when they are, they represent forms of diverse size and habits which are unlikely to have entered into direct conflict. While the reptiles were treading the ground with heavy step in search of fodder or living prey, the mammals were scurrying through the underbrush after insects and other invertebrates. Since the little mammals could not have attacked the grown reptiles, it has been proposed that they might have destroyed them by eating their eggs. That the emergent mammals did sometimes dine on reptilian eggs is almost certain, but their extinguishing the entire group by this means is improbable. Animals of any sort that lay eggs have always produced a far larger number than is necessary to ensure the survival of the species. The discovery of row upon row of *Protoceratops* eggs in nests in the compacted sands of Mongolia showed that this dinosaur was not inferior to modern reptiles in reproductive capacity, and nothing argues that its relatives were any less fertile. The late Cretaceous mammals, unable to defeat the reptiles in any way, may simply have occupied territory as the reptiles abandoned it. Once established, the mammals evolved larger forms which could have resisted the return of the reptiles to their old habitat and challenged them effectively in the warmer regions to which they had retreated.

The further progress of the beleaguered reptiles toward extinction can be readily imagined. Although the animals might have survived if this or that modification in their structure or their behavior had occurred, the genetic mutations that would have effected the changes either did not take place or were not successfully selected for. As the archosaurian reptiles became more and more restricted in their distribution, their variety declined and their number decreased. Their abrupt disappearance from the fossil record need not be explained by some final calamity like an epidemic or an unusual barrage of cosmic rays. Once their populations shrank below a certain minimum, subsequent deterioration would have accelerated under normal circumstances. Segregated into isolated groups as the number of individuals became fewer, the remaining archosaurs would have been progressively weakened by inbreeding. Eventually, when the position of the animals became precarious, some small alteration of the flora, the fauna, or the local climate which they might have withstood at an earlier time would have been sufficient to bring about their end. On only one group the final blow did not fall—or has not yet fallen. The crocodiles, conservative amphibious hunters rather than land forms, were reduced but not swept entirely away.

The surviving four-footed reptiles are the last animals reminiscent of the archaic forms that first came out of the water to lay their eggs during the Paleozoic era. The crocodiles and the lizards, especially, are in their outward appearance not far removed from those ancient tetrapods that moved with undulating body and long tail upon spraddled legs. The crocodiles seem to have escaped extinction by continuing to exploit the habitat that was safest for the reptiles from the beginning, and the lizards, also, in remaining largely insectivorous, have achieved longevity through retention of an old and conservative reptilian life style. Although the turtles and the snakes have moved much farther from the form of their ancestors, they share with the thriving lizards the qualities generally characterizing organisms that manage to live through harsh and changing times. Animals which are small, numerous, and collectively adaptive have the advantage, it seems, irrespective of their degree of specialization or the age of the group to which they belong. No other explanation accounts for the diversity of the reptiles that survived into the Cenozoic era and made their way in a world dominated by the mammals and the birds.

8

The Legacy of the Reptiles: The Birds

Evolution of flying vertebrates

Birds were described by de Maillet in the eighteenth century as having arisen from flying fishes that were chased or blown to land, but the passage of the vertebrates from water to sky was in reality an indirect and very lengthy process. The 100 million years or more of evolution that took place between the appearance of the first known fishes in the Ordovician period and the emergence of the ichthyostegids from the water in late Devonian time prepared the vertebrates only to breathe in air and to rest their bodies on legs. Another 150 million years of change were required before backboned animals acquired the structures necessary for even the most rudimentary kind of flight. By the late part of the Triassic period, gliding was at last achieved by the lizards *Kuehneosaurus* and *Icarosaurus* and perhaps by the forerunners of the pterosaurs and the birds as well. This form of flying, which involves no expenditure of energy, can be exercised by any small-bodied animal that develops an expanded surface against which the air can press. Although gliding flight cannot be long sustained, it allows fast escape from enemies and rapid progress from one place to another. For these reasons, the ability to glide has been selected for several times among vertebrates of different classes since Mesozoic times. There exist today, besides lizards with rib-supported flight membranes like those of the fossil species, flying frogs, flying geckos, flying snakes, squirrels, "lemurs," and opossums, all of which parachute over short distances buoyed by outspread webs, frills, or folds of skin. To this list should be added also the flying fishes which impressed de Maillet. Actually teleosts of an advanced type, they propel themselves through the water until they gain speed enough to leave it and glide parallel to the surface on expanded but rigidly held pectoral fins. It is certain that they arose long after the birds.

Flight requiring greater activity and motor skill than gliding has been far less common among the vertebrates. Forms that could soar or lift themselves against the pull of gravity have evolved only three times: pterosaurs and birds appeared during the Jurassic and bats near the beginning of the Tertiary period. The rarity of vertebrates possessed of true flight is understandable because the necessary modification of the body is extreme and the probability of all the requisite changes taking place is small. That as many as three groups of true fliers evolved independently against great odds is a measure of the selective pressure in favor of animals that could take to the air. Besides escaping from predators, animals capable of long, controlled flights could widen their search for food more effectively than ground-dwellers and could pass over obstacles that regularly limit the dissemination and migration of nonvolant forms.

The feathered wing

Of all the flying vertebrates, the birds alone evolved an alternative to the flight membrane, or patagium, to keep themselves aloft. Their production of feathers spared them the disadvantages inherent in the use of an extension of the body wall as a sail. Since a fold of skin has to be stretched taut to catch the air, a tear in one part of it destroys the effectiveness of the whole structure. Intact, it requires attachment to the hind legs as well as to the front ones and so interferes with the freedom of the former to function in locomotion. No one knows exactly how the pterosaurs moved on the ground, but bats are reduced to crawling because the pelvic appendages are twisted outward and upward to carry the lower edge of the wing membrane. Feathers are stiffer and lighter than the living tissue of the patagium (Fig. 8.1). The long feathers of the bird's wing are affixed to the bones of the forearm and hand and need no other support. The hind

Vertebrate History: Problems in Evolution

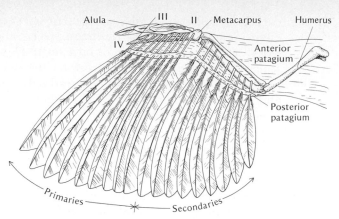

Figure 8.1. Left wing of a sparrow outstretched, in dorsal view; drawn to show the relationship of primary and secondary flight feathers to patagium, or flight membrane, and bones of the wing. *II, III, IV,* remaining digits. (From Storer.)

limbs, not being associated in any way with the wing, have become adapted for walking, running, or swimming, and the birds consequently have been able to assume habits of life impossible for other fliers. Besides leaving the hind limbs free, the feathered wing is superior to the patagium in other ways. Since it is constructed of separate elements instead of a continuous membrane, it is less likely to be incapacitated by injury. If a feather is damaged, the others function unaffected and a replacement for the spoiled one eventually grows in. During flight, the feathers can be turned or twisted by muscles at their base so that the wing changes the resistance it offers to the air.

Feathers may have been selected for originally because they increased their possessors' ability to fly, but they soon became indispensable to birds for other reasons. They provided an insulating cover without which the animals could hardly have maintained the high body temperature that made their activity possible. In living birds, the color as well as the substance of the feathers is important for survival. Many birds rely upon the hue of their feathers for concealment, whether they are white like the snow, speckled like the dry vegetation underfoot, or brightly mottled like sunlit tropical foliage. Recognition of members of their own species for the purpose of mating and other essential activities also depends upon the color pattern of the feathers.

The origin of feathers

It is not difficult to imagine how feathers, once evolved, assumed additional functions, but how they arose initially, presumably from reptilian scales, defies analysis. Scientists of an earlier generation attributed the appearance of feathers to environmental factors. G. Heilmann, who published an exhaustive discussion of the origin of birds in 1927, suggested that the scales of an arboreal avian ancestor lengthened in response to increased air pressure and then gradually frayed at the edges and metamorphosed into typical feathers as a result of friction generated between the air and the body of the leaping animals. Heilmann's quaint, Lamarckian explanation is unacceptable today, but no other has been put forth. The problem has been set aside, not for want of interest, but for lack of evidence. No fossil structure transitional between scale and feather is known, and recent investigators are unwilling to found a theory on pure speculation. Their supposition that feathers were derived from the scales of reptiles is based upon the fact that both are nonliving, keratinized structures generated from papillae on the surface of the body. Since reptiles and birds are closely related, it seems more likely that their papillae are homologous than that those of birds arose *de novo* and replaced the reptilian scale-producing tissues.

The way a feather grows suggests that it is a scale much modified. It develops as a scale does from the epidermal cells of the papilla, but instead of forming in a flat plate at the surface it takes its origin from a collar of cells at the base of the papilla and extends outward (Fig. 8.2). Its substance subdivides into numerous hollow barbs, which are fringed with barbules and conjoined at a central shaft. If the shaft is short and the barbules smooth-walled, the feather is of a type called down. Its barbs form a fluffy, insulating cover for the adjacent skin. Flight feathers and contour feathers that give the body its shape have longer, stronger shafts and barbules equipped with hooks. The hooks on each barbule catch the barbules farther forward, so that the barbs radiating from either side of the shaft are held in a flat, wind-resistant vane. It seems, from the complex construction of feathers, that their evolution from reptilian scales would have required an immense period of time and involved a series of intermediate structures. So far, the fossil record does not bear out that supposition. The oldest bird known, *Archaeopteryx,* still exhibited skeletal characters reminiscent of reptilian ones, but its feathers gave no hint of primitive features. The imprint they left in the rock, clear and sharp, makes it evident that the feathers of *Archaeopteryx* were already in Jurassic time exactly like those of birds flying today (Fig. 8.7, page 365).

Adaptive similarities in birds and pterosaurs

Although the birds evolved, in feathers, a unique modification

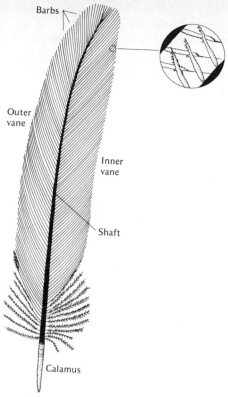

Figure 8.2. Contour feather with inset showing a small area of the vane magnified. Enlargement shows parts of two barbs, each bearing three interlocking barbules. (From Rawles, in Marshall.)

for flight, in several of their other adaptations for aerial life they paralleled the pterosaurs. The lightening of their skeleton was effected in the same way. Initially slender, their long bones became thin-walled cylinders. Instead of being filled with heavy marrow, the interior cavities housed air sacs that outpocketed from the lungs. The weight of the skull was minimized by selection for thin or spongy bone, and its strength was retained by the fusion of its elements when the animal was full-grown. Like the pterosaurs, the birds showed a tendency toward loss of the teeth, which culminated in their becoming wholly edentulous. The horn-covered beak that replaced the heavier, mineralized reptilian dentition was as serviceable for cutting and holding as the latter had been and ultimately proved as versatile in adapting to a variety of food-getting habits. The ability to fly, in birds as in pterosaurs, involved an increase in the complexity of the nervous system, particularly in the brain, where sensory and motor activity is coordinated. Casts of the brain of Mesozoic fliers show that the cerebrum and cerebellum had enlarged and overlapped the midbrain. By the beginning of the Tertiary period, the two structures in the avian brain had converged, pressing the optic lobes of the midbrain laterally and covering the entire brainstem in dorsal view (Fig. 8.3). Much of the sensory data sent to the coordinating centers undoubtedly came from the eyes in pterosaurs, and the large orbits in the skull of *Archaeopteryx* indicate that vision became the primary sense in birds, also.

While birds were evolving traits parallel to those of pterosaurs, they were accumulating a suite of new structural characteristics that would make of them more efficient flying machines than had ever existed before. Not only did the body become compact and rigid through the shortening and fusion of the trunk vertebrae, as in the airborne reptiles, but the wing musculature increased enormously in bulk, making forceful flight possible (Fig. 8.4). The pectoralis muscle, which brought

Modification of the musculature in birds

Figure 8.3.   Brain of **(a)** an alligator and **(b)** a goose in dorsal view. Note that in the reptile, the cerebrum and cerebellum are relatively small and the optic lobes of the midbrain are visible between them. In the bird, the cerebrum and cerebellum have enlarged, covering the optic lobes. (From Romer.)

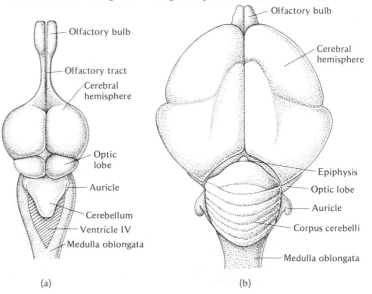

(a)

(b)

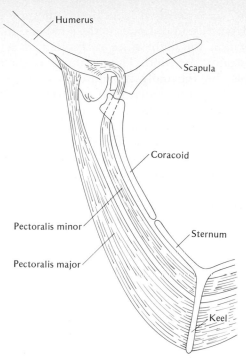

Humerus

Scapula

Coracoid

Pectoralis minor

Sternum

Pectoralis major

Keel

Figure 8.4.   The location of the muscles which raise and lower the bird's wing. Although it raises the wing by pulling on the upper surface of the humerus, the pectoralis minor lies against the underside of the body beneath the depressor of the wing, pectoralis major. With the bulk of the flight musculature situated ventrally, the center of gravity is located low in the body and the stability of the bird in flight is enhanced. (From Storer.)

about the downstroke of the wing, expanded over the surface of the sternum, and that bone developed a keel, or carina, for the attachment of its fibers. The great weight of the pectoralis on the underside of the body gave the bird added stability in the air. The arrangement of the rest of the appendicular muscles contributed to the maintenance of the center of gravity between and beneath the wings. Instead of being located above the shoulder, the fibers which opposed the pectoralis were also situated ventrally. Originating upon the sternum beneath the flexor muscle, they inserted upon a tendon that slipped upward through a notch in the pectoral girdle to pull upon the humerus dorsally and raise it. Little musculature remained distal to the humerus, where its weight could cause the bird to roll as it flew. The reduced mobility of the pectoral appendage below the elbow, which resulted from the diminution of the musculature and from fusion of bones in the wrist and hand, was in itself an adaptation for flight since it resulted in better bracing of the limb against the wind. The distal part of the hind legs remained flexible, but virtually no muscle extended downward to the ankle. Most of the musculature lay on the upper leg, close to the trunk, and pulled upon the bones of the feet and toes through long tendons. Since the legs were carried directly beneath the body, their weight contributed to keeping the center of gravity low. The axial muscles, which would have added weight to the dorsal region of the body, were reduced to insignificance because of the fusion of the trunk vertebrae and the loss of the long reptilian tail.

Mechanics of flight   Because the wings are flapped in active flight, the interplay of forces involved in the process is extremely complicated. In beating the wings downward and forward and then drawing them upward and backward, the bird must manipulate the bones and feathers so that its lift is not lost. This feat is accomplished by keeping at least the proximal part of the wing at a certain angle to the airstream. Since the wing is cambered like that of an airplane (convex on its upper surface and tapered from the leading to the trailing edge), when tilted into the wind, it divides the oncoming air into two currents which act upon the wing differently. The lower of the two passes under the wing smoothly, pressing upward against it. The upper one rises over the blunt leading edge of the wing and then slips down behind the high point of its curvature, exerting a downward force. The stream passing over the convexity of the wing, having a longer distance to go, flows faster and pushes less strongly upon the feathered surface than the air moving below. The difference in pressure constitutes the lift which, when equal to the weight of the bird or greater, sustains the animal in flight. The ease with which a bird can fly and the speed of which it is capable depend upon the relative size and shape of its wings, but the lift generated at a given moment is determined to a large extent by the speed at which the bird is traveling and the angle at which it holds its wings. To compensate for the decrease in lift that accompanies a decrease in speed, a bird twists the leading edge of the wing upward so that its undersurface is presented more directly to the oncoming air. The rotation of the wing must be carefully controlled, however, because, beyond a certain point, increasing turbulence above and behind the wing will cause the bird to stall. Birds adapted for relatively slow flight are frequently able to separate the feathers at the wing tip, where turbulence tends to be greatest, and to raise feathers mounted on the first digit, thereby opening slots for small, smooth streams of air. These airstreams reduce the turbulence somewhat and thus allow the bird to tilt its wings more sharply against the air than would otherwise be possible. Control of turbulence by slotting has also permitted widening of the wings for maximum lift in birds that fly in dense cover or those, like hawks, whose speed is reduced by the weight of the small animals they snatch and carry away. Since wing size itself has a limit, wings that are broad tend to be proportionately short. Long wings are characteristic of birds that can accelerate fast enough to develop the lift they require from a wing of restricted breadth.

Once aloft, the work a bird has to do to remain in flight varies with the air currents it encounters and the ratio of its wing surface to its total weight. Light forms with relatively large wings, like swifts and swallows, expend less energy in flapping than birds, like ducks and geese, which are heavy in proportion to the size of their wings. No matter how easily or how gracefully a bird travels through the air, however, the actions involved in flying create strong stresses which its body must absorb. The structural modifications that have evolved to withstand them include, besides the rigidity of the vertebral column, the development of especially stout coracoid bones which project the thrust from the wings against the side of the heavy sternum (Fig. 8.5). The ribs, which are pulled upon indirectly during flight, have ossified rather than cartilaginous connections to the sternum, and each rib produces from its midregion a finger-shaped uncinate process that overlaps the rib behind, reinforcing the thoracic basket further. Whereas the expansion of the iliac bones and their fusion with a long series of vertebrae to make the synsacrum undoubtedly contributed to the rigidity of the body necessary in flight, the pelvic region assumed its extraordinary strength in response to the stresses generated in landing. Even though a bird slows itself by braking with its wings and outspread tail feathers, the force projected upward through the legs is considerable when the animal comes to rest. The pelvic girdle, open ventrally as an adaptation for egg laying, is adequately resistant, nevertheless. Although the pubis is a delicate bone twisted back beneath the ischium, the latter element is broadened and fused to the expanded ilium that lies above it. The ischia, the ilia, and the vertebrae between them make a bony arch into which the femur of each leg is fitted. A hole in the

Adaptation of the avian skeleton to stresses of flight

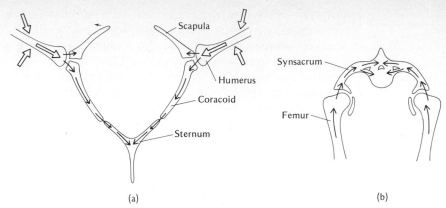

Scapula

Humerus

Coracoid

Sternum

Synsacrum

Femur

(a)

(b)

Figure 8.5. Adaptation of the avian skeleton for resistance to stress projected from the appendages. **(a)** The pectoral region, showing major stress thrust through coracoid bone to sternum. (After Storer.) **(b)** The pelvic region, showing stress projected through vault formed by the bones of the synsacrum.

base of the acetabulum that allows the head of the femur to be deeply inset and a ridge of bone projecting over the socket assure that the leg will not be dislocated from the girdle by a sudden thrust.

Advantages of efficient metabolism and thermoregulation

The structural adaptation of the skeletal and muscular system for flight was accompanied in birds by physiological changes which enabled them to produce continuously the enormous amounts of energy they required. It is certain that in early members of the avian line a higher metabolic rate was selected for. Not only were the animals with enhanced powers of energy production capable of the most sustained muscular and nervous activity, but they would have been best able to resist the cooling effect of the air, which, by lowering the temperature of their body, tended to make them sluggish. The small size that was necessarily retained in animals becoming airborne created selective pressure for the most efficient metabolic processes possible. Since the large surface-to-volume ratio characteristic of small forms resulted in an especially rapid loss of body heat through the skin, the evolution of the birds as active fliers continued, most likely, only because the increase in the metabolic rate that did occur coincided with the appearance of feathers. With feathers to provide insulation, the energy produced in the cells of emerging birds was sufficient to support the demands of flight and also to keep the temperature of the body high enough to assure the stability of the metabolic rate. Once the birds could maintain their activity in the face of low environmental temperatures, they were free not only to soar into the cool air at high altitudes but also to spread into regions of the world where diurnal and seasonal fluctuations were marked or where cold weather prevailed the year round.

Development of avian thermoregulation is obscure

How temperature control gradually manifested itself as the birds evolved remains unknown. Some investigators have looked to the reptiles, hoping to find exhibited among them primitive forms of thermoregulation that could represent the first steps in the development of the avian mechanisms. Others have pointed out that since none of the living reptiles belongs to the line from which the birds sprang, there is little justification for supposing that the two kinds of animals shared a common approach to temperature control. Indeed, the evidence suggests that they did not. Such ability as modern reptiles have for keeping their body temperature at an optimum level for metabolic activity depends more heavily upon behavioral adaptations—sunning, seeking shade, and enduring periods of torpidity—than upon adjusting their production of energy. Studying the development of temperature control in young birds just after hatching has been recommended as a better method of retracing the steps by which avian thermoregulation evolved. The value of this kind of research is also questionable, however, because ontogeny is now recognized as an unreliable guide to phylogeny. It is quite possible that the early stages of thermoregulation observable in nestlings are the result of juvenile adaptation rather than reprises of the type of temperature control found in the first birds. The hope that different thermoregulative abilities would be discovered in adult birds, permitting the construction of a graded series of stages approximating the evolutionary ones, has proved unavailing. In all living birds, apparently, the mechanisms of temperature control are equally sophisticated.

The avian digestive system

Study of the internal organs of modern birds does make clear the physical basis for the efficient metabolism which makes thermoregulation possible. The avian digestive, respiratory, and circulatory systems are modified in such a way that the oxygen and nutritive materials necessary for energy production are moved to the cells more quickly and in greater quantity than in poikilothermic vertebrates. The structure of the foregut allows birds to ingest a large volume of food rapidly. Fish, insects, and other small animals may be swallowed whole, and seeds, grain, and algae are often downed by the beakful. In grain eaters, the lower end of the esophagus is expanded into a storage organ, called the crop, which can be stuffed prodigiously. Uncracked grain is passed little by little from the crop to the muscular region of the stomach, termed the gizzard, where powerful contractions of the gut wall crush the fruit and seed coats, exposing the digestible material within. Birds that are carnivorous lack the crop but secrete such highly acidic gastric juice that food coming in contact with it disintegrates speedily and virtually completely. The gizzard in many of these forms compresses the small amount of material not dissolved, principally teeth and hair or feathers, and sends it forward for discharge through the mouth. The digestive and absorptive activity of the intestine and its blind outpocketings, or caeca, is so efficient that an unusually large proportion of the material taken from the environment into the gut is made available for use by the cells. The amount of nutrient matter supplied to the body within a given time is great, because as well as being efficient the digestive process is rapid, and birds are able to consume a sizable quantity of food in proportion to their body weight.

The avian respiratory system

The construction of the respiratory system permits the intake of oxygen and the expulsion of carbon dioxide at an especially high rate (Fig. 8.6). Instead of being blind sacs which retain a considerable volume of stale air, the lungs are continuous with spacious air sacs which fill the body cavity and extend into the interior of many of the bones. Since incoming air passes through the lungs and into the sacs, the gaseous content of the lungs is changed completely with each breath. All spent air is driven even from the minute spaces where oxygen is given up and carbon dioxide accepted from the bloodstream, because the spaces interconnect and rejoin bronchial channels rather than terminating as closed alveoli. The evolution of respiratory organs of this design was advantageous not only in making available large volumes of oxygen for metabolic reactions but in dissipating excess heat generated by the muscle fibers during flight. Cooling both by escape of heated air and by evaporation are fostered by the thorough ventilation of the air sacs. Birds' best defense against overheating is a rise in the respiratory rate, and a center exists in the brain which responds to an increase in the temperature of the blood by initiating more rapid breathing. The system of air sacs as it is developed in birds is unique among living vertebrates, but there is evidence that it had evolved independently in the pterosaurs. If the openings found in pterosaurian long bones are, as they are assumed to be, pneumatic pores for the entrance of air sacs into the interior, it may

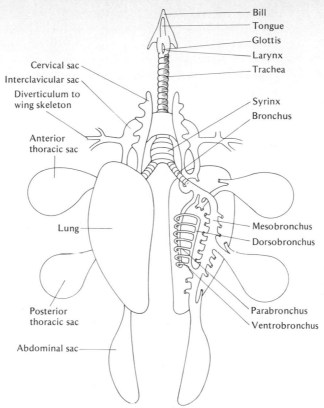

Bill
Tongue
Glottis
Larynx
Trachea

Cervical sac
Interclavicular sac
Diverticulum to
wing skeleton

Syrinx
Bronchus

Anterior
thoracic sac

Lung

Mesobronchus
Dorsobronchus

Posterior
thoracic sac

Parabronchus
Ventrobronchus

Abdominal sac

Figure 8.6. The avian respiratory system, showing direct channels to the several air sacs and one of the many sets of thin-walled parabronchi within the lung, through which excurrent air passes. (From Welty.)

mean that flight even in the pterosaurs was facilitated not only by the lightness of the bones but also by the high metabolic rate and stability of temperature for which the saccular extensions of the lungs are in good part responsible.

Efficiency of the avian circulatory system

The oxygen and nutritive materials which enter the body reach its cells quickly because the blood that carries them flows under great pressure. The pumping action of the heart, remarkably strong and fast in modern birds, doubtless accelerated as the respiratory and digestive systems of early avians improved, because the capacity of cardiac muscle for work depends upon its own supply of metabolites. That an increase in the rate of the heartbeat could raise the pressure of the blood significantly was due to changes that had taken place in the circulatory system long before, when vertebrates were acclimating themselves to life out of the water. As the gills disappeared at that time, their capillary network, which had been interposed between the heart and the main distributive artery, vanished also. Instead of losing pressure by passing through countless capillaries whose combined volume was larger than the ventral aorta that fed them, the blood in gill-less tetrapods flowed from the heart through major arteries directly to the various tissues of the body where the substances it carried were used. The location of the respiratory capillaries in the lungs, where they did not interrupt the flow of blood to other organs, made possible greater arterial pressure and thus speedier delivery of materials, but it introduced a new limitation. Because oxygenated blood returning from the lungs mixed with unoxygenated blood from the rest of the body as the two streams passed through the heart, the blood pumped by the heart into the aorta always contained less oxygen than it had the potential of holding. In amphibians and reptiles, the heart became partially divided so that much of the unoxygenated blood was diverted into the pulmonary artery but the archosaurian ancestors of the birds, if they were like living reptiles, still had one large aortic vessel which conveyed unoxygenated blood to the systemic circulation. As the birds evolved, maximum oxygen tension in the arterial blood was assured by the loss of this vessel and by completion of the partitions within the heart that separated unoxygenated blood bound for the lungs from the oxygenated blood on its way to other parts of the body.

Reproduction and breeding behavior in birds

As birds became capable of producing enough heat to keep their body temperature stable and above that of the environment—in other words, as they become homoiothermic, or warm-blooded—their reproductive habits changed somewhat from those of reptiles. Whereas reptile embryos can, and presumably always could, develop within eggs laid in holes, covered, and left subject to the changing weather, avian embryos require for their growth the same constant and elevated temperature that is maintained by the adult birds. Although the embryos could have been protected from cooling by being retained within the body of the female, a tendency toward retention of the eggs either never expressed itself in birds or was selected against because it increased the weight of the airborne animals. Instead of paralleling mammals in delaying the release of their offspring, birds continued to lay their eggs in the reptilian manner. Rather than abandoning the clutch of eggs, however, they remained with it, brooding the eggs to keep them warm, or if the temperature of the air was too high, shading the nest to prevent them from overheating. Caring for eggs in this way and feeding the hatchlings, as is done by a large number of modern birds, requires the attention of at least one and usually both of the parents. The behavioral modifications which rendered the birds careful of their young stemmed from the evolution of nervous pathways within the brain which are poorly understood. Neurophysiologists have traced impulses forward through the central nervous system to a greatly enlarged region in the floor of the forebrain called the basal nuclei and suppose that neural circuits originating in that specially developed area are involved in producing the birds' breeding behavior.

The connections among the nerve cells in the brain are very precisely determined, for the patterns of activity that emerge are instinctive and extremely rigid. Particular stimuli from the environment elicit whole series of responses which, once begun, are continued even when their objective is in some way nullified. Thus, species of birds which do not distinguish between their eggs and those of others can often be made to brood a variety of substitutes, sitting on a ball or even a light bulb simply because it fills the nest at a time when the eggs are expected to be there. This kind of invariant behavior has helped ensure the survival of birds, because it renders them persistent in their reproductive activities in the face of all but the most dire environmental conditions. Birds are not unusual among the vertebrates, certainly, in having their essential reproductive functions assured by powerful instincts. Their instinctive behavior is remarkable only because of its elaborate development. The birds are brought together initially by a courtship ritual which in some species is extraordinarily complex, involving patterned flights or dances on the ground and even the presentation of food or other objects by one bird to the other. Although some birds lay their eggs in simple depressions in the ground, protection of the young is often augmented by provision of a nest which both parents help to prepare. Instinct is responsible for the finest details of all these activities, from the movements used in courtship displays to the twisting and knotting of fibers by which some types of nests are secured to the branches of trees. Instinctual behavior of a complex nature also governs the migrations made by some birds to areas favorable for breeding as well as a number of activities not connected with reproduction, like care of the feathers, associating in flocks, and defense of their chosen territory.

When investigators turn from the study of modern birds to the fossil record for information about avian evolution, they find their research severely limited by the scarcity of informative remains. The delicacy of the avian skeleton, as well as the habitat of the majority of birds, makes their preservation a rare occurrence. Birds that are killed by predators are either eaten completely and reduced to a wad of indigestible feather fragments or left in a mangled condition to be macerated still further by bacterial action. Scavengers of the field and forest soon dismember a bird that dies without being attacked, and eventually its broken, disarticulated bones are scattered by wind and rain. Skulls, which would be valuable to the paleontologist, seldom survive. Collectors find most often splintered pieces of long bones, isolated ribs, bits of the pelvis, and parts of vertebrae, all of which are far less helpful than the skull in making clear the nature of the bird from which they come. Because death in shallow water led most readily to fossilization, the record is weighted in favor of swimming and diving birds and those living near the shore. Fliers based farther inland were preserved especially well only on the rare occasions when they were blown away from their usual haunts into a lake or a sea. Most commonly, whatever the circumstance of fossilization, no more than a few bones belonging to the same animal are found in association with each other. In places like the tar pits of Rancho La Brea in California and the ancient lake bed famous for its avian fossils in Oregon, the remains of so many individuals of the same species are preserved that paleontologists have been able to collect enough material to make composite reconstructions of the extinct forms. Both the California tar pits and the Oregon site date from the Pleistocene epoch, however, and so contain birds less than million years old. No comparable reservoirs of fossils are known from the Tertiary period or from Mesozoic times, when the transition from the reptilian to the avian condition occurred.

Since the preservation of birds is infrequent and, when it does take place, almost always incomplete, the discovery of two excellent specimens of *Archaeopteryx* in the Upper Jurassic limestones of Bavaria was an incredible piece of good fortune. In 1861, some months after finding a single fossilized feather, workmen in a quarry near Pappenheim came upon the partially disarticulated remains of an individual which had been buried belly down, wings outspread, and legs extended under fine sediments in the shallow sea water that once covered the region. Recognizing the fossil for the prize it was, they brought it to the medical officer of the district, Dr. Friederich Karl Häberlein, who had an extensive collection of plant and animal remains discovered in the local quarries. He sold it to the British Museum, where it was examined by Richard Owen and found to be missing only a few cervical vertebrae, the right foot, and the lower jaw. The second specimen, even more perfectly preserved than the first, was removed in a slab of limestone from a site 10 miles from Pappenheim in 1877 and presented to Häberlein's son, who disposed of it to the Museum für Naturkunde in Berlin (Fig. 8.7). Although fragmentary remains of two more of the animals came to light subsequently, study of the London and Berlin specimens has provided all that is known of the Jurassic feathered forms.

Few investigators have hesitated to call *Archaeopteryx* a bird because its feathers conform in structure and arrangement to those of modern members of the class Aves. The impressions in the limestone make it plain that the hand bore primary, and the forearm secondary, flight feathers, their quills shielded by shorter wing coverts. The posture of the animal was bird-like as well. *Archaeopteryx* was bipedal, resting its weight on long hind legs with three-toed feet braced behind by a short, backward-directed first digit. The wings were carried folded against the rib cage, and the head was held high on a long sinuous neck. At first, only the toothed jaws and the long, drooping tail of the animal seem out of keeping with its avian nature, but scrutiny of the skeleton reveals a number of differences between it and that of a modern bird.

Structure of the skull of Archaeopteryx

Although the skull was relatively small and large-eyed, the elements of the snout, cheek, and roof bore an unmistakable resemblance to those of pseudosuchian reptiles (Fig. 8.8). Heilmann, whose analysis of the anatomy of *Archaeopteryx* is less questionable than his explanation of the evolution of feathers, pointed out that the premaxilla in the Jurassic form was short and in no way similar to a beak or bill. Instead of being horn-covered, this bone and the maxilla behind it supported thecodont teeth. The nasal opening lay far forward, separated from the eye by a large preorbital fenestra in typical archosaur fashion. Whereas the palate and cheekbones in birds are flexibly articulated with a mobile quadrate, as far as Heilmann could tell, this was not the case in *Archaeopteryx*. The palate was not visible in the Berlin specimen, which he studied, but the stoutness of the jugal element and its union with the bones in front of and behind the orbit suggested that the jugal arch and the quadrate posterior to it were immovable. It was difficult to discern the

(a)

structure of the temporal region in *Archaeopteryx* because the bones in that part of the skull had been compressed and somewhat crushed. It seemed to Heilmann that the lower of the two archosaurian temporal fenestrae was still present but that the upper one was either small and slit-like or lost entirely. The expansion of the rear of the skull characteristic of birds had not yet taken place. Since the cerebellum of *Archaeopteryx* and the

(b)   (c)

Figure 8.7. Two fossil specimens of the Upper Jurassic bird *Archaeopteryx*. **(a)**
The London specimen. (From De Beer, courtesy of the Trustees of the British
Museum.) **(b)** The Berlin specimen, formerly designated as *Archaeornis*. (Photo-
graph, Museum für Naturkunde an der Humboldt-Universität zu Berlin.) **(c)**
Feather of *Archaeopteryx* found in 1860. (Photograph, Bayerische Staatssam-
mlung für Paläontologie und historische Geologie.)

bones which covered it were still of reptilian proportions, the
roof of the skull was not dome-shaped, as it is in existing birds.

*The postcranial
skeleton of Ar-
chaeopteryx*

The postcranial skeleton also showed traits that were more
reptilian than avian. The cervical vertebrae, instead of having

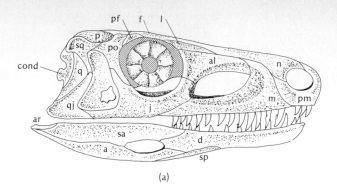

(a)

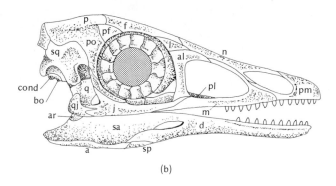

(b)

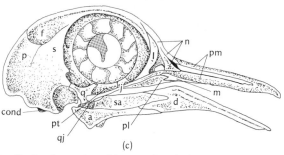

(c)

Figure 8.8. Skulls of pseudosuchian reptile, *Archaeopteryx*, and modern bird in lateral view. (**a**) *Euparkeria*, a pseudosuchian. (**b**) *Archaeopteryx*, reconstructed from the Berlin specimen. (**c**) *Columba*, the modern pigeon. Abbreviations: *al*, adlacrimal; *cond*, occipital condyle. Other abbreviations as in Fig. 6.7. (After Heilmann.)

the heterocoelous centra peculiar to birds, had centra with unmodified biconcave surfaces like those of primitive archosaurs. The posterior part of the column lacked the compactness and rigidity found in birds. The vertebrae were not fused with one another in the trunk region, nor were they braced by extremely elongated iliac bones. Rather than being supported by a solid synsacrum, the weight of the front of the body was counterbalanced, as in pseudosuchians and bipedal dinosaurs, by the long tail. The pelvic girdle of *Archaeopteryx* was much smaller than that of an extant bird of comparable size (Fig. 8.9). It had an avian appearance chiefly because the pubes were twisted backward below the ischia and the latter were not joined in the midline. The ilia abutted far fewer sacral vertebrae—from

Figure 8.9. Pelvic girdles of **(a)** *Archaeopteryx* and **(b)** a modern hen in dorsal view, both natural size. Although the Jurassic bird was as large as a small hen, the bones of its pelvic girdle were proportionately far smaller. (From Heilmann.)

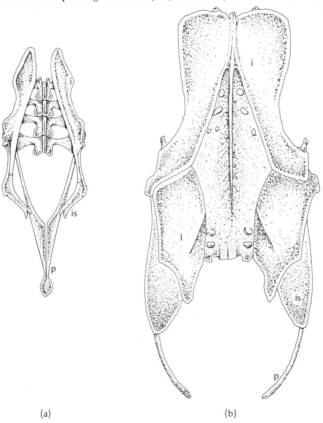

(a)   (b)

four to six in contrast to a minimum of eleven in modern birds—and did not extend posteriorly along the dorsal edge of the ischia or fuse with those bones as they do in avian fliers of today. The three bones on each side of the girdle articulated at the edge of the perforate acetabulum, as they did in advanced archosaurs, rather than ankylosing about the socket for the femur, as they do in mature birds. The hind limb came closer to the avian condition than the girdle, but it remained reptilian in keeping a fibula equal in length to the tibia and in maintaining its three long metatarsals as independent bones.

To the degree that the pectoral appendage, its girdle, the sternum, and the ribs fell short of the avian state, the ability of *Archaeopteryx* to fly must have suffered. Since the sternum of *Archaeopteryx* was small and without a keel and the coracoid bone was short, the animal had little surface for the origin of flight muscles. That *Archaeopteryx* could not fly far is also suggested by the weakness of the skeleton in the thoracic region (Fig. 8.10). The ribs were slender and lacked the bony uncinate processes by which the ribs of living birds are braced against one another. Most if not all of the ribs tapered to a fine end in the body wall rather than being secured ventrally to the side of the sternum. The latter bone did not extend posteriorly beneath the ribs to form a rigid support for the viscera. The belly wall was stiffened only by gastralia, serially arranged slivers of dermal bone that appear in several kinds of reptiles as vestiges of the armor of ancient tetrapods. The wing skeleton in *Archaeopteryx* was not fully adapted to resist the air currents encountered in soaring or flapping flight. The hand and wrist had not yet become an inflexible blade. The fusion of the carpal elements had barely begun, and the metacarpals still remained independent of each other. The retention of claws at the ends of the three digits makes it probable that *Archaeopteryx* used its forearms not only in flying but also in grasping branches as it scrambled about in trees. There is a living bird, the South American hoatzin (*Opisthocomus cristatus*), in which the nestling possesses clawed digits and exhibits this kind of behavior (Fig. 8.11). In the hoatzin, the claws disappear as the bird matures. Carpal and metacarpal fusions occur, limiting the mobility of the hand, the primary feathers grow to full size, and the bird abandons the juvenile habit of crawling for flight. Although the condition of the hoatzin is undoubtedly a secondary development, the behavior of the young birds proves that flexible, clawed forelimbs can be used to advantage by feathered forms. It is possible that

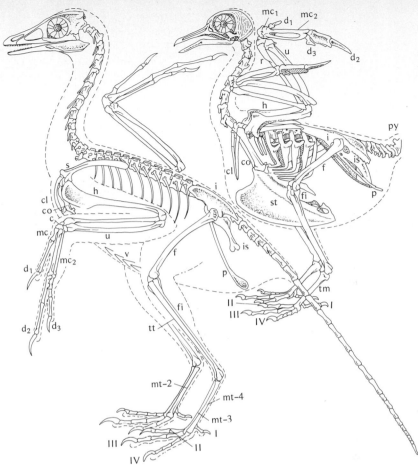

Figure 8.10. Skeletons of *Archaeopteryx* (left) and the modern pigeon (right). The skeleton of the pigeon is reduced in size to a greater degree. Abbreviations: *c*, carpal; *cl*, clavicle; *co*, coracoid; *d*, digits; *f*, femur; *fi*, fibula; *h*, humerus; *i*, ilium; *is*, ischium; *mc*, metacarpals; *mt*, metatarsals; *p*, pubis; *py*, pygostyle; *r*, radius; *s*, scapula; *st*, sternum; *tm*, tarsometatarsus; *tt*, tibiotarsus; *u*, ulna; *v*, ventral ribs; *I–IV*, toes. (From Heilmann.)

Archaeopteryx survived for a time because, in addition to making short flights, it was able to crawl agilely through the treetops with the help of its wing fingers.

Since *Archaeopteryx* occupies an isolated position in the fossil record, it is impossible to tell whether the animal gave rise to more advanced fliers or represented only a side branch from the main line. The fossil has been less significant, in fact, for what it

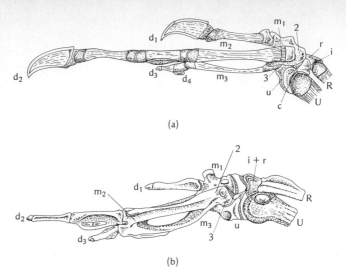

(a)

(b)

Figure 8.11. Skeleton of the left hand of the hoatzin. **(a)** Hand of the embryo, × 4. **(b)** Hand of the adult, × 1. In the adult, claws have disappeared from first and second digits and fusions of the carpal bones have taken place. Abbreviations: *c*, central carpal; d_1-d_4, first to fourth fingers; *i*, intermedium (a proximal carpal); m_1, m_2, m_3, metacarpals; *R*, radius; *r*, radiale (a proximal carpal); *U*, ulna; *u*, ulnare (a proximal carpal); *2, 3*, distal carpals. (From Heilmann.)

The evolution of flight

has implied about avian evolution from the Jurassic period onward than for the evidence it offers concerning the earlier history of the birds. Although study of modern members of the class Aves has linked them certainly to the reptiles, examination of *Archaeopteryx* has enabled investigators to specify the pseudo-suchians as the group from which the first avians came. Not every one agrees, however, about the way in which the primitive terrestrial thecodonts were transformed into flying forms. Heilmann, who pictured the hypothetical proavian as an arboreal animal (Fig. 8.12), voiced the opinion, still favored by the majority, that flight developed as the result of persistent hopping, leaping, and gliding among the branches. The light bones and backward-turned first toe on the hind foot Heilmann regarded as reptilian adaptations for life in the trees which facilitated the evolution of fully bipedal fliers. F. Nopcsa believed that avian flight, dependent as it is upon feathers and bipedal posture, would not have evolved in animals that had initially taken to the trees on all fours. He pointed out that arboreal tetrapods which jump and glide spread their four appendages wide and inevitably develop some sort of patagium that prevents their descendants from becoming bipedal. Nopcsa asserted that flying birds evolved from cursorial predecessors, bipedal forms which held their pectoral appendages outstretched to act as airfoils as they ran. Heilmann knew of Nopcsa's idea and disagreed with it, arguing that the skeletal changes which occurred as birds evolved were the opposite of those which took place in archosaurs and other terrestrial animals adapted for running on two legs. In bipedal terrestrial runners, the pectoral appendages became progressively smaller and the clavicles tended to disappear. Also, the posteriorly directed first digit dwindled as the cursorial habit became more pronounced. Although he was aware of the striking similarity between certain parts of the skeleton in the advanced archosaurian bipeds and in birds, Heilmann maintained that the two groups had specialized along divergent paths.*

Ratites: primitive or specialized forms?

In supporting his hypothesis, Nopcsa called attention to the early existence of birds called ratites, which are good runners but poor fliers. Included in this category are the living ostriches, rheas, cassowaries, emus, and kiwis and the extinct moas and elephant birds. Since remains of ratites are known from the beginning of the Tertiary period, Nopcsa regarded these terrestrial birds as primitive forms, closely related to the cursorial avian stock in which he thought flight had first appeared. He was not alone in judging the ratites to be archaic types. Huxley had upheld this idea in the 1860s, saying that the structure of the ratite palate was more primitive than that of flying birds. In ratites, he pointed out, the pterygoid bones extend straight forward to meet large vomers, whereas in other avians the pterygoids articulate only with the palatine bones, the vomers having fused completely and diminished in size. So convinced was Huxley of the significance of the structure of the palate that he based his classification of the birds upon it. The ratites he placed in the superorder Palaeognathae, and the birds without the pterygoid-vomerine articulation he housed in the superorder Neognathae. The separation of the ratites from the supposedly more advanced birds was broadly accepted. More than 60 years after Huxley published his theory, P. R. Lowe reviewed and defended the thesis that the flightless birds were

---

*J.H. Ostrom commented recently (1970) upon the origin of flight in connection with his discovery of a hitherto unrecognized specimen of *Archaeopteryx* in a museum collection in the Netherlands. His study of the clawed manus of the fossil led him to conclude tentatively that the animal had sprung from terrestrial (and therefore cursorial) rather than arboreal ancestors.

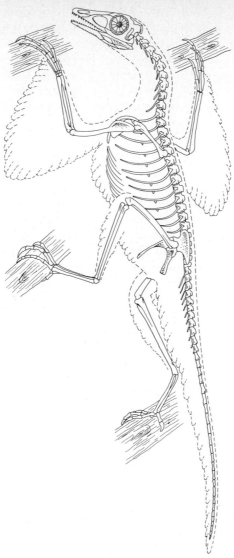

Figure 8.12. Heilmann's reconstruction of the skeleton of *Proavis*, a hypothetical form intermediate between a thecodont reptile and *Archaeopteryx*. (From Heilmann.)

relics of avian populations whose members had not yet evolved the ability to sustain themselves in the air. He argued that in addition to having a primitive palate, ratites like the ostrich fail to develop certain of the traits that characterize the advanced birds. They produce only downy plumage, akin to that which in neognathous birds is hidden by the contour feathers as the animals mature, and they continue to show the sutures between the skull bones that vanish in neognathous forms as growth is completed. Lowe believed that the absence of the clavicles in the ostrich and the short coracoid elements in all the ratites proved their direct relationship to the flightless, bipedal archosaurs.

The presence of *Archaeopteryx* in Upper Jurassic times was an embarrassment to investigators, like Lowe, who thought that arboreal, flying birds evolved only after the cursorial flappers had succeeded in leaving the ground. Lowe dismissed *Archaeopteryx* as a gliding dinosaur which anticipated the advanced birds but did not give rise to them. B. Petronievics, thinking it improbable that feathers had evolved twice, resolved the problem differently. After studying the Berlin and London specimens of *Archaeopteryx*, he declared that the two were not alike. Although the sternum was badly preserved in both and not entirely exposed, he stated that he had found indications of a keel in the Berlin animal but not in the London form. He set the Berlin specimen at the base of the lineage of flying birds, or carinates, stating that it differed in so many traits from the individual in London that it should be classified not as *Archaeopteryx lithographica* but rather as *Archaeornis siemensi*, the keel-less *Archaeopteryx* he believed to be the ancestor of the flightless ratites.

De Beer disagreed with Lowe and with Petronievics. In his opinion, the conclusion of the latter that the Berlin specimen had a keeled sternum was unjustified by the evidence and the raising of a separate genus and species for that fossil was an error. De Beer argued that both specimens belonged to *Archaeopteryx lithographica* and that the Jurassic avian led not to earthbound birds but to forms whose powers of flight were superior to its own. He held that the ratites were not primitive forms which originated directly from flightless avian precursors but specialized descendants of *Archaeopteryx's* carinate successors. Owen had taken the same stand years before. He considered the ratites to be birds which had lost the ability to fly and doubted that they constituted a natural group. Since birds, especially large birds, find most of their food on the ground, it seemed that natural selection would promote the evolution from several avian stocks of forms which were adapted for an entirely terrestrial life. The stout legs and reduced wings of these birds would be the result of convergent modification rather than a sign of common ancestry. Further study of the palate has supported the polyphyletic origin of the ratites. Workers who followed Huxley found that the structure of the palate was far from uniform among the flightless birds. There existed so many variations in the arrangement of the bones, including some which approached the neognathous condition nearly, that derivation from a single source seemed highly unlikely. To De Beer it seemed more probable that the condition of the palate in the ratites was neotenous. W. P. Pycraft had discovered that the palate of many carinate birds passes through a "paleognathous" stage during its development. De Beer reasoned that the differences among ratite palates as well as the common possession of a pterygoid-vomerine articulation could be best explained by supposing that birds of diverse origins, in becoming wholly terrestrial, had retained embryonic characters to a varying degree. Other traits which Lowe had interpreted as primitive in arguing that the ancestors of the ratites had never flown, De Beer also interpreted as neotenous and regarded as additional evidence in favor of a link between the ratites and flying birds. Besides exhibiting similarities to the carinate embryo, ratites also retain structures which arose in birds as adaptations for flight. If the precursors of the ratites had not yet become airborne, De Beer maintained, the presence in flightless birds of the large cerebellum, the abbreviated tail, and the typically reduced and immobilized avian hand would be inexplicable.

History of the flightless birds

The ratites are currently regarded not as primitive relicts which failed in competition with more modern birds but as specialized forms that appeared early and prospered until the emergence of the advanced mammalian carnivores. The appearance of the fast, aggressive, sharp-toothed tetrapods put an abrupt end to the contest between the birds and the mammals for the territory which had been relinquished by the great reptiles. Although the fossil record is slim, it reveals that during the Eocene epoch the archaic mammals had been forced to share their habitat with a number of avian forms whose pectoral appendages had already undergone reduction or disappeared entirely. Some of these birds were related to the paleognathous ratites which still survive. Others exhibited more direct neognathine affinities. Most of them were large enough to prevent all but a few of their competitors from challenging them effectively. *Diatryma*, the most completely known of the Eocene flightless birds, stood at least 7 feet tall (Fig. 8.13). The proportions of its body were similar to those of an ostrich, but its neck was stouter, to support the enormous, rounded beak with which it scissored its food. Remains of *Diatryma* have been found in

Figure 8.13. Skeleton of *Diatryma*, giant flightless bird of the Eocene. (From Matthew and Granger.)

both North America and Europe. On the latter continent there lived at the same time a flightless goose-like bird of great size, *Gastornis*, and another form, *Eleutherornis*, known only from a pelvic girdle, which seems to have been a forerunner of the ostriches. Africa had its terrestrial avians, too, in the elephant birds, or aepyornids, which spread across the northern region, perhaps from southwest Asia, and gradually penetrated southward.

The carnivores which arose in Oligocene and Miocene times extinguished all but a few of these birds wherever they had access to them. The ostriches, which had distributed themselves over parts of Europe, Asia, and Africa, resisted the onslaught of the mammals longest but after the close of the Pleistocene epoch were confined to restricted areas of Africa and Arabia. Ten-foot-tall elephant birds (Fig. 8.14) survived until recent times on the island of Madagascar, leaving behind fragments of what surely must be the largest eggs ever laid. Islands elsewhere

Figure 8.14. Restoration of *Aepyornis*, the elephant bird, which survived in Madagascar beyond the end of the Pleistocene. (After Wilson, from Swinton.)

provided havens for other flightless birds, since carnivorous tetrapods were stopped at the continental shore. Emus and cassowaries exist still on New Guinea and Australia, and the relatively small kiwi remains in New Zealand. South America, long uncolonized by placental mammals of carnivorous habits, supported a series of flightless birds throughout the Tertiary period. Older forms like the long and slender-legged, heavy-beaked *Phororhacos* (Fig. 8.15) are gone, but the rheas are found yet in the unforested southern part of the continent. A few other flightless forms might have survived had it not been for the depredations of man. The flightless pigeon, *Raphus*, commonly known as the dodo, was wiped out by sailors of the seventeenth century who landed on the island of Mauritius, and the large and stately moas were hunted for their feathers and flesh by the Maoris of New Zealand until there were no more.

It seems that the inability to fly at all was the fatal flaw in these birds. Whereas no extant group of flightless forms contains more than a half dozen species, the paleognathous tinamous of Central and South America, which can rise from the ground if pressed, number more than thirty. A flight of a few hundred feet, no matter how awkwardly performed, is apparently enough to frustrate all but the most determined predators, and often flapping from the ground to the lowest branches of a tree is sufficient to remove a bird from some impending danger. Two orders of neognathous birds with representatives worldwide, the Galliformes and the Gruiformes, include many species that feed, nest, and run on the ground but save themselves by short flights in moments of dire emergency. Unlike the birds which became completely flightless, the fowls that constitute the former order and the rails, coots, and bustards in the latter remained small enough to conceal themselves from their enemies. By being able to run, hide, or fly, these birds managed to survive in the presence of the rapacious mammals, whereas the huge ratites and other flightless forms that, once spotted, had no choice but to flee in full view of their pursuers largely failed.

Figure 8.15. Restoration of *Phororhacos*, a large, flightless bird that inhabited South America during the Miocene. (After Wilson, from Swinton.)

While some birds were leaving the trees for the ground, many others were becoming adapted for life in and near the water. Selection for those that could swim was rapid, for lakes and oceans offered to birds, as to other kinds of higher vertebrates, both an ample supply of food and safety from attack by terrestrial animals. Although various workers have cautioned that water birds and shorebirds predominate in the fossil record because of the greater likelihood of their preservation, they do not deny that these birds were of real importance early in avian history. The variety and number of birds which travel exclusively in or over the water today is an indication that a large avian population in this environment is a phenomenon of long standing. Isolated and fragmentary bones dating from the Cretaceous period attest to the existence of the diving loons and grebes before the end of the Mesozoic era. Despite the absence of skulls among the remains, birds ancestral to the ducks, geese, and swans have been identified from that time and the first of the flamingos as well. By the Eocene epoch, cormorants and relatives of the albatross were in evidence, and penguins were swimming in the southern seas. These birds, the first representatives of modern orders, coexisted with birds of an older radiation, which, though more primitive in some structures, were as specialized for their habitat as the later forms. The two best known of the ancient avians (Fig. 8.16) have been reconstructed from fossil material found in the Cretaceous Niobrara chalk of Kansas in association with bones of pterosaurs, mosasaurs, plesiosaurs, and marine fishes. Called *Ichthyornis* and *Hesperornis* by Marsh, who described them, they proved to differ greatly from the earlier *Archaeopteryx* and from each other.

Ichthyornis

Ichthyornis was a far better flier than *Archaeopteryx*. Unlike the Jurassic form, its bones were lightened by spaces for air sacs, its tail was short, its ribs braced by uncinate processes, its wing bones modern, and its sternum equipped with a deep keel for attachment of the pectoralis muscles. From the proportions of its body and the nature of its feet, Marsh presumed that *Ichthyornis* was rather like a tern in its habits. That it might actually be related to the terns, gulls, or other members of the modern order Charadriiformes seemed out of the question to Marsh. Impressed by the primitive traits exhibited by *Ichthyornis*, he considered it to be more closely related to the reptilian precursors of the birds than to any Recent genera. *Ichthyornis* still had vertebrae with biconcave centra, he pointed out, and those in the thoracic region of the column had not yet fused with

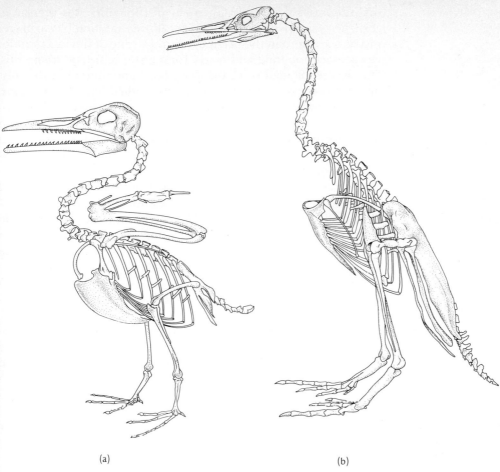

(a)   (b)

Figure 8.16. Skeletons of Cretaceous birds. **(a)** *Ichthyornis* as reconstructed by Marsh, with heavy, toothed mandible. **(b)** *Hesperornis*, a primitive toothed bird with reduced wings, apparently specialized for swimming and diving. (From Marsh.)

one another. The sacral vertebrae were co-ossified but only ten in number. Marsh noted that the brain of *Ichthyornis* was smaller than that of modern birds and more reptilian in its structure. The most strikingly primitive characteristic of *Ichthyornis* was, in Marsh's opinion, its toothed jaws. Although he had apparently had some doubts about the matter at first, Marsh had finally accepted as belonging to *Ichthyornis* a mandible bearing thecodont teeth which had been found beside the avian bones. The jaw, which was well preserved, looked very much like that of a small mosasaur. The arrangement of the bones was the same, and the two halves were connected in the midline by cartilage, just as they were in the marine lizard. In spite of the fact that the jaw seemed too long and heavy for a bird the size of *Ichthyornis*, Marsh fitted it against the quadrate bone and thereafter thought of *Ichthyornis* as a primitive bird still toothed like *Archaeopteryx*. Because *Hesperornis* also had teeth, Marsh regarded the Mesozoic birds that he had discovered as significantly different from living forms. He set *Ichthyornis*, *Hesperornis*, and a few genera known from more fragmentary remains apart in the subclass Odontornithes.

Marsh's classification, published in 1880, stood unquestioned until 1952, when J. T. Gregory reexamined the toothed mandible attributed to *Ichthyornis*. He agreed with Marsh that the element was so similar to that of a mosasaur that it would not have been associated with *Ichthyornis* had not it been near the bones of the bird. Gregory insisted, however, that the proximity of the toothed jaw and the avian bones was accidental and that the mandible looked mosasaurian because it had really belonged to one of the lizards, albeit a very young one in view of its size. Since *Ichthyornis* had little in common with *Hesperornis* except the toothed jaw that Marsh had attributed to it, investigators who accepted Gregory's conclusion removed *Ichthyornis* (and the related form *Apatornis*) from the Odontornithes and placed it in its own order, the Ichthyornithiformes. The primitive characters of *Ichthyornis* notwithstanding, it has seemed to these workers preferable to set the order Ichthyornithiformes near the Charadriiformes than to ally it with the more archaic Mesozoic orders.

Hesperornis

A category for toothed birds was kept in the classification scheme for *Hesperornis* because the presence of teeth in that form was indisputable. The jaws found in articulation with other bones of the skull contained grooves in which the teeth were anchored. Obviously a fish catcher, *Hesperornis* was armed with the uniform, sharp, slightly recurved teeth common in piscivorous animals. The teeth were reptilian in structure but unique in their distribution. Although the lower jaw bore teeth along its whole length, only the maxilla of the upper jaw was so equipped. The two long premaxillae were edentulous and probably sheathed by a horny bill. Gregory, who studied the jaws of *Hesperornis* as well as those attributed to *Ichthyornis*, remarked, as Marsh had, that the mandible of *Hesperornis* was also astonishingly like that of the mosasaurs. Besides bearing teeth and having its left and right halves flexibly joined at the midline, it had a movable joint between the dentary bone and the elements behind. Unlike the supposed mandible of *Ichthyornis*, that of *Hesperornis* could be identified as avian, according to Gregory, by the absence of the reptilian coronoid bones, by the shape of the articular facet for the quadrate, and by the relations of several of its other elements. The appearance of specializations similar to those of the marine lizards he ascribed to convergence, since both animals would have benefited from modifications that allowed the ingestion of larger prey.

The structure of the appendicular skeleton in *Hesperornis* proves that the bird could not fly but swam and dove after fish, as the mosasaurs did. The pectoral girdle and limbs were vestigial. The scapula was thin, the coracoid short, the sternum flat, and the wing reduced to a spindly humerus. *Hesperornis* depended for locomotion entirely upon the hind legs, which were powerfully developed. Although the girdle that supported them was still primitive in certain ways, it was braced against no fewer than fourteen ankylosed sacral vertebrae and was expanded to afford a large surface for muscle attachment. The legs were long and stout. Marsh pictured the bird as standing upright upon them, but Heilmann argued that they were situated in such a manner as to render them useless for walking. By describing carefully the fit of the femur against the girdle, Heilmann made it clear that the thighbone could not have been oriented downward and forward as Marsh showed it. When the head of the femur was inserted into the acetabulum and the articular surface on the trochanter was set against the iliac flange above, the femoral shaft projected directly outward from the body. The shortness of the femur kept the thigh from bulging and so spoiling the streamlined form advantageous in swimming. The leg gained its length chiefly from the tibiotarsus, which extended backward from the knee to the level of the tail. Heilmann thought that this bone, like its homologue in the loon, was tied tightly to the body by muscles and that the feet alone extended laterally and moved through the water like a pair of oars. They propelled the animal forward as they swept to the rear, pulled upon through the tendon of the gastrocnemius muscle (Fig. 8.17). That muscle in birds originates in large part upon the femur, but, the shaft of the femur being extremely short in *Hesperornis*, most of the fibers affixed themselves, apparently, to the dagger-like patella, which jutted anteriorly from the knee. Heilmann noted that a similar arrangement

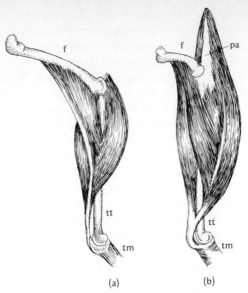

Figure 8.17. Skeleton and gastrocnemius muscle of the leg in **(a)** the pheasant, a terrestrial bird, and **(b)** the loon, a swimming form. In both birds, the gastrocnemius originates on the femur (*f*) and on the patella (*pa*) and inserts through a tendon on the tarsometatarsus (*tm*). In the loon, the femur is short, and surface for the attachment of the muscle is provided in great part by enlargement of the patella. *tt*, tibiotarsus. (From Heilmann.)

existed in the loon, a bird which has not lost the ability to fly but which also swims underwater, as he imagined *Hesperornis* used to do.

The resemblance between *Hesperornis* and the living loons and grebes, noticeable in the skull as well as the legs, struck Heilmann as indicative of genetic affinity. H. Howard also thought that a relationship among these diving birds was possible and suggested that *Hesperornis* might have been an early, nonvolant offshoot from the line of Mesozoic fliers that produced the modern genera. The teeth and small brain of *Hesperornis* still require that that bird be set apart in its own order, but Cretaceous forms like *Baptornis*, known from postcranial bones similar to those of the loons and grebes, have been classified in the same order as the extant birds by several workers. Marsh, less impressed by *Hesperornis'* adaptation for diving than by its inability to fly, postulated that the bird was more closely akin to the ostrich. In supporting his theory that *Hesperornis* was a carnivorous, swimming ratite, he emphasized the underdevelopment of the pectoral appendages, similarities to the ostrich in the cranium and palate, and the retention of primitive reptilian traits. Although he acknowledged the possibility that the ancestors of *Hesperornis* and the ostriches might have been flying forms, it seemed more probable to him, as it did later to Lowe, that these ancient and primitive birds had come directly from flightless reptilian stock.

Penguins

   The belief of Marsh and others that the antiquity of certain flightless birds made more probable their direct descent from nonflying reptiles underpinned the idea that the penguins, too, had evolved from forms which had never left the ground. This hypothesis was upheld by M. Menzbier, a contemporary of Marsh's, and seemed borne out by reports of fossil penguins more terrestrial than the living birds. G. G. Simpson, who reviewed the problem of penguin origins more recently, declared that the bones on which the supposed terrestrial genera were based did not, in fact, belong to penguins at all. Since Simpson believed that the arboreal *Archaeopteryx* was intermediate between the reptiles and all the birds which followed, he opposed the theory of Menzbier and the later version of it put forth by Lowe. He espoused instead the idea expressed by M. Furbringer in the 1880s that the penguins were derived from the stock of flying birds that produced the diving petrels and other members of the order Procellariiformes. Simpson admitted, however, that the fossil evidence for this supposition is scanty. Although penguin remains have been found in Australia, New Zealand, the Galapagos Islands, and South America, no complete skeletons are known. Most extinct species have been based upon a few bones, a single element, or even a lone fragment. The one skull that has been found, belonging to the Miocene *Paraptenodytes*, seems to be closer in structure to the skull of a procellariiform bird than the modern penguin skulls are, but convergence between the ancient penguins and the procellariiforms is not readily apparent from comparison of the postcranial bones.

   In the absence of strong anatomical evidence, Simpson built his argument on other grounds. He suggested a parallel between the evolution of the penguins in the Southern Hemisphere and the now extinct flightless auks in the north. Investigators are quite sure that the northern birds were the specialized and ultimately unsuccessful descendants of forms like the surviving auks, which can fly as well as swim. Apparently the wings of these birds gradually became stubbier and more useful in propelling the animals through water until the agility of the birds when submerged was more important for their survival than the ability to fly. Thereafter, selection of characters that facilitated swimming and diving proceeded rapidly at the expense of those which made flight possible. The wings, shortened and flattened, served efficiently as flippers but could no longer bear the body aloft. Once the weight of the body was no longer a crucial factor, some of the birds increased enormously in size. The great auk, *Pinguinus*, was more than 2 feet tall, and the Pliocene form, *Mancalla*, was even larger. Simpson thought it likely that the penguins evolved in a similar manner from ancient procellariiforms which behaved like the diving petrels now living in the regions below the equator. The wings of penguins are really not far removed in structure from those of the seabirds. They are specialized rather than underdeveloped or degenerate, as they would have been if the ancestors of the penguins had been flightless. The lack of air sacs in their bones is not a sign of affinity to terrestrial forms but an adaptation which reduces the buoyancy of the penguins, enabling them to dive and swim underwater more easily. The loss of the broad flight feathers was also an advantage, because they were more of a hindrance than a help when the birds were submerged. Only the downy plumage remained, providing the body with a smooth, streamlined surface and, in combination with the subcutaneous fatty tissue, minimizing the loss of heat. Although the necessity of retaining heat when exposed to cold ocean water doubtless mitigated against the evolution of very small forms, even among the flying seabirds, the flightless penguins—in which the relationship between wing size and body weight was far less important—could produce varieties large enough to reduce the surface-to-volume ratio, and consequently the escape of heat, substantially. The larger species of penguins, protected against the cold by their fat, feathers, and size, could then extend their range as far south as the frozen shores of Antarctica.

Simpson has defended this theory of the origin and spread of the penguins against attack by investigators other than Lowe who still favor the idea of flightless terrestrial ancestors for these birds. B. de Meillon, for instance, has maintained on the basis of the kind of parasites carried by penguins that the precursors of the modern animals must have been earthbound South American forms. He pointed out that the only two genera of fleas found on penguins belong to a subfamily whose other members are confined to seabirds and to South American rodents. In the opinion of de Meillon, this distribution of fleas can be explained only by assuming that the parasites spread from the rodents to flightless birds in South America and thence to their descendants, the seagoing penguins. Simpson argued in opposition that the penguins could have originated as he suggested they did and then, while ashore, caught the fleas either from rodents or from the seabirds which carried them. The presence of the fleas proved nothing about the home of the earliest penguins or their ancestors, for the restriction of penguin parasites to the two genera could mean only that no fleas capable of living upon penguins happened to arise elsewhere. Although he acknowledged the incompleteness of the fossil record, Simpson added that the earliest known penguin remains come from Australia and New Zealand rather than South America but that, in any event, the penguins seem to have arrived on the latter continent long before the rodents did. Such fossil evidence as there is also fails to support the older theory of Matthew's that flightless ancestors of the penguins lived on Antarctica. If Matthew were correct in his idea that flying birds, based on a warmer ancient Antarctica, eventually became flightless and later took to the sea to feed when the continent became barren and cold, the distribution of penguin remains should reflect the southern origin of the birds. Actually neither the fossil record nor the distribution of the Recent animals weighs in favor of Matthew's hypothesis. Now and in the past, penguins have settled in both cold and temperate regions. Since no site has yet given up a specimen recognizable as protopenguin, it is impossible to guess where or even in which climate penguin history began.

Evolution of ter-
restrial flying birds
 The radiation of the birds in progress at the beginning of the Tertiary period produced not only swimming and running birds but those which remained in the arboreal habitat and diversified. Following the evolution of these terrestrial flying forms has been most difficult, for many of them were small and left no fossil trace. Few were accidentally swept into the water where their bodies might be covered quickly with protective sediments. In general, these birds came to rest on the ground after death and suffered the same fate as other kinds of small upland vertebrates. As a result, virtually nothing is known of the differentiation of the birds found at present in the world's jungles, forests, and grasslands. More than half the species living in these environments seem to be variants of a single stock and are thus regarded as members of a single order, Passer-

iformes. These are the perching birds, those with three toes in front and one behind, which characteristically build nests above ground to house the featherless young that hatch from their pale-colored, often speckled eggs. The success of these birds in exploiting the arboreal habitat can be gauged by their diversity: the living passeriforms belong to sixty-nine families, each of which contains forms adapted for a somewhat different niche. In the absence of fossil evidence, paleontologists can say little about the date at which these families appeared. The ancestors of the passeriforms, if not the passeriforms themselves, were almost certainly present in Eocene times, because a sparrow-like bird, *Palaeospiza bella*, is known from the Oligocene epoch immediately following. Unfortunately, the condition of the specimen has prevented investigators from learning very much from it. The bird fell into a lake and was preserved almost entire, but pressure flattened it into a thin plate and rendered the boundaries of many of the bones indistinct. In addition, the beak, a valuable aid in the classification of passeriforms, is missing. A. Wetmore, who studied *Palaeospiza*, thought that it might have been an early songbird and so referable to the suborder Oscines. If *Palaeospiza* were an oscine, it would mean that the evolution of the passeriforms was already well advanced in the Oligocene, for that suborder includes fifty-three families of the most highly specialized passeriform birds.

Study of the living species has not shed any light on the sequence in which these different songbirds emerged. Three major assemblages can be distinguished within the suborder Oscines, but which of these groups is the most primitive or the most advanced is far from clear. Some workers would put the crow-like birds in the highest position, because they seem to display the greatest avian intelligence and the most complex behavior. Others think that the songbirds which have only nine primary flight feathers, like the vireos and the tanagers, should be considered the most progressive, since they may have diverged most lately from the large assemblage of oscines which retain the more conservative number of ten such feathers. It is possible that the degree of specialization is not a clue to the relative age of these birds and that the corvine, nine-primaried, and ten-primaried forms emerged almost simultaneously near the beginning of the Tertiary period, during the early and perhaps explosive radiation of the passeriforms, and have been evolving separately ever since. The three assemblages of songbirds would then bear to each other what Wetmore described as a three-dimensional relationship rather than the linear one that any listing in a classification scheme implies.

Difficulty of ascertaining avian relationships   A surer understanding of the evolution of arboreal birds may be slow in coming because workers in the field are few and many of the anatomical data necessary for comparative study are lacking. Most ornithologists have interested themselves more in the outward appearance and behavior of birds than in their skeletal structure. As a consequence, paleontologists often have difficulty in determining the modern group to which genera based upon fossil bones should be assigned. Ignorance of the skeletal anatomy of living birds led early investigators to identify as new species specimens which belonged in fact to existing ones. Later workers, wary of increasing the number of species in this way, sometimes elected to postpone the description of the passeriform material in their collections. Wetmore, for instance, left unnamed the bones of perching birds included among avian remains from a Pliocene deposit in Kansas because he had a hypothesis concerning their identity that he could not test. He suspected that his Kansas passeriforms belonged to species now resident in South America, which in warmer Pliocene times had ranged farther north. Since the skeletons of the modern birds were not available for comparison with the Pliocene bones, Wetmore was forced to abandon the question.

Although difficulties encountered in tracing the history of the passeriforms and classifying them are aggravated by the size of the group, they actually differ little in kind from those which impede the study of other types of arboreal birds or, for that matter, of the swimming and running forms. Although research has been undertaken in other fields to supplement the morphological evidence of relationship, no results have been obtained—from comparative studies of behavior, geographical distribution, physiological traits, chromosomal structure, or protein chemistry—that clarify the phylogenetic connections among the various avian orders. Since the degree of difference between birds in separate orders is much less than that between vertebrates distinguished at the ordinal level in other classes, investigators are forced to depend upon differences or peculiarities in relatively minor characteristics for clues to the branching of the avian family tree. For this reason, it is especially hard to distinguish between convergence and signs of real affinity. Where nothing argues to the contrary, similar adaptation has often been assumed to indicate the existence of a common ancestral stem or at least the attainment of a similar level of development. Hence the loons of the order Gaviiformes and the grebes of the order Podicipediformes are placed next to one another in the classification scheme because both are water birds with strong, posteriorly set hind feet specialized for swimming. The two orders are set first in the list of neognathous flying birds on the supposition that the similarity of their modified hind legs to those of *Hesperornis* marks them as primitive if not actually related to the Cretaceous form. Just as birds with similar modifications are juxtaposed, so those with extraordinary morphological traits are usually regarded as less closely related to the central avian stock and are given a special position in the classification scheme. Simpson's explanation of penguin origins notwithstanding, some investigators still argue that the penguins merit inclusion in a superorder separate from the rest of the modern birds. Simply listing the penguins in an order beside the Procellariiformes seems to these workers insufficient to mark the degree of divergence these birds exhibit.

Classification of the birds

The classification scheme for the birds, like that for other groups of vertebrates, changes little by little as theories of phylogeny are revised. The scheme in its present form suggests certain ideas which may prove quite accurate and others that are of doubtful validity. The isolation of *Archaeopteryx* in its own subclass, Archaeornithes, marks its undeniable position as an intermediary between the reptiles and the birds. All the other feathered vertebrates are housed in the subclass Neornithes and are considered to be (quite surely correctly) much farther removed from the ancestral archosaurian stock. Although Marsh's subclass Odontornithes has been dropped and the Cretaceous birds accorded membership in the Neornithes, a separate superorder Odontognathae has been reserved for the toothed *Hesperornis*. Workers who still believe, despite Gregory's arguments, that the toothed jaw found with *Ichthyornis* really belonged to that bird place *Ichthyornis* in the Odontognathae as well. Segregation of the orders of Tertiary birds depends upon an author's opinion of Simpson's work on the penguins and of De Beer's theory of the evolution of the ratites. If he agrees with Simpson, he drops the superorder Impennes and distinguishes the penguins only by assigning them to their own order, Sphenisciformes. Adoption of De Beer's conclusion that the ratites are neither a natural group nor a primitive one leads to the abandonment of Huxley's category Palaeognathae. All the orders of Tertiary birds are then listed seriatim in the super-order Neognathae, the ratites first and the flying birds afterward. This arrangement can be interpreted as implying that the ratites did diverge earlier from the avian main stream or at least that they are the most conservative of the modern birds. Although it can be argued that all living birds are highly specialized rather than primitive, placing the orders of water and shorebirds before those which consist largely of arboreal species perpetuates the view, based perhaps upon their preponderance among Eocene fossils, that the former are in general less advanced forms. The orders of arboreal birds are also arranged in a manner that suggests an evolutionary hierarchy. The diurnal birds of prey, the Falconiformes, and the weak-flying fowl of the order Galliformes are listed before orders containing such forms as pigeons, parrots, owls, woodpeckers, and swifts; and the large and extremely diversified order Passeriformes is set last, a placement which implies that its members are the most advanced and progressive of all the birds.

Avian evolution during the Tertiary period Despite their uncertainty concerning the interrelationship of the various orders, paleontologists are fairly sure of the course of avian evolution since the close of the Mesozoic era. As sparse as it is, the fossil record reveals that even before the Cretaceous period had ended, the rapid radiation which led to the establishment of the modern birds was under way. By Eocene times, at least half of the modern orders were represented, and birds belonging to most of the rest emerged during the Oligocene years which followed. Differentiation at the ordinal level was succeeded by diversification within each group, so that, by the middle of the Tertiary period, the number of families and genera had increased substantially. The birds reached the high point of their development just before the Tertiary period ended and the Pleistocene ice sheets began their descent. At that time, when subtropical conditions extended far north of their present boundaries, species now living coexisted with others which have since become extinct.

The richness of the avifauna of this time can be inferred from the slightly later Pleistocene fossils reclaimed from the tar pits of La Brea and two other less famous Californian deposits of the same type, those at McKittrick and Carpinteria. In each of the three areas, birds alighting upon what seemed to be a solid surface sank into the black ooze or tarry sand. No articulated skeletons were preserved, because gases that bubbled through the tar eventually separated the bones of the buried birds, but single elements were often left intact. The problem over the years has not been to find these remains but to obtain enough hands to clean and sort them from the bones of other kinds of vertebrates which also died in the asphalt. Workers have recovered, so far, more than 100,000 avian elements. Since the environment at the three sites was different, the bones represent a great variety of species. At Carpinteria, which in Pleistocene times was near the coast, sea- and shorebirds predominate. McKittrick, where the tar appeared in an open valley dotted with ponds and marshes, produced bones belonging to ducks, geese, and other freshwater forms. Birds of prey make up more than half of the avian collection from La Brea. The latter was in an inland area populated by many kinds of mammals. These animals, struggling in the tar, apparently attracted large numbers of falconiform birds—vultures, hawks, falcons, eagles, and condors—which became stuck when they descended to the attack. A count of the species from the three sites and comparison with the number that remains today make it plain that the history of the birds in the area since the Pleistocene has been one of decline. Similar comparisons of species elsewhere indicate that the trend was widespread.

Causes of decline in variety of modern birds

The increasing severity of the climate and changes in the mammalian population were apparently the chief factors contributing to the impoverishment of the avifauna that occurred during Pleistocene and Recent times. As the glacial intervals succeeded one another and parts of the Northern Hemisphere became frigid wastes, species of birds which could not withstand the cold either died out or shifted their range southward. Of the birds that were able to wrest a place for themselves where warm weather still prevailed, some took up permanent residence and others evidently evolved the habit of migrating seasonally as far northward toward their former breeding or feeding grounds as the temperature permitted. Although at the present time the climate in the Northern Hemisphere is not so rigorous as it once was, the avian population has not regained its former variety. Despite their larger land mass, the continents of North America and Eurasia support far fewer families of birds than exist in the regions below the Tropic of Cancer. Even in the southern part of the world, however, avians have been under duress since the Pleistocene began. The gradual disappearance of the great terrestrial birds as the mammalian carnivores arose has already been described. Eventually, the failure of the largest of the mammals entrained the loss of other birds as well. The birds of prey suffered, in particular, as the animals on which they fed became scarcer. With the extinction of huge beasts like the mastodons and the saber-toothed cats, giant condors, vultures, and eagles vanished. Birds like *Teratornis*, a condor which must have weighed between 40 and 50 pounds (Fig. 8.18), and *Wetmoregyps*, a long-legged eagle, were replaced by smaller hawks and owls, better suited to pursuing the rodents that were fast becoming the most populous mammals.

Figure 8.18. Two birds found fossil in the La Brea tar pits. **(a)** *Teratornis*, a vulture which attained a wingspan of 12 feet. **(b)** *Polyborus*, a carrion-eating falcon. (After Howard.)

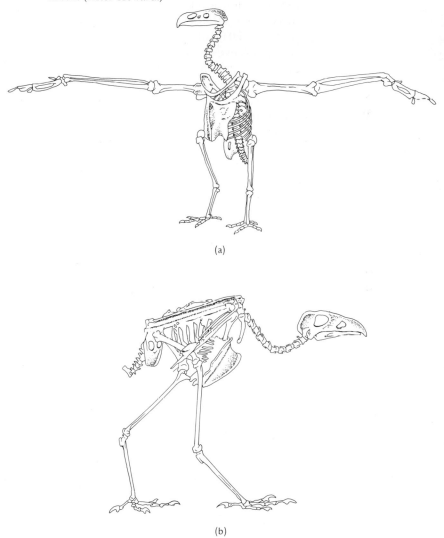

(a)

(b)

In historic times, and doubtless in the Pleistocene also, large birds of all kinds and many smaller ones have been endangered or extinguished by man. The kitchen middens of ancient tribes and modern ones bear witness to the numbers taken with arrow, net, and spear, and the use of guns has tipped the odds further against hunted species. Man's newly developed skill in chemistry has now given him weapons with a far greater range than any of his firearms. The near extinction of the peregrine falcon has proved that poisonous compounds can cut down even those species which live in relatively inaccessible areas. This bird, though it nested in rugged regions, preyed upon smaller birds which ate insects that fed upon domesticated plants sprayed with the pesticide DDT. The chemical accumulated within the bodies of the falcons eventually disturbed their calcium metabolism. As a result, the eggs laid by the female had inadequately calcified shells and broke long before the young were ready to hatch. Against this kind of assault, the avian instinct to fly from danger was no defense. It is not improbable that an increasing number of species will succumb to this insidious sort of attack, but the impact of man's newly expanded powers of destruction upon the avian population as a whole is harder to evaluate. The variety, distribution, and adaptability of birds make it unlikely that the dire predictions of the pessimists will be realized. It seems that the activities of man are more apt to reinforce the current trend in avian evolution than to change it drastically. Unless the birds should evolve a suite of new and advantageous traits enabling them to embark upon another major radiation, it is conceivable that they will make no inroads upon the territory of other animals but continue to hold their position in the sky and the marginal place on the ground necessary to sustain it.

9

The Legacy
of the Reptiles:
The Mammals

Vertebrate fauna
of the Mesozoic
era

The Mesozoic era, justly remembered as the time when fantastic reptiles arose and triumphed, was also the period during which the modern terrestrial vertebrate fauna began to take shape. The labyrinthodont amphibians vanished then and were replaced by frogs and salamanders. Turtles, lizards, and primitive snakes crept beside the dinosaurs, and the earliest birds flew. Although they played a minor role in the terrestrial environment, the first mammals appeared and underwent considerable diversification. The task of tracing the roots of the vertebrate lineages of the present day back into Mesozoic time has been most difficult because of the small size of the animals, the extreme rarity of their remains, and the limited number of localities in which terrestrial deposits of Mesozoic age are exposed. By dogged perseverance, however, and the use of improved collecting techniques, paleontologists studying the evolution of the mammals have made significant progress, especially within the last 30 years.

Evolution of the
mammals and
birds compared

In certain ways, the history of the mammals paralleled that of the birds. The tetrapods that were to replace the archosaurian reptiles as dominant forms at the end of the Cretaceous period had, like the birds, departed from reptilian stock much earlier in the Mesozoic and achieved a more advanced grade of organization. Changes in the visceral organ systems and in the muscles, skeleton, and teeth had brought about internal temperature regulation in the mammals and with it the ability to remain active over a wider range of climatic conditions than their cold-blooded forebears. The establishment of homoiothermy demanded increased metabolic efficiency, just as it had in avians, but since the ancestors of the mammals and the birds were only distantly related, the requirements for increased energy production were differently met. The mammals, obliged to eat as regularly, if not as continuously, as the birds, did not lose their teeth but came to depend upon a variety of specialized dentitions to pierce, cut, crush, and grind their food. Relatively late, a few mammals became secondarily toothless, but the group as a whole was characterized by ever more complex tooth structure and precise occlusion. The evolution of mammalian teeth and the habit of chewing were accompanied by the appearance of a secondary palate that separated the air passage completely from the area of the mouth cavity in which the food was held. As a result, prolonged chewing, which enlarged the number of materials usable as food, and the continuous and rapid respiratory exchange necessary for the maintenance of a high metabolic rate developed simultaneously. The capacity of the lungs increased, not by the outward extension of air sacs, as in birds, but by the multiplication of alveoli within each lung and the evolution of a muscular diaphragm that worked with the muscles of the chest wall to induce a regular and forceful respiratory current. The oxygen introduced into the bloodstream reached the tissues in the greatest possible concentration, since the blood leaving the lungs passed through the left side of the heart, completely separated from the venous flow on the right. Supplied copiously with oxygen and nutrients, the cells metabolized fast enough to assure the rapid exercise of their special functions and the production of sufficient heat to raise the temperature of the body above that of the environment. The advantages of a constant internal temperature were secured by the evolution of several regulatory mechanisms. A covering of hair insulated the body against excessive heat loss by surface radiation, and in the absence of air sacs sweat glands appeared and augmented cooling in warm weather. Sweating and manipulation of the hair as well as panting, shivering, and change in the diameter of the surface blood vessels came under the control of the autonomic, or involuntary, portion of the nervous system so that delicate adjustments to thermal change were made quickly and without conscious effort on the part of the animal.

Although the mammals paralleled the birds in evolving homoiothermy, they diverged sharply from avians in the manner in which their activities were controlled. The basal nuclei in the brain, which in birds are highly elaborated and presumably responsible for the rigid patterns of behavior exhibited by those animals, are in mammals far less prominent. The dominant region of the mammalian brain is the neocortex, an area of the cerebral hemispheres that first appeared in reptiles but remained undeveloped in their avian descendants. The neural connections within the neocortex in living mammals are so complicated that few of them have been traced, but observation and experiment have made it plain that the neocortex governs voluntary actions directly and exerts an influence indirectly over many involuntary processes. In contrast to the avian brain centers, which permit few deviations from a predetermined set of responses, the mammalian neocortex enables its possessor to react more variably to changes in the environment. A mammal can perceive, discriminate among, remember, and adjust its actions to a much larger number of stimuli than a bird is able to do. In short, mammals have gained flexibility through the development of intelligence.

Development of viviparity and parental care in mammals

The dependence of the mammals upon intelligence as well as instinct for survival apparently grew concomitantly with the evolution of viviparity and continued postnatal association between the female and her young. Retention of the fertilized egg within the body of the mother may have been selected for initially as an adaptation for proper maintenance of the temperature of the developing embryo and as a way of protecting the embryo without reducing the mobility of either parent, but prolonged foetal life and provision of sustenance for the newborn also afforded a long period of time in which the intricate mammalian brain could establish itself and store its first impressions. If the condition in monotremes and marsupials—living mammals supposedly more primitive in their reproductive processes than placental forms—gives an accurate clue to the evolution of viviparity and maternal care, it seems that the suckling of the young was an early development and the extension of the time spent *in utero* a gradual process. Since the mammary glands are simply modified sweat glands, the secretion of milk might have followed soon upon the appearance of the latter structures. Holding an actively growing embryo within the body longer than it can sustain itself upon yolk material, however, required the evolution of an arrangement which would allow service of the embryonic tissues by the body of the mother. The uterine wall in both monotremes and marsupials affords nutriment to the resident embryo, but since the monotreme embryo is still surrounded by a shell and that of the marsupial associates itself only in a rudimentary way with the uterine wall, the period each spends within the mother is relatively short. Not until the advent of the chorioallantoic placenta, a structure in which the capillaries of the embryonic membranes are brought into intimate association with those of the uterine lining, did extended development *in utero* become possible. The prolongation of the developmental period led, in the most advanced mammals, to the establishment of familial and social relationships far more complicated than those in any other group of vertebrates. In man, of course, the lengthening of the maturation process and the supervision of the young have reached an extreme point, but other kinds of modern mammals also exhibit helplessness at birth, dependence upon learning for survival, and considerable ability to adjust their behavior in response to changed conditions.

Reptilian precursors of the mammals

There have been occasions in the history of vertebrates when the relatively sudden appearance of a new adaptive complex precipitated the expansion of a group at a new level. The transition to the mammalian state, however, even though it did not involve invasion of a different medium, seems to have been as lengthy a process as the evolution of the amphibians from the rhipidistian fishes. The deep-seated changes that transformed poikilothermic, often sluggish, egg-laying reptilian forms into warm-blooded, active, intelligent animals that gave birth to their young alive depended upon countless mutations and the conservation and combination of a multiplicity of new traits in viable populations. It has long been clear that specifically mammalian characteristics began their formation in very ancient reptilian stock. By early Triassic time, therapsid reptiles had made sufficient progress toward the mammalian condition to be recognized without question as the group from which the new class ultimately sprang. The carnivorous theriodonts— gorgonopsians, therocephalians, bauriamorphs, and cynodonts —all showed the incipient differentiation of the teeth, widening of the temporal fossa, and enlargement of the dentary bone in the mandible that constituted the first steps in the evolution of the mammalian type of jaw apparatus. Bones of the skull which are absent in mammals were already in these reptiles quite small (Fig. 9.1), and development of the secondary palate, at least in cynodonts, was far advanced. The limbs and girdles of these therapsids were assuming a mammalian orientation, and the vertebral column had also become more mammal-like by losing the intercentral elements and developing an atlas and axis that permitted greater mobility of the head. The extreme reduction of the lumbar ribs in some therapsids suggested that these forms may already have developed diaphragmatic respiration. The shortening of the ribs attached to the cervical vertebrae, in addition, foreshadowed the restriction of separable ribs to the thoracic region in mammals.

How nearly the theriodonts approached the mammalian level in the structure of their soft organs, in their reproductive habits, and in their behavior cannot be determined directly. Watson and several other investigators have argued that foramina in the

Figure 9.1. Skulls of mammal-like reptiles and a placental mammal. **(a)** *Dimetrodon*, a Permian pelycosaur. **(b)** *Scymnognathus*, a later Permian therapsid (suborder Theriodontia). **(c)** *Thrinaxodon*, a more advanced, Lower Triassic theriodont therapsid. **(d)** *Canis dirus*, a wolf-like mammal of the Pleistocene. Abbreviations: *b*, auditory bulla; *fio*, infraorbital foramen; *oc*, occipital; *so*, supraoccipital; other abbreviations as in Fig. 6.7. (From Romer.)

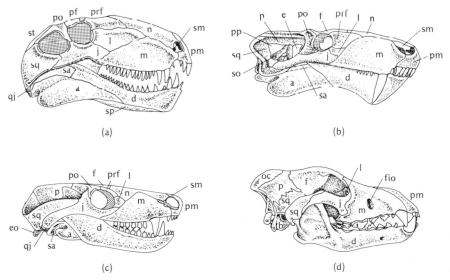

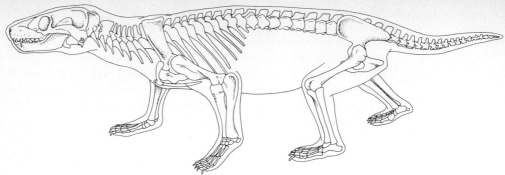

Figure 9.2

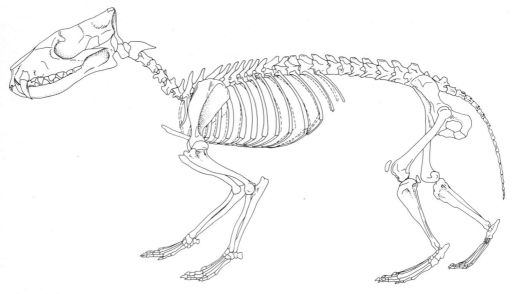

Figure 9.3

Figures 9.2 and 9.3. Skeletons of **(9.2)** *Thrinaxodon*, a mammal-like reptile, and **(9.3)** *Hyaenodon*, a carnivorous mammal of the Oligocene. Compare the posture and skeletal structure of these forms with that of the more primitive mammal-like reptile, the pelycosaur *Dimetrodon* in Fig. 7.20. (9.2 from Jenkins; 9.3 from Scott.)

facial bones for the escape of blood vessels and nerves signify that the mammal-like reptiles may have evolved loose vascularized skin, hair, facial muscles, and the movable lips necessary for suckling. Brink interpreted the association in one nodule of two specimens of the cynodont *Thrinaxodon*—one large and one very small—as suggestive either of internal development of the embryo or of maternal attention to the young. The strengthening of the vertebral column and the pulling of the limbs underneath the body have persuaded others that the therapsids were capable of relatively rapid locomotion and an active terrestrial life (Figs. 9.2 and 9.3). Casts of therapsid skulls which reveal the size and form of the brain reinforce the possibility that the animals were agile and vigorous. The cerebellum, the region in which motor coordination is effected, seems to have been quite large, and the fibers leading to and from it were sufficiently numerous to constitute a mass beneath the brainstem, termed, in mammals, the pons. The cerebral hemispheres in the two theriodonts in which their size is known were no bigger than those of other kinds of reptiles. If their condition is typical, the neocortex was not yet the predominant center in the brain and the animals did not evince even the first signs of the intelligence that was to characterize the mammals.

The affinity of the therapsids to the mammals was determined by comparison of the Triassic species to the modern forms. For a long time nothing was known of the animals intermediate between the reptiles and the mammalian genera that appeared at the beginning of the Cenozoic era, less than 70 million years ago. It was believed until the second quarter of the nineteenth century that true mammals did not exist in Mesozoic time, for no bones recognizable as mammalian had ever been distinguished in deposits of that age. When Buckland proposed that two lower jaws retrieved from the Middle Jurassic Stonesfield Slate in England might have belonged to mammals, the possibility was dismissed by the French biologist Blainville on the supposition that the warm-blooded tetrapods did not arise until the heyday of the reptiles was over. Agassiz, when he was consulted, sidestepped the question by saying that the material was too fragmentary to decide so crucial an issue, but Cuvier agreed with Buckland that the bones and teeth might be the remains of some primitive mammalian form. Owen concurred because the mandible of each specimen contained no bone other than the dentary and the cheek teeth were multicusped and double-rooted. Further exploration of the Stonesfield Slate produced three additional kinds of Mesozoic mammals, and rocks of somewhat later Jurassic age on the English Isle of Purbeck yielded a large collection of jaws and teeth, among which specimens assignable to the class Mammalia were found.

Discovery of Mesozoic mammals

Although the search for remains of more of the newly discovered animals was pressed, during the ensuing century relatively few were brought to light. In Europe, three teeth came from the Wealden deposits of Lower Cretaceous age and a small number from Upper Triassic sediments that had settled into fissures in older limestones. Collectors who went into the American West in the 1870s in search of dinosaurs stumbled upon the now famous pocket of Jurassic mammalian material at Como Bluff, Wyoming, and also recovered remains from late Cretaceous strata. All these fossils consisted of teeth isolated or set in fragments of maxillary or dentary bones. The first skulls came, in the 1920s, from the dry upland Cretaceous deposits at Djadochta in Mongolia. Although none of the seven discovered was complete, these skulls were for a long time the only clue that paleontologists had to the cranial anatomy of any of the Mesozoic mammals. Some 40 years later, more skulls were found at the same site, and in China and South Africa a few skulls and some postcranial bones belonging to much older mammalian forms were unearthed. In recent years, paleontologists have redoubled their efforts to add this kind of material to their collections, using techniques that were designed to retrieve even the smallest fragments. It became standard practice at certain sites to dig out hundreds of pounds of compacted sediment, carry it in sacks to the nearest stream, and put it into screen-bottomed boxes, which were then set to wash in the current. After the fine particles of sand and clay were swept away through the mesh into the water, the concentrate left in the boxes was dried, spread out in the sunlight and carefully examined. The chief result of this method was that the number of teeth known was multiplied enormously.

At present, the fossil record includes dozens of genera—perhaps more than are actually warranted—based upon tooth-bearing jaw fragments or even isolated teeth. Since there is no way of knowing whether upper and lower teeth belonged to the same or different animals, each is given a separate generic name. The apparent number of forms may also have been exaggerated by assigning different names to teeth which were in reality not from different species but from different positions in the jaw of the same form. The multiplicity of genera now in the record is thus not an accurate reflection of the extent of the Mesozoic mammalian fauna. A better understanding will result when discovery of more complete remains makes it possible to improve the correlation of material belonging to the same species and to prune synonymous names from the list of genera.

Structure of the
teeth in Mesozoic
mammals

Because of the nature of the fossil evidence, paleontologists have been forced to reconstruct the first two-thirds of mammalian history in great part on the basis of tooth morphology. Their first assumption was that at the end of the Triassic and in the early Jurassic periods, several new groups of terrestrial forms appeared which could be readily distinguished from the better known reptilian animals. Whereas most of the latter were characterized by conical or peg-like single-rooted teeth, the newcomers had a dentition of far more complex design. Although the front teeth and the canines remained relatively simple, the cheek teeth bore crowns with a number of cusps supported by two or three roots. Cusped crowns had appeared among Lower Triassic cynodonts, but in those animals the pattern of the cusps varied widely from one genus to another. The teeth assigned to the Mesozoic mammals showed some specific differences, but they could be sorted into five or six major taxonomic categories on the basis of their general structure.

The cheek teeth of the Mesozoic mammals called triconodonts were most similar to the cusped teeth of the carnivorous cynodonts (Fig. 9.4). Each cheek tooth in animals of this type bore three main cusps, a large center one with a smaller one in front of it and another behind. Below the cusps, there arose from the base of the crown a slight ridge, or cingulum, that edged the inner side of a lower tooth and girdled an upper one almost completely. On many teeth of triconodonts, a little cusp projected from the posterior part of the cingulum and abutted the tooth behind in a firm interlocking arrangement. Sufficient remains of teeth and jaw fragments of triconodonts have been found to make it clear that the postcanine tooth row could be divided into anterior premolars and posterior molars. Both kinds of cheek teeth displayed the anteroposteriorly aligned three cusps on the crown and double roots below, but they differed somewhat in size and shape. Teeth of triconodonts, like those of carnivorous cynodonts, were straight, cutting blades. When brought together, the large center cusp of each upper tooth sheared down behind and slightly external to the principal cusp of the corresponding lower tooth. The worn surfaces on the teeth of triconodonts suggest that the triconodont dentition was a more effective cutting device than the reptilian one. Whereas reptilian teeth were constantly replaced and newly erupted teeth would have occluded imprecisely against older, worn ones, the teeth in triconodont mammals, where a series in the same jaw is known, seem to have erupted at about the same

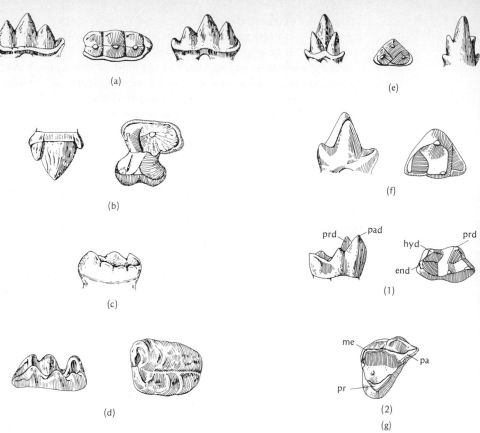

Figure 9.4. Structure of the cheek teeth in Mesozoic mammals. **(a)** A tricodont lower molar; internal, crown, and external views. **(b)** A docodont upper molar, external and crown views. **(c)** A haramyid lower molar. **(d)** A multituberculate lower molar, external and crown views. **(e)** A symmetrodont lower molar; internal, crown, and external views. **(f)** A pantothere lower molar, internal and crown views. **(g)** Tribosphenic molars: (1) lower left molar, internal and crown views; (2) upper right molar, crown view. Abbreviations as in Fig. 9.5. (From Simpson.)

time and worn down uniformly, permitting the maintenance of a close fit between the upper and lower teeth. The reduction of tooth replacement seems to have taken place early in mammalian evolution and to have made possible complicated and highly efficient occlusal relationships, which proved highly advantageous.

As well as improved shearing, more precise occlusion permitted better crushing dentitions. In teeth of early mammals called docodonts, expansion of the crown on the inner, or lingual, side resulted in the formation of rectangular lower molars which met the inner half of the upper molars directly. Another type of tooth seemingly adapted for crushing is that assigned to mammals termed microcleptids or, more recently, haramyids. The crown is broad with a shallow depression on the occlusal surface rimmed with a ridge of fine cusps. Although teeth of haramyids have been known for more than a hundred years, paleontologists are still not sure that they are properly classified as mammalian because they are rare and have never been found associated with any bone. Far better known and certainly mammalian are the crushing molars belonging to animals known as multituberculates. These teeth are broad and usually much elongated. They bear two and, in some upper molars, three parallel series of cusps. They have been found not only isolated but implanted in lower jaws and partial skulls in company with anterior teeth. It is obvious from the complete dentition that the animals possessing it were similar to modern rodents in their eating habits. They had one large pair of gnawing incisors; behind a space, or diastema, left by the absence of canines, were the premolars and the robust molar teeth. Besides their crushing function, the premolars exerted a shearing force, and a single (often huge) premolar in the lower jaw usually became knife-edged and specialized entirely for cutting (Fig. 9.10*b*).

Two other kinds of teeth attributed to Mesozoic mammals seem to stand closer than any of the ones described above to those which characterized the first marsupial and placental mammals of later times. The simpler of the two, characteristic of forms called symmetrodonts, were like teeth of triconodonts in having a single large cusp flanked by two lower ones but differed in that the cusps were arranged in a trigon, or triangle, rather than in a straight line. Since the apex of each triangular-shaped upper tooth faced inward and that of each lower tooth outward, the teeth, when brought together, interdigitated tightly. The total shearing surface was greater than in the dentition of triconodonts, because instead of a single straight blade, each tooth had two shearing edges, one that worked against the opposing tooth in front and another that scissored the opposing tooth behind. The increase in shearing capacity was apparently advantageous, because in teeth which paleontologists presume belonged to descendants of the first symmetrodont animals the length of the shearing edges continued to grow. In advanced symmetrodont teeth, the angle at the apex is more acute, and, as a result, the shearing sides of the tooth are longer and more transversely oriented.

Other teeth, from mammals termed pantotheres, which seem also to have derived from teeth of the original symmetrodont type, not only have long, nearly transversely oriented shearing crests but an added shearing crest and cusp at the back of each lower cheek tooth. As it moved into occlusion between the triangular surfaces of adjacent lower teeth, the innermost cusp of each upper tooth cut downward and slightly inward, coming to rest against the new heel, or talonid, of the more anterior lower tooth. Because the pantotherian jaw moved somewhat sideways as well as vertically, food seems to have been ground between the teeth as well as cut. Whether the addition of the grinding function or a progressive change in some other part of the body (which is completely unknown) was responsible for their success, the pantotherian mammals did diversify during Jurassic time. There are, according to Clemens, at least four different types of pantothere teeth, representing four divergent families of these animals. Among them are teeth which are clearly antecedent in form to those of the earliest therian mammals, the progenitors of the marsupials and placentals which replaced the archaic Mesozoic mammalian groups before the end of the Cretaceous period.

Tribosphenic teeth of early therian mammals

The first therians retained triangular occlusal surfaces on the cheek teeth and the talonid which developed as an extension on the lower molars of the pantotheres. The talonid, which was larger than that of the pantotheres, was basined for the reception of the tall cusp on the inner corner of the upper tooth. Teeth of this structure, described as tribosphenic, enabled their possessors to eat insects and other small animals as well as softer vegetable material. When the teeth were brought together, the sharp cusps pierced the food, then cut it, and finally ground and crushed the pieces caught between the upper cusps and the talonids. The tribosphenic teeth of the earliest therians are easily comparable to the teeth of the later marsupials and to those of the placental mammals that became dominant in the first years of the Cenozoic era (Fig. 9.5). Although, in the upper teeth of these animals, an outer, or stylar, shelf with one or more cusps developed to a variable degree, it is always possible to recognize the three cusps of the original triangle. The large inner one, called the protocone, stands alone, but the outer two—the paracone anteriorly and the metacone behind—sometimes are

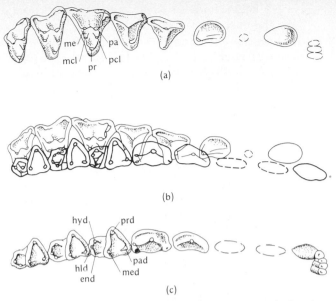

Figure 9.5. Dentition of a primitive placental mammal, showing tribosphenic cheek teeth and their relative position in occlusion. **(a)** Upper teeth, right side. **(b)** Upper and lower teeth in occlusion. **(c)** Lower teeth, left side. Abbreviations: Upper tooth cusps—*mcl*, metaconule; *me*, metacone; *pa*, paracone; *pcl*, protoconule; *pr*, protocone. Lower tooth cusps—*end*, entoconid; *hld*, hypoconulid; *hyd*, hypoconid; *med*, metaconid; *pad*, paraconid; *prd*, protoconid. (From Romer.)

crowded so close together that their bases merge. The triangular portion of each lower cheek tooth is termed the trigonid to distinguish it from the upper. Over the objections of some later workers, the three cusps on the trigonid are still called by the names given them by Osborn in 1888. Since Osborn believed that the cusps of the upper and lower triangles were equivalent, he created corresponding terms for the latter: protoconid for the outer, apical cusp, and paraconid and metaconid for the cusps that form the base of the triangle on the lingual side. In addition to the three cusps on the trigon and trigonid, a number of accessory cusps appear on tribosphenic teeth. Besides the stylar cusps already mentioned, small cusps commonly develop on ridges trailing downward from the protocone toward the outer part of the tooth, and, posterior to the protocone, a more important cusp, the hypocone, may arise. In the lower molars, a hypoconid appears as a prong on the outer edge of the talonid and faces an entoconid on the lingual side (Fig. 9.5).

Paleontologists have been able to follow the permutations in the cusps of the tribosphenic teeth of therian mammals as they diversified after the end of Mesozoic time, but tracing homologies between cusps and crests in these teeth and those of earlier mammals has been much more difficult. Despite some uncertainty concerning the correspondence of the major cusps on therian and pantothere teeth, workers do agree that the immediate ancestors of the modern mammals sprang from some group of pantothere stock. They are willing to say, further, that the ancestors of the pantotheres were surely closely allied to the known symmetrodonts. About the relationship between the symmetrodonts, the triconodonts, and the other Mesozoic mammalian tribes, there has been more divergence of opinion. Over the years, paleontologists who supported different interpretations of the interrelationships among these earliest mammals often had conflicting ideas about the origin of the class Mammalia as a whole.

Mammals: a monophyletic or a polyphyletic group?

To many paleontologists, the differences in tooth structure among the earliest mammals suggested that they evolved polyphyletically from the therapsid reptiles. According to this view, the mammals, long accepted as constituting a natural group, represented instead a structural and functional grade. The proponents of this theory supported it by pointing out that mammalian traits appeared in representatives of several therapsid lines. Bauriamorphs and various kinds of cynodonts all seem to have been evolving in a similar direction, perhaps because the genetic complex they shared tended to change in the same way and selective pressure upon all these animals favored the development of homoiothermy. Presumably, therapsids of different groups exhibited different mosaics of reptilian and mammalian characters. It seemed not at all improbable that eventually members of more than one group attained mammalian status. Certainly it was hard to imagine that the several lines of Mesozoic mammals shared a common ancestor. Osborn had tried to tie a number of them to a central triconodont stem, arguing that the triangular arrangement of cusps seen in symmetrodonts and pantotheres was arrived at by rotation of the first and third triconodont cusps around the principal one. Simpson, and later Patterson, spoke out strongly against Osborn's idea and declared themselves in favor of a separate origin of the triconodonts and the mammals with triangular teeth. Multituberculates with their highly specialized dentition, in Simpson's opinion, must have arisen from a third reptilian stock and the monotremes, primitive egg-laying forms without a fossil history, from a fourth. Reed, who opposed the retention of a polyphyletic assemblage as a class, proposed to save the validity of the class Mammalia by extending it to include all the therapsids and the pelycosaur ancestor from which they came. Simpson bridled at the prospect of accepting so reptilian an animal as a pelycosaur as a mammal and, in addition to the small- and medium-sized carnivorous therapsids, the whole flock of large, lumbering herbivorous ones as well. He pointed out that a classification scheme cannot express phylogeny exactly by including at the base of each taxon the particular form from which all its other members sprang. He believed that the requirements of monophyly were reasonably satisfied if it could be demonstrated that a group had evolved from a single taxon of lower rank. Having arisen from the order Therapsida, the class Mammalia was acceptable as a natural unit.

Criteria for the mammalian condition

Since progress toward the mammalian level of organization was gradual, it is important that participants in a discussion of the origin of the class Mammalia agree on their criteria for the attainment of the mammalian grade. Reed justified his proposed enlargement of the class on the basis that a new major category may be considered established with the appearance of a new adaptive trend. Since he regarded high energy production and thermoregulation as the key adaptive innovations in mammals and believed that these functions were developing in therapsids, he thought it proper to recognize the latter as primitive members of the mammalian assemblage. Van Valen, who also attempted to introduce the therapsids into the class Mammalia, explained his criteria in a somewhat different way. He visualized the mammals as being distinguished from the reptiles by viviparity, greater intelligence, and a higher level of activity. Aware that these characteristics were not directly determinable from the fossil record, he suggested defining a mammal as a synapsid tetrapod in which the skull lacks a supratemporal bone and has only a small quadrate and in which the innermost toe on each foot has two phalangeal bones and the others three. Simpson, reviewing Van Valen's definition, objected that his criteria would really be of little help in deciding whether an animal had achieved mammalian grade. The phalangeal formula is almost always impossible to ascertain because of the absence of postcranial remains, and since skulls are also rare, the status of the quadrate and the supratemporal usually cannot be used as a guide. Simpson stressed the practical value of the characteristics traditionally regarded as diagnostic of the mammalian state. As long ago as Owen's time, a mammal had been recognized as a vertebrate animal that had a lower jaw consisting solely of the dentary bone, a jaw joint between the dentary and squamosal elements rather than the articular and quadrate, and three small bones in the middle ear. Two of these traits could be determined and the third inferred if only the mandible of a specimen was known.

Drawing a line between reptiles and mammals on the basis of the traditional criteria became problematic only after new studies of certain Triassic forms made it apparent that the several characters in question did not appear simultaneously. A. W. Crompton, who traced the origin of the mammalian jaw step by step from that of therapsids, pointed out that the dentary-squamosal joint became established before the postdentary dermal bones had vanished completely from the mandible and before the quadrate and articular had lost their suspensory function (Fig. 9.6). *Diarthrognathus*, a form which emerged from therapsid stock near the end of the Triassic period, had a skull that resembled the mammalian one except for the retention, in close conjunction with the new dentary-squamosal articulation, of the old reptilian articular-quadrate hinge. The dentary bone, to which virtually all the jaw musculature had transferred its attachment, bore on its medial side a shallow groove for the last remnants of the postdentary bones, and to these the articular was still attached. Lying just beneath the condylar process of the dentary, it continued to function in the jaw joint rather than assuming the role of an ossicle in the middle ear. A similar arrangement existed, apparently, in the tritylodonts, therapsid derivatives which were contemporary with *Diarthrognathus* but not closely related to it. These animals were once accepted as mammals on the basis of their teeth, which were somewhat like those of multituberculates, and the general structure of the skull. Scrutiny of the jaw joint in the tritylodonts *Bienotherium* and *Oligokyphus* revealed that although contact between the dentary and squamosal bones was in the process of evolution, the articular-quadrate hinge was still definitely present (Fig. 9.7). Paleontologists who believed, as Simpson did, that the existence of an articulation between the dentary and squamosal elements could serve as a practical, though admittedly arbitrary, criterion of mammalian status, were willing to recognize as mammals not only the forms from whose lower jaw the articular had disappeared entirely but also *Diarthrognathus* and the latest

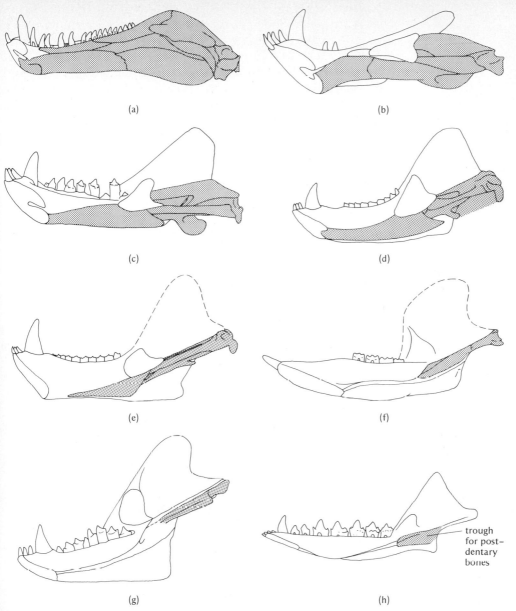

Figure 9.6. Lower jaws of mammal-like reptiles and an early mammal in medial view, showing expansion of the dentary bone and diminution of the postdentary elements. Dentary, unshaded; postdentary bones, shaded. **(a)** *Dimetrodon*, a pelycosaur. **(b–g)** Therapsids: **(b)** a gorgonopsian; **(c)** *Thrinaxodon*; **(d)** *Diademodon*; **(e)** *Cricodon*; **(f)** *Oligokyphus*; **(g)** *Diarthrognathus*. **(h)** *Morganucodon*, an early mammal. (Modified from Crompton.)

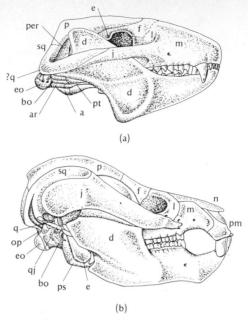

(a)

(b)

Figure 9.7. Skulls of Upper Triassic therapsid reptiles in which evolution of the mammalian type of jaw articulation was far advanced. **(a)** *Diarthrognathus*; **(b)** *Bienotherium*. Abbreviations: *per*, periotic; other abbreviations as in Fig. 6.7. (From Romer.)

tritylodonts, in which contact between the dentary and squamosal bones was presumably established. Since these animals surely had diverse origins within the order Therapsida, the acceptance of this view implied that the reptilian-mammalian boundary had been crossed independently several times.

An argument for monophyly of the mammals

By the late 1960s, Crompton and his colleagues F. A. Jenkins, Jr., and J. Hopson had formed a new opinion concerning the origin of the Mammalia. They think now that the triconodonts, docodonts, symmetrodonts, pantotheres, and perhaps even the multituberculates can be traced to a single line which emanated from one specific family among the Therapsida and thus that the class Mammalia is monophyletic in a much narrower sense than that understood by Simpson. Their definition of a mammal continues to depend upon the presence of the dentary-squamosal jaw joint rather than the absence of the articular from the mandible. However, by requiring that a mammal possess teeth that are not repeatedly replaced, that are (or primarily were) differentiated posteriorly into premolars and molars, and that are (or were at some time in their history) characterized by a primary cusp set between accessory ones in front and behind, they have excluded from the class Mammalia the tritylodonts, the cynodonts with crowned molars, and *Diarthrognathus* and its immediate relatives.

The conviction on the part of Crompton, Jenkins, and Hopson that the mammals did constitute a monophyletic group stemmed from studies of newly discovered Upper Triassic fossils and reexamination of others of approximately equivalent age. The new material, which came from red beds in Lesotho in southern Africa, consisted not just of teeth but of skulls and postcranial bones belonging to animals eventually named *Erythrotherium* and *Megazostrodon*. From the structure of their teeth these animals proved to be mammals, rather than reptiles like the majority of forms at the site. Although Crompton speculated from the groove on the medial side of the dentary of *Erythrotherium* (Fig. 9.8) that postdentary bones and an articular had been present in the lower jaw, the little animal was in its other skeletal characters a perfectly good mammal. A reconstruction

Figure 9.8. Reconstruction of *Erythrotherium*, an Upper Triassic mammal. (Drawing by R. T. Bakker courtesy of A. W. Crompton.)

based upon the postcranial remains shows *Erythrotherium* to have been rather similar to a rat in its appearance and far too small to have attracted the attention of the prosauropod dinosaurs that shared its territory.

The teeth of *Erythrotherium* and *Megazostrodon* resembled those found many years earlier in Upper Triassic fissure fillings in Great Britain and first assigned to the genus *Morganucodon* but now referred by most workers to *Eozostrodon*. Eozostrodont teeth bore cusps arranged linearly as in triconodonts but differed from those of the Jurassic and Cretaceous forms in the way they interlocked and in the details of the cusp design. The suggestion was made that the eozostrodonts might have been ancestral to the triconodonts, and after weighing the differences in tooth pattern and in the structure of the lower jaw in the two groups of mammals, Hopson and Crompton stated that such a relationship seemed probable. Other workers had proposed that the eozostrodonts also gave rise to the docodonts, because docodont teeth, which date from late Jurassic time, seem to have evolved from cusp-in-line molars that broadened lingually. The discovery of a Middle Jurassic form, *Haldanodon*, in which the teeth were morphologically intermediate between those of *Eozostrodon* and *Docodon*, made that theory seem reasonable as well.

Since the eozostrodonts seemed to be emerging as a basal mammalian group from which others radiated, it was logical to explore the possibility of their relationship to the symmetrodonts. It appeared immediately that the known Upper Triassic eozostrodonts could not also have been ancestral to the mammals with triangular teeth because a contemporaneous symmetrodont had been described by Kühne. This symmetrodont, ultimately called *Kuehneotherium*, had been advanced by K. A. Kermack and his colleagues as the forerunner of all the later forms with an angular arrangement of the cusps and perhaps of the triconodonts as well. To derive triconodont teeth from those of *Kuehneotherium* Kermack was obliged to suggest that the accessory cusps of the latter had been displaced in a direction opposite to that imagined by Osborn, from a position internal or external to the main cusp to one in line with it. Crompton and Hopson agreed that *Kuehneotherium* could have sired the later symmetrodonts and pantotheres, but they rejected a symmetrodont origin for the triconodonts as far less probable than an eozostrodont one.

Comparisons that he and Jenkins had made between the teeth of *Kuehneotherium* and the eozostrodonts had convinced Crompton that these two kinds of Upper Triassic mammals, though they gave rise to divergent descendants, were themselves not very different from one another. He remarked, as Parrington had earlier, that the cusps on the teeth in each were clearly homologous despite their different arrangement. The teeth interlocked in a similar manner, and although in *Eozostrodon* the upper teeth closed almost beside the corresponding lower ones, those of *Megazostrodon* fitted between the lower teeth, as they did in *Kuehneotherium.* The strong resemblance between the eozostrodont and earliest symmetrodont dentitions caused Crompton and Jenkins to abandon the older theory that mammals with cusp-in-line teeth and with triangular ones had evolved from separate therapsid groups. They proposed, instead, that the Upper Triassic forms had had a common ancestor and that its teeth, if they were ever found, would prove to be of the eozostrodont type. The latter conclusion was based upon the similarity of eozostrodont teeth to those of the galesaurid therapsid *Thrinaxodon* and upon their conviction as a result of studies of the wear patterns in therapsid and early mammalian teeth that the triangular teeth of symmetrodonts were not primitive but the product of a long sequence of evolutionary changes leading to precise occlusion without excessive attrition of the tooth crowns. The ancestral eozostrodonts, according to Crompton and Jenkins, inherited teeth with crowns shaped like those of their carnivorous cynodont predecessors but lost the ability to replace them continuously. Their definitive set of teeth wore severely as the animals grew older and, by doing so, developed triangular abraded surfaces which fitted against each other more tightly than had the original, unworn occlusal surfaces of the newly erupted teeth. The arrangement must have been advantageous, because the eozostrodonts were, besides the tritylodonts, the only therapsid offspring with reduced tooth replacement to survive beyond the end of the Triassic period.

To explain the origin of the symmetrodonts from the hypothetical eozostrodont stem, Crompton and Jenkins returned to the theory of cusp rotation which had been formulated by Cope and restated by Osborn. Animals that appeared with teeth on which the accessory cusps were set slightly to one side of the main one would have been selected for, they believed, because the opposing teeth would have been so shaped as to fit together more accurately upon eruption than those with cusps in line. With very little wear, the occlusal surfaces of forms like *Kuehneotherium* developed the obtuse-angled triangular outline that allowed them to interdigitate precisely. Eventually, as cusp rotation continued in the Jurassic descendants of the early wide-angled symmetrodonts, the apices on the occlusal surfaces of the teeth became sharper. In advanced symmetrodonts and pantotheres, the pattern of reversed triangles that permitted accurate closure and maximum shearing capacity was almost entirely dependent upon genetic factors rather than upon wear.

If the common ancestry of the symmetrodonts and the cusp-in-line forms was accepted, only one challenge to the monophyly of the class Mammalia remained. To maintain that the Mesozoic mammals derived from a single crossing of the reptilian-mammalian line it would have to be shown that the multituberculates and the monotremes were also traceable to the primitive eozostrodont group. Such a demonstration could not be based securely upon dental morphology because the multituberculates enter the fossil record in Jurassic time with a unique, highly specialized dentition and the almost completely toothless monotremes have no fossil record at all. Hopson and Crompton did argue that the postcanine teeth of the multi-tuberculate mammals and the few teeth possessed by the young duck-billed platypus, *Ornithorhynchus*, could have evolved from precursors of the eozostrodont type, but they acknowledged the tenuous nature of their speculations. The possibility of a rela-tionship between the multituberculates and monotremes and the eozostrodonts was strengthened, however, by evidence from the structure of the skull (Fig. 9.9). The skull of *Erythrotherium* revealed that the eozostrodont had inherited from the cy-nodonts their tendency to ossify the braincase anterior to the inner ear, a region usually left membranous in reptiles. Study of *Thrinaxodon* and other cynodonts showed that in these animals the epipterygoid bone, which had been a narrow element lateral to the braincase proper, enlarged until it covered the unpro-tected surface and abutted the periotic bone posteriorly. An expanded epipterygoid, called the alisphenoid bone in mam-mals, remains in marsupial and placental therians, but in the living monotremes it has diminished markedly, yielding before the advance of an anterior flange of the periotic. The difference in the development of the alisphenoid suggests that the mono-tremes are not, like other Recent mammals, descendants of the symmetrodont-pantothere line. Although the skulls of symme-trodonts and pantotheres have not been found, it would seem that the alisphenoid in those animals must have been transitional in size between that of cynodonts and the still larger one of the

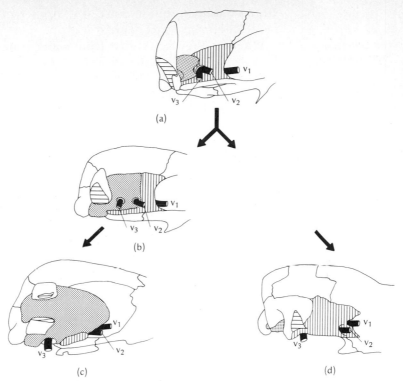

Figure 9.9. The cranial region of the skull in three kinds of mammals, arranged to suggest hypothetical lines of descent of the mammals from therapsid reptiles. **(a)** A therapsid. **(b)** A primitive nontherian mammal such as *Eozostrodon*. **(c)** A monotreme. **(d)** A therian, or placental mammal. Branches of the trigeminal nerve (*V*) indicated in black; periotic bone, shaded; cut surface of squamosal, lined horizontally; alisphenoid bone (epipterygoid), lined vertically. (From Hopson and Crompton, in Dobzhansky et al.)

modern therians. Decrease in the size of the alisphenoid and enclosure of the side of the braincase by an extension of the periotic seems to have begun among the eozostrodonts, perhaps after the symmetrodonts separated from them. A flange from the periotic bone has already spread forward in *Erythrotherium* to cover part of the area occupied by the alisphenoid in *Thrinaxodon*. Kermack pointed out that the flange existed in *Trioracodon*, the triconodont he studied, and new fossil discoveries have confirmed his belief that it was present in the multituberculates also. It is possible that multituberculates and monotremes paralleled the eozostrodonts and their descendants in expansion of the periotic element, but the most economical explanation,

Hopson and Crompton think, is that the two groups emanated from an eozostrodont source.

First appearance of therian mammals

Whether they descended from a single stock or several, the Jurassic mammals radiated widely and held their ground well into Cretaceous time. Their superior anatomical and physiological attributes made them more than a match for the Mesozoic reptiles, and when they finally gave way, it was to a new group that advanced from their own midst. A solitary tooth signaled the existence of one of the new forms in the early part of the Cretaceous period. A lower molar found in the Wealden strata of England and assigned to the genus *Aegialodon* shows for the first time a basined talonid behind the trigonid and is thus a true tribosphenic tooth. Paleontologists assume that *Aegialodon* belonged to the common therian stock from which the marsupial and placental mammals sprang. Lower Cretaceous deposits near Forestburg, Texas, which seem to be somewhat younger than the rocks in which *Aegialodon* was found, have yielded more extensive material of similar forms. Patterson, who first described the Forestburg teeth, suggested that they belonged to animals still not differentiated into separate marsupial, or metatherian, and placental, or eutherian lines. New discoveries in the same deposits have convinced R. H. Slaughter, however, that such a division had already occurred. Besides *Pappotherium*, whose teeth Patterson described as being simply of metatherian-eutherian grade, Slaughter has recognized three molars that, in his opinion, are definitely marsupial and two other teeth that he believes could have belonged only to a placental mammal. Nowhere else in the world have remains of such animals come to light. The only other specimen with tribosphenic teeth known from the Lower Cretaceous—*Endotherium* found in a coalfield in southern Manchuria—cannot be classified specifically as a marsupial or a placental form.

Late Cretaceous mammals

There is a great gap in the history of the mammals between the Forestburg years and late Cretaceous time. No fossil evidence at all is available from the period during which, while the angiosperm flora supplanted the cycads and conifers, the older mammalian tribes failed and the newest therians proliferated. The late Cretaceous fauna, like the early one, is known from a relatively small number of localities. Except for one tooth from Europe, all the mammalian material that is definitely late Cretaceous in age comes from the upland deposits in Djadochta, Mongolia, and from North American sediments laid down in what is now northwestern Canada, Montana, Wyoming, and New Mexico when those regions were lowlands bordering a shallow sea in the Great Plains area. Although ecological conditions in upland and lowland environments cannot have been the same, the climate seems to have been mild and the vegetation plentiful on both continents.

The mammalian fauna at Djadochta and at the late Cretaceous North American sites contrasts sharply with that of Forestburg. Triconodonts are rare, and the symmetrodonts have disappeared completely. Only the multituberculates of the older Mesozoic assemblage remain numerous (Fig. 9.10), and these animals had evolved new varieties that apparently thrived on the grasses and broad-leaved flowering plants. The Djadochta fauna included several undoubted placental forms. One of these, *Kennalestes,* could be described as a very primitive insectivore, possibly ancestral not only to animals of the order Insectivora but also to some of those which branched away and became more carnivorous in habit. *Kennalestes* may have been a survivor of an already obsolete placental stock, because some of the Djadochta mammals exhibit long canines and cheek teeth which identify them as deltatherids, early members of the carnivorous group known as the Creodonta. A third line of placental mammals was represented by *Zalambdalestes,* an animal that seems to have been a rather specialized insectivore but which might possibly have been close to the ancestry of the rabbits whose remains are found in Mongolian deposits of a slightly later date.

The nature of the fauna from the North American sites has led some investigators to suggest that, despite the existence of sea channels separating North America and Asia, there may have been some movement of animals from one continent to the other. The multituberculates found in the American Northwest are similar to those from the slightly older Mongolian strata, and the placental forms, though fewer in number, are apparently not far removed in structure from the primitive insectivores and nascent creodonts of Djadochta. The spread of animals from continent to continent could not have been as free as it was earlier in the Mesozoic era, however, because the marsupials, which are the most common mammals in the North American deposits, do not appear at all in Asia. The existence of somewhat different faunal assemblages in distant locations implies that in the late Cretaceous period the development of mountain chains and seaways was separating hitherto contiguous populations into isolated breeding groups and thus fostering the emergence of new forms, each characteristic for its area.

Early marsupials in North America

The apparent restriction of the marsupials to the Western

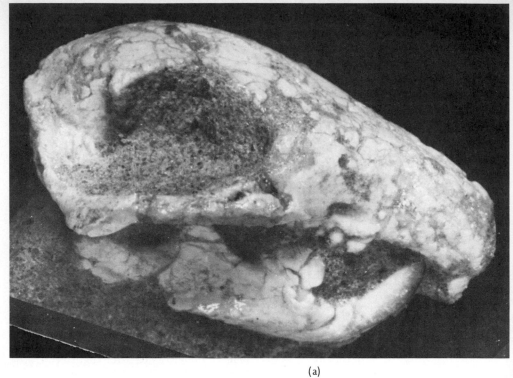

(a)

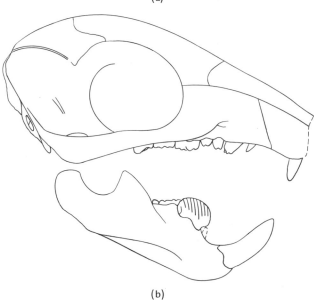

(b)

Figure 9.10. Skull of *Sloanbaatar mirabilis*, a multituberculate from Mongolia. **(a)** Fossil before final preparation. **(b)** Reconstruction from the fossil specimen. (From Kielan-Jaworowska.)

Hemisphere throughout Cretaceous time suggested to Clemens that the metatherian line originated in the Americas. Before Slaughter reported marsupial teeth from Forestburg, Clemens had pointed out that the molars of *Alphadon*, the most primitive of the late Cretaceous marsupial forms, were not far removed structurally from the early Cretaceous tribosphenic teeth found at the Texas site. The rise of the marsupials seems to have taken place unhindered by competition from placental mammals because not many of the latter appear in the Western Hemisphere until the very end of the Cretaceous period. After Forestburg time, a generalized metatherian population, of which *Alphadon* was a late representative, must have flourished in the growing angiosperm vegetation and gradually replaced the triconodonts and symmetrodonts. Whether the marsupials triumphed over the older insectivorous forms because their teeth were more effective or because of some other advantageous traits cannot be said, but by the late Cretaceous period they were the predominant terrestrial animals of small size at all the localities known from New Mexico to Canada. Virtually nothing remains of these marsupials except teeth, but the dental evidence makes it plain that considerable diversification occurred within the group. Clemens identified three species of *Alphadon* from the Lance Formation in Wyoming as well as five genera of more specialized forms derived from the central stock. Most of the animals were no bigger than mice, but some grew as large as housecats. The hegemony of these marsupials in North America was brief, for in the last years of the Cretaceous period, all but a few were displaced by placental mammals. Those that remained managed to survive another major faunal revolution early in the Cenozoic era but finally became extinct by Miocene time, about 17 million years ago. The marsupials living today, including the North American oppossum, are descendants of Cretaceous forms that migrated southward from their place of origin and underwent extensive diversification in Australia and South America, where, for most of the Cenozoic era, they were isolated from the most formidable members of the placental assemblage.

Origin of placental mammals

When Patterson diagnosed teeth from the early Cretaceous deposits at Forestburg as being of metatherian-eutherian grade, he implied that the marsupial and placental mammals arose by the division of a common stock. It follows from that assumption that the placentals as well as the marsupials could have originated on the shores of the American Cretaceous sea. Although

Slaughter's discovery of teeth at Forestburg that seem distinctively eutherian strengthens this hypothesis, other factors argue against it. Simpson commented upon the abruptness of the shift in the mammalian fauna at the end of the Cretaceous period and expressed doubt that the numerous kinds of placental mammals which appeared then would have had time to evolve *in situ*. It is possible that they did so but through most of the late Cretaceous remained so rare that they left no fossil record. This explanation is regarded as unsatisfactory by most investigators, however, because it hardly seems likely that the eutherians, which were to extinguish the North American marsupials completely, would have been repressed by those animals for so long. Simpson suggested, and Lillegraven and others who have studied the problem more recently agree, that the marsupials were overcome not by a group with which they had coexisted for years but by invaders from elsewhere with superior anatomical and physiological characteristics.

If the placental mammals did not arise on the North American continent, they must surely have come from Asia. The northwestern American region, where they appeared first, was completely cut off from the eastern part of the same continent by the broad inland sea but separated from the land masses to the south and to the northwest by shallow waters that from time to time were evidently crossed by a few animals at least. There is no fossil evidence to suggest that South America might have harbored mammals nearing the eutherian state, but the record proves that northeast Asia certainly did. The presence of *Endotherium* confirms the existence there of early Cretaceous mammals with tribosphenic teeth, and the diversity of the placental forms at Djadochta shows that the area (now Mongolia) was an important center of eutherian evolution. Although the formation of the Rocky Mountain chain was under way in western America by late Cretaceous time, migration from Mongolia across the Bering region and down into the North American continent was still possible, as the history of the ceratopsian dinosaurs illustrates. That the placental mammals traveled from Asia to America like the dinosaurs rather than in the opposite direction is strongly suggested by the fact that they appear earlier and in greater variety in Mongolia than on the American continent. The case for the Asian origin of the eutherian group is weakened only by the existence at Forestburg of the mammals that had approached or even reached the placental level. Clemens has speculated that these forms might have paralleled the early placentals but left no descendants or, if they were true eutherians, have come from Asia themselves at an earlier time. He does not ignore the possibility, however, that the Forestburg animals may represent the true root of the later placental mammals and that the hypothesis of an Asian source for the group, as probable as it seems, will prove erroneous after all.

The early history of the placental mammals is known almost entirely from fossils obtained in Montana, Wyoming, and the Canadian province of Alberta. This region is the only one in the world in which sequential deposits have been found covering the latest part of the Cretaceous period and the first years of the Cenozoic era. In the earliest fauna represented, ornithischian dinosaurs are numerous and placental mammals rare. Later assemblages found in the uppermost Cretaceous strata include fewer dinosaurs but contain a much larger eutherian population and one which clearly foreshadows that of the opening Cenozoic years. Above the Mesozoic rocks lie Cenozoic layers that are older than most of the early Cenozoic deposits elsewhere. Assigned to the Paleocene epoch of the new era, they have yielded primitive and specialized mammals, which gradually disappeared during the following Eocene epoch as the ancestors of the modern placental forms evolved.

When the great horned *Triceratops* still grazed and multituberculates and marsupials shared the underbrush, the eutherians emerged already divided, as they had been in Mongolia, into insectivores and forms that were developing more carnivorous habits. Their teeth still showed a simple tribosphenic design not widely different from that of the contemporary metatherians, but instead of having three premolars and four molars, like those animals, they had four premolars and only three molars behind. The teeth of *Gypsonictops*, one of the earliest known placentals from the *Triceratops* zone, were so similar to those of living shrews that some paleontologists regard *Gypsonictops* as the basal member of the superfamily to which the shrews belong (Fig. 9.11). More conservative investigators place the Mesozoic form among the leptictoids, a group of primitive insectivores from which several others are thought to have sprung. The upper molars of *Gypsonictops*, with paracone and metacone set well apart from each other near the outer edge of the tooth, were distinct from those of the Cretaceous placentals that gave rise to noninsectivorous descendants. *Cimolestes*, the earliest of the latter, produced upper teeth like those of the Mongolian deltatherids: paracone and metacone arose from a point near

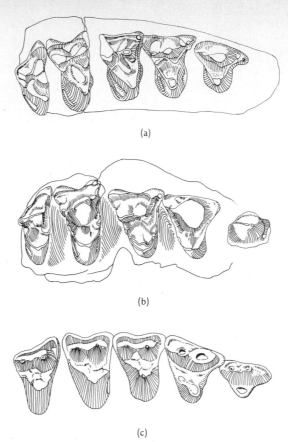

(a)

(b)

(c)

Figure 9.11. Upper cheek teeth of early placental mammals. **(a)** *Gypsonictops*, a shrew-like insectivore with molar paracone and metacone well separated. **(b)** *Cimolestes*, possibly a somewhat more carnivorous animal with paracone and metacone set closer together. **(c)** *Procerberus*, a form with cheek teeth morphologically intermediate between those of *Gypsonictops* and *Cimolestes*. (a, b from Lillegraven; c from Sloan and Van Valen, copyright 1965 by the American Association for the Advancement of Science.)

the center of the crown and stood so close together that the bases of the cusps merged. Both *Gypsonictops* and *Cimolestes* could have been (and almost surely were) derived from ancestors structurally like *Kennalestes* from Djadochta. Lillegraven believes that the two lines which they represent did actually originate in Mongolia and that *Gypsonictops* and *Cimolestes* or their immediate forebears crossed into North America from Asia. The point is not settled, however, because there were possibly some structurally intermediate forms on the American continent from which they could have come. *Procerberus*, found in uppermost Cretaceous rocks in company with *Gypsonictops* and *Cimolestes*, shows dental characters suggestive of both but, of course, lived too late to have served as the common ancestor. It is still possible that an older relative of *Procerberus* will be discovered in the Western Hemisphere and prove to be a suitable precursor for the late Cretaceous leptictid and deltatherid genera. Should such a form appear, it would be necessary to suppose that closely appressed, centrally placed cusps had evolved twice, once in the Mongolian carnivores and again in the lineage of *Cimolestes* and its allies.

Diversification of the early placental mammals

As the dinosaurs declined and the reptilian fauna took on a more modern aspect, the placental mammals diversified rapidly. The degree to which the group expanded can be gauged by the report of Sloan and Van Valen that in 10 weeks of collecting at a site in Montana near Bug Creek they found 26,000 mammalian teeth, about 1,000 fragmentary jaw elements, and hundreds of postcranial bones. Despite the large quantity of fossil material, paleontologists have had difficulty in defining the interrelationships among the emergent groups. Many of the animals are still known only from isolated teeth, and the available skeletal remains are too sparse to support extensive comparative studies. In addition, identifying incipient branches from the central placental stem is problematic, because the latest Cretaceous forms had not yet diverged far enough from the primitive eutherian type to show clearly the traits characteristic of the later Cenozoic orders. When investigators prepare a classification scheme, the similarity of the early placental mammals is somewhat masked because the authors remove from the leptictid and deltatherid assemblages the forms they believe are close to the ancestry of later orders and place them at the base of the more advanced taxa. As a result, animals which in life would have resembled each other strongly are presented as differing at the ordinal level. The fact remains that mammals of latest Mesozoic time referred to as primitive primates, ungulates, and carnivores were all small, clawed animals that scampered after insects and perhaps supplemented their diet with soft-bodied invertebrates and some plant food.

The descendants of *Gypsonictops* and its fellow leptictids seem to have retained their small size and insectivorous habits longer than the offspring of the deltatherids. By Paleocene time there existed a half dozen or more families of insectivores whose ancestry paleontologists trace more or less certainly to animals similar to *Gypsonictops,* and isolated groups occur later which they believe may also have come from a leptictid source. The primates were apparently a very early branch from the leptictid stock. Van Valen and Sloan describe and designate as *Purgatorius ceratops* a primate tooth recovered from a Cretaceous stream-channel deposit that also held part of a large dinosaur. Another species of *Purgatorius* appeared in earliest Paleocene rocks, and from sediments laid down later in the epoch there come at least a dozen more primate genera, including one with enlarged incisors, called *Plesiadapis* (Fig. 9.12), that has been mentioned as possibly allied to the rodents.

While small, insectivorous animals diversified from the leptictid group, *Cimolestes* and other deltatherids were giving rise to forms that would eventually produce herbivores and carnivores to fill the niches left empty by the departed dinosaurs. Recognizing in the deltatherid dentition the potential for improved shearing and grinding action, Simpson had speculated in the late 1920s that the deltatherids, or the paleoryctids as he called them, were ancestral to the condylarths, animals antecedent to the modern hoofed mammals, and to two distinct lines of carnivores that preyed upon them. Simpson's theory seems borne out in part by the discovery in uppermost Cretaceous rocks of teeth that had paracone and metacone set quite close together yet showed in their wear pattern similarities to the teeth of the earliest Paleocene condylarths, the arctocyonids. Although Sloan and Van Valen, who described the teeth and assigned them to the genus *Protungulatum* (Fig. 9.13), did not

Figure 9.12. Skull of *Plesiadapis,* a specialized primate of the late Paleocene possibly descended from a stock which gave rise to the rodents. (From Simpson. Reprinted by permission, *American Scientist,* journal of The Society of the Sigma Xi.)

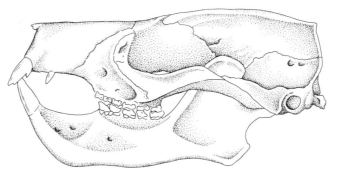

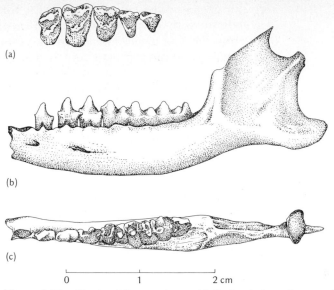

(a)

(b)

(c)

0	1	2 cm

Figure 9.13. Teeth of *Protungulatum*. **(a)** Occlusal view of upper right premolars and molars. **(b)** Lateral view of the teeth and mandible. **(c)** Occlusal view of lower left cheek teeth implanted in mandible. (From Sloan and Van Valen, copyright 1965 by the American Association for the Advancement of Science.)

find among the known primitive eutherians a form they believe might have been an actual ancestor of the Cretaceous specimen, Lillegraven and other workers are of the opinion that *Protungulatum* shows undoubted deltatherid affinities. The Paleocene successors of *Protungulatum* evolved less sharply cusped and more squarish cheek teeth, better suited to masticating plant material than the teeth of the deltatherids had been. The majority of the condylarths did not compete with the ubiquitous small herbivorous multituberculates but grew as large as some of the modern grazing animals and adopted an entirely different mode of life.

The creodonts: primitive carnivores

From the deltatherids that retained their shearing teeth came forms specialized for hunting game bigger and tougher than insects. In these animals, postcanine teeth developed as carnassials, with lengthened shearing crests that could scissor through the muscle and connective tissues of other vertebrates. The carnivores which became most common in the Paleocene and Eocene epochs were the creodonts, predators in which carnassial function was concentrated toward the posterior part of the tooth series. There were two families of these animals, the oxyaenids and the hyaenodonts (Fig. 9.14). The former were rather heavily built, with short, deep jaws and carnassial teeth developed from the first upper molar and the second lower one. The latter, lighter and more fleet of foot, had long-snouted faces and carnassial teeth even closer to the jaw joint. Although the creodonts of both families became formidable animals by the Eocene, they were apparently not as efficient hunters as members of the modern order Carnivora, whose first representatives, the miacids, appeared in Middle Paleocene time. Simpson thought that the miacids, which were small and perhaps somewhat arboreal, were another branch from the deltatherid stock,

Figure 9.14. Skulls of early carnivorous mammals. **(a)** *Oxyaena*, an oxyaenid creodont of late Paleocene and early Eocene times. **(b)** *Sinopa*, a hyaenodont creodont of the Eocene. **(c)** *Vulpavus*, a miacid of the early and middle Eocene. The skulls are, respectively, about 8, 6, and 3 inches long. (From Romer.)

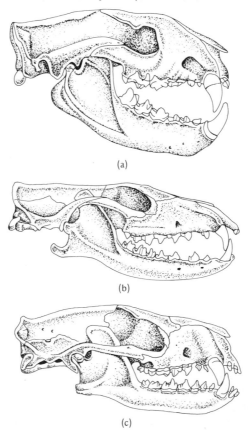

(a)

(b)

(c)

Vertebrate History: Problems in Evolution

but this relationship has been disputed. Not even the most primitive known miacids showed the closely appressed paracone and metacone that distinguished the upper molars of the deltatherids, and miacid carnassials developed more anteriorly than those of the deltatherid-descended creodonts, from the last upper premolar and the first lower molar. Van Valen, who regarded the similarities which did exist between miacids and creodonts as the result of parallel adaptation, thought that the miacids evolved from a leptictid-arctocyonid source rather than from animals of the deltatherid-creodont line. Despite differences in the dentition, however, Lillegraven still holds that the miacids, like the creodonts, stemmed from deltatherid ancestors and sees in small and large species of *Cimolestes* the source of the two diverging lines.

Mammalian fauna of the Paleocene epoch

The great variety of placental mammals that existed in the Paleocene, the epoch following the Cretaceous period, has suggested to some paleontologists that their colleagues may be incorrect in trying to trace every placental group to either leptictid or deltatherid ancestors. Believing that time was too short to allow the evolution from small leptictids or deltatherids of such animals as the large, herbivorous Paleocene taeniodonts and the bizarre, pig-sized, rat-toothed early Eocene tillodonts, these workers suppose that there may be, yet undiscovered, other kinds of late Cretaceous primitive eutherians. It is a large assumption, they point out, that terrestrial regions worldwide were all populated by the same kind of placental mammals as existed at the few localities from which fossil forms are now known.

The explosive radiation of the placental mammals during the later part of the Cretaceous period produced Paleocene faunas that would look strange to the modern observer. The glimpse that paleontologists have of them from fossils found in deposits in North America, Central Asia, and Europe shows a myriad of animals, few of which were directly ancestral to those living today. Since rodents did not appear in large numbers until the beginning of the Eocene, the niches available to small herbivores were filled in earliest Cenozoic time by aberrantly specialized primates and the persistent, nontherian multituberculates. Whether the placentals first inhabited the trees or the forest floor has been the subject of much debate. If the eutherians were not originally arboreal, some of them soon became so, it seems, because the oldest postcranial remains suggest the early evolution of feet suitable for grasping branches. The appearance of bats in the Middle Eocene is further evidence that the insectivorous mammals must long have been at home in the trees.

The first forms to emerge from the cover of the forest may have been the condylarths. The earliest of these animals, the arctocyonids, kept their claws, but by the end of the Paleocene epoch herbivores appeared with hoofs on the toes. The ungulate condylarths never became highly specialized for grazing and running, as did the modern forms that replaced them. Their cheek teeth remained low-crowned, and the canines were often as robust as those of carnivores. Short-legged and long-backed, the animals could have run with a gait little different from that of the creodonts which presumably chased them. The condylarths and most of the creodonts which followed them out into open country were of modest size, but during the late Paleocene some peculiar herbivores evolved that grew much larger than any placentals had before (Fig. 9.15). Initially, these big animals, called pantodonts and uintatheres, roamed from one northern continent to another, challenged only by the largest oxyaenids. Like many outsized herbivores, however, they were exceedingly heavy and probably not very fast. They survived the leaf-eating taeniodonts and the tillodonts, but they and the genera of creodonts that hunted them vanished almost completely in the last years of the Eocene. Some creodonts survived in Africa and India through Miocene times, but pantotheres and uintatheres failed everywhere far earlier.

The early Cenozoic fauna also included animals which were rarer than the archaic insectivores, carnivores, and herbivores so far described but which were more closely related to the placental mammals that became predominant in post-Eocene times. Some of these forms paleontologists know from a single specimen or from a few found at only one locality. Others, they assume, must have existed as the ancestors of the new groups that burgeoned suddenly in the fossil record during the Eocene years. Although they must have been present, the immediate forebears of the modern ungulates (with the exception of a late Paleocene horse) have not been found or perhaps have not been recognized among the fossils already in hand. Similarly, among the small forms of the earlier Paleocene, there were animals still not identified which gave rise to the paramyid rodents whose descendants gradually replaced the ancient multituberculates. Besides rodents, rabbits should perhaps be counted as an element of the early Eocene fauna, for they appear in the fossil

(a)

(b)

Figure 9.15. Restorations of two archaic herbivores of Eocene time. **(a)** *Uintatherium,* an animal the size of a large rhinoceros. **(b)** The pantodont, *Coryphodon,* about the size of a large pig. (From Colbert.)

record before the end of the epoch. Several genera from North America represent what must have been a small group of invertebrate-eaters with reduced teeth, collateral relatives, most probably, of the edentates which were more successful south of Panama. By Lower Eocene time, the primates antecedent to the lemurs, tarsioids, and New World monkeys were certainly established on the American continent, and it may be that the ancestors of the whales were wading offshore. In short, while eutherians that had evolved before the end of the Mesozoic era still survived, the progenitors of the modern placental orders were emerging as a potential threat to the older animals.

Although a number of the archaic mammals lived beyond the Eocene into Oligocene time, by the latter epoch the newer forms were predominant and the fauna was fast coming to resemble that of the present day.

Significance of geographical distribution of mammals

Post-Paleocene deposits from continents other than North America make it clear, by inference if not directly, that mammals of all kinds spread widely over the globe as soon as they became differentiated from their reptilian forebears. Since the Triassic and Jurassic mammals found in Africa, Europe, Asia, and North America were not markedly divergent, their distribution tells paleontologists little except that they must have moved outward easily and relatively quickly from their place of origin. Beginning in late Cretaceous time, however, the mammals of one region are readily distinguishable from those of another, and investigators are able to draw conclusions about the evolutionary history of the various forms from their relative location and the degree of difference between them. Data from geographical distribution are especially helpful in determining whether animals of similar structure are closely allied phylogenetically or merely convergent. The former supposition is reasonable if it can be shown that the groups to which the animals belong could have migrated from a common source. If, on the contrary, it is demonstrable that one of the groups was cut off by some sort of barrier from the others throughout the period during which its special characters evolved, an investigator is forced to conclude that the resemblance of its members to the distant animals is the result of similar adaptive change.

There was a time when some paleontologists sought to make the facts of geography fit phylogenetic interpretations they favored on the basis of their morphological studies. They insisted upon the hypothetical land bridge across the deep waters of the Atlantic Ocean, for instance, to explain the existence in Africa and South America of rodents that showed similar characteristics and suggested the migration of certain marsupials to Australia from South America by way of Antarctica. Recent advances in marine geology and new evidence for continental drift have virtually ruled out almost all these highly speculative theories. Although most geologists now think that the southern continents were once a single land mass and that North America and Eurasia were united as well, they believe that the separation of the various continents was far advanced by the end of the Mesozoic era. By the time the marsupials and the placental mammals evolved, direct passage from one southern continent to another was no longer possible and the only links between adjacent terrestrial blocks were those that are exposed or under shallow water today.

Opportunities for intercontinental migration

The opportunity for migration from one continent to the next has been intermittent since the Cretaceous period. When the sea level has fallen or the land has risen, dry land has connected the northeastern tip of Asia with North America and, perhaps once, the eastern part of North America with northern Europe (Fig. 9.16). Entrance into Africa, South America, and Australia, possible only from the north, has been prevented or restricted by the presence of marine barriers during much of the Cenozoic era. Africa was long divided from Europe and southern Asia by a sea, called the Tethys, which extended from the Atlantic to the Indian Ocean; South America was cut off from North America by water covering northwestern Colombia south of Panama; and Australia was set apart from the southeast Asian mainland by even broader marine straits than exist in the area now. Although the presence of water constituted an important barrier for the

Figure 9.16. Map of the world with land areas exposed during early Eocene times shown in white. A broad connection between Asia and North America, a narrow bridge between North America and Europe, and a seaway running north and south through Asia are consistent with the faunal distribution known for the Paleocene and early Eocene years. (Modified from Kurten, after Schaffer.)

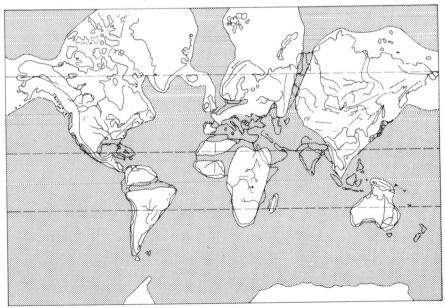

terrestrial mammals, it was not the only factor that inhibited their spread. Animals adapted to warm lowlands were stopped by mountain ranges or by the cold climate that developed in northern latitudes toward the end of the Cenozoic era. Temperate forms sometimes moved gradually southward or northward into the hotter regions close to the equator, but tropical species rarely migrated into cooler areas. Rain forest was impenetrable to animals of the savannah and vice versa, and deserts were impassable by all but a few mammalian forms. The availability of suitable food, the nature of the terrain, and the presence or absence of competitors and predators have also played a part in determining the animals' range.

On the basis of the amount of faunal interchange that has taken place between neighboring continents, Simpson has characterized the passageways from one land mass to another as corridors, filter bridges, or sweepstakes routes. He counts as a corridor the broad land connection that has long bound China and Siberia to western Europe. This path is so wide that a considerable variety of ecological niches exist upon it, giving most of the animals in one region an opportunity to reach the other. That a corridor allows continuous migration in both directions is proved by analysis of the faunas of the adjacent areas. The same orders are represented in each, and the percentage of shared genera is generally high. Migration over a filter route is much more restricted. On the passage that has linked Asia and North America from time to time and on a narrow land bridge like that which now connects the two American continents, environmental conditions are suitable for some animals but not for others. As a result, some forms succeed in crossing, but so many fail to do so that the faunas of each continent remain far more distinctive than those connected by a corridor.

A sweepstakes route, as the name suggests, is characterized by such serious obstacles that animals move over it infrequently and then only by chance. For terrestrial mammals the most common pathway of this type has been along island chains such as exist now between Indo-China and Australia. Large mammals almost never cross the strait that separates the mainland from the first island offshore, but small species can drift on logs or other plant material and arrive unscathed on a beach many miles from where they started. Dispersal over a sweepstakes route is a slow and unpredictable process. Transport of a live land animal across a marine barrier does not in itself make the advance of the form to an outlying island possible. To establish a new colony, the individual transported must be either a pregnant female or accompanied by an animal of the opposite sex. Even if, improbably, these conditions are met, the success of the species requires that the animal find an available niche and reproduce before any calamity befalls it. Continued advance through a chain of islands demands that the extraordinary sequence of events be repeated several times. Under the circumstances, it is easy to understand why passage from one continent to another over a sweepstakes route is a rare occurrence. Although, theoretically, animals could move in either direction, study has shown that most of the chance migration has taken place from the larger land areas, where a greater number of the existing niches are filled, to regions of sparser mammalian population. Islands like Madagascar and Australia, which have received almost their entire fauna via a sweepstakes route, exhibit, because of the random nature of sweepstakes migration, a mammalian assemblage that is relatively poor in variety and unique in kind. The few mammals that do arrive diversify in a manner different from the mainland group from which they separated and often survive long after their parent stock has been extinguished. Their radiation, though it may be unopposed, is limited by genetic factors, and thus a number of niches available on the island are left unoccupied.

Geographical distribution and evolutionary change

It should be obvious that the migrations by which mammals reached different areas of the world were not, like the familiar seasonal migrations of the birds, journeys of individuals from one place to another. The terrestrial animals distributed themselves by gradually extending their range over a period of many generations. Successful forms spread farther afield, and others, defeated on their original ground by competitors or inimical changes in the climate, survived by multiplying in contiguous areas where the environment remained or was becoming favorable to them. The rate at which mammals expanded or shifted their territory varied, but their progress was usually slow enough to permit differentiation of new species and even new genera to occur along the way. Thus, when zoologists compare animals that have emanated from the same source but now live far apart, and when paleontologists make similar studies among fossil forms, they usually find that the degree of difference between the animals bears some relation to the distance which separates them.

Since diversification from a parent stock requires time, it is possible to judge from the variety that exists within a group approximately how long it has lived isolated from its fellows elsewhere. This sort of analysis is useful in estimating the date of entry of certain kinds of mammals into a region separated by a filter or a sweepstakes route from the adjacent one. Study of the entire fauna of such an area may reveal the presence of several groups in different stages of differentiation. Investigators conclude that an assemblage has sprung from ancestors only recently arrived if its members are similar to one another and not far removed in type from their relatives on the neighboring continent. If, on the other hand, a group contains numerous distinctive forms and differs at the familial or ordinal level from those elsewhere, it can be assumed that it has been in residence for a very long time.

Interaction between successive waves of immigrants differed in detail on every continent but followed the same general pattern. The first mammals to reach an area established themselves without having to struggle against animals equally active or intelligent. They carved new niches or occupied those from which the ruling reptiles had departed. Their radiation produced a native fauna of considerable diversity, often characterized by highly specialized forms. The mammals that arrived next, coming from different stock at a later time, initially enriched the assemblage. Those adapted for a way of life completely unlike that of the older animals disturbed the established forms very little. They slipped into available niches and began their own radiation. Competition ensued between the new immigrants and the native species when their requirements were similar. The entry of a competitive form into the region proved that it was to some degree superior to the old inhabitants; had it not been, it would not have been able to penetrate the area at all. The most resistant of the older immigrants or the species descended from them were forced to share their territory with the newcomers. Subdivision of niches rendered the mammalian population more diverse and usually more dense. The introduction of forms very much more efficient than the established ones led to a succession of extinctions and a rebalancing of the fauna. More advanced carnivores, for instance, first caused the disappearance of the older hunters by killing their prey to feed themselves. The species preyed upon frequently vanished next. Although they had existed in equilibrium with their habitual predators, they usually lacked adequate defenses against the attack of the newcomers. Their place was often taken by more progressive animals that had survived beside the immigrant carnivores in the area of their origin. After a time, the oldest elements of the fauna were reduced to a relatively small number of forms unique in their habits and often restricted in range, and the newer immigrants became predominant.

History of the marsupials in the Cenozoic era

By such a sequence of events, apparently, the marsupials and the oldest placental insectivores were first eclipsed and then extinguished in the area of North America where they had radiated during the Cretaceous period. The subsequent history of the marsupials has been reconstructed largely from the geographical distribution of the living animals, with some clues from the fossil record, mostly from post-Eocene times. Like all successful groups, the Cretaceous marsupials spread outward from the area of their origin. The eastward migration of these animals must have been made difficult by the shallow continental sea that covered much of the Great Plains area at the end of the Mesozoic era. By the time the waters ebbed completely, the ranks of the marsupials in western North America had been depleted, and few forms were left to expand into the newly accessible regions. Since one marsupial, *Peratherium*, a descendant of the primitive North American didelphoid *Alphadon*, has been found in early Eocene deposits in Europe, it is apparent that some metatherians did finally reach lands east of the Atlantic. Whether they arrived there by moving eastward over the land bridge that connected the American continent with Europe at the beginning of the Eocene or by going westward through Asia is a matter of speculation. It is certain that they did not fare any better in Europe during the Cenozoic than they did in North America. The placental mammals of the Paleocene–early Eocene fauna apparently inhibited radiation of the *Peratherium* stock, and the more advanced eutherian assemblage that emerged in the Eocene destroyed it entirely by the Miocene.

The marsupials that spread into the Southern Hemisphere entered areas where competition from more progressive mammals either did not exist or was less intense. Having arrived in South America and perhaps also Australia as early as the late Cretaceous, the marsupials underwent a diversification that produced species to fill almost as wide a variety of niches as were filled by the placentals in the north. The route by which the marsupials entered South America is certainly known. The Central American isthmus was exposed during the Cretaceous period, and since it lay so close to the region in which the marsupials presumably originated, it was probably invaded early by those animals. How the metatherians reached Australia is not as well understood. The complete absence of marsupial remains in the Cretaceous fauna sampled from Asia led Ride and other investigators to keep alive the suggestion that the animals had migrated to Australia from South America via Antarctica when the three continents were still part of a single land mass. Raven and Axelrod, convinced primarily by studies of floral distribution that land connections remained between Australia, Antarctica, and South America into the early years of the Eocene, believe that marsupials surely entered Australia by this southern route. This idea is not supported by the many advocates of continental drift, who think that the vast territory of Gondwanaland had begun to break apart long before the end of Mesozoic time. Audley-Charles has stated, on the basis of ecological studies, that Australia had reached its present location near the southeast edge of Asia before the Cretaceous period began and from that time forth lay off the part of the continental shelf which bears the islands of Borneo and Java. Although during some periods Borneo and Java were more broadly connected to the Asian mainland than they are now and Australia extended northward to the area that currently exists isolated as New Guinea, there has always been some sea water between the large continent and the smaller one. Paleontologists who believe that the marsupials did cross to Australia from the mainland explain the apparent lack of marsupial fossils in northern Asia, through which the animals would have had to pass, by saying that the Cretaceous sediments there in which mammals have been found were all deposited in an upland environment unfavorable to metatherian forms. They point out that the earliest marsupials were adapted to the lowlands in North America and would doubtless have edged their way along the shoreline of the Bering bridge and then moved southward through the Asiatic coastal regions. When an Asian fauna characteristic of well-watered Cretaceous lowlands is found, they expect it to include the missing marsupials.

Australian mono-tremes

Although not a scrap of fossil evidence exists to document the entry of the metatherians into Australia, investigators infer that the immigrants encountered a population of primitive monotremes when they arrived. Because monotremes, living or fossil, have never been discovered elsewhere in the world, it is suspected that they represent an archaic branch of early mammals which evolved in isolation on the island continent. The extant forms (Fig. 9.17), the echidnas *Tachyglossus* and *Zaglossus* and the platypus *Ornithorhynchus*, are highly specialized, like most relict animals, but in many aspects of their structure and physiology they are much closer to the reptilian level than the most primitive of the marsupial and placental mammals. The monotremes retained from their reptilian predecessors not only their well-known habit of laying eggs but also their minimal powers of thermoregulation and such skeletal traits as the presence of two independent coracoid bones. Since no certain remains of monotremes have been found in Australian Tertiary rocks, the history of these animals is speculative. Advocates of mammalian polyphyly derive them from a separate reptilian stock, while supporters of a single origin for mammals suppose, as has been mentioned, that monotremes sprang either from the same general cynodont group as other hair-covered forms or arose

Figure 9.17. Recent monotremes. **(a)** *Tachyglossus*, the echidna, or spiny anteater. **(b)** *Ornithorhynchus*, the duck-billed platypus.

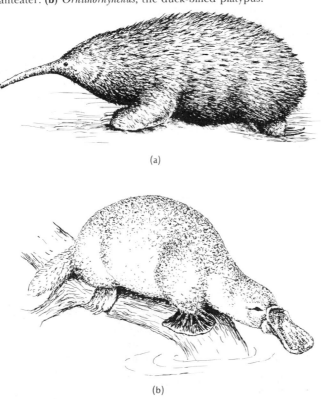

(a)

(b)

from an eozostrodont source. The widely different habits of the terrestrial ant-eating echidna and the aquatic mud-grubbing platypus suggest that the monotremes underwent extensive radiation, like many newly emergent groups, when they first appeared in Australia. Their physiological inferiority evidently caused them to give way to the marsupials wherever the latter advanced. The survival of the platypus may be due to the failure of any of the marsupials to take to the water and that of the echidna to the fact that none of the newcomers became so expert a myrmecophage or was able to tolerate so wide a range of environmental conditions.

Entrance of therian mammals into Australia

The displacement of the hypothetical monotreme fauna, if it did take place, and the establishment of marsupial insectivores, herbivores, and carnivores was accomplished long before the late Oligocene or early Miocene epoch, from which the oldest known marsupial fossils in Australia date. Paleontologists are quite certain from the untrammeled expansion of the marsupials that they did not have any competition on the island continent from placental forms. The absence of placentals during the Paleocene, when diversification of the marsupials most probably occurred, did not seem strange to the nineteenth-century investigators who studied the history of the mammals in Australia. Since they believed that the marsupials were the ancestors of the more advanced, eutherian animals, they assumed that the former radiated before the latter evolved. After it was discovered that marsupials and placentals diverged during early Cretaceous time from a common stock, a new explanation was necessary. Simpson's theory of sweepstakes migration provided one. According to Simpson, both kinds of mammals were probably residents of the Asian mainland at the end of the Mesozoic era, and neither was better equipped than the other to cross the sea barrier that kept them from the islands to the east and the Australian shore. The fact that marsupials rather than placentals happened to reach the southern continent was a matter of chance, not a reflection of competitive advantage. The successful migration of the marsupials could have been effected simply because one species belonging to that group habitually lived in a nearshore environment where individuals were frequently set adrift.

The high odds against negotiating the whole length of the sweepstakes route from Asia to Australia makes it probable that few animals accomplished the feat and thus very likely that the marsupial population of Australia was derived from a single immigrant source. Before this guiding principle was recognized, investigators had postulated, on the basis of their morphological studies, separate ties between different groups of Australian marsupials and various South American forms. Their reason for doing so was understandable. Because no evidence from the fossil record was available to indicate that the modern marsupials on the two continents had evolved separately, biologists mistook for evidence of close relationship the structural similarities which had arisen through adaptive convergence.

Australian mar-supials

The extant Australian marsupials (Fig. 9.18) can be classified in one of three groups. The carnivorous forms, placed in the superfamily Dasyuroidea, are with few exceptions small species ranging in length from a few inches to a foot or two. Although some of the extinct dasyuroids were much bigger, the largest now remaining, *Thylacinus*, the Tasmanian wolf, is about the size of a collie dog. The aberrant mole-like *Notoryctes* and the anteater *Myrmecobius* are usually included with the dasyuroids, being considered peculiar offshoots from the ancient stock. The living herbivores, generally a more specialized assemblage than the carnivores, have been bracketed in the superfamily Phalangeroidea. The most diverse group in this category consists of the phalangerids, animals of arboreal habits termed possums, cuscuses, gliders, and koalas. To this superfamily belong also the burrowing wombats and the saltatory wallabies and kangaroos. The Phalangeroidea once included rhinoceros-sized animals, but these beasts became extinct in the Pleistocene epoch less than 100,000 years ago. Despite their variety of form, the phalangeroids agree in peculiarities of tooth and foot design. They have in the upper and lower jaws from one to three pairs of incisors, among them a pair enlarged in a somewhat rodent-like manner. This diprotodont condition, as the arrangement is called, distinguishes the phalangeroids from other Australian marsupials which exhibit three, four, or five pairs of uniform incisors and are described as being polyprotodont. The hind foot of the phalangeroids is characterized by the reduction and partial fusion of the second and third toes. The two syndactylous digits balance the fifth toe, and the fourth, extended between them, carries most of the animal's weight. Small, omnivorous forms, known as bandicoots, are syndactylous like the phalangeroids, but not being diprotodont as well they are housed in their own superfamily, the Perameloidea. Since the bandicoots still have the primitive number of incisors, it was thought that they might represent the ancestral group from which the more

Figure 9.18. Recent marsupials. **(a–d)** Polyprotodonts; **(a)** *Didelphys*, the American opossum; **(b)** *Dasyurus*, a small Australian carnivore; **(c)** *Thylacinus*, the Tasmanian "wolf"; **(d)** *Myrmecobius*, an anteater. **(e)** A peramelid: *Peragale*, a bandicoot. **(f, g)** Diprotodonts; *Phascolarctus*, the koala; **(g)** *Petrogale*, a wallaby.

specialized diprotodont, syndactylous phalangeroids sprang. Serological tests indicate that the bandicoots are more closely related to the dasyuroid carnivores than to the phalangeroids, however, so the problem of phalangeroid origins is far from solved.

Relationship between Australian and American marsupials

Some investigators, noting the degree of diprotodonty in South American marsupials called caenolestids, postulated that the phalangeroids might be derived from that group. Although there is no fossil evidence to support the theory, they thought that caenolestid animals, which were already present in South America in the Paleocene epoch, might have migrated around the periphery of the Pacific Ocean, finally entering Australia and giving rise to the fully diprotodont herbivores. Paleontologists who favored uniting the caenolestids and phalangeroids as a natural group usually assumed that the same kind of relationship obtained between the South American polyprotodont carnivores, the borhyaenids, and the Australian dasyuroids. In their opinion, the diprotodonts and the polyprotodonts diverged originally from primitive didelphoid forms, somewhat like the opossum native to South America. This theory has fewer advocates than it once had, not only because repetition of the long trek around the Pacific to Australia seems improbable, but also because recent research on marsupial anatomy and physiology has produced evidence against it. Ride has pointed out that the enlarged lower incisors of the caenolestids and the phalangeroids are not homologous but convergent, since those of the former develop from teeth adjacent to the midline of the jaw and those of the latter from teeth more laterally placed. Blood tests performed upon the South American didelphoids and upon the phalangeroids and other Australian forms confirm Ride's conclusion that the metatherians of the two continents represent two quite separate radiations.

Apparently, the common ancestor of the South American and Australian assemblages existed among the primitive marsupials that lived in North America during Cretaceous time. From the group that spread northward and rounded the Pacific Ocean, animals perhaps already verging on the dasyuroid condition entered Australia. In the course of their diversification, they produced, beside descendants with ordinary feet, a population with partially fused second and third toes. The syndactylous branch soon divided in two. One lineage, the perameloids, evolved conservatively, but the other, the phalangeroids, developed diprotodont incisors and diverged far from the primitive dasyuroids as they became herbivores adapted variously for open country and forest life. The grazing kangaroos and large wallabies, in many ways the most advanced of the phalangeroids, evolved specializations that closely paralleled those of the modern grass-eating placentals. They lost the canine teeth and developed high-crowned molars well suited for cutting and crushing coarse grass. Like ruminants, they became able to regurgitate their food and swallow it again, thus assuring breakdown and bacterial digestion of the plant cell walls. Their body form evolved quite differently from that of the larger placental herbivores, however, since they acquired speed by developing the ability to jump rather than run.

Bats of the Australian region

The placental mammals that finally did reach Australia during the Tertiary period presented no serious challenge to the reigning marsupials. The bats came first, sweeping eastward out of Asia soon after their origin in the Eocene. Since ribbons of sea water were no barrier to them, they migrated in continuous waves along the island chain. From the distribution of the living forms, investigators conclude that later immigrants replaced some of the earlier ones on small islands but that all of them found room to coexist on New Guinea. There they evolved new genera and species, and these, as well as the older types, moved southward to colonize Australia and eastward to solitary islands in the South Pacific inaccessible to other terrestrial mammals. The uniformity of the climate along the entrance route into the Australian region facilitated the spread of the bats, and their relatively free movement inhibited the development of a fauna completely distinct from that on the Asian mainland. The seven families represented in Australia and its environs all originated in Eurasia, and more than two-thirds of the existing genera live on both sides of Wallace's line, an imaginary boundary extending between Borneo and Celebes that separates the faunal assemblages of the Australian and Oriental regions. Without disturbing the marsupials, the immigrant bats sought and claimed a wide variety of food materials. Large forms, known colloquially as "flying foxes," feasted on tropical fruit and flowers, while other sizable species survived on the blooms alone. Most of the small bats took insects on the wing, but *Macroderma*, one of the few types unique to Australia, became a carnivore with a taste for its own kind.

Australian rodents

The rodents, which passed over the sweepstakes route long after the entry of the bats began, surely depended upon the same sort of chance transportation as the marsupials before them. From the extent to which the rodents which reached Australia have radiated it is apparent that their acclimatization after reaching that continent was rapid. Paleontologists agree that the rodents invaded the Australian region in several waves, but they disagree concerning the time of arrival of the first immigrants. Simpson places the initial entry of rodents into the area in the Miocene. Many of the animals, which he thinks stemmed from the Muridae, or Old World rats, migrated only as far as New Guinea. There they produced a peculiar assemblage that included one subfamily with aquatic members. Of the forms that did spread from New Guinea to Australia, one is presumed to have given rise to the subfamily Pseudomyinae, the group to which a major portion of the rodents on the southern continent belongs. The broad adaptive radiation of the pseudomyines, providing species that filled niches occupied in other regions by rats, mice, jerboas, squirrels, and rabbits, was what convinced Simpson that many millions of years must have elapsed since the ancestral murids crossed Wallace's line. Keast has objected to the Miocene date of entry that Simpson suggests, pointing out that available evidence does not support it. The radiation of the Australian rodents was accomplished without the kind of profound structural changes that would have required a long period of time, Keast argues, and there are no rodents known in the fossil record of the region until the Middle Pliocene. The Middle Pliocene remains come from New Guinea and consist of a single tooth. A flourishing rodent fauna does not appear until the Pleistocene.

Sometime after the arrival of the ancestors of the pseudomyines, Keast agrees with Simpson, rodents of the genus *Rattus* or closely related forms made their appearance in New Guinea. No fossils of these invaders have been found, but Simpson infers the presence of such animals from the existence on the island today of four genera of rats that almost surely trace their ancestry to *Rattus* or a rodent very much like it. Because the modern forms are distinct from their forebear only at the generic level rather than at the subfamilial level like the Pseudomyinae, Simpson has supposed that their ancestor entered the Australian region later than the rodents that gave rise to the Pseudomyinae and the more archaic New Guinean genera.

A fresh (or perhaps a continued) invasion of *Rattus* occurred after the first members of the group had secured a place. Enough time has elapsed since the entry of these animals into New Guinea and Australia for native species to have emerged on both islands. They coexist with a more recent arrival, *Rattus exulans,* which seems to have evolved somewhere in the South Pacific area and to have spread from place to place on native boats. In the last few hundred years, European ships have brought into the Australian region species of rats and mice that are now worldwide. *Rattus rattus, Rattus norvegicus,* and *Mus musculus* have crept off sailing ships and ocean liners and spread out of the port cities into the countryside in the East Indies and Australia, as they have everywhere.

Competition between marsupial and placental mammals

Simpson mentions the possibility that this army of rodents did displace marsupials that occupied the same niche, but fossil evidence is not available to prove his hypothesis. Interaction between immigrant placental and resident marsupial animals is indicated by such evidence only in the case of the dingo and the thylacine wolf. The dingo is a variety of domestic dog that arrived with an aboriginal people and then escaped to the wild. Since the earliest remains of the dingo in Australia are approximately contemporaneous with the last traces of the marsupial wolf on the continent, it does seem that the arrival of the placental carnivore may have brought about the downfall of the native form. The persistence of the thylacine wolf on the island of Tasmania, an area never reached by the dingo, reinforces this hypothesis. The disappearance in the Pleistocene epoch of some of the larger marsupials, however, cannot be attributed to competition from placental tetrapods. Since all the animals known to have become extinct at this late date were herbivores, it is possible that a change in the climate and a consequent reduction of the vegetation on which they depended was responsible for the loss. It is also quite likely that man hastened the disappearance of the larger forms in Australia, as he did elsewhere. The niches these animals occupied were not immediately filled. The room still available for herbivores is evident from the rapidity with which the rabbits multiplied when they were introduced into Australia by settlers from Europe.

The marsupials that migrated directly southward from their place of origin into South America did not enjoy isolation from placental mammals to the extent that the Australian marsupials did. The earliest metatherian immigrants were accompanied over the sweepstakes route between North and South America in Cretaceous time by at least two groups of the more progressive mammals, condylarths and ancestral edentates. In Pliocene time, when the present isthmus appeared, a wave of placental immigrants far larger than any that reached Australia broke in upon the continent to the south. As a result of the more intimate connections of the continents in the Western Hemisphere, the fauna of South America was never as impoverished as that of Australia. The marsupials were forced from the beginning to share the territory and then to give ground more extensively than they did in the Australian region.

Mesozoic reptiles and mammal-like forms in South America

The nature of the terrestrial vertebrates that inhabited South America when the northern mammals arrived is not demonstrable. With the exception of some deposits of the Jurassic and late Cretaceous age, there are no Mesozoic deposits containing remains of land animals younger than the Triassic. As in South Africa, the South American Triassic strata record the fading away of the therapsid reptiles and the rise of the archosaurian thecodonts and dinosaurs. There is as yet no evidence to suggest that the late Triassic fauna included, besides large archosaurs, small eozostrodont mammals like those that Crompton described from Lesotho. Except for one possible fossil condylarth and a few marsupial teeth, mammals are also unrecorded from the late Cretaceous deposits which contain reptiles similar in general to those living elsewhere in the world at the same time. The conclusion that no mammals existed in South America before the marsupials, condylarths, and edentates came, resting as it does on negative evidence, may yet prove erroneous. Primitive mammals had spread widely in other parts of the world before the end of the Triassic period, and (particularly if the South American continent was in contact with Africa at that time, as many scientists now believe) there is no reason why creatures near the eozostrodont level should not be found there. There is already evidence on hand to indicate that the Middle Triassic fauna included forms very near the reptilian-mammalian boundary. Since the mid-1940s collectors have recovered remains of carnivorous cynodonts, called chiniquodonts, that are even closer in structure to the mammals than Lower Triassic galesaurids like *Thrinaxodon.* One of their number, *Probainognathus* (Fig. 9.19), exhibits, besides a skull quite mammal-like in the facial, palatal, and occipital regions, an articulation between the dentary and squamosal bones which reinforces the reptilian articular-quadrate joint. The quadrate seems to have been rather loosely attached to the squamosal element, an arrangement which logically might have preceded its dissociation from the jaw and its conversion into an ossicle of the middle ear in mammals. Hopson and Crompton regard the

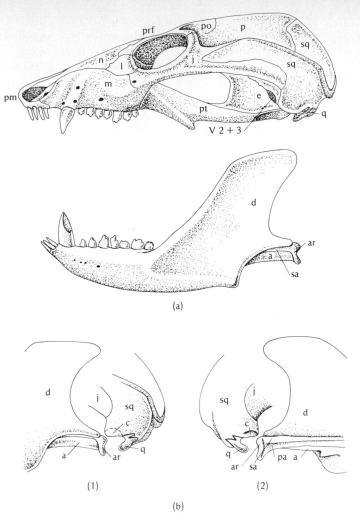

Figure 9.19. *Probainognathus*, a chiniquodontid mammal-like reptile from the Middle Triassic of South America. **(a)** Lateral view of skull and lower jaw. **(b)** Jaw articulation in detail: (1) lateral view; (2) medial view. Abbreviations as in Fig. 6.7. (From Romer.)

chiniquodonts as a side branch of the galesaurids, but Romer, who has described several of the animals, finds no reason to exclude them from the direct ancestry of the class Mammalia. He suggests that the chiniquodonts may eventually be found outside of South America because no evidence points to that continent as the birthplace of the mammals. It is possible, however, that during the 100 million years for which there is no fossil record, some sort of mammalian animals might have appeared there and undergone at least a modest radiation.

Cenozoic mammals of South America

Since the first known terrestrial fauna of the Cenozoic era in South America dates from the late Paleocene, the mammals it contained belonged to groups that had long ago replaced the animals living on the continent when the metatherians and placentals entered it. Diversification from the immigrant stock had already produced the beginnings of a distinctive assemblage of herbivores, insectivores, and carnivores. The early condylarths, more specialized than the other invaders for feeding upon vegetation, virtually preempted the first role. At Rio Chico in Patagonia and Itaborai in Brazil, the Paleocene localities, they are still in evidence, but by that time, it seems, they had given off divergent groups, which, by the Eocene epoch, became separate at the ordinal level from the older ungulates.

Litopterns

Fossil evidence for the transition from the condylarth condition is clearest in the case of the litopterns (Fig. 9.20). These hoofed animals, some of which came to resemble horses closely in becoming adapted for running, are traceable to Eocene forms similar enough to the archaic condylarths to be classified as members of that group. The litopterns achieved their greatest variety in the Miocene. Three families existed then, at least two of which contained animals with feet narrowed through the loss of the first and fifth toes. Although the litopterns projected their weight through an enlarged third digit, most of them retained the flanking second and fourth toes, as the early horses did. One form, *Thoatherium*, had evolved a single-toed foot by Miocene time (long before such a structure appeared in the equid line) but failed to survive despite this innovation. Its early extinction was due, perhaps, to its retention of primitive, low-crowned teeth. The litoptern that persisted longest, the three-toed animal called *Macrauchenia*, had developed wear-resistant, high-crowned molars similar to those that characterize successful grazers in other groups. The survival of *Macrauchenia* into the Pleistocene epoch may have depended also upon its occupation of a peculiar ecological niche. The animal seems to have been less fleet than its Miocene ancestors and, with its nostrils located far posteriorly above the orbits, may have been accustomed to feeding, snout under water, upon partially submerged marsh grasses.

Notoungulates

Throughout the Tertiary period, hoofed forms, termed notoungulates, were far more numerous and various than the litopterns. Although investigators suspect that these animals

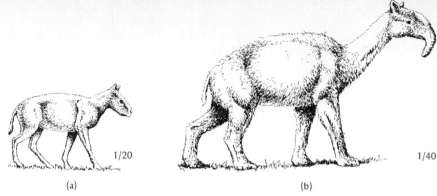

1/20

1/40

(a)

(b)

Figure 9.20. Restorations of litopterns. **(a)** *Thoatherium*, a single-toed form from the Miocene. **(b)** *Macrauchenia*. The last and largest of the litopterns, this animal flourished during the Pleistocene. (From Patterson and Pascual.)

were, like the litopterns, descendants of the condylarths, they are not sure that they arose in South America. Primitive notoungulates appear there in Paleocene time, but early members of the group are known also from contemporary deposits in North America and Asia. Whether the animals originated in Asia, tropical North America, or South America is impossible to ascertain. In the Northern Hemisphere, the notoungulates evidently failed in competition with the ancestors of the modern hoofed herbivores. In South America, where there were no members of the modern ungulate orders until the end of the Tertiary period, the notoungulates flourished and diversified. A number of the notoungulates remained small and evolved a rodent-like dentition, but the rest of them became large grazing or root-eating animals. For reasons not well understood, the notoungulates dwindled after the Miocene epoch. By the time the land bridge between the two Americas was reestablished late in the Pliocene, their number was much reduced. Some genera survived into the Pleistocene, among them the heavy-headed and ponderous *Toxodon* (Fig. 9.21). That animal was quite common, despite the presence in its territory of modern herbivores and carnivores from the north, and its large size assured its frequent preservation. Its remains were made known in 1833, the year after the young Charles Darwin reached South America in the H.M.S. *Beagle*. He bought a *Toxodon* skull which had been dug out of an earth bank near Montevideo for the equivalent of 18 English pence. Owen later described the specimen as belonging to a group intermediate between the rodents and the elephants because the dentition showed a rodent-like gap behind the incisors and the bulk of the skull suggested advance toward the graviportal pachyderms. Modern paleontologists have, of course, rejected this view.

Although the notoungulates and the litopterns made up a large part of the herbivore assemblage, there were, especially early in the Tertiary period, a variety of other plant-eating forms. The ungulates among them tended toward gigantism like the uintatheres and pantodonts that evolved at the same time in the north. The ancestry of these big hoofed animals is not apparent. One of them, *Carodnia*, was found fully differentiated in the late Paleocene deposits, and others, called astrapotheres and pyrotheres (Fig. 9.22), were already highly specialized when they prospered briefly in the Oligocene and Miocene. The rare Eocene specimens that can be related to the pyrotheres show a superficial resemblance to the elephants, because the animals' robust, elongated incisors and molar cheek teeth seem similar to those of the early proboscideans. In addition, the posterior position of the nostrils suggests that the pyrotheres may have had a trunk-like snout. Since the elephants are known to have arisen in Africa and the pyrotheres in South America at a time when there was no connection between the two continents, a common ancestry for the two groups is considered by most paleontologists to be unlikely. The astrapotheres are too singular in structure to be linked with any other kind of ungulate. The front of the upper jaw in these animals was vestigial and lacking incisor teeth. It bore a pair of huge, ever-growing

Astrapotheres and pyrotheres

Figure 9.21. A restoration of *Toxodon*, a huge notoungulate that survived into the Pleistocene. (From Patterson and Pascual.)

1/40

Figure 9.22. Restorations of **(a)** *Pyrotherium*, an Oligocene pyrothere, and **(b)** *Astrapotherium*, a Miocene astrapothere. (From Patterson and Pascual.)

canines which opposed a toothless area in the long mandible below. The lower incisors, which jut out beyond the end of the skull, must in the living animal have closed against toughened soft tissue as they do in the cow. The source of these peculiar characteristics is unimaginable, so that the astrapotheres, like most of the large and specialized Paleocene and Eocene ungulates, can be linked with other hoofed forms only indirectly through the ancient condylarths.

Edentates   The array of mammalian herbivores in South America included, besides ungulates, animals assigned to the order Edentata. The plant-eating members of this group, the sloths, do not appear in the fossil record until the Oligocene epoch, but they surely existed long before that time. It has been supposed that they evolved from the primitive armadillos which are known to have been present in South America in the Paleocene and that those forms stemmed from degenerate-toothed immigrants from the north called palaeanodonts. Those who object to deriving the soft-skinned, hairy sloths from the already armored Paleocene species propose instead that the sloths might have come from older animals in the edentate line, forms that were still arboreal or at least semiarboreal. The residence of such ancestors in a forest environment would explain the lack of fossils from the early part of the Tertiary period. The sloths that make their appearance in the Oligocene and Miocene epochs had become terrestrial in habit and so presumably wandered into areas where preservation of their bones was more likely to take place. An arboreal origin for the sloths is suggested further by the lightly built frame and clawed feet of some of the first known members of the group (Fig. 9.23). Later sloths exhibited a tendency to grow extremely large. In each of the three families that emanated from the early stock, giants 10 to 20 feet long evolved. The heavy limb bones of these animals and their habit of walking on the outer edge of the feet and the knuckles of the hands implies that they were awkward and slow-moving. They proved viable, nevertheless, invading North America when the land bridge reemerged in late Pliocene times and surviving there, in South America, and in the West Indies until the Pleistocene was far advanced. The only sloths alive today are smaller South American forms arboreal in habit. A group separate from the extinct ground sloths, they hang quiescent from the branches by day and at night move slowly through the trees browsing on the leaves. Since these tree sloths have no fossil history, their exact relationship to the old ground sloths is unknown. While some investigators declare on the basis of structural similarities that the tree sloths and certain of the terrestrial forms are phylogenetically close, others maintain that the modern arboreal animals are less likely to have been derived from terrestrial ancestors than to have descended from an archaic lineage whose members never left the trees.

Edentates other than sloths were at first eaters of invertebrates or scavengers. The oldest among them, the armadillos, plodded about during Paleocene time as they do now, their body safe from attack under a series of bony plates. In the Eocene, they gave off a side branch, a group of animals called glyptodonts that seem to have been adapted for grazing. The glyptodonts developed even heavier armor than the armadillos. The plates over the back fused into a solid carapace, and those encasing the tail in some genera became adorned with spikes. In the post-Eocene years, armadillos grew large, and glyptodonts grew larger. By the Pleistocene epoch, when both kinds of animals had spread northward into the southern part of North America, there were forms in each group as large as the great ungulate beasts for which the time is famous. Although the giant armored edentates seem to have had few enemies, they disappeared like the ground sloths and the outsized animals of other orders at the close of the Pleistocene. The largest representative of the group now extant, the "giant" armadillo of Brazil, is less than 5 feet long, and the common nine-banded armadillo, *Dasypus novemcinctus* (Fig. 9.24c), is much smaller.

Besides the armadillos, one other tribe of invertebrate-eating edentates survives today in South America, the anteaters (Fig. 9.24a) assigned to the infraorder Vermilingua. Because these animals are soft-skinned forest forms, their fossil record is

scanty and dates only to the Miocene. The only truly toothless edentates, they tear apart rotting logs with their claws or climb through the tree-branches in search of termites and ants, which they lick up with a long, thin, sticky tongue. Zoologists were once inclined to lump with the South American anteaters, the scaly anteater, or pangolin, of the Old World tropics and the aardvark of Africa, forms similarly specialized for the same diet. Since neither the pangolin nor the aardvark shows the peculiar structural traits that link the South American anteaters, the armored edentates, and the sloths, however, they have been housed in orders of their own. The South American edentates, although they appear diverse, share such unusual characteristics that their common ancestry is not in doubt. All of them possess an auxiliary pair of articulating processes below the normal zygapophyses through which the neural arches of the vertebrae join one another, and in the sacral region of the column the ischiac bones abut the vertebrae behind the normal sacroiliac joint. Their teeth, if not missing entirely, are much reduced.

Figure 9.23. Restorations of edentates which survived into late Pleistocene time. Left: Three giant ground sloths of the genus *Megatherium*. Right: Two glyptodonts. (The Field Museum of Natural History, Chicago; from a mural by C. R. Knight.)

The incisors are absent, and the cheek teeth are generally undifferentiated, rootless pegs of enamel.

Early carnivores in South America

The primitive, opossum-like marsupials that entered South America with the first wave of immigrants were apparently prevented by the rapid radiation of the placental ungulates from expanding into the niches available for herbivores. The Paleocene forms kept to the trees, as many of the edentates seem to have done, and continued their insectivorous or omnivorous habits. One group, the caenolestoids, evolved by Eocene time a dentition with projecting incisors and rectangular molars adapted for cutting and crushing, and some genera even produced fan-shaped premolars like those of the herbivorous multituberculates; but the marsupial assemblage as a whole seems to have provided the predaceous members of the South American

(a)

(c)

(b)

Figure 9.24. Recent edentates. **(a)** *Tamandua*, the anteater. **(b)** *Choloepus*, the two-toed tree sloth. **(c)** *Dasypus*, the armadillo.

fauna (Fig. 9.25). From the sharp-toothed little didelphoid creatures that reached the southern continent before it was cut off from North America at the end of the Mesozoic era came the borhyaenids, terrestrial animals that pursued the condylarths and their kin. Although they paralleled the northern placental carnivores in evolving shearing teeth and clawed feet moderately well adapted for running, the borhyaenids had passed their peak by the Pliocene epoch. Patterson and Pascual have suggested that the predatory marsupials may have suffered in competition with the large, cursorial, carnivorous birds, like *Phororhacos*, which were common in South America in the Miocene and Pliocene. Certainly reduced in number, the borhyaenids disappeared completely when the carnivores from the north invaded their territory at the close of the Tertiary period.

Mid-Tertiary invaders of South America

Before the reestablishment of the land bridge that allowed the entry of the placental carnivores, a few small, arboreal animals reached South America by island-hopping, and some of them gave rise to groups still important in the fauna there today. The most recent of these immigrants were raccoon-like forms, which, if they are not ancestral to the modern coatimundi and kinkajou, as some workers have supposed, apparently disappeared without issue. The date of their entry is not certain, but it seems to have taken place as late as Miocene or Pliocene time. Earlier, probably during the late Eocene years, the progenitors of the New World monkeys and the rodents arrived. In the vast tropical forests, these animals prospered beside the arboreal edentates and marsupials and spread rapidly. The rodents may have displaced some caenolestoids that had evolved teeth for gnawing and munching, but the monkeys seem not to have caused the extinction of any group.

Because the monkeys remained arboreal inhabitants of the warm, dank forests, their fossil record is meager. Only a few genera have been found in Tertiary deposits from Oligocene time onward. The animals represented had realized the potential (already partially expressed in northern lemuroid and tarsioid primates of the Lower Eocene) for development of the visual sense at the expense of the olfactory powers. The eyes had enlarged and rotated forward, thus developing overlapping fields of vision and endowing the animal with improved depth perception. This faculty was undoubtedly selected for as an adaptation to arboreal life, since better stereoscopic vision would have permitted the early monkeys to gauge distances more accurately as they moved from branch to branch. As a secondary effect, more acute sight may have fostered use of the hands for more intricate tasks. The growth of the brain as interneuronal connections evolved between the visual centers and other areas forced the expansion of the braincase into a hemispherical shape. Anterior to it, the facial region became shorter as the olfactory epithelium and the snout declined in importance. The skull of the monkeys was further distinguished by the walling off of the orbit from the space for the temporalis muscle by an extension of the alisphenoid bone and the formation from the frontal and zygomatic elements of a postorbital bar.

Figure 9.25. Restorations of two South American marsupial carnivores. **(a)** *Borhyaena,* a Miocene form. **(b)** *Thylacosmilus,* a Pliocene animal which paralleled the placental saber-toothed cats in evolving large upper canine teeth. (From Patterson and Pascual.)

(a)   (b)

With large eyes and large brains, the South American monkeys became active and social animals that rarely left the treetops. The little marmosets ran about, gripping the branches with claw-like nails, in search of insects and fruit, while the cebids, bigger but still lightly built forms, picked leaves and fruit and sometimes supplemented their diet by taking eggs or small birds from nests. As they grew increasingly specialized for arboreal life, the cebids evolved slim limbs and elongated hands and feet with digits that could be wrapped around the boughs. Although the innermost toe remained opposable to the others, as it had been in the earlier primates, the thumb showed a tendency toward reduction in some genera. In the living spider monkey, the hand has only four long fingers, which are held bent and used to hook the branches as the animal swings rapidly from place to place. These monkeys, the capuchins, the woolly monkeys, and the howlers have developed a prehensile tail, a structure which gives them an extra handhold when sitting or moving through the forest canopy.

Relationship between New and Old World monkeys

The resemblance of the New World monkeys to those of Africa and Eurasia caused paleontologists to assume for a long time that one group might have been derived directly from the other or that both might have come from a common stock of primitive monkeys. These hypotheses are now controverted by evidence from every quarter. Since monkeys appear in South America and Africa in the Oligocene, it is thought that these animals must have evolved during the Eocene epoch, a time at which scientists are now convinced the two southern continents had no link with one another. Some fragments of ape-like primates have been found in late Eocene strata in Burma, but there is no evidence that any anthropoid form existed farther north or spread eastward or westward through land bridges in nontropical regions. Comparative anatomical studies have shown that although the South American monkeys are more primitive in having kept the third premolars lost by other higher primates, they are too specialized in many ways to have fathered the Old World forms. Similarly, it is hardly likely that the Old World monkeys could have regained the third premolars and other primitive characters to give rise to the cebids and the marmosets.

The fossil record suggests an alternative theory and one that is now generally supported. The Eocene primates that did pass east and west from continent to continent in the Northern Hemisphere, leaving their remains in North America and

Eurasia, were not monkeys but lower forms classified variously as lemurs or tarsioids. It seems possible that it was these animals which filtered into South America and Africa and that from them the two groups of monkeys evolved independently and in parallel. The attainment by the New and Old World monkeys of an equivalent level of structure and function is explicable on the basis of their common genetic heritage from ancestors already recognizable as primates. This theory seems correct to many investigators, because there are among the northern Eocene forms several that qualify structurally as possible progenitors of the South American and African monkeys (Fig. 9.26). The animal first advanced as a possible common ancestor was *Notharctus*, a lemur known both from North America and Europe. The chief qualification of *Notharctus* was its generalized skeletal and dental structure. Unlike some of the other Eocene primates, it did not have enlarged, rodent-like incisors or elongated tarsal bones which would have eliminated it from the lineage of both groups of monkeys. It still retained a full complement of teeth, a long snout, and an unexpanded braincase. Although its structure was in these features quite primitive, the pattern of the molar cusps in *Notharctus* seemed proper for a form ancestral to monkeys, and the animal had already developed a postorbital bar. As study of the Eocene primates continued, it became apparent to Gazin, and later to Simons, that forms termed omomyids might be closer than *Notharctus* to the ancestry of the monkeys. None of the omomyids is known from remains as complete as those of *Notharctus*, but they seem to have been a large and diverse group nearly related to the primitive lemur. *Hemiacodon*, of which the facial part of the skull has been found, had orbits facing more anteriorly than those of *Notharctus*, and the dentition of other omomyids shows a trend toward the loss of premolar teeth, a trend which existed also in the evolution of the monkeys. Most omomyids still had three premolars, a characteristic which, Gazin observed, qualified them as forebears of the marmosets and cebids. Further reduction of the premolars to two would have produced the dental formula typical of the Old World monkeys. That certain omomyids did enter Africa and change in this way seemed possible to Simons because he found a resemblance between the omomyids and *Oligopithecus*, the earliest higher primate known from that continent. Investigators who are convinced that the Old and New World monkeys have descended separately from the omomyids or similar forms have ceased to bracket the two

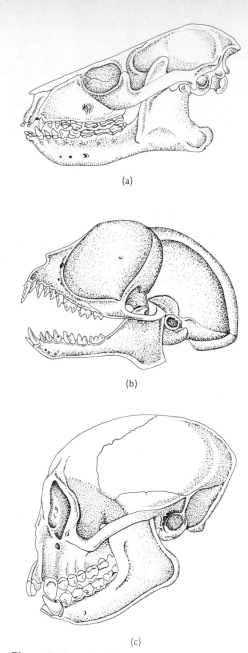

(a)

(b)

(c)

Figure 9.26. Skulls of lemuroid, tarsioid, and platyrrhine primates. **(a)** *Notharctus*, a North American lemur of the Eocene. (From Romer.) **(b)** *Tarsius*, the modern tarsier of the East Indies. (From Parker.) **(c)** *Cebus fatuellus*, a Recent platyrrhine monkey from South America. (After LeGros Clark.)

groups in the suborder Anthropoidea, as has long been done; instead, they have raised to subordinal status and listed *seriatim* the taxa Platyrrhini and Catarrhini, in which the South American and Old World higher primates respectively have been housed.

South American rodents

In contrast to the primates, the rodents, which seem also to have arrived in South America by island-hopping during the Eocene, radiated beyond the confines of the forest and diversified in a number of different niches. They encountered little opposition and proliferated throughout the remainder of the Tertiary period unchallenged by other animals with similar adaptive traits. The living descendants of the Eocene immigrants are ubiquitous on the continent and are widely variant in their habits (Fig. 9.27). The porcupine is still at least semiarboreal, but the chinchilla is found in barren, mountainous country, where it burrows or hides in holes among the rocks. The vizcacha, a larger relative of the chinchilla, lives in colonies on the grassy plains, burrowing like its cousin, the Patagonian cavy, a form better adapted for running. The 4-foot-long, 100-pound capybara, the biggest rodent extant, has become a swimmer and eats water plants as well as terrestrial vegetation. These animals and their many relatives, living and fossil, have been divided among four superfamilies, all of which are characterized by the same arrangement of the jaw musculature. Described as hystricomorphous, it is peculiar in that the deep part of the masseter muscle passes upward from its insertion on the mandible through a greatly enlarged infraorbital canal and obtains an origin on the face in front of the eye. The other parts of the masseter are unspecialized and originate as usual on the zygomatic arch.

Ancestry of the South American rodents

That the South American rodents with the masseter in this condition are derived from a common stock is almost certain. Although rodents from the early Oligocene, the oldest known on the continent, are recognizable as members of the four superfamilies, they are also very similar to one another. Their mutual resemblance has been interpreted as evidence of their having only recently begun to diverge from the same stem, presumably the group of rodents whose representatives succeeded in reaching South America by the sweepstakes route. Paleontologists have argued not about the existence of such a common ancestral form but about the kind of rodent it was. The problem faced by investigators of the question is very much like that which confronts workers studying the origin of the New

Chinchilla

Viscacha

Cavy

Capybara

Porcupine

Figure 9.27. Recent South American rodents.

Vertebrate History: Problems in Evolution

World monkeys. In the case of the South American rodents, also, there are animals so similar to them in Africa that the possibility of the origin of one group from the other or of both groups from a common ancestor at the same level of development had to be carefully examined. The African rodents, which include forms unfamiliar to dwellers in the Northern Hemisphere like mole and cane "rats," "flying squirrels," and gundis, are hystricomorphous and often are characterized by an outwardly flaring mandibular angle like that of the South American animals. Investigators were for a long time so sure that rodents sharing these peculiar traits must constitute a natural group that they lumped them in the suborder Hystricomorpha. Once it was established that Africa and South America had been widely separated from each other since pre-Eocene days, preservation of this suborder depended upon demonstrating the possible entrance into the two continents of related hystricomorphous rodents from the north. At first it seemed that paleontologists might be able to do so, since they had found in Europe hystricomorphous rodents of Eocene age. None of these forms, called theridomyids, were discovered in North America, however, territory through which they would have had to spread before island-hopping to the South American continent. Investigators such as Wood, Patterson, and Lavocat soon found what they considered positive evidence for disqualifying the theridomyids as ancestors of either group of southern hystricomorphs in the complicated pattern of crests on the molar teeth and in other characters. Landry and Schaub, on the other hand, still persuaded that theridomyids or unknown forms similar to them must have served as the common stock, theorized that such animals were absent from the Eocene strata of North America because they had migrated and flourished there earlier, in Paleocene time. Wood and Patterson objected to this hypothesis on several grounds. They pointed out that rodents of the theridomyid type seem not to have existed, even in Europe, until the Eocene; that if theridomyids had lived in North America in the Paleocene, it is unlikely that they would have been replaced there by the primitive paramyid rodents known to have inhabited the region in the succeeding epoch; and lastly, that the mutual resemblance of the South American Oligocene hystricomorphs made it improbable that their common ancestor arrived on the continent as far back as Paleocene time. Wood and Patterson accepted the possibility that the hystricomorphous condition, as peculiar as it is, could have evolved two or even more times independently. Whereas they did not speculate upon its origin in the African rodents (which, it seems, may not all have emanated from the same source), they did have an idea about its evolution in the South American animals. Through *Platypittamys* (Fig. 9.28), a South American Oligocene form in which the infraorbital canal was not large enough to allow passage of the deep masseter muscle, they tied the southern rodents to a group within the northern paramyid assemblage. Since, among the northern genera there was one, *Rapamys*, in which a tendency toward enlargement of the infraorbital foramen was expressed, Wood and Patterson supposed that similar animals must have migrated southward and evolved the hystricomorphous condition just before or after they reached South America. Because all the other South American Oligocene rodents were fully hystricomorphous, according to this theory, *Platypittamys* would represent a relict of the primitive island-hopping group. Convinced by the fossil evidence that the South American hystricomorphous rodents were a natural group not closely related to the African forms, Wood followed Simpson in abandoning the old suborder Hystricomorpha and substituting, for the South American hystricomorphs, the suborder Caviomorpha. Not being able to link

Figure 9.28. Skull of *Platypittamys*, an Oligocene rodent from Patagonia, showing an enlarged infraorbital foramen approaching in size that of the later South American caviomorphs. (After Wood.)

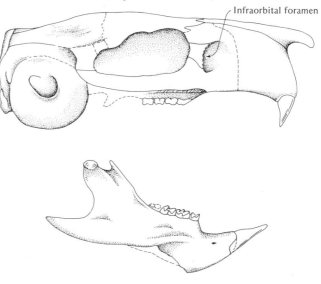

Infraorbital foramen

the African rodents to a common ancestor, he took the extraordinary step of listing them in his classification scheme in superfamilies and families without any subordinal assignment.

The caviomorph rodents were not obliged to share the South American continent with any other members of their order until the end of the Tertiary period, when the seawater barrier disappeared between Central America and the land to the south. Immediately, field mice and squirrels arrived, and sometime later the herbivorous rabbits (not rodents, but members of their own order Lagomorpha) invaded and, like the field mice, spread beyond the tropics to the temperate grasslands far below the equator. The entrance of many small gnawing animals coincided with the extinction of the giant-sized caviomorphs but apparently did not cause it. The medium-sized and small hystricomorphous rodents seem to have been little affected by the newcomers. The field mice, squirrels, and rabbits either shared the niches of the native rodents or slipped into unoccupied ones, and the caviomorphs continued to flourish as they had before.

Later invaders
from North
America

Despite the variety of the South American fauna, there were open places for many animals other than rodents and lagomorphs that came from the north. Because the native ungulates had been in a decline since early Pliocene time, the northern hoofed forms encountered little opposition. The last of the litopterns and the toxodontids faded away in the Pleistocene epoch, leaving the mountains, forests, and grasslands to the camelids, deer, peccaries, tapirs, and horses that made their way through the Panamanian isthmus. All the herbivores except the horses have survived in tropical or temperate South America to the present day. The only herbivorous immigrants besides the horses which ultimately proved unsuccessful on the southern continent were the mastodonts. The great tusked beasts entered with the other ungulates, enjoyed a brief radiation, and then vanished, as did most other large mammals, at the end of the Pleistocene. The herbivores and the omnivorous peccaries were accompanied in their southward movement by predatory species, all members of the order Carnivora, that quickly overcame the native marsupial flesh eaters. None of the borhyaenids was a match, apparently, for the cats, dogs, weasels, and bears that had emerged in the Northern Hemisphere through competition more intense than any experienced by the older, carnivorous mammals while their territory was cut off from other lands.

Differentiation of
the mammalian
faunas of North
and South America

The invasion of South America by mammals from the north did not result in the creation of similar faunas in the two continents of the New World. The immigrants in the south soon evolved new genera and species that set them apart from the animals which remained in North America, and each region was further distinguished by residents that did not spread across Panama. For reasons that are still not clear, the North American moles, pronghorns, bison, and mountain sheep did not reach the southern continent, and the shrews, although they colonized its northwestern part, never advanced further. In addition to the caviomorph rodents, advanced primates, and older inhabitants of South America that did not establish themselves in temperate North America, the tapirs and camelids gave a different character to the southern fauna after their northern cousins became severely limited in range or disappeared.

Although the Central American land bridge could, theoretically, have allowed as much migration in one direction as another, it did not prove as ready a passage into the neighboring continent for the southern animals as it did for the northern ones. In early Pleistocene times, toxodonts, opossums, several families of edentates, and a few caviomorph rodents moved entirely across it, but the bridge acted as a filter to stop the primates, anteaters, tree sloths, most of the caviomorphs, and even, in later years, descendants of the latest immigrants from spreading to the north. The explanation of the imbalance between the passage of animals north and south lies in the climatological history of Central and North America and in the relationship of the old marine barrier to the boundary between the tropic and temperate zones. Throughout the Tertiary period, when the earth's climate was warmer than it is now, the tropics spread far to the north and south of the straits that connected the Atlantic and Pacific oceans. To the north of the straits, not only was the Central American peninsula tropical, but the greater part of what is today the United States was either hot or at least mild and humid. For some 60 million years, families of mammals that had entered North America over the northern land bridge from Asia or evolved in the cooler regions of the continent produced genera which, little by little, had moved into the spacious warmer regions, acclimating themselves first to subtropical and then to tropical conditions. Toward the end of the Tertiary period, when the Panamanian land bridge was emerging, the environment grew generally drier and colder and the tropical portion of the North American continent shrank. The mammalian populations shifted southward, and in the Ice Ages which occurred periodically during the Pleistocene, forms unable to move into new territory or to adapt to the colder weather became extinct. The Panamanian isthmus allowed the tropical animals of northern ancestry that inhabited Central America to pour into the vast regions of South America which, because they straddled the equator, were never reached by the encroaching cold. At the same time many mammals still adapted to the northern temperate climate were able, by migrating southward through the Rocky Mountains, to circumvent the hot and steamy areas and to gain access to the cool slopes of the newly heightened Andes and the southern temperate zone beyond.

The South American mammals that spread onto the land bridge were almost exclusively tropical forms from the northwestern part of the continent. They colonized the warm, wet regions of Central America readily, but most of them were unable to advance into the highlands of northern Mexico and to the forests and grasslands at higher latitudes, where the climate in late Pliocene times was already becoming quite severe. Of all the South American forms that moved northward, only the opossum and the Canadian porcupine penetrated deep into the temperate part of North America and survived. The modern opossum ranges from the southern United States to the Canadian border, but the Canadian porcupines adapted so narrowly to the northern woods that they can no longer live in warmer regions. The other South American mammals that reached the temperate zone soon lost their toehold there. The water-loving capybaras were pushed back to the tropics and the armadillos to the southern fringe of the continent, where the climate is continuously mild. The giant ground sloths and glyptodonts might have had better success, for they survived the periodic intervals of cold associated with the advance of ice sheets in the far north, but these animals encountered inimical forces other than climatic ones and succumbed. It is still possible, of course, that migration of South American mammals into the northern continent will increase as sufficient time elapses for the tropical forms to produce descendants adapted for a colder environment. Investigators who emphasize this point note that the northern mammals had the entire Tertiary period to acclimate themselves to tropical conditions before the land bridge appeared, allowing them passage from Central America to the larger region like it to the south. In contrast, the South American mammals came in contact with subtropical and temperate climates little more than 3 million years ago. Whether the magnitude of movement to the north would ever equal that to the south is questionable, however, because it appears historically that animals enter the tropics with greater facility than they leave.

Difficulty of tracing spread of early Tertiary mammals

The general tendency of mammals to migrate southward was argued strongly by W. D. Matthew in his now classic treatise on climate and evolution. In that work, he maintained that new groups of mammals arose, almost without exception, somewhere in the vast expanse of northern Eurasia and lived there until forced to spread toward warmer regions by the appearance of more progressive competitors. Although few would hold now, as Matthew did in 1915, that tropical mammals are in virtually every case descendants of primitive forms driven from their place of origin by more advanced members of their line, most paleontologists agree with his contention that northern Eurasia was the area in which many of the orders of mammals first emerged. Support for this hypothesis is based in good part on inferences from the distribution of Tertiary mammals, because direct fossil evidence for the evolution of the earliest members in each order is extremely scarce. The record of mammalian history in Asia is especially poor. What is known of the fauna on that continent between Paleocene and late Miocene time has been gleaned from material found at less than a dozen localities in Mongolia, a few in China, and a single Eocene deposit in Burma. A continuous series of deposits dating from the end of the Miocene epoch to the Pleistocene exists in the Siwalik Hills of India, but the mammals that lived elsewhere in Asia at that time are known only from a few sites in China, a Pliocene deposit in Iran, and Pleistocene beds in Burma. The conclusion that many mammals spread outward from the Asian land mass rests upon the nearly simultaneous appearance of numerous closely related forms in post-Paleocene Europe and North America and upon the discovery that the fauna in each of those regions often has more in common with the fauna of Asia than with that of the other. The migration of specific groups across the continents of the Northern Hemisphere is not easy to trace, however. The broad corridor between Asia and Europe and the frequent availability of land routes between the Old World and the New has allowed such a complicated interchange of animals that the place of origin of each one can only be estimated approximately.

Study of the evolution and spread of the mammals native to the Northern Hemisphere is complicated further by the rapidity with which the animals of a new order undergo diversification. When a highly advantageous character, or complex of characters, appears within a population, its members may quickly enter niches from which they were barred before and develop, besides their distinctive ordinal traits, a variety of adaptive ones. The paleontologist is faced then with a multiplicity of forms, in part primitive and in part specialized, of whose exact interrelationships he cannot be sure. Since the animals have evolved only a short time earlier from a common stem, they are all very much alike, but their unique adaptive characters often require their classification in separate families. When the evolution of a group follows this pattern, the source and early dissemination of the animals may be difficult or impossible to trace. Only later, when many of the older forms have become extinct and the members of the surviving families have diverged farther in structure from each other, does it seem easier to discriminate among the different kinds of animals and to chart their movements.

Evolution of the insectivores

The early history of the modern insectivores is unclear because their ancestors did experience an explosive radiation of this type. As mentioned earlier, investigators have rather good evidence that the progenitors of the order Insectivora appeared first in Mongolia or possibly on the other side of the Bering bridge in western North America, but the proliferation of families within the group at the beginning of the Tertiary period resulted in a fossil record from which it has proved difficult to derive an evolutionary scheme. By the Oligocene epoch, there were about a dozen families of animals that had retained the insectivorous habits of the primitive eutherians and become specialized at that level. Of these, only three, or possible four, can be linked to the insectivores of the present day (Fig. 9.29). The adapisoricids, probably direct descendants of the early leptictids, spread throughout the Northern Hemisphere and apparently gave rise to the hedgehogs, shrews, and moles. The hedgehogs, more primitive than the other two, died out in North America during the Pliocene but survived in Eurasia and eventually extended their range southward into Africa. On the basis of tooth morphology alone, the early Tertiary plagiomenids of North America have been regarded as possible early allies of a curious animal found today in the East Indies called *Galeopithecus* or the "flying lemur." This squirrel-sized, gliding form, which seems to have no living relatives and is not a primate as its name suggests, could conceivably be derived from ancient insectivore stock, but the long gap in the record between

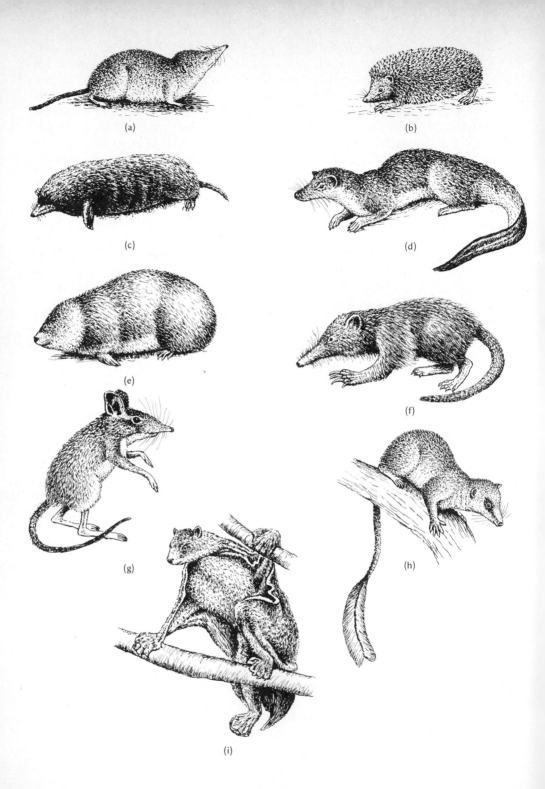

(a)

(b)

(c)

(d)

(e)

(f)

(g)

(h)

(i)

the plagiomenids and the modern animal makes a connection between them highly speculative. Investigators who prefer not to found a liaison on such tenuous evidence acknowledge the isolation of *Galeopithecus* by excluding it from the Insectivora and housing it alone in the order Dermoptera.

The only other family of early Tertiary insectivores that may be antecedent to living forms is the North American Apternodontidae. *Apternodus*, which, unlike most of the extinct insectivores, is known from the skull as well as the teeth, seems rather too specialized to have given rise to any of the extant animals, but the molars in this form and its relatives resemble those of a heterogeneous group of modern insectivores called zalambdodonts (Fig. 9.30). Whereas the upper cheek teeth in most insectivores are characterized by laterally placed paracones and metacones set wide apart, as they are in the leptictids, those teeth in *Apternodus* and the living zalambdodonts are more sharply wedge-shaped with the outer cusps close together or fused in a more central position. Although it may be possible that apternodontids were near the ancestry of the New World insectivores with zalambdodont molars, namely, *Solenodon* and the recently extinguished *Nesophontes* of the West Indies, there is no hint in the fossil record of the spread of apternodontids or their descendants to the African region to which the Old World zalambdodonts are now confined. Inspection of the African zalambdodonts themselves does not provide evidence pertinent to their ancestry. From their first appearance in the Miocene, the golden moles of South Africa have been so extremely specialized for burrowing that the more primitive state from which they evolved is completely masked. The tenrecs of Madagascar, long-snouted and tailless nocturnal animals, show similarities in structure and behavior to the most generalized northern insectivores, as does the otter shrew of West Africa, a form peculiar in its habit of swimming and eating fish.

Despite the resemblance that does exist between the African zalambdodonts and the non-zalambdodont northern insectivores, Van Valen has suggested that the former are not descendants of the leptictids like the latter but offspring of the deltatherids, which had narrow V-shaped upper molars with conjoined paracones and metacones centrally placed. By offer-

Figure 9.29.   Recent insectivores. (a) *Sorex*, a shrew. (b) *Erinaceus*, a hedgehog. (From Young.) (c) *Scapanus*, a mole. (From Storer.) (d) *Potamogale*, the African otter shrew. (e) *Chrysochloris*, the golden mole of Africa. (f) *Solenodon*, a West Indian zalambdodont. (g) *Macroscelides*, the African elephant-shrew. (h) *Ptilocercus*, a tree shrew showing some primate-like characters. (i) *Galeopithecus*, the East Indian "flying lemur."

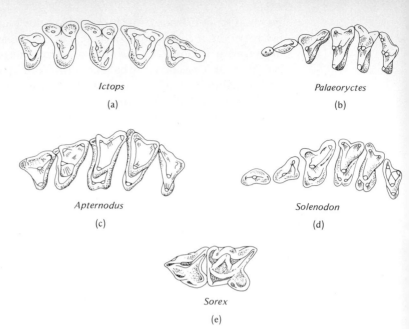

Figure 9.30. Upper teeth of fossil and Recent insectivores in occlusal view. **(a)** *Ictops*, a leptictid. **(b)** *Palaeoryctes*, a deltatherid. **(c)** *Apternodus*, an early Oligocene form with teeth similar to those of modern zalambdodonts. **(d)** *Solenodon*, a West Indian zalambdodont. **(e)** *Sorex*, the shrew, a modern non-zalambdodont insectivore. (*a, b, c, d* from Gregory; *e* from Repenning.)

ing a classification scheme in which the zalambdodonts were removed to the order Deltatheridia and so associated with the carnivorous creodonts, Van Valen split the group of animals that has formed the nucleus of the order Insectivora since it was erected by Bowdich in 1821. Van Valen recognized that his resolution of the problem of zalambdodont relationships was tentative and in many ways unsatisfactory. In addition to separating families of similar animals, exclusion of the zalambdodonts did not improve the cohesiveness of the insectivores as a group. The order still embraces, besides the hedgehogs, shrews, and moles, the living non-zalambdodont African elephant shrews of uncertain origin; the tupaiids and apatemyids, long regarded as primitive primates; the "flying lemur," *Galeopithecus*; and a host of diverse archaic genera, including, in the view of some paleontologists, the first eutherians from Forestburg. Subdividing this assemblage accurately into natural subordinal units cannot be accomplished until far more fossil evidence is at hand. Present understanding of insectivore relationships is perhaps best expressed by associating shrews, moles, and hedgehogs in one suborder, placing the zalambdodonts with them or in their own suborder, segregating the elephant shrews and *Galeopithecus* in separate taxa, and, finally, housing all the early Tertiary families that cannot be linked to modern ones in a suborder of their own.

Paleontologists have had equal difficulty in constructing a phylogenetically correct classification scheme for the rodents and the primates. These animals, like the insectivores, radiated rapidly in the second half of the Paleocene epoch and in the Lower Eocene. The scarcity of deposits of Paleocene age and the small size of the forms in question have restricted the amount of fossil material available, so that investigators find themselves unable to verify relationships among the several families that can be identified. Because the primates did not diversify to the extent the rodents did, it has proved somewhat easier to erect suborders that reflect at least the general pattern of divergence within the order. The primitive primates common by mid-Paleocene time, thought to be descendants of *Purgatorius* or animals like it, are still difficult to distinguish from early insectivores. They are classified as primates chiefly because their molar teeth have changed from the tribosphenic form to a more quadrate one and, in most genera, the incisors have been reduced from three pairs to two, the number typical for later members of the order. These early primates, of which three families are known, have been assigned by E. L. Simons to the same suborder, the Plesiadapoidea, because, like *Plesiadapis*, most of them evolved long, forward-jutting incisors that functioned like those of multituberculates and rodents in cutting plant food. The primates that enter the fossil record in the Lower Eocene, the adapids, the anaptomorphids, and the omomyids, showed no trace of rodent-like incisors. Since the adapids, like the modern lemurs and lorises (Fig. 9.31), preserved the long snout, laterally facing eyes, and unenlarged braincase that the primitive primates inherited from their insectivore ancestors, they have been bracketed with the modern animals in the suborder Lemuroidea. The anaptomorphids and omomyids, more progressive in brain and eye development and in the shortening of the snout, seem to have resembled the living tarsier and so have been assigned with that form to the suborder Tarsioidei.

The plesiadapoids had a short history. They became extinct

Diversification of the early primates

Figure 9.31. Recent lower primates. **(a)** *Lemur catta*, the ring-tailed lemur. **(b)** *Nycticebus cinereus*, the gray loris. **(c)** *Tarsius spectrum*, the tarsier.

without issue before the close of the Eocene, presumably as a result of competition from the more efficient rodents and perhaps because of the increasing aridity of the climate in the Northern Hemisphere at that time. A change in the climate is implicated in the failure of the plesiadapoid primates to survive, as members of the Eocene families not competitive with the rodents also disappeared from Europe and North America during the late Eocene and early Oligocene years. The adapids, anaptomorphids, and omomyids seem to have continued in some other region, for the lemurs, lorises, and tarsiers of the

Old World and the first anthropoid primates are certainly descended from animals very much like them. There is little direct fossil evidence to link the early Tertiary primates with the lower primates of the present day, but the absence of such evidence is explicable on the basis of the primates' gradual specialization for a tropical environment. When the climate of the Northern Hemisphere became too rigorous for the lemuroid and tarsioid animals, they surely withdrew southward. In the New World, except for the omomyid animals that escaped by island-hopping to the rain forests of South America and gave rise to the platyrrhine monkeys, the primates perished. In the Old World, numbers of genera apparently spread into Africa and southern Asia. The catarrhine primates that evolved thereafter in Africa maintained themselves well, but their less progressive relatives died out or became relics. All the surviving lemurs live on the island of Madagascar, where they are protected from the most voracious continental carnivores and the more advanced primates. The galagos and lorises are confined to the trees in restricted areas of sub-Saharan Africa, India, and Malaysia, and *Tarsius*, the last of the tarsioids, is found only on islands in the Philippines and the East Indies.

Classification of the rodents

Initially, it seemed that the rodents could also be divided into suborders that would correspond at least in a broad sense to natural groups. In 1855 Brandt had pointed out that the living rodents were separable into three categories on the basis of the structure of their jaw muscles (Fig. 9.32). It was he who described the native South American and African forms as having a deep masseter that passed forward through the infraorbital foramen and thus as belonging together in the Hystricomorpha. Similarly, he housed the squirrels, beavers, gophers, and pocket mice in the Sciuromorpha, because in all these animals the deep masseter originated on the zygomatic arch and the middle portion of the muscle extended anteriorly below the arch to obtain an origin on the bones of the face. The remaining rodents—a vast assemblage including the rats, mice, jerboas, gerbils, lemmings, hamsters, and their relatives—he classified in the Myomorpha, since these forms exhibited the extension forward of both parts of the masseter muscle. The deep masseter curved upward to pass through a relatively small, dorsally located infraorbital foramen, while the middle masseter followed a course beneath the zygomatic arch, as it did in the sciuromorphs. Brandt's system of classification proved invaluable to zoologists, since all but one of the living rodents fitted

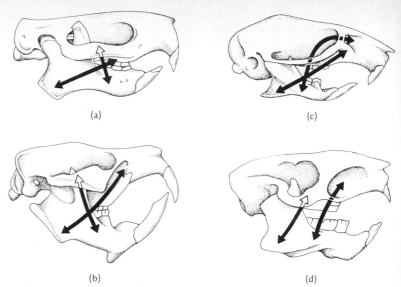

(a)

(c)

(b)

(d)

Figure 9.32. Skulls of various types of rodents in lateral view to show arrangement of the middle and deep portions of the masseter muscle. **(a)** A primitive rodent: both parts of the masseter originate from the underside of the zygomatic arch. **(b)** A typical modern sciuromorph: the middle masseter extends anteriorly below the zygomatic arch to obtain an origin on the face. **(c)** A myomorph: the middle masseter originates on the face, and the deep masseter passes forward through a small, dorsal infraorbital foramen. **(d)** A hystricomorph: the middle masseter is similar to that in primitive rodents, but the deep portion passes through a large, low-set infraorbital foramen to originate anterior to the orbit. (From Romer.)

neatly into it, but paleontologists soon found it inadequate for fossil forms and incorrect in some of the phyletic relationships it implied. In addition to their discovery that the Hystricomorpha was not a natural group, investigators unearthed whole families of extinct rodents which, like the living mountain beaver *Aplodontia,* had an unspecialized masseter muscle that originated entirely upon the zygomatic arch. No revision of Brandt's scheme accommodated the ever growing number of known rodents happily, and no one seemed to be able to devise a substitute that was more satisfactory. Some of the systems that were suggested contained "wastebasket" taxa—categories erected for miscellaneous genera or families that fitted nowhere else—and others, by dividing the rodents into two large groups, consisted of suborders which critics thought could not possibly be monophyletic. When, for instance, Schaub offered a scheme in which he divided the rodents into those with five crests on each molar and those with four, Wood objected that similar tooth patterns were known to have evolved convergently in rodents of different ancestry and so could not be used to identify natural assemblages within the order. Wood was not convinced, as Schaub was, that the order Rodentia could be divided more or less equally into taxa of subordinal rank. Like Simpson, he suspected that the explosive radiation of the early rodents had produced the bases of numerous lines, some of which continued to diversify and some of which did not. Such a pattern of evolution would result in the appearance within the order of monophyletic groups of unequal taxonomic rank. If this reasoning were correct, the phyletic history of the rodents would be conveyed most accurately by listing isolated families and superfamilies beside the valid suborders that could be defined.

Origin of the rodents

The disagreement between Wood and Schaub concerning the classification of the rodents, as well as their different opinions, alluded to earlier, about the ancestry of the South American hystricomorphs, stems from their conflicting theories of the evolution of the entire rodent group. Wood is quite certain that the paramyid animals (Fig. 9.33) which appeared first in the late Paleocene rocks of western North America were the basal members of the order and thus the source, direct or indirect, of

Figure 9.33. Restored skeleton of *Paramys delicatus*, a primitive rodent from the late Paleocene and early Eocene of North America. (From Wood.)

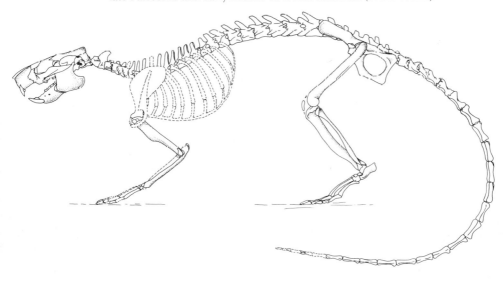

all the later rodents. He believes, further, that the picture of rodent evolution given by the fossil record is the true one: rodents arose in the middle Paleocene and by the late Paleocene and early Eocene had embarked upon their first great radiation, one that was to produce the forerunners of the diverse groups of the present day. Schaub has found, among the Oligocene squirrels, forms with teeth which he believes are more primitive than those of the paramyids and so refuses to accept the paramyids as the primary members of the order Rodentia. The Oligocene squirrels, in Schaub's opinion, are relics of a group of rodents older than the paramyids and ancestral to them. Although the fossil record gives no indication of their presence before late Paleocene times, Schaub presumes that the rodents must have emerged much earlier, possibly in the Cretaceous period, and undergone radiations of which there are no longer any trace. Because he regards inferences drawn from tooth structure as more compelling than negative evidence from the fossil record, Schaub has stood firm in his belief that new paleontological discoveries will eventually verify his hypothesis.

Wood has responded to Schaub's argument by expounding in detail his reasons for proposing the paramyids as the earliest rodents. These animals had developed the continuously growing pair of incisors, the diastema, and the broad-crowned cheek teeth that are characteristic of the order but still showed dental traits that mark them as more primitive than any other rodents, even the Oligocene squirrels. The paramyid upper molars, although flanked fore and aft by cingula that widened them, retained a vestige of the three-cusped, triangular design present in the rodents' insectivorous forebears (Fig. 9.34). The lower cheek teeth were already square, having lost the anterior cusp of the trigonid, but they too showed cusps instead of the crests that appear in more advanced rodents. There was clearly a transition from the paramyid cuspidate tooth to the crested type, however, because paramyid molars exhibit low ridges on the crown which in some genera join the bases of the cusps to produce elongated cutting blades in the pattern that the crests later assume. Although in paramyids the most posterior and only remaining premolars, like those of other rodents, are molarized to increase the length of the grinding battery, the change in the form of the teeth seems to have been a recent one. In one genus of paramyid, *Franimys*, the upper fourth premolar is still unmodified, and in none of the early Eocene members of the family is the lower premolar completely transformed. From paramyids in

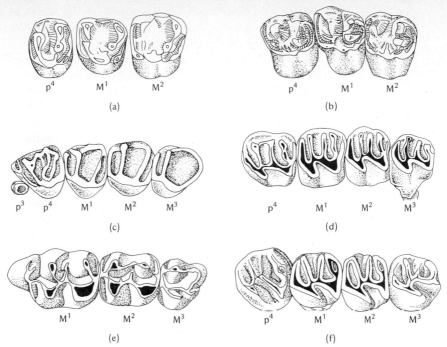

Figure 9.34. Upper cheek teeth in rodents in occlusal view, showing characteristic patterns of cusps and crests. **(a)** *Franimys*, a Lower Eocene paramyid with an incompletely molarized fourth premolar. **(b)** *Paramys*, an Upper Paleocene rodent with molarized fourth premolar but little development of crests or ridges joining cusps. **(c)** *Sciurus*, a Recent squirrel, with four crests. **(d)** *Theridomys*, a European Oligocene form with five crests. **(e)** *Eumys*, an Oligocene myomorph, with five crests. **(f)** *Erethizon*, the modern New World porcupine, a hystricomorph, with five crests. Abbreviations: *P*, premolar; *M*, molar. Teeth are numbered from front to rear. (*a, b* from Wood; *c, d, e, f* from Stehlin and Schaub.)

which the skull is known, it is apparent that the fossa for the mandibular condyle was shallow enough to allow the lower jaw to move forward and backward as well as up and down, but the jaw musculature was not yet specialized to effect the full range of motion that the design of the jaw joint allowed. The masseter was relatively short, arising on the zygomatic arch and extending nearly vertically to its insertion on the mandible. Assisted by the anterior fibers of the large temporalis muscle, the masseter lifted the jaw but did not impart to it the strong forward thrust necessary for vigorous gnawing. Wood has found evidence in certain genera of paramyids that elongation of the masseter was impending. In one case, the extension of the fibers portended a unique pattern of muscle attachments, but in others the sciuromorph or the hystricomorph condition was foreshadowed. The postcranial skeleton of some of the paramyids showed specialization for burrowing or jumping, but *Paramys* and several other forms had a generalized frame that could have been antecedent to that of a wide variety of later rodents.

Wood has found series of fossils which he believes are phylogenetically as well as morphologically transitional from the paramyid level to the aplodontid and the sciuravid states. Aplodontid and sciuravid rodents, which retained the masseter in its primitive condition, were replaced (the latter immediately and the former, with the exception of *Aplodontia*, during the Miocene) by forms with more progressive jaw musculature. Members of the Sciuridae, or squirrel family, spread thoughout the Northern Hemisphere and followed the ancestors of the caviomorphs into South America when the Panamanian isthmus appeared. Another group of paramyid derivatives, the Pseudosciuridae, was less successful, being confined to Europe and dying out in the Oligocene, but these animals apparently gave rise to the theridomyids, the hystricomorphous rodents with five-crested teeth, which Schaub and other European paleontologists have viewed as forerunners of the later hystricomorphs of the Southern Hemisphere. Although Wood cannot link the paramyids to the other rodents with five-crested teeth, the castorid beavers and the myomorphs, he does not find, in the lack of a direct connection between these forms and the paramyids, grounds for agreeing with Schaub that there must have been another and presumably earlier group of ancestral rodents. In Wood's opinion, not only the absence of such animals from the fossil record makes their existence unlikely but also the survival through the Paleocene and early Eocene of numerous multituberculates, archaic mammals that could hardly be expected to flourish in competition with physiologically superior placental rodents.

In the continents of the Northern Hemisphere during late Eocene time, there were other groups of animals besides the paramyids whose diversification was important in shaping the mammalian fauna of the future. The small, flesh-eating miacids, which had been overshadowed throughout the Eocene by the oxyaenid and hyaendont creodonts, were by the end of the epoch giving rise to the first members of the modern families in the order Carnivora. At first glance, the reason for the replace-

Miacids: ancestors of the modern carnivores

ment of the creodonts by descendants of the forest-dwelling miacids is not apparent. The creodonts had evolved powerful canine teeth and, toward the rear of the tooth row, specialized carnassials that were efficient in cutting and slicing. They ranged in size from rat to bear, some species being adapted for life in the forest and others, with digitigrade feet, for running in the open fields. Although the miacids may initally have no advantage over the creodonts in body form, they may have been cleverer hunters, for their descendants are among the most intelligent of mammals. In the competition for prey, animals that were beginning to stalk their intended victims or to work together in twos and threes would eventually have displaced carnivores incapable of such calculated behavior. The miacids had also in their favor a dentition that was potentially more versatile than that of the creodonts. Since their carnassial teeth developed farther forward than those of the oxyaenids and hyaenodonts, they retained the use of crushing molars at the end of each tooth row. These teeth, primitively tribosphenic in structure, showed a tendency to disappear in several lines descended from miacids, but only in the cats did they lose their function entirely. In other carnivores, though the carnassial blades elongated and became parallel with the long axis of the jaw for more efficient cutting, the rear teeth and to some extent the carnassials themselves served to masticate food material. The retention of crushing as well as shearing teeth allowed the early carnivores a varied diet. When the group began to diversify, some of its members, like the hyaenas and cats, became more completely dependent upon flesh; others, like the dogs, maintained the original balance between crushing and slicing action; and still others—the raccoons and bears, for example—broadened the posterior cheek teeth and enlarged the vegetable component of their diet.

History of the modern carnivores
    Although the animals of the order Carnivora became almost as cosmopolitan as the rodents, their evolutionary history is easier to follow. Instead of splitting into a profusion of separate lines with distinctive and convergent traits, the carnivores subdivided in a relatively simple manner. One, or possibly two, groups of the flesh eaters became adapted for life in the sea. For these animals, as for secondary marine forms that belong to other classes, fossils documenting the transition from the ancestral tetrapod state are lacking. Enough structural evidence links the seals and walruses to the terrestrial carnivores, however, to prevent their inclusion in the order Carnivora as members of

the suborder Pinnipedia from being seriously challenged. The terrestrial carnivores have been assigned to the suborder Fissipedia, an assemblage which is clearly divisible, above the miacid level, into two groups (Fig. 9.35). Good fossil evidence exists for the gradual segregation of the advancing fissipedes into feloids and canoids or, as some prefer to call them, aeluroids and arctoids. The branch first named includes the viverrids, hyaenas, and cats and the second, the dogs, bears, raccoons, and more generalized forms like otters, badgers, and skunks, weasels, and wolverines.

Since the modern carnivores radiated conservatively rather than explosively and left numerous fossil remains, Matthew was able to demonstrate that their evolution followed the pattern set forth in his thesis. The appearance of the miacids in Paleocene deposits of North America was consonant with his theory that the territories of the Northern Hemisphere served as the cradle of the mammalian orders in Tertiary time. Evidently, the miacids evolved the traits that were to characterize the more advanced carnivores before they spread outward from their place of origin, because the oldest felids from Europe and the canids which appeared almost simultaneously in North America shared such structural specializations as the fusion of several bones in the wrist and the complete ossification of the bulla protecting the middle ear. The cats, the most successful members of the aeluroid group, evolved rapidly, grew in size, and migrated to the farthest points of the northern land mass before the end of the Oligocene epoch. There were apparently several lines of felids, of which at least two culminated in great saber-toothed beasts and another led to the surviving genera *Acinonyx* and *Felis* (Fig. 9.36). Many kinds of cats entered South America and Africa when terrestrial bridges were available and coexisted there and elsewhere until the Pleistocene, when man doubtless hunted them and some of the large herbivores on which they fed disappeared. Thereafter, the greatest of the cats fared less well. The sabertooths became extinct, and the bigger felines dwindled. The tigers, lions, leopards, and cheetahs in the Old World and the cougars and jaguars in the New linger in the few remaining areas of wilderness or live now protected by law. In Matthew's opinion, the initial success of the cats was responsible for the present restriction of the less highly specialized aeluroids to the tropics. The civets and mongooses, living representatives of the viverrid stock from which the cats came, were driven southward in the Old World, he believed, by the expansion of

Figure 9.35. Fissipede carnivores. Aeluroids: **(a)** *Herpestes*, the mongoose; **(b)** *Hyaena striata*, the striped hyaena; **(c)** *Felis pardalis*, the ocelot. Arctoids: **(d)** *Procyon*, the raccoon; **(e)** *Mustela*, the weasel; **(f)** *Lycaon*, the Cape hunting dog, **(g)** *Ursus americanus*, the black bear.

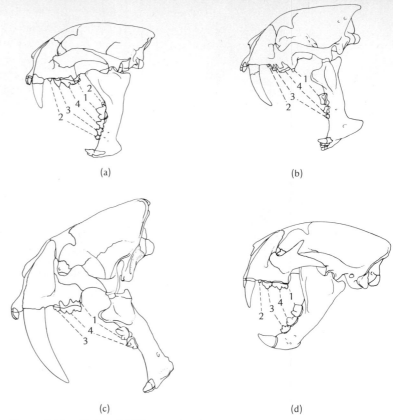

Figure 9.36. Skulls of felids. **(a)** *Dinictis*, an Oligocene form with medium-sized upper canine teeth. **(b)** *Hoplophoneus*, an Oligocene form with incipient saber-sized canines. **(c)** *Smilodon*, a Pleistocene saber-toothed cat. **(d)** *Felis*, the modern felid, with large lower canine teeth but upper canines of modest size. Note that in saber-toothed forms, the jaw articulation is depressed below the floor of the braincase, allowing the mouth to open more widely. (From Matthew.)

the more advanced felids. Hyaenas, which arose from viverrid ancestors in Miocene time, sheltered themselves somewhat from competition with the cats by adapting to a diet of carrion as well as live prey, but these animals also failed eventually in the northern regions and survive today only in Africa and in adjacent parts of Asia.

That the forces which determined the distribution of the carnivores and, in fact, of other mammals were more complex than Matthew supposed is illustrated by the dispersal of the arctoid fissipedes. The least progressive members of that group, known collectively as mustelids, have tended to remain in the cool forests of the north while the dogs roamed worldwide. Rather than competing in the Northern Hemisphere, the mustelids and the dogs have occupied different niches. The dogs, most of which have become excellent runners, usually chase their game in open country, whereas the short-legged mustelids, like weasels and skunks, hunt within the woodlands or subsist largely upon the fruits, eggs, and insects to be found there. The radiation which produced this distribution seems to have been well under way by Oligocene time. The fossil record of the forest forms is thin, but it seems that as the mustelids diversified, they separated from the basic canid stock, which in turn enjoyed a radiation of its own. One group of canids, distinguished by the peculiar crested talonid on the lower carnassial tooth, left only a few survivors in southern regions, but another, with a more normal, basined heel on the tooth, gave rise to several families with living representatives.

The raccoons and their relatives, which constitute the family Procyonidae, are essentially Western Hemisphere in distribution and to a large extent have remained arboreal. Whether they evolved from a primitive canid stock or a mustelid group is not clear. Unlike the dogs, the procyonids and a second arctoid group, the Ursidae, or bears, lost the long, sharp ridge on the carnassial teeth and developed, behind the huge canines, cheek teeth specialized for crushing all kinds of food (Fig. 9.37). Although the bears have become big and rather ponderous animals, many of them have retained the ability to climb trees shown by primitive carnivores. The ancient hunting instinct has been suppressed, however, and no ursid except the polar bear depends primarily upon flesh for its food.

Marine carnivores   The prey of the polar bear consists largely of its distant cousins, the pinnipeds. The sea lions, seals, and walruses enter the fossil record during the Miocene, when the bears evolved, but it is certain that they split away from the central carnivore stock far earlier. Their possession of the fused wristbones characteristic of carnivores more advanced than miacids implies an origin from progressive members of that group or possibly from some emerging arctoid strain. In the absence of fossil evidence, nothing certain can be said about the way in which they evolved, but it does seem that their rate of change from land to sea creatures must have been rapid. Even the time that elapsed between the Eocene epoch and the Miocene is short for the great number of mutations required to effect the structural

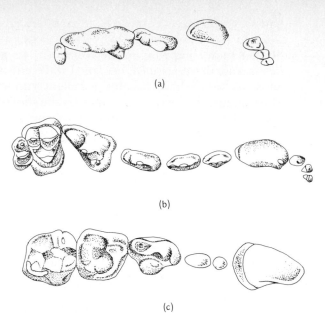

Figure 9.37. Right upper teeth of fissipede carnivores. **(a)** *Pseudaelurus*, a Pliocene cat, showing extreme specialization for shearing. **(b)** *Temnocyon*, a mid-Tertiary relative of the modern dog, with the two most posterior teeth (molars) somewhat adapted for crushing. **(c)** *Agriotherium*, a bear of the Pliocene and Pleistocene with molars strongly modified for crushing. (From Romer.)

and physiological transformation that made of terrestrial animals forms that habitually move, feed, and mate in the water. The Miocene fossils attest to the existence, then as now, of three families of pinnipeds differently specialized for marine life (Fig. 9.38). All the animals are streamlined for swimming and are obliged, for lack of a muscular tail, to propel themselves through the water with enlarged, webbed feet. The hair seals and the sea lions, of the family Otariidae, and their relatives, the walruses, which constitute the Odobaenidae, can still twist the feet forward to support their hindquarters when on land, but the fur seals, belonging to the family Phocidae cannot. The phocids, which may have arisen from terrestrial carnivores independently of the otariids, are the pinnipeds most highly modified for swimming. Their vertebrae interlock less tightly than usual so that the body flexes freely in the water, and the respiratory, circulatory, and nervous systems function cooperatively to allow the animals to dive and stay submerged for periods up to half an hour. The head of the phocid seal is more perfectly streamlined than that of the otariid since it has lost the projecting external ears, but the vibrissae are still in evidence on the snout, giving the face a resemblance to that of the terrestrial carnivores. All the pinnipeds retain the strong canine teeth of their ancestors but have lost the carnassials. In seals, which feed on fish, the cheek teeth have been reduced to uniform holdfasts and, in the mollusk-eating walruses, to flat-topped pegs for cracking shells. Only the reproductive habits of the pinnipeds tie them to the shore. Unable to give birth in the water, the animals must emerge once a year to produce their offspring. Since they move awkwardly on the ground, it is the isolation of their breeding places rather than their ability to fight that has assured their survival to the present time.

Evolution of the whales

When the ancestors of the pinnipeds entered the sea, it already harbored mammals far more extremely adapted for marine life than any of the pinnipeds, even the phocids, were

Figure 9.38. Recent pinnipeds. **(a)** *Otaria*, the sea lion. **(b)** *Odobaenidae*, the walrus. **(c)** *Phoca*, the fur seal.

(a)

(b)

(c)

ever to become. There is strong evidence in the absence of hind limbs in the mid-Eocene *Protocetus* that when the forebears of the seals were perhaps still semiterrestrial, the whales, or cetaceans, had already developed the structural and physiological prerequisites not only for swimming with the agility of fishes but also for bearing their young in the water. *Protocetus* belonged to a group of archaic whales called archaeocetes that radiated during the Eocene to fill the adaptive zone in the ocean left empty by the disappearance of the ichthyosaurs and the mosasaurs. From the skeleton of the archaeocetes and the form of other fossil and living whales, it is obvious that in becoming obligate swimmers the cetaceans evolved traits convergent with those of the earlier tetrapods that divorced themselves completely from the land. In the early whales (Fig. 9.39) the head developed an elongated snout with simple teeth that functioned both as cutwater and fish trap. The vertebrae of the neck became shorter and in large, advanced species fused with one another, so that the head no longer moved independently of the body. Like the ichthyosaurs, the whales retained a heavy, tapering tail which acted as the locomotor organ. It flexed in a dorsoventral direction, however, rather than laterally and enlarged its surface by developing horizontal flukes rather than a fish-like caudal fin. As in the swimming reptiles, the projecting upper part of the forelimbs disappeared in the process of streamlining by shortening of the long bones, and the hands grew to paddle form by the addition of extra phalangeal elements. In many smaller whales, a single dorsal fin supported by connective tissue arose, selected for, as dorsal fins always have been, because it prevented excessive rolling as the animals swam.

As with most tetrapods secondarily modified for aquatic living, ascertaining the terrestrial stock from which the whales came is exceedingly difficult. Since the anatomy of the archaeocetes does not reveal any unique structures, like the fused carpal bones of the pinnipeds, which link the whales to animals of a

Figure 9.39. Restoration of the skeleton of the Upper Eocene whale *Zygorhiza kochii.* (From Kellogg.)

particular order, paleontologists have had to rely upon indirect reasoning to identify the mammalian group most probably ancestral to the first whales. Observing that the archaeocetes were fully adapted for swimming by the Eocene epoch, investigators recognized that similarities between whales and animals which appeared then and later, like the modern ungulates, sirenians, and pinnipeds, would have little or no phylogenetic significance. R. Kellogg pointed out, and others agreed, that the extent of the change involved in the transformation of a terrestrial mammal into a completely oceanic one was so great that the process must have begun at least as long ago as the early Paleocene and possibly even before that time, at the end of the Cretaceous period. Because the archaeocetes were carnivorous, workers tended to look toward the insectivore-creodont assemblage as a likely ancestral stock. Van Valen objected to the descent of whales directly from an insectivore group on the grounds that the insectivores tended to remain small. He eliminated the possibility of origin from oxyaenid creodonts because although they did tend to grow large, they showed dental specialization divergent from that of whales. The teeth of the hyaenodontid creodonts seemed more readily convertible to those of early archaeocetes like *Protocetus*, but Van Valen believed that there was an even closer likeness between the teeth of archaeocetes and early Tertiary forms known as mesonychids. The mesonychids, now regarded as primitive ungulates because of their hoofed toes and blunted tooth cusps, had teeth with enough of an edge for them to have been once considered members of the Creodonta. The advanced Eocene mesonychids were large and heavy-headed animals with feet modified for running, but their cheek teeth were not quadrangular like those normally found in herbivores. If they did not feed upon vegetable matter, they may have used the strong cusps of the tribosphenic molars to break mollusk shells. Van Valen, asserting that the wear pattern and the structure of certain teeth in Paleocene mesonychids like *Dissacus* (Fig. 9.40) were similar to those in middle Eocene archaeocetes, speculated that some ancient mesonychids may have shifted from a diet of mollusks to fish and become aquatic forms in the process. Van Valen's hypothesis is unacceptable to many paleontologists, however, because they find in the teeth of mesonychids certain characters, like the prominent protocone in the upper molars, which they believe debar those animals from the ancestry of the early cetaceans.

Where, geographically, the archaeocetes first appeared is as

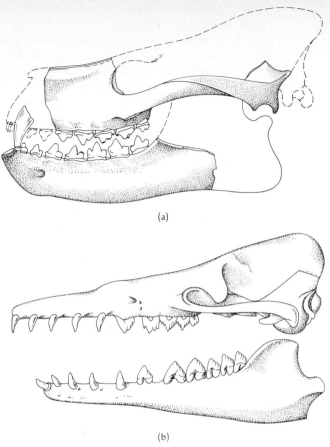

(a)

(b)

Figure 9.40. The skull of **(a)** *Dissacus*, a Paleocene mesonychid, and **(b)** *Prozeuglodon*, an Eocene whale. (*a* from Matthew; *b* from Kellogg.)

puzzling as the identity of the group from which they sprang. Finding *Protocetus* and two other middle Eocene genera of whales in Africa has led investigators to think that the cetaceans may have emanated from that continent. The occurrence in England of an Eocene specimen older than the African forms has shown, however, that the African discoveries may be misleading. Although the African archaeocetes still showed traces of the former tribosphenic construction of their posterior teeth and apparently retained a vestigial pelvic girdle jointed to the vertebral column, the animals' forebears had surely been in the ocean long enough by Eocene time to have dispersed themselves widely. A single vertebra tentatively referred to *Protocetus* found in upper middle Eocene rocks in Texas suggests that whales existed on both sides of the Atlantic during the mid-Eocene and thus could have originated in the New World as well as the Old. The absence of cetacean remains from Eocene sediments bordering the Pacific and Indian oceans hints that the whales were confined at first to Atlantic waters, but new fossil discoveries could prove that supposition untrue.

The later Eocene archaeocetes, known from numerous remains found in the southeastern United States and from others in Europe and Africa, were a more advanced and specialized group than *Protocetus* and its contemporaries. The small forms, *Dorudon* and *Zygorhiza*, and the monstrous 60-foot *Basilosaurus* had lost the sacroiliac articulation, lengthened the lumbar vertebrae, and eliminated the bracing of one neural arch upon the next so that the long trunk and tail region attained maximum flexibility. The serpentine form of the body and the peculiar serrated cheek teeth made it plain that these archaeocetes could not possibly have been ancestral to any of the

Modern whales

modern whales. Living cetaceans belong to two groups, the toothed whales, or Odontoceti, and the whalebone whales, or Mysticeti, both of which must have separated from the archaeocete line long before the appearance of its end forms. The lumbar vertebrae in the whales of these two suborders are not attenuated, and the body is thus more compactly built than that of the Upper Eocene archaeocetes. Also, the structure of the skull in the odontocete and mysticete forms shows a strange modification not present, even in a rudimentary way, in *Basilosaurus* and its smaller relatives: in conjunction with the backward migration of the nostrils to form a blowhole on the dorsal surface of the head, the nasal bones have been reduced and carried upward and the premaxillary and maxillary elements have expanded to the rear to cover the original braincase roof. In the mysticetes, the nostrils remain paired (Fig. 9.41), and the arrangement of the bones adjacent to them symmetrical, whereas the odontocetes exhibit a single blowhole and a variable degree of asymmetry in the growth of the neighboring bones. The several kinds of odontocete whales have maintained the carnivorous habits of the first cetaceans, although only the porpoises and dolphins have retained teeth in both the upper and lower jaws. The sperm whales (Fig. 9.42*a*), which have teeth only in the mandible, and the bottle-nosed whales, which are virtually toothless, depend nevertheless upon a diet of fish and squid, passing these animals backward into the gullet without

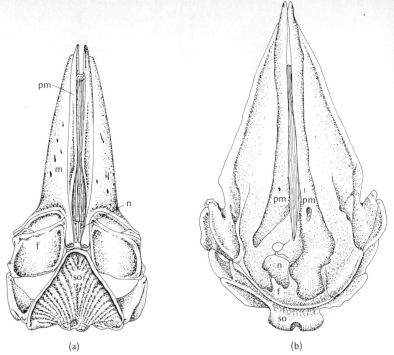

(a)   (b)

Figure 9.41. Whale skulls in dorsal view. **(a)** *Balaenoptera*, a mysticete, showing symmetrical arrangement of bones roofing the skull. (After Kellogg.) **(b)** *Physeter*, an odontocete, showing asymmetry. (From A. B. Howell, "Aquatic Mammals," Charles C Thomas, 1930.) Abbreviations: *so*, supraoccipital; other abbreviations as in Fig. 6.7.

chewing, as, in fact, even the toothed marine carnivores do. The mysticetes, such as the right whales (Fig. 9.42c) and the rorquals, have evolved an entirely different style of feeding. These forms have relatively enormous mouths into which they admit quantities of water filled with plankton. Tons of microscopic organisms are strained out by the long shreds of tough skin (called baleen or "whalebone") (Fig. 9.42d) that hang downward from the palate and then are maneuvered by the muscular tongue to the entrance of the gut.

The origin of whales: mono-phyletic or di-phyletic?

The structural differences between odontocetes and mysticetes led some investigators to suggest a diphyletic origin for the modern whales, but the majority of workers, including Weber and Kellogg, disagree. Aware that close convergence does occur in different tetrapods secondarily modified for aquatic life, they maintain, nevertheless, that certain peculiar traits shared by the two kinds of whales are not likely to have arisen twice. The structure of the ear, for instance, is the same in the animals of both groups and unique among the mammals. Instead of being fused to the braincase, the periotic bone containing the inner ear is isolated from the rest of the skull, to which it is bound only by fibrous tissue. Thus insulated from vibrations transmitted through the bones of the head, the auditory ossicles are set in motion only by pressure waves which reach the bulla and the air within it directly from the exterior via the heavy tympanic membrane and the narrow external ear canal. Both odontocetes and mysticetes depend heavily upon this sensory mechanism, as large auditory pathways in the central nervous system show, since they have relatively poor vision and a reduced sense of smell. It has not been possible to trace the origin of the modern cetacean ear to its source, presumably in some early archaeocete stock, but the discovery of Eocene and Oligocene whales in which the bones of the snout have begun their backward migration is a clue to the antiquity of the existing cetacean lines. The oldest of these forms, *Archaeodelphis* and *Agorophius*, contemporaries of *Basilosaurus*, apparently represent already divergent groups, for the former

Figure 9.42. Recent whales. **(a)** *Physeter*, the sperm whale, an odontocete. **(b)** Skull of *Physeter*, showing teeth in the lower jaw. **(c)** *Balaena*, the right whale, a mysticete. **(d)** Skull of *Balaena* with "whalebone" in place. (*a, c* after Slijper.)

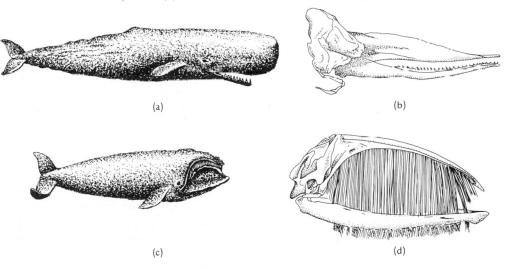

(a)

(b)

(c)

(d)

exhibited the symmetrical arrangement of the bones around the blowhole characteristic of the mysticetes and the latter a pattern more likely to have been antecedent to that of the toothed whales. With the exception of two late Oligocene relatives of *Archaeodelphis* which survived beside the first of the more modern whales, no fossil record of the immediate precursors of the later Tertiary cetaceans exists. The specimens in hand indicate that the odontocetes and the less diversified mysticetes enjoyed their greatest prosperity during the Miocene epoch.

The decline of the whales

Toward the end of the Tertiary period, important families in each suborder became extinct, and the remainder were diminished by the disappearance of numerous genera. The decline in the cetacean population has continued to the present time, accelerated in the last hundred years by the efficient hunting techniques of modern man. The small dolphins and porpoises, which have no commercial value, are not in danger of vanishing in the near future, but the larger whales in both suborders are on the verge of extinction. Recognizing the gravity of their situation, the United States government has placed these animals on its list of endangered species and brought whaling by resident companies to an end. Representatives of the nations still actively engaged in hunting whales have met to discuss conservation measures, but, before they agree upon the amount of "harvesting" the threatened species can withstand, it is possible that the world will lose another one or two of its most fascinating mammals.

Appearance of the modern ungulates

When the whales were still new to the sea, there were evolving in the Northern Hemisphere two groups of animals which were to be the mainstay in the diet of the large carnivores that remained on land. Both groups consisted of advanced ungulates, forms that enlarged rapidly and tended to become either extremely heavy-bodied or highly modified for running. Evidently descended from some Paleocene condylarth stock, the two kinds of ungulates developed distinguishing characteristics but paralleled each other in many adaptive traits. They are segregated on the basis of the structure of their feet into two orders, Perissodactyla and Artiodactyla (Fig. 9.43). The former, which includes the tapirs, rhinoceroses, and horses, as well as several kinds of creatures no longer extant, unites the forms that thrust their weight through the strongly developed third toe on each foot and possess an astragulus with a pulley-shaped articular surface for the lower end of the tibia and a less deeply grooved one for its joint with the more distal tarsal bones. The

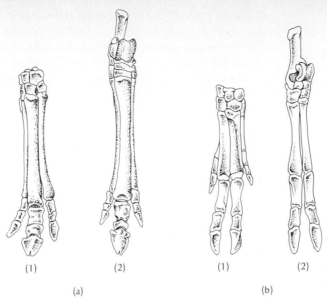

(1)   (2)   (1)   (2)

(a)   (b)

Figure 9.43. The structure of the feet in a perissodactyl and an artiodactyl. **(a)** The manus (1) and pes (2) of *Miohippus*, an Oligocene horse (Perissodactyla). **(b)** The manus (1) and pes (2) of *Mylohyus*, a Pleistocene peccary (Artiodactyla). (From Romer.)

animals of the order Artiodactyla, the swine, camels, deer, cattle, and their extinct relatives, support themselves equally upon the third and fourth toes and have an astragulus pulley-shaped at both ends. Although primitive members of both groups appeared in the early Eocene, the perissodactyls immediately became the predominant forms. They failed to hold their ground, however, and soon took second place to the artiodactyls, which remain the most numerous ungulates today.

Evolution of the perissodactyls

Thanks to a fortunate set of circumstances, paleontologists have been able to follow quite closely the rise, radiation, and decline of the perissodactyls. They know fossil forms not only from Europe and Asia but also from a series of North American deposits almost continuous from Eocene time onward. They have enough material, particularly from North America, to demonstrate the ramification that took place within each family of the order, and the availability of living species representing different levels of development attained by the perissodactyls allows investigators to understand the characteristic modifications in soft tissues as well as skeletal structure. Because all extant members of the group have an enlarged intestinal caecum, for instance, it is presumed that the appearance of this organ was a basic step in the adaptation of the gut in perissodactyls for the more efficient digestion of their cellulose-rich plant food. From the living animals, also, some idea can be gained of the defenses their ancestors evolved against the many carnivores which preyed upon them. The large forms may have developed, besides great size combined with fleetness of foot or sharp horns, thick skin, like that of the rhinoceroses, to protect them, and the primitive species, which are smaller, perhaps had striped or particolored coats similar to those of tapirs to camouflage them from their enemies.

Hyracotherium

In *Hyracotherium* (Fig. 9.44) a little dog-sized ungulate that inhabited Europe and North America in the late Paleocene and early Eocene, paleontologists believe they have an animal nearly related to the stock from which all of the perissodactyls evolved. Although classified as the most primitive member of the horse family, *Hyracotherium* (or *Eohippus*, as it was once called), exhibited the distinctive traits of its order in their least advanced form. Its dentition was definitely that of a herbivore, but only the molars were broadened and flattened for grinding. The premolars were still relatively primitive rather than fully molarized as they were to become in most later, more specialized perissodactyls. The number of teeth typical of primitive placental mammals still remained, but the canines had decreased in size and a diastema had appeared in front of the cheek teeth. Since the teeth of *Hyracotherium* were low-crowned and not especially resistant to wear, the animal evidently subsisted upon the leaves it pulled from low branches of trees and bushes rather than upon more abrasive grasses. The presence of a bony crest on the roof of the braincase for the origin of the posterior part of the temporalis muscle signifies that the jaw musculature was as yet little modified from its condition in the condylarths. Later, in the descendants of *Hyracotherium*, the temporalis was to diminish in bulk, and the crest, which afforded it additional area for attachment, was to disappear as the masseter enlarged and provided the force for the grinding action of the jaws.

The structure of the postcranial skeleton of *Hyracotherium* indicates that the animal was already better adapted for running than any of the carnivores which pursued it. The high neural spines on the vertebrae at the shoulder make it clear that the head and chest balanced each other over the forelegs as the body was thrust forward by the strong hind limbs. The stride was lengthened by the elongation of the legs below the knee and

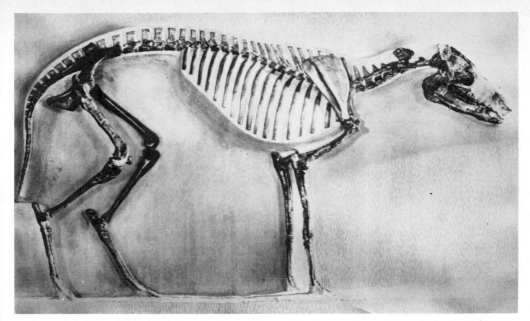

Figure 9.44. Skeleton of the earliest known horse, *Hyracotherium* (= *Eohippus*). (Photograph, D. G. Stahl.)

elbow. Not only were the bones distal to those joints proportionately longer than they had been in earlier mammals, but the metapodials and phalanges were added to the vertical portion of the legs, since the animals stood upon the hoofs and the small pads behind them. From the configuration of the articular surfaces on the limb bones it seems that, as in more advanced ungulates, movement at the joints in the legs of *Hyracotherium* was restricted to bending in the direction of the stride. Since there was little thrusting of weight to the side, the digits flanking the central third one were reduced. In the hind limb, the first and fifth toes had disappeared entirely, but, in the foreleg, four toes remained, the outermost, fifth one so small that it surely touched the ground only when the animal walked or ran on soft terrain. The action of the legs was fast, because the muscles chiefly responsible for moving them back and forth were inserted upon the short proximal segments of the limbs. The advantage of the muscles' contracting quickly through a small arc seems to have compensated for the extra expenditure of energy required to swing the leg from a position so near its pivot.

Tapirs   In the modern tapirs (Fig. 9.45) students of the perissodactyls

Figure 9.45. *Tapirus*, the modern tapir, a browsing perissodactyl now confined to the tropics.

can see forms advanced only a little beyond the level of *Hyracotherium*. Tapirs had emerged by Oligocene time from one of a number of primitive Eocene stocks and spread over North America and Eurasia. Since that time, they have changed hardly at all. Specialized only in having evolved a short, trunk-like proboscis, the tapirs are browsing animals of the forest. Shy and nocturnal in habit, they frequent wet regions and feed upon water plants, young shoots, and fruits as well as leaves. Although their teeth are still low-crowned, the grinding surface has been enlarged by the molarization of the premolars, and the gap between the front and rear teeth has become more distinct. They are not large animals, the biggest of the living forms being about 3 feet high at the shoulder, but the body is relatively heavy and the limb bones are stout to support it. Like *Hyracotherium*, the tapirs have four toes on the forefeet and three on the hind. Despite their archaic structure, the tapirs competed successfully with more progressive ungulates in the temperate regions until the early part of the Pleistocene epoch. Thereafter, they became confined to the tropics and survive today in Malaysia and in Central and South America.

Rhinoceroses   The rhinoceroses, which arose from tapiroid ancestors during the Eocene, became more specialized than the tapirs and radiated in a much more complex manner but finally came to a similar end. After having roamed all the lands of the Northern Hemisphere, they failed everywhere during the Pleistocene except in tropical areas of southeast Asia, India, and sub-Saharan Africa. Unlike the tapirs, which are still quite numerous, the rhinoceroses are on the brink of extinction. Since they have always tended toward large size, it is rather remarkable that their group has endured so long. Only the primitive running rhinoceroses of late Eocene and Oligocene times remained lightly enough built to retain slender legs and a graceful gait. Even while these animals, the hyracodontids, lived, there were other rhinoceroses, the amynodonts, that became graviportal and apparently behaved like hippopotamuses, wading in lakes and stream channels, where their bones are usually found. Both the running rhinoceroses, which had only three toes on the front feet, and the heavy amynodonts, which kept four, died out before the Miocene epoch began.

The ancestors of the later rhinoceroses (Fig. 9.46) were members of a third branch of the family which seems to have emanated from the same source as the running rhinoceroses. During the Oligocene years, the forebears of the later rhinoceroses evolved molarized premolars and reduced the toes on the front feet from four to three, as the running forms had done. At the same time, they increased in bulk and developed shorter, rather than longer, legs. This group radiated widely in the Oligocene and Miocene epochs. From the available fossil material, paleontologists have been able to identify a half dozen divergent lines but have been unable to define their mutual relationships. The most aberrant forms were the giant rhinoceroses of Asia like *Baluchitherium*, which had a skull more than a yard long and could reach higher than 20 feet into the trees to strip away the leaves. Even though these animals were the largest terrestrial mammals that ever lived, they must have been especially vulnerable, for they vanished before the middle of the Miocene. A number of smaller but still large and ponderous rhinoceroses which coexisted with the giants and survived them were distinguished from those and earlier forms by having borne horns. Had not living rhinoceroses been available for study, the nature of those horns might have been something of a puzzle. Although the fossil skulls exhibit scars where the horns were attached, the horns themselves are never found. Examination of the structures of existing animals makes the reason obvious: the horns have no bony core but are composed solely of compressed hair, which decays upon burial.

An early horned form, *Diceratherium*, lived in Europe at the end of the Oligocene and in the first half of the Miocene wandered the length and breadth of North America. Its bones have been discovered in deposits of that age from South Dakota eastward to the island of Martha's Vineyard, off the coast of Massachusetts, and southward in Florida and the states in the

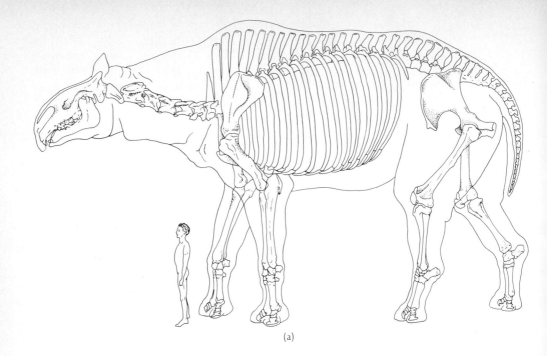

(a)

Gulf coastal plain. The animal spread into Central America as far as Panama but, because of the marine barrier that existed then, never reached South America. There was another lineage of horned rhinoceroses which had North American representatives during the Miocene, but its last member, a squat and perhaps amphibious form called *Teleoceras*, died out in Pliocene times before the elevation of the Panamanian isthmus. If rhinoceroses had survived in North America at the end of the Tertiary period, they would undoubtedly have moved over the Panamanian land bridge and found refuge in tropical South America. The fortunes of the rhinoceros family declined in the New World after the Oligocene epoch, however, and by the late Pliocene and Pleistocene the animals were confined to Eurasia and Africa. There, various kinds of one- and two-horned rhinoceroses evolved. In *Elasmotherium*, a huge beast which disappeared during the Pleistocene, the horn was mounted on the forehead above the eyes: in the surviving Indian and Javanese rhinoceroses, the single horn grows over the nasal bones at the end of the snout. The woolly rhinoceroses depicted by the cavemen were two-horned forms of the cold north. They are gone, but a small population of animals possibly related to

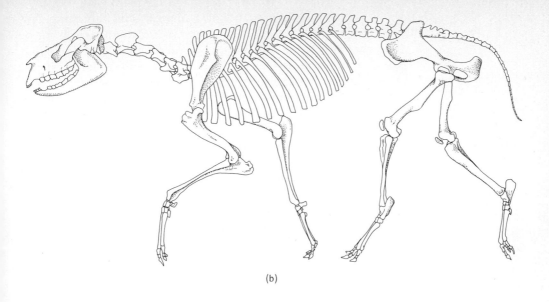

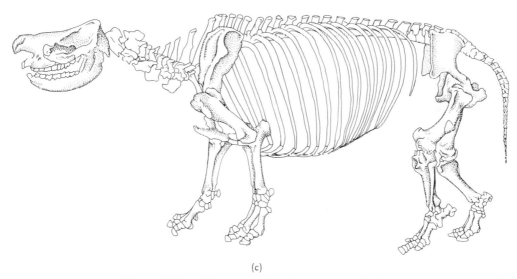

(c)

Figure 9.46. Skeletons of rhinoceroses. **(a)** *Baluchitherium*, giant Asian rhinoceros of the late Oligocene and early Miocene, shown beside a 6-foot man (From Gregory and Granger.) **(b)** *Hyracodon*, a running rhinoceros of the Oligocene. (From Scott.) **(c)** *Diceros*, the living black rhinoceros of Africa. (From Piveteau.)

them remains on the island of Sumatra. The Sumatran rhinoceroses, which have a thin rather than a woolly covering of hair, retain a pair of tusk-like incisors in each jaw, structures that are absent in the two-horned rhinoceroses still extant in Africa. Although the African species have the horns arranged like those of the Sumatran animals (a large one on the snout and a smaller one behind it), the animals differ from the eastern forms in other traits and seem to have had a long separate history.

Titanotheres and chalicotheres

Besides the rhinoceroses, other heavy-bodied perissodactyls appeared during the Tertiary period (Fig. 9.47). The titanotheres, or brontotheriids as they are properly called, were derived from small, Lower Eocene forms that were contemporary with *Hyracotherium* and similar to it in structure. These animals evolved quickly to large size and spread from North America, where they apparently originated, eastward through Asia. Some of the Oligocene titanotheres developed fearsome horns on the face, but the structure of their teeth and skeleton remained so primitive that the group failed in competition with other herbivores before the middle of the epoch. Possibly their small brains made them incapable of outwitting their enemies, or perhaps their teeth, which remained low-crowned and never developed cross ridges like those of rhinoceroses, restricted them to some special kind of soft vegetation that eventually became unavailable.

The chalicotheres were less numerous but more persistent animals than the titanotheres and much more perplexing to the paleontologist. From the pattern of crests and cusps in the molar teeth it is apparent that the chalicotheres were more closely related to the titanotheres and horses than to the rhinoceroses and tapirs, but their habits were surely not like those of any other perissodactyls. Although they paralleled other members of their order in increasing in size between Eocene and Miocene times, they evolved stout legs that terminated not in hoofs but claws. How they used their strangely constructed feet has never been satisfactorily determined. Since they were tall animals with front legs somewhat longer than the back ones and a head rather like that of a horse, some investigators believe that they browsed as early horses did, employing their claws to rake high branches downward toward the mouth. Others suppose that the chalicotheres fed at least in part upon tubers and roots which the claws were adapted to unearth. However they behaved, the animals had evidently found a unique and stable niche, for they endured with little change from the Miocene epoch to the Pleistocene.

(a)

(b)

Figure 9.47. Two extinct perissodactyls. Restoration of **(a)** the Oligocene titanothere *Brontotherium* and **(b)** the Miocene chalicothere *Moropus*. (From Colbert.)

Like the rhinoceroses, the chalicotheres declined in North America during the Miocene and eventually became restricted to the Old World. Their shift of habitat may have been caused in part by the increasing aridity of the climate in their original home, but it is also evidence, surely, of the intense competition that existed among the different kinds of perissodactyls and the growing array of artiodactyl herbivores. The only perissodactyls that held their ground in the Western Hemisphere in the post-Miocene years were the horses. These animals had been more conservative in evolving to larger size than other members of their order. In the Eocene, only European relations of the horse family, the palaeotheres, grew gigantic, and they, like the titanotheres, became extinct in the Oligocene. The immediate descendants of *Hyracotherium* in North America underwent minimal modification in the Eocene and Oligocene epochs. They did enlarge, but only to the dimensions of a good-sized dog. The body remained light, and the legs continued to grow more slender and longer in the shank and foot. During the early and middle Oligocene, the North American horses were represented by *Mesohippus*, a form which had three toes on each foot and premolar teeth that were molariform but still low-crowned.

Ever since Marsh assembled the series of fossil horses intermediate between *Hyracotherium* and the modern *Equus* from abundant remains collected in the American West, there have been those who have used it to illustrate the concept that change in a group of animals proceeds relentlessly in a single direction to produce highly specialized forms. The proponents of this kind of evolution—for which the name "orthogenesis" has been coined—have explained it in a number of different ways. Those whose reasoning was teleological declared that animals evolved in a "straight line" because they were moving along a predetermined path toward certain structural and functional goals. Lamarckians preferred the theory that vertebrates progressed in a particular direction, impelled, they believed, as all living things were, by an inner drive to reach a higher or more perfect level of adaptation. Investigators who looked with disfavor upon supernatural or intangible internal forces as explanations for evolutionary change supposed that rigidly oriented progress resulted from the simultaneous operation and controlling influence of a variety of natural laws whose nature would eventually be made clear through future research. All arguments about the cause of orthogenesis became irrelevant, however, when close inspection of supposed examples of straight-line evolution cast doubt upon the reality of the phenomenon. The work of Simpson and others on the evolution of the horse demonstrated that the history of that form had been far more complicated than Marsh had imagined when he made his brilliant studies a hundred years ago (Fig. 9.48). Certainly the animals had gradually become better adapted for their particular mode of life

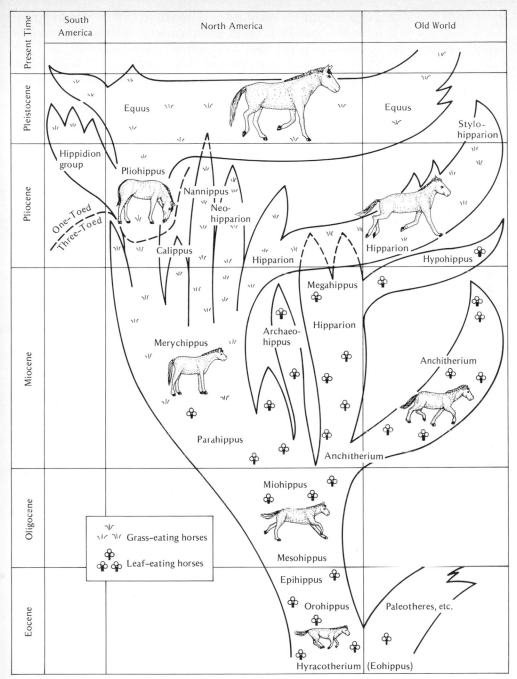

Figure 9.48. Radiation of forms within the horse family. The earliest horses, successful as three-toed browsers, gave rise to a number of separate lines. Evolution of the ability to graze and invasion of the continent of South America resulted in a further multiplication of equine genera. (From Wendt.)

through the operation of natural selection, but they could not be described as having progressed in a direct and uniform manner to the state in which they exist today.

Their rate of change, slow at first, suddenly increased in the early Oligocene years. During the last half of that era and the first part of the next, the three-toed horses had diversified in North America to produce five or six genera, each distinctly divergent from the others. While *Miohippus* and *Anchitherium* became somewhat larger than their Oligocene ancestors, *Archaeohippus* grew smaller and *Hypohippus* and its descendant, *Megahippus*, evolved toward gigantic size. Although the teeth, and therefore presumably the diet, of these horses had not changed from those of *Mesohippus*, most of them survived longer and spread farther than the earlier form. *Anchitherium*, besides ranging throughout North America, crossed into Asia and reached Europe, where it existed into Pliocene time. *Hypohippus* also invaded Asia and remained extant there beyond the end of the Miocene. Among this assemblage of forest-dwelling, browsing animals there was another and more progressive horse, *Parahippus*, which probably supplemented its diet of soft vegetation by cropping grass on the open plain. The teeth of *Parahippus* were better suited for grinding grasses than those of other early Miocene horses, for, in addition to being larger, they were partly covered by a protective layer of cement.

By the middle Miocene, although the other genera continued unchanged, the descendants of *Parahippus* were so altered in structure as to merit generic separation from their ancestor. The modifications exhibited in these animals, designated *Merychippus*, carried on the trends observable in the evolution of *Parahippus*. The teeth, which had continued to enlarge, attained the high-crowned, completely cement-covered condition characteristic of modern horses (Fig. 9.49). The crowns of these teeth consisted of tall ridges of enamel-covered dentin separated from one another by a filling of cement. Soon after the teeth erupted, the cement and the enamel at the crest of each ridge wore away, exposing the dentin within. As chewing continued, all three materials wore down, but, because of the height of the crowns, the teeth served their possessor for a great length of time. The skull in the various species of *Merychippus* also became more modern in appearance, because it lengthened and deepened in front of the eye as the cheek teeth elongated and grew heavier. The orbit, partially bordered behind by a process of the frontal bone in *Parahippus*, was finally entirely enclosed by a bar which

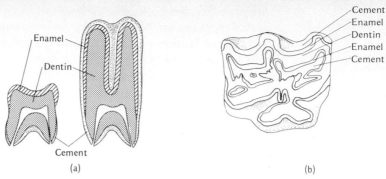

Figure 9.49. The structure of the hypsodont tooth. **(a)** A normal molar (*left*) and a hypsodont molar (*right*), showing the difference in the height of the crown and extent of the cement covering. **(b)** Occlusal view of a worn upper molar of the modern horse, *Equus*, showing platforms of dentin and cement separated by ridges of enamel. (From Romer.)

provided increased area for the attachment of jaw muscles. It was plain from the structure of the legs of *Merychippus* that these animals lived on the dry grasslands rather than under the forest canopy. The large middle toe was now supported entirely by its greatly strengthened hoof, and the digits flanking it, which used to brace *Parahippus* on the soft woodland floor, were so reduced that they could no longer touch the ground.

In contrast to the browsing horses, which ceased to diversify and began to dwindle toward the end of the Miocene, *Merychippus* continued to evolve and radiate at a rapid rate. While the population of these animals maintained itself, the stock also gave rise to a strain of smaller horses, *Nannipus,* and to an immensely successful genus, *Hipparion,* which replaced *Anchitherium* in Eurasia and migrated southward into Africa. Another branch from the *Merychippus* stem produced a group of heavier-bodied horses than *Nannipus* and *Hipparion,* one of which was the progressive *Pliohippus.* When the Pliocene epoch began, an array of horses existed that would have astonished those who viewed the evolution of the horse family as orthogenetic. The giant, functionally three-toed form, *Megahippus,* still browsed in the forests of Asia; light-footed animals with vestigial side digits populated the grasslands of both hemispheres; and *Pliohippus,* a horse verging on the modern form, roamed beside the other species on the North American plains.

Of all the Pliocene horses, only *Pliohippus* produced descendants that endured beyond the early years of the Pleistocene.

Members of this genus were almost as large as the living horses and well adapted for running by the slimming of the legs through reduction of the ulna and the fibula. *Pliohippus* spread into Central America, and from it there arose to inhabit South America several forms which were eventually displaced by the modern *Equus*, the offspring and successor of *Pliohippus* on the North American continent. Just as *Hipparion* had done in the Pliocene and *Anchitherium* even earlier, *Equus* moved quickly eastward over the Bering land bridge and established itself in all the contiguous areas of the Old World. For a while it flourished, but toward the close of the Pleistocene epoch, its numbers diminished. *Equus* eventually disappeared from the Western Hemisphere entirely, leaving the northern plains devoid of equids for the first time in millions of years. It fared little better in Eurasia, where only a few small wild populations of horses, onagers, and kiangs now remain in isolated regions. Like many of the large mammals, *Equus* is making its last real stand in Africa. There, in the form of asses and zebras, it seems to be moving slowly but inevitably toward extinction in the wild. Should these animals vanish, the family Equidae would be represented only by the domesticated species of *Equus*, whose existence is subject to the whims of man.

Early artiodactyls   The decline of the horse and other perissodactyls has been much easier to trace through the fossil record than the rise and proliferation of the artiodactyls. One of the first of these ungulates to appear was *Diacodexis*, a contemporary of *Hyracotherium* in North America during the Lower Eocene. It was an extremely primitive mammal, undoubtedly derived from Paleocene condylarth stock, advanced only in having evolved the double-pulleyed astragulus characteristic of the artiodactyls and in having lost the innermost toe on each foot. Since the molars of *Diacodexis* were tribosphenic, like those of the early eutherians, and bore pointed cusps, it is assumed that the animal was not yet a herbivore. A small form, it may have subsisted upon a mixed diet of soft plant material and invertebrates which it found in the forest.

The successors of *Diacodexis* and its relatives dispersed themselves in North America and Europe, eventually forming isolated populations from which many different groups of more advanced artiodactyls evolved. By the late Eocene, there were well under way in the Old World and the New independent radiations of artiodactyls at the level represented today by the wild and domestic pigs, the peccaries of the warmer regions, and the hippopotamuses. Before the end of the epoch, the first members of the more highly specialized camel family had appeared, as had the immediate forebears of the modern pigs and peccaries. By the end of the Oligocene, the ancestors of the antlered and horned ruminants had arisen. The latter animals became the predominant ungulates by late Miocene time, diversifying so extensively that zoologists studying the antelopes and oxen have had difficulty in dividing the vast continuum of surviving forms into strictly defined subgroups.

Classification of the artiodactyls

The fossil remains of these animals are so numerous that no one has yet been able to review them all and draw a clear picture of the radiation of the modern lineages from their common stem. The classification scheme now in use imposes a measure of order within the tangled assemblage and, though far from perfect, is helpful to investigators studying the group. The most advanced artiodactyls have been housed in the suborder Ruminantia, a category which almost everyone agrees should include the deer, giraffes, and cattle, animals further associated in the infraorder Pecora. Some workers also place in the suborder Ruminantia the North American pronghorns and the camels and their extinct relatives, because these animals qualify as ruminating forms. Others, however, prefer to set the camels and their kin aside in the suborder Tylopoda and to isolate the pronghorns also, pointing out that these artiodactyls evolved separately and, at least in the case of the camels, have a somewhat different ruminating mechanism from that of the deer, giraffes, and cattle. Forms that do not chew the cud have been placed in the suborder Suina. Although some systematists include here (beside the pigs, peccaries, and larger amphibious forms) *Diacodexis* and its most primitive descendants, other workers set the earliest artiodactyls apart in their own suborder, Palaeodonta.

Reasons for the success of the artiodactyls

The reason for the triumph of the artiodactyls over the other medium- and large-sized herbivores is not easy to explain. Their greater success cannot be attributed entirely to the progressive traits of the ruminants, because the pigs, which, like the tapirs among the perissodactyls, remained at a relatively primitive level, have withstood quite well the rigors of the Pleistocene and the depredations of the carnivores and man. Initially, it seems, the modification of the astragulus, which at once prevented sideward bending at the ankle and facilitated movement fore and aft, gave a significant advantage to the animals in which it appeared. Certainly, the changes that took place in the tarsal elements and in the more distal part of the foot must have been continuously adaptive, because the arrangement of the bones had reached an optimum point mechanically by the time *Diacodexis* entered the fossil record. With few exceptions, the descendants of *Diacodexis* evolved square-crowned upper molars and enlarged lower ones that masticated plant materials efficiently. The pigs and peccaries, which retained both upper and lower incisor teeth as well as strong canines, never became restricted to feeding upon vegetation but maintained their ability to utilize a wide variety of foodstuffs. The canine teeth, huge in some forms, allow the animals to dig for roots and other succulent underground plant parts, and, more important perhaps, provide them with an effective means of defense. The tusks of the hippopotamus, combined with its adaptations for an amphibian existence, have apparently contributed to the survival of that form at a time when the large ungulates have been hard-pressed.

Structural adaptations for running

Although the ruminants benefit from the special structure of the astragulus and their specialized stomach, they enjoy the advantage of a different complex of characters from that which sustains the Suina. Most of them have abandoned canine tusks as instruments of defense and have come to depend instead upon their legs to carry them out of the way of danger. In becoming adapted for running, they evolved, in addition to the double-pulleyed astragulus, legs remarkably convergent in structure to those of the horse (Fig. 9.50). The limbs retained a short proximal segment for attachment of the major muscles that drive the leg while their lower portions lengthened, increasing the size of the stride. The fibula disappeared, as it did in the equids, and the shaft of the ulna blended with that of the radius, leaving only the olecranon process, or elbow, projecting to provide anchorage for the muscle that extends the lower leg. In several families of ruminants, independently, the side toes and their supporting metapodials vanished, narrowing the foot to the third and fourth digits. These toes always remain separate and slightly divergent, but the metapodials proximal to them frequently unite to form a single cannon bone. Except in the camels and giraffes, which move the two legs on the same side of the body forward at the same time, the running gait of the ruminants involves sequential movement of the four feet, as in horses. Because their vertebral column has a longer lumbar region, however, the ruminants display a bounding motion and have produced forms adapted for jumping in rocky or rough terrain.

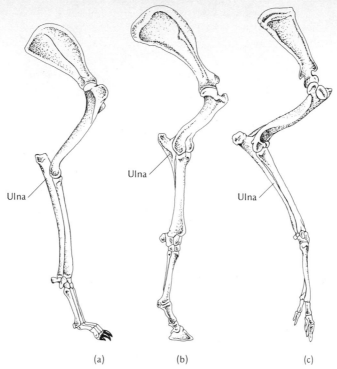

Ulna

Ulna

Ulna

(a)   (b)   (c)

Figure 9.50. Skeleton of the forelimb of three mammals adapted for running.
(a) The dog, showing digitigrade stance and no loss of digits or long bones.
(Adapted from Sisson and Grossman.) **(b)** The horse, showing unguligrade
stance, reduction or loss of all digits except the third, and virtual disappearance
of the shaft of the ulna. **(c)** The deer, showing modifications similar to those in
the horse except that the third and fourth digits remain.

Modifications as-
sociated with
plant-eating in
ruminants

The ruminants have become obligate herbivores but highly
specialized and efficient ones (Fig. 9.51). In both camelids and
pecorans, the anterior teeth in the upper jaw have been gradually replaced by a tough pad against which the lower cutting
teeth crop foliage or grass. The camels, in which this cropping
mechanism developed more slowly, still retain small but typical
canines and a single pair of upper incisors. In pecorans, the
upper incisors disappeared entirely at an early date, and the
canines behind them vanished as the most advanced forms
differentiated during the Miocene. The lower canines in all
pecoran ruminants became incisiform, lengthening the row of
cutting teeth that worked against the pad. Although the premolars never broadened as they did in perissodactyls, the molars
increased in size, bringing about a lengthening of the facial

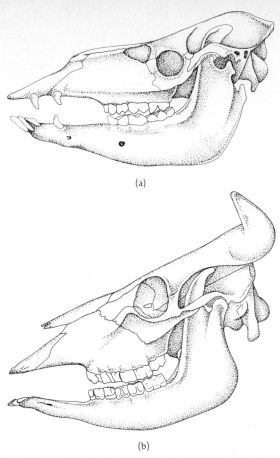

(a)

(b)

Figure 9.51. Skulls of two ruminating artiodactyls. **(a)** *Camelops*, a Pleistocene camel (Tylopoda), showing loss of upper anterior incisors in development of cropping mechanism. (From Romer.) **(b)** *Bos*, the modern ox (Pecora), showing loss of all upper incisors as well as upper and lower canine teeth. (From Sisson and Grossman.)

region similar to that which occurred in horses. Even in the browsing ruminants in which the cheek teeth remained low-crowned, the tooth cusps expanded into crescentic ridges that improved the ability of the teeth to chop and grind. As in the horses, the masseter muscle enlarged, its fibers inserting upon the wide angular region at the rear of the lower jaw. The combined action of the masseter and pterygoid muscles in ruminants is able to produce a rotary motion that enhances grinding, because the shallowness of the depression into which the mandibular condyle fits allows the lower jaw to slide back and forth and from side to side. The food the animals swallow passes first from the esophagus into a large outpocketing of the gut, called the rumen, where it is moistened and invaded by bacteria and ciliated protozoans which partially decompose it. The advantage of having such a chamber near the anterior end of the gut rather than in the more posterior position occupied by the caecum of the perissodactyls is twofold. The ruminants are able to utilize the organ, much as grain-eating birds employ the crop, for storage of food eaten hastily. They return the food, or, as it is now termed, the cud, to the mouth some time later to chew it more slowly and thoroughly. By exposing ingested plant matter first to microorganisms and then, when the cellulose plant cell walls are disrupted, mashing the material between the teeth once again, the animals derive the greatest amount of nutriment possible from a given quantity of herbiage. Although certain substances, including useful ones produced by the microorganisms; are absorbed directly through the walls of the rumen, others do not become available until the bolus of food, swallowed a second time, passes by the entrance to the rumen, moves through two chambers beyond it, the reticulum and the abomasum, and reaches the omasum, or true stomach, where it is subjected to the first stages of chemical digestion by the animal's own enzymes.

Complex behavioral patterns in ruminants

Finally, the ruminants depend to a greater extent than the pigs and their relatives upon complex behavioral patterns for their survival. Their organization into small bands or great herds affords increased protection for the members of the group and by increasing contact and competition among the males assures selective reproduction of well-favored offspring. The intricate social habits of the ruminants as well as their territorial claims are sustained by a rather sophisticated system of communication which depends upon signals received through every sense. Special scent glands beneath the skin on the face, back, and feet have evolved as an important adjunct to this system. Neither these glands nor the kinds of social activity they help promote are unique in the ruminants. What distinguishes the ruminants and makes them a thriving group is the high level of development of these and their other adaptive characters.

Geographical distribution of the early artiodactyls

Although the ruminating and nonruminating artiodactyls have long been the most numerous ungulates, their populations have shifted in place and kind. The finding of *Diacodexis* and all its known relatives in North American deposits suggests that the first members of the order may have lived exclusively in the New World; but if early Eocene sediments of northern Asia ever come to light, it is possible that similar animals will be discovered to have existed at the same time west of the Bering region. There is evidence from the distribution of other kinds of mammals that the northern lands of the Eastern and Western Hemispheres were broadly connected during the early part of the Eocene epoch by corridors north of both oceans, and it is reasonable to suppose that primitive artiodactyls spread from one continent to another as they began to multiply. Palaeodonts with more advanced teeth than *Diacodexis* are known from Eocene strata of Europe and southern Asia. These animals, typified by *Dichobune*, already have the squarish molars with blunted, or bunodont, cusps characteristic of later pigs, but the reduction in the size of their canine teeth makes it unlikely that they were ancestral to the modern suids. They may have given rise, however, to a group of pig-like artiodactyls called anthracotheres.

Amphibious arti-odactyls

The first anthracotheres, which entered the fossil record in the Middle Eocene, resembled *Dichobune* and its relatives in being primitive in their general structure yet having bunodont molars and canines of moderate size. Two trends appeared as the anthracotheres continued to evolve which were to be repeated subsequently in several artiodactyl lines: the animals grew to large size, and their molars began to develop crescentic ridges in place of the bunodont cusps. Despite the transformation in the pattern of their molars, the animals seem not to have become ruminants. Short-legged forms that walked on four toes, they evidently took to wading after vegetation rooted in the shallow water of lakes and rivers, as the amynodont rhinoceroses did (Fig. 9.52). They must have adapted well to amphibious living, for they flourished in Eurasia, where they originated, far into Pleistocene time. A branch of the family which migrated to North America in the Oligocene survived only briefly, but it seems that the anthracotheres may have left descendants in the Old World in the form of the hippopotamuses. First represented in the fossil record by a foot bone from Miocene deposits in Kenya, the latter animals seem in several ways not too distant from certain exceptionally broad-snouted anthracotheres of Lower Pliocene age. The hippopotamuses coexisted with the anthracotheres in Africa and Eurasia and, like them, became extinct in Eurasia late in Pleistocene time. They remain firmly

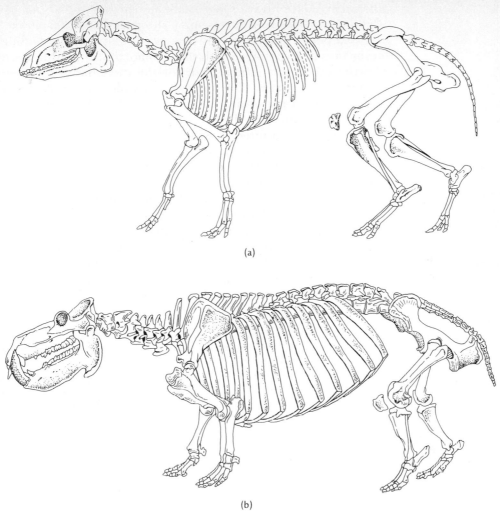

Figure 9.52. Skeletons of two amphibious members of the Suina. **(a)** *Bothrio-don*, an Oligocene anthracothere. (From Scott.) **(b)** *Hippopotamus*, a living form. (From Gregory.)

established in Africa, however, where they occupied the niche left empty by the disappearance of the sub-Saharan anthracotheres.

Caenotheres and entelodonts

By the end of the Eocene, several additional groups of artiodactyls had emerged and spread in areas within the Northern Hemisphere. In Europe, an aberrant family of small forms called caenotheres developed with elongated, slender hind legs and other skeletal traits superficially similar to those of rabbits. Although the teeth of these animals were not at all rabbit-like, the idea persists that they may have been jumping or bounding herbivores that competed with the lagomorphs. If they did so, it was the rabbits that triumphed, for the caenotheres died out during the Miocene. The entelodonts, which can best be described as giant pigs, ranged more widely than the caenotheres but became extinct at the same time. They evolved powerful jaws and teeth that fitted them well for digging the quantities of roots they required to sustain their hulking bodies, but the increased size of the skull that the development of their feeding apparatus entrained must have made them ultimately very awkward animals. By the Lower Miocene, forms, like *Dinohyus*, that reached more than 10 feet in length and had a skull almost a yard long may have been more vulnerable to carnivores than smaller pigs that could slip away more easily. The entelodonts

Evolution of the pigs and peccaries did have smaller relatives which spread as widely as they, and it is from these animals that paleontologists believe the modern

Figure 9.53. Skulls of swine. **(a)** *Archaeotherium*, an Oligocene entelodont. (From Peterson.) **(b)** *Platygonus*, a Pleistocene peccary. (From Gregory.) **(c)** *Sus*, a modern pig. (From Sedgwick.)

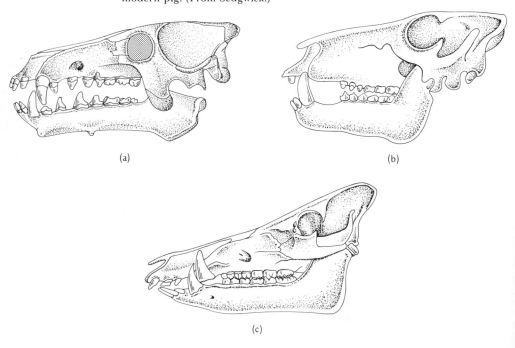

(a)

(b)

(c)

Figure 9.54. The babirusa, a modern wild pig of the East Indies.

pigs may have come (Fig. 9.53). The European form, *Choeropot-amus*, which is known to have existed from late Eocene to early Oligocene time, had a long facial region similar to that of the suids and no specialized structures that prevented it from being close to the base of the new line.

Once the first of the modern pigs appeared in the Oligocene, all other pig-like animals except the amphibious anthracotheres vanished from the Old World. As the years passed, the surviving pigs became more formidable animals than they had been originally. Although they never rivaled the entelodonts in size, they did enlarge, and they developed in the upper jaw continuously growing, outward curving canine tusks. Their molars stayed low-crowned but lengthened as the face did and evolved a peculiar wrinkled enamel that effectively increased the number of cusps. With the exception of the warthog, which wanders into the savannah in Africa, the pigs remained inhabitants of the forest and wet lands beside the rivers, where their pair of short side toes helped to support them on the spongy ground. They spread everywhere in the Eastern Hemisphere and even crossed the deep strait beyond Borneo to reach Celebes. How the ancestors of the horrid-looking babirusa (Fig. 9.54), which lives on that island today, reached it will never be known, but it is certain that other kinds of suids were carried over stretches of water and into the New World by man. Perhaps because the suid pigs did not enter North America during the Tertiary period, a related group of animals, the peccaries, persisted on that continent. The peccaries, which may have sprung from a North American branch of the Eocene stock that gave rise to the suids, were more advanced than the latter forms in the structure of the legs and thus better adapted for running. They did not become as large as the Old World pigs, however, and their canine and molar teeth remained unspecialized. The peccaries proved less hardy than the suids, because, unlike the latter animals, they failed to hold their ground in the temperate region as the Pleistocene progressed. They survive today only in the warm, southwestern corner of the United States and in Central and South America.

When *Perchoerus*, the earliest of the peccaries, ranged throughout the northern continent and anthracotheres and giant entelodonts existed there, other kinds of artiodactyls were also establishing themselves. The group of these animals most successful in the Oligocene, the oreodonts (Fig. 9.55), exhibited a mosaic of characters, resembling at once archaic artiodactyls, ruminants, and pigs. Assemblages of this sort constitute a problem for the systematist because they fall across the neatly drawn taxonomic boundaries within the classification scheme. They do demonstrate, however, that populations radiating from the same base carry on unique combinations of ancestral traits and evolve others which appear segregated in divergent but related stocks. In fitting animals like the oreodonts into a

Figure 9.55. Skeleton of the Oligocene oreodont, *Merycoidodon* (= *Oreodon*). (From Scott and Jepsen.)

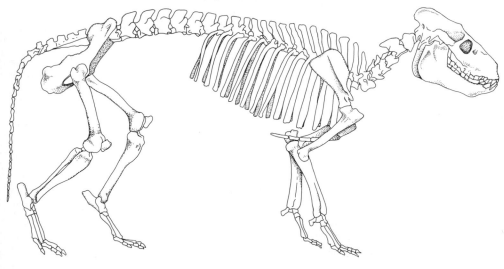

conventional suborder, the systematist must select what he believes is a key character and use it as a guide. Since investigators have differed concerning the key character of these artiodactyls, the oreodonts appear in the Suina in some schemes and in the Ruminantia in others. The development of crescentic ridges on the molars and incisiform lower canines provides the basis for classifying them as ruminants. Like ruminants, also, the oreodonts evolved a complete postorbital bar and, in advanced forms, exhibited high-crowned cheek teeth that imply adaptation to grazing. The great number of individuals found fossilized in the American Northwest strongly suggests that the animals moved in herds, as ruminants do in open land. If the oreodonts were ruminants, they were rather small ones, for the largest were no bigger than good-size pigs. Their long body, short legs, and four-toed feet made them similar in appearance to pigs, and as a result some paleontologists prefer to regard them as closely allied to the swine. Although the oreodonts remained confined to North America throughout their existence and never lost their archaic body form, they did diversify to a considerable degree. Most species were conventionally hoofed, but one group, the agriochoerids, reacquired claws. As with the chalicotheres, paleontologists have had to guess what the habits of these clawed ungulates might have been, since there are no such forms extant to observe. The agriochoerids, never a large group, died out early in Miocene time, but the main body of oreodonts persisted until the pronghorns and the deer displaced them in the late Miocene and early Pliocene.

Besides the oreodonts, there were other artiodactyls in North America which evolved traits characteristic of ruminants. Although these animals also began as short-legged four-toed forms, they were more progressive than the oreodonts, developing long limbs and two-toed feet as well as the dentition typical of artiodactyl herbivores. The longest enduring of these animals were the camels and their relatives. The first members of this tylopod assemblage appeared in the late Eocene, already subdivided into two lines. One, which included *Camelodon* and *Oromeryx*, consisted of animals that retained a great many primitive traits and soon vanished without issue. The other advanced faster and led eventually to the modern camels. From *Poëbrotherium*, an Oligocene member of the more progressive group, it is apparent that the rate of change in the structure of the limbs was more rapid than that of the dentition. *Poëbrotherium* did have molars with crescentic ridges (Fig. 9.56), but its upper incisors were all in place and the bony bar behind the orbit was still incomplete. The legs, on the other hand, were already two-toed and considerably longer proportionately than those of nonruminant artiodactyls. By the Miocene, when the climate in North America had become rather arid, the camels had evolved the almost toothless upper gums that rendered them capable of cropping hard, dry, and even spiny vegetation. They still bore hoofs, but these were to be lost in the Pliocene, when the camels relapsed into a digitiform stance, a position that gave them great stability despite the absence of side toes. The two remaining digits then diverged from one another and rested upon a newly evolved pad, and each became covered with a nail. Even before the development of the modern camelid foot, however, the animals were evidently extremely well adapted for the environment of the time. In the Miocene, although the camels remained in North America, they became a numerous and diverse group. The tendency toward elongation of the neck that accompanied the lengthening of the legs culminated in the production of *Alticamelus*, a camel which, like a giraffe, could reach high into the trees for foliage not accessible to smaller forms. Of the latter, there were many, including the light, gazelle-like *Stenomylus* and the ancestors of the modern llamas, guanacos, alpacas, and vicuñas of South America (Fig. 9.57). The camelid ruminants entered South America when the Panamanian isthmus emerged at the end of the Tertiary period but never advanced into the tropical lowland regions. The animals, which had evolved thick coats of hair, became permanent residents of the cold Andean plateaus.

At some time in their history, the camels that migrated late in the Pliocene northwestward over the Bering bridge to Asia developed the hump, a structure that increased their powers of endurance and ultimately made the animals especially useful to man. The existence of camels in the Near East and North Africa

Figure 9.56. Skull of Poëbrotherium, an early camelid from the Oligocene of North America. (From Romer.)

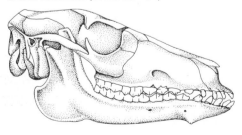

Vertebrate History: Problems in Evolution

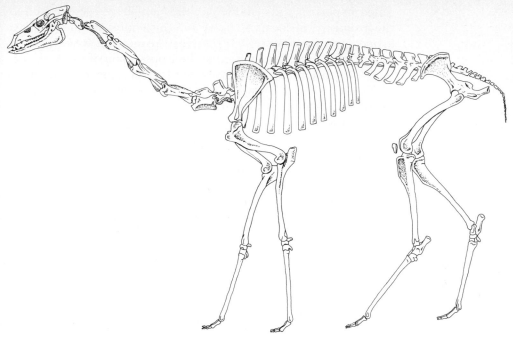

Figure 9.57. Skeleton of *Oxydactylus*, a Lower Miocene camel. (From Gregory.)

is due entirely to man's having kept them there as the only beasts of burden able to withstand the desert conditions in that region. Doubtless because they could metabolize materials stored in the hump when food was not available and retain water in specially constructed pockets in the rumen, the animals were able to establish a population in the Gobi Desert in Mongolia, an area in which they have remained relatively undisturbed. Why the camels should have survived equally well in North America throughout the Pleistocene and then become extinct at the very end of that epoch is still not clear. Some paleontologists have guessed that the coming of the bison resulted in their displacement. Others believe that the camels of North America succumbed belatedly to the forces or conditions that caused the extinction of many large North American mammals in the Pleistocene.

Precursors of the pecorans

Another group of ruminants, more progressive even than the camels, was also represented in North America during the Oligocene and early Miocene. Like other early Tertiary artiodactyls, these animals were small, but they were already quite long-legged forms with crested molars and upper incisors either reduced or completely absent. The structure of these ruminants, called hypertragulids, shows that they may have been related to the primitive pecoran stock from which most of the modern artiodactyls have come, but the North American forms proved to be a sterile line. Specialized in their possession of a caniniform lower first premolar, they gave rise to archaic horned ruminants of North America known as protoceratids and then, during the Miocene, became extinct. The hypertragulids had originated during the Eocene epoch in Asia, and the evolution of the pecorans continued to be centered there.

The source of the pecoran group in the Old World is not clearly demonstrable from the fossil record. As in North America, it seems, there were several populations of artiodactyls, descended directly from forms at the level of *Diacodexis* perhaps, that approached the advanced ruminant condition independently. Middle Eocene animals called anoplotheres seemed to be in the process of developing crested molars from those of bunodont design, but the retention of three, rather than two, large toes on each foot has made their relationship to later ruminants questionable. The late Eocene xiphodonts were somewhat more advanced than the anoplotheres, but they appear from the structure of their legs and teeth to have been evolving parallel to the camels of the New World. Two genera of more generalized late Eocene ruminants, *Amphimeryx* and *Pseudamphimeryx*, could have been precursors of the pecorans, but there are no forms intermediate between these animals and those that can be included more certainly within the pecoran assemblage (Fig. 9.58). The only well-known artiodactyl of the late Eocene that can be assigned definitely to the Pecora is the hypertragulid *Archaeomeryx*, and this little herbivore already possessed the specialized traits that debarred it from being ancestral to any of the existing pecoran lines.

The best representatives of the common stock from which the most advanced ruminants arose are, strangely enough, not fossils but living forms. The chevrotain, *Tragulus*, that lives in Asia and its African cousin, *Hyemoschus*, have preserved almost without change the traits that paleontologists presume are primitive for the Pecora. The animals, often called "mouse deer," are small, solitary, forest forms. They browse and chew the cud, but the development of the special chambers at the base of the esophagus has not progressed as far as in true deer and cattle. The legs in chevrotains are characteristically pecoran in the fusion of the navicular and cuboid tarsal bones and in the simplification of both the tarsal and carpal regions by the loss of certain other elements, but in other ways they approach but do

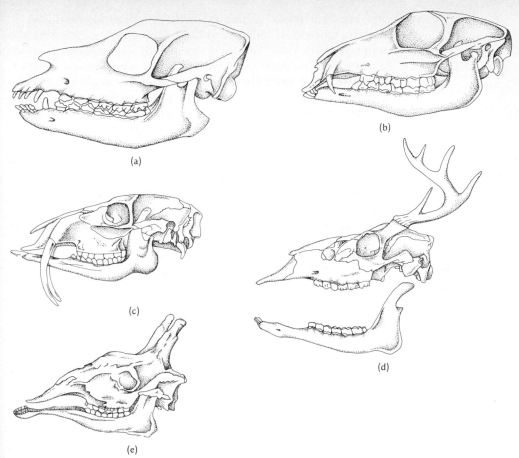

Figure 9.58. Skulls of pecorans. **(a)** *Archaeomeryx*, a primitive traguloid of the Eocene. **(b)** *Tragulus*, the modern Asian chevrotain. **(c)** *Moschus*, the modern Asian musk deer. **(d)** *Dama*, the Recent American white-tailed, or mule, deer. **(e)** *Giraffa*, the Recent African giraffe. (*a* after Colbert. *b, e* from Gregory. *c* from Owen. *d* after E. R. Hall and K. R. Kelson, "The Mammals of North America," vol. 2, The Ronald Press Company, New York, 1959.)

not reach the pecoran condition. The fibula, which is vestigial in the more advanced forms, is united with the tibia in chevrotains but still entire. A cannon bone exists in the hind foot, but in the front one the metacarpals supporting the two principal toes are either separate or incompletely fused. Unlike even the most primitive of the true deer, the chevrotains, or the tragulids as they are more properly termed, still bear hoofs on the short side toes. Although the teeth of *Tragulus* and *Hyemoschus* are similar to those of many early deer, it is possible that the dentition of the chevrotains is not as primitive as that of the unknown Eocene stock from which the Pecora sprang. In the two chevrotains, the upper incisors have vanished completely and enlarged upper canines have evolved, selected for, no doubt, because of their defensive value since the animals have no horns. When Eocene tragulids are discovered, they may prove to have teeth of this sort, but it is also possible that they will have very small incisors and normal-sized canines in the upper jaw like *Archaeomeryx*. Because *Archaeomeryx* and *Tragulus* are similar in their general structure, paleontologists have speculated that the hypertragulids and the tragulids diverged from a common ancestral stem in the Eocene. The tragulids, they think, then separated into the conservative line of which *Tragulus* and *Hyemoschus* are the surviving members and a more progressive one, which by Oligocene time produced the first true deer.

Evolution of the deer and giraffes

The deer that appeared at the end of the Oligocene were already a diverse and widespread group. Both the Eurasian and North American continents were inhabited by archaic forms without antlers and by others which were developing protuberant growths on the top of the head. These animals merited inclusion with the modern deer and the giraffes in the superfamily Cervoidea because the structure of their legs had advanced beyond the tragulid level and, it is believed, the modification of the gut for rumination had reached the modern condition. The latter supposition is based upon the anatomy of the gut in *Moschus*, the Asian musk deer, the only living cervoid that seems similar to the archaic mid-Tertiary forms. *Moschus*, like the archaic Oligocene and Miocene deer, has continued to rely for defense upon long, sharp upper canines, but the more advanced deer lost these teeth entirely. As in other groups of artiodactyls, the more progressive genera among the cervoids became larger and thus stronger and faster. They must have depended upon their legs to deliver them from their enemies, for their new "horns" could hardly have been as effective weapons as the true horns of cattle. As far as paleontologists can tell, the growths from the skull of Miocene cervoids were not yet modern antlers, because they seem not to have been shed. Straight prongs in some forms and slightly branched in others, the "horns" may have been permanently covered with skin, like those of giraffes or the bases of the antlers in modern deer.

Deer in which bony antlers, stripped of skin, are shed annually after the breeding season first became numerous in Pliocene time. Although the black- and white-tailed deer of the genus *Odocoileus* were native to North America, eventually migrating through Panama to reach the continent to the south, most of the modern cervids* evolved in the Old World. The muntjacs and a few other varieties remained confined to Eurasia, but the wapiti, the moose, and the caribou spread eastward into North America during the Pleistocene. None of the Old World forms ever penetrated the lands of the Southern Hemisphere. Either these cervids could not move into the warmer regions adjacent to the north temperate zone, or they arrived on the scene so late that the multiplication of the bison in North America and the antelopes in sub-Saharan Africa prevented their progress southward. The only cervoids that established themselves broadly in Africa were the giraffes. These animals emerged as a separate line in Miocene time, before the parent stock had evolved deciduous antlers extending from the skin-covered knobs on the skull. The giraffes developed their "horns" no further than their Miocene forebears, but they advanced farther than other cervoids in losing all trace of the side toes. The elongation of the neck and forelegs that was to assume extreme proportions in the modern *Giraffa* was already apparent in *Paleotragus*, a form which roamed Eurasia and Africa late in the Miocene era. *Paleotragus*, or early giraffids like it, left two groups of descendants in the Pleistocene. One, the sivatheres, consisted of animals that became heavy-bodied rather than tall and grew elaborate sets of "horns." The other included *Giraffa* and allied forms, animals better adapted for running and browsing among the trees. At the beginning of the Pleistocene the sivatheres and the giraffes inhabited large parts of Eurasia and Africa. The sivatheres, whose niche perhaps partially overlapped that of the large antelopes and other kinds of cattle, did not survive, but the giraffes have maintained a remnant of their population in sub-Saharan Africa despite the radiation there of the bovoid artiodactyls. The only other extant giraffid is the rare okapi that lives deep in the forest of central Africa. This lone form, in which the lengthening of the neck and forelegs is far less pronounced than in *Giraffa*, seems to be a relict derived with little change from *Paleotragus* or a close ally.

Origin of the bovoids

In contrast to the browsing ruminants, which increased in number and diversified gradually over a long period of time, the grazing bovoids entered the fossil record late in the Miocene and radiated explosively as the Pliocene began. Apparently, the evolution of high-crowned cheek teeth lifted this group of

---

*"Cervid" refers specifically to deer—animals in the family Cervidae—whereas "cervoid" applies to any form housed in the superfamily Cervoidea.

pecorans to a new adaptive level and enabled them to enter a niche in which they were challenged later only by horses like *Hipparion.* The early Tertiary ancestors of the antelope, oxen, sheep, and goats that appeared in the Old World and of the pronghorns that emerged in the New are not certainly known. Stirton suggested that they may have been primitive cervoids because, like those animals, the bovoids have in all four limbs cannon bones whose distal ends are keeled from front to rear, allowing movement of the tip of the foot through a wide arc fore and aft. Other paleontologists, like Colbert, have supposed that the cervoids and the bovoids shared a common ancestor at the traguloid level and that each evolved its foot structure independently. Since evidence of a diverse artiodactyl assemblage in the Miocene of sub-Saharan Africa has been discovered, some investigators have even speculated that the Old World bovoids may have arisen on that continent from one of the more advanced cervoid forms like *Paleotragus.*

Although the bovoids surely owe their rapid rise to their newly developed ability to eat grass and run speedily over open ground, the change in the form of their horns must have contributed to their success (Fig. 9.59). Unlike the cervoids, in which bony growths not covered by skin are usually confined to male animals and shed annually, the bovoids have permanent horns that are present (except in a few small antelopes) in females as well as males. The horns of Old World bovoids consist of unbranched prongs of bone covered closely by a sheath of compact, fibrous keratin. In early bovoids, they were short and extended straight backward, but as the animals diversified, the horns evolved in various spiral or curving forms. Since they were strong, sharp, permanent, and present in both sexes, the horns of bovoids were far more useful in fighting than those of any cervoid. It seems that the inclination to employ them in this fashion was selected for, because many of the most advanced bovoids have become exceedingly belligerent animals.

Pronghorns

The horns of the New World pronghorns, or antilocaprids, are somewhat different in structure, suggesting that these animals, often called antelopes, are only distantly related to the antelopes of the Eastern Hemisphere. In the pronghorns, the horny outer covering is of a softer keratin than in other bovoids and is replaced yearly. In the last living genus, *Antilocapra,* the central bony core, which is permanent, curves slightly as it extends upward but has a small anterior branch at its base. The horns of extinct pronghorns were even more elaborately divided and for that reason superficially resembled antlers. The antler-

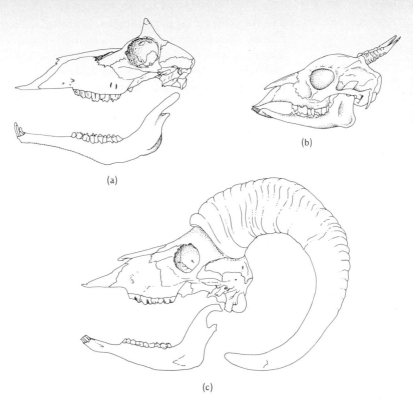

Figure 9.59. Skulls of bovoid pecorans. **(a)** *Antilocapra*, the modern pronghorn of western North America. **(b)** *Cephalophus*, the duiker, an African antelope. **(c)** *Ovis dalli*, Dall's sheep, of western North America. (*a, c* from E. R. Hall and K. R. Kelson, "The Mammals of North America," vol. 2, The Ronald Press Company, New York, 1959; *b* from Gregory.)

like form of the horns in early pronghorns has not been a significant clue to the ancestry of the animals, which appeared suddenly in North America in mid-Miocene time and became extremely numerous and diversified. In the absence of a series of fossils transitional to the antilocaprids, there is no way of deciding whether these forms evolved on the American continent independently of other bovoids from some native stock or whether they split from the assemblage of Old World pecorans and came alone over the Bering bridge. If the pronghorns were immigrants, they were exceptional among the grazing ruminants, very few of which spread to the Western Hemisphere.

Radiation of the bovoids

During the Pleistocene, the shaggy-coated musk oxen and bison traversed the northern land bridge and a mountain sheep and a goat-like relative of the chamois worked their way southward through the Rocky Mountains, but, like the prong-horns, these animals never reached South America. Investi-gators suppose that by the Pliocene epoch, when the great radiation of bovoids began in Eurasia, the climate was already growing cold enough in the northern latitudes to constitute a barrier against the movement of most animals from the Asian plains northward to the Bering region. Thus prevented from leaving the Old World, the antelopes and the heavier-bodied ox-like forms derived from them spread east and west through the parts of the continent still covered with warm grasslands and forests. There, the group continued to diversify, producing small sheep and goats adapted for living in the mountains and, by the beginning of the Pleistocene, larger animals antecedent to the cattle domesticated by early man.

The repeated glaciations of the Pleistocene and the later rapid growth of the human population put an end to the expansion of the bovoids in Eurasia. The species that escaped extinction during the glacial periods were hunted by men of the Stone Age and then displaced by the spread of agriculture and further reduced by the use of firearms. By the seventeenth and eigh-teenth centuries, when men began to think about preserving endangered species, most of the wild bovids* had disappeared. Many of those that remained, like the auroch, *Bos primigenius*, were so few in number that even when protected they could not maintain themselves. Others, like *Saiga*, an antelope of Central Asia, responded well to protection but then again competed with man for space. The saiga, having doubled its population within 5 years after killing it was prohibited, intruded upon the wheat fields that had been planted on the Russian steppes. Only in Africa have the bovids continued to flourish. Undisturbed by cold or (until relatively recently) by dense human settlement, the animals have diversified to an extraordinary degree. Ranging in size from the diminutive dik-dik to the huge eland, the bovids have produced forms adapted to forest, grassland, mountain, and desert environments. Since several genera seem to occupy the same niche, it is possible that, in Africa, the radiation of the bovids is perhaps only just past its height. Even if the activities of man were not becoming increasingly inimical to the multiplica-tion and free movement of these animals, they might, like other groups that reach a similar point in their evolution, undergo a

---

*"Bovid" denotes members of the family Bovidae (goats, sheep, bison, musk oxen, cattle, Old World antelopes). "Bovoid" refers collectively to these animals and the pronghorns of the family Antilocapridae, all of which are housed in the superfamily Bovoidea.

thinning of their ranks as competition among their many species continues.

Early Tertiary mammals in Africa

When the bovids appeared in Africa, that continent already supported a thriving mammalian fauna about which paleontologists know comparatively little. Deposits yielding remains of terrestrial mammals of the Miocene age are rare in sub-Saharan Africa, and those of earlier Tertiary years are still undiscovered there. Even in the northern part of Africa no fossils have been found in sediments laid down before late Eocene time. At Fayum in Egypt, strata dating from the Upper Eocene and Lower Oligocene have given investigators a peephole view of an early mammalian assemblage of remarkable diversity but one whose origin is largely a matter of speculation. The older strata at Fayum were formed where river water entered the sea and contain, besides the remains of precursors of the elephants, remnants of two genera of whales and early sirenians, or sea cows. The younger, Oligocene rocks, formed under more terrestrial conditions, have given up mammalian species belonging to nine orders. Some of the animals, the primates, bats, rodents, creodonts, artiodactyls, and the single insectivore, were representative of orders established elsewhere, but others, mostly herbivores, were apparently peculiar to Africa.

Terrestrial connections between Africa and Eurasia

Since the fossil record of the African mammals of the Tertiary period is scanty after the Fayum years and blank before them, paleontologists seeking clues to the migrations of mammals into and out of Africa have consulted the geologists concerning the history of terrestrial connections between Africa and adjacent lands. From the maps they draw, it seems that those mammals which arose in the north may have reached Africa later than they reached the other continents of the Southern Hemisphere. During the Cretaceous, when marsupials and other archaic forms were evidently entering Australia and South America, Africa was isolated and, in fact, transected by broad swaths of sea. The absence of marsupials from Africa, although it is negative evidence, reinforces the probability that the African continent was an unattainable island when the therian mammals replaced the more primitive Mesozoic types worldwide. By the Paleocene epoch at the beginning of the Cenozoic era, the entire mass of Africa, with Arabia attached, was uplifted and perhaps connected by dry land to the Iberian peninsula of Europe. There may also have been terrestrial links at this time to southwestern Asia, but for most of the Tertiary period Africa and Arabia were cut off from lands to the north and east by the broad Tethys Sea. The sea water flooded the northern part of the continent again during the Eocene, and faunal interchange seems to have ceased until the late Oligocene, when the Atlas Mountains were rising higher and elevations appeared as islands in the strait separating Africa from Spain. Passage between Europe and Africa became easier again for a brief time at the end of the Miocene epoch, when a temporary land bridge emerged where the Strait of Gibraltar now exists, but the Red Sea appeared then, severing Africa almost completely from the Arabian peninsula. By Pliocene time, the Red Sea, which had been confluent at first with waters to the north, opened southward into the Indian Ocean, and, thereafter, migration of terrestrial animals to and from Africa took place only through the Suez region and via Gibraltar during short periods when the strait was closed. The drying of the Sahara eventually created a barrier to the movement of animals southward within the continent in the Pleistocene, so that late immigrants to Africa, which gave the fauna of the northern region a Eurasian cast, never reached the sub-Saharan lands.

Entry of mammals into Africa

The geologists' sketch of the sporadic and tenuous links that have existed between Africa and Eurasia has served as a guide to the times when access to the southern continent was easiest, but from their study of the faunas of Fayum and the Miocene deposits near Lake Victoria and Fort Ternan in Kenya, investigators infer that mammals must have entered Africa in limited numbers by sweepstakes routes during years when direct terrestrial bridges were not available. In commenting upon the rodents at Fayum, for instance, Wood pointed out that the animals seem just to have begun their radiation from an immigrant stock. He assumes, therefore, that the rodents must have arrived in Africa not very long before early Oligocene time, presumably late in the Eocene epoch. By the Miocene, although many of the rodents could still be assigned to the Phiomyidae, the single family represented at Fayum, there were also rodents of other kinds on the continent. Eventually all the phiomyids except two—the modern *Thryonomys*, or cane "rat," and *Petromus*, the rock "rat"—were extinguished by the rodents which invaded Africa after the early Oligocene years. Like the phiomyid rodents, the Fayum artiodactyls must also have spread to Africa during the Eocene, when the submergence of the present northern coast contributed to the isolation of the remaining land mass. The identification of a short-faced pig among them corroborates this hypothesis, because the family to which the animal belongs seems to have originated in the Northern Hemisphere in Middle Eocene time. The existence in the Lower

Oligocene strata at Fayum of six different kinds of anthracotheres suggests either that conditions were especially favorable in northeast Africa for the development of these water-loving artiodactyls or that their adaptation for moving through swampy terrain facilitated the migration of these animals to the continent. The absence from Fayum of many mammals that existed in similar environments in the Northern Hemisphere does bear out the geologists' contention that Africa was quite isolated in the Eocene, however. The fauna from Kenya and the vicinity of Lake Victoria, which dates from the years following the late Oligocene uplift of North Africa, is far more varied than the Fayum assemblage, indicating that passage to the continent probably did become easier during mid-Tertiary time. In addition to the diverse rodents, remains of rhinoceroses and chalicotheres and viverrid, felid, and canid carnivores have been discovered at the Lake Victoria localities. Rabbits appear, too, beside an aardvark, several insectivores, and a number of primates that probably evolved on the African continent from older immigrants. The Fort Ternan deposits are late Miocene in age; they include antelopes which, if they did not arise in Africa, were still newer arrivals.

Archaic African subungulates

Although the identity of the earliest immigrants into Africa may never be known, it seems from the presence of the peculiar and highly differentiated herbivores at Fayum that they may have been condylarths of some sort. That the Fayum forms appear related yet sufficiently diverse to be assigned to four orders argues in favor of their long and divergent evolution in Africa from similar or common ancestors. Since all the animals had unguals intermediate in structure between a hoof and a nail and were apparently herbivores of long standing, their descent from condylarths that reached Africa during Paleocene time is not unlikely. Collectively called subungulates, the animals (Fig. 9.60) by the late Eocene and early Oligocene had not only evolved a variety of land forms but also some that had become adapted for life in the water.

The most diversified subungulates at Fayum were the hyracoids, terrestrial animals ranging in size from hare to hog. Although the hyracoids were in many ways the least specialized of the subungulates, being quadrupeds of moderate proportions, the differences in body size and in cusp design and crown height of the cheek teeth in the six genera identified make it apparent that these animals had already radiated considerably. They commanded a niche, evidently, that they were later forced to share with more advanced and specialized herbivores like

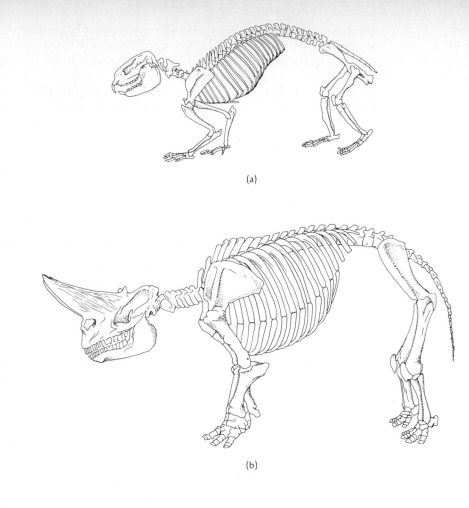

(a)

(b)

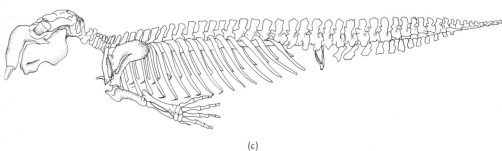

(c)

Figure 9.60. Skeletons of subungulates. **(a)** *Procavia* (= *Hyrax*), the modern cony of Africa and the Middle East. **(b)** *Arsinoitherium*, a giant form from the Oligocene of Egypt. **(c)** *Halicore*, the Recent dugong. (*a, b* from Gregory; *c* from Owen.)

rabbits and artiodactyls. The larger forms were replaced by the latter animals, but a few small species, misleadingly called "dassie rabbits" or "conies," remain in Africa and the Middle East today.

The hyracoids must have been extremely conservative in their evolution because other subungulates at Fayum had undergone radical change in size and in skeletal and dental characteristics. The sea cows in the Eocene strata, although more primitive than their modern descendants, the manatee and the dugong, had already begun to lose their hind legs and to develop highly specialized cheek teeth. Besides anticipating the living sirenians in the structure of the skull, the Eocene forms exhibited a tendency to enlarge that was common among the subungulates. The Lower Oligocene terrestrial deposits contained remains of one animal, *Arsinoitherium*, that had reached the size of a rhinoceros. This great herbivore bore huge bony horns on the nasal bones and had paralleled other kinds of heavy mammals in evolving graviportal, columnar legs and spreading feet. Although its cheek teeth resembled those of some of the hyracoids in their pattern of crests, *Arsinoitherium* lacked the prominent pair of incisors present in those and other subungulates. Because it had evolved divergently and no animals related to it appear in the fossil record, paleontologists have housed *Arsinoitherium* alone in the order Embrithopoda. How long it survived after the early Oligocene is unknown, but it seems to have been the sort of archaic, ponderous herbivore that was rapidly displaced by the more advanced hoofed mammals.

Origin of the proboscideans

Moeritherium, a smaller subungulate from the oldest Fayum rocks, was once thought to have given rise to the proboscideans—the gomphotheres, mastodonts, and elephants (Fig. 9.61)—animals that eventually became far larger than *Arsinoitherium* and far more important in the world fauna than the hyracoids and the sirenians. Although it was only 3 feet tall at the shoulder, *Moeritherium* seemed to foreshadow the proboscideans in the development of long stout upper and lower second incisors and in the appearance of a toothless space between those incipient tusks and the premolars. Continued study of *Moeritherium* convinced paleontologists, however, that the form was, after all, only an aberrant subungulate and that the true ancestors of the tusked animals that flourished in the Miocene were the animals that appeared in the early Oligocene strata at Fayum, *Phiomia* and *Paleomastodon*. *Phiomia*, a form twice the size of *Moeritherium*, and *Paleomastodon*, which grew more than three times as large, seem both to have been adapted

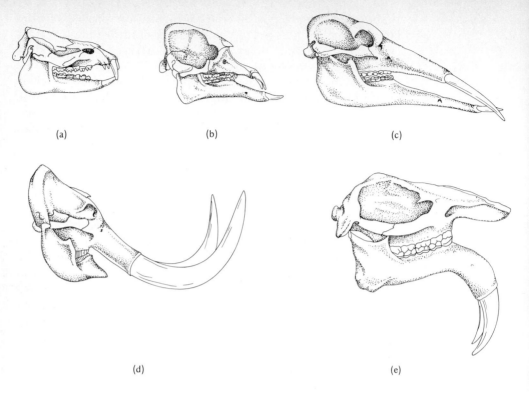

(a)

(b)

(c)

(d)

(e)

Figure 9.61. Skulls of proboscideans. **(a)** *Moeritherium*, a primitive proboscidean of the Upper Eocene and Lower Oligocene of Fayum, Egypt. **(b)** *Phiomia*,
an early gomphothere from the Lower Oligocene of Fayum. **(c)** *Gomphotherium*
(= *Trilophodon*), a long-jawed Miocene gomphothere. **(d)** *Mammuthus primigenius*, the woolly mammoth of the Pleistocene, showing cement-covered lamellar
teeth characteristic of elephants. **(e)** *Deinotherium*, a singular proboscidean of the
Miocene. (From Romer.)

for digging with their enlarged straight lower incisors, a method
of food gathering that preceded the evolution and use of the
trunk. From one or both of these subungulates sprang several
lines of gomphotheres, proboscideans which, like the mastodonts descended from them, had heavily cusped cheek teeth,
all or most of which erupted and functioned simultaneously.
The elephants, in which three elongated, ridged molars succeeded one another during the life of the animal, were relatively
late offshoots from gomphothere stock, appearing near the close
of the Miocene.

The radiation of the gomphotheres during the middle and
late Oligocene years cannot be traced, because fossiliferous

rocks of that time have not been found in Africa and proboscidean remains of that age are not known from other continents. Gomphotheres appear again in the record at the beginning of the Miocene epoch, differentiated into more than a half dozen genera and spread throughout the Old World. From the Miocene onward, bones and teeth of the large animals are plentiful and whole specimens of their Pleistocene elephantine descendants have been discovered frozen in Siberia and the arctic region of North America. Despite the wealth of fossil material, paleontologists have had great difficulty in arranging the known genera properly in phylogenetic series. Since all the derivatives of the Fayum proboscideans grew larger, they evolved in parallel the complex of characters that permit the support and feeding of a body of huge size. Unlike animals in many of the other mammalian orders, the gomphotheres seem not to have developed, in combination with their common traits, an admixture of special, adaptive features that would allow investigators to separate the members of different lines. In becoming taller and heavier, all the gomphotheres acquired appendages in which shearing stress was reduced to a minimum. The weight of the hindquarters was projected straight downward, passing from the vertically aligned iliac bones to the upturned heads of the femora beneath and thence through the long, stout femoral shaft on each side to the short shank and wide, padded foot. In walking, knee and elbow bent very little, and the vertebral column, which supported a long series of ribs, barely flexed.

Structural adaptations for feeding in gomphotheres

The changes that occurred in the head and teeth of the gomphotheres were virtually all related to the improvement of food gathering and mechanical digestion. Since a large head and a short, strong neck to support it continued to be selected for in every line, as the animals grew taller, reaching down to root for food with the lower tusks would have become more and more difficult. Under the circumstances, the gomphotheres probably became increasingly dependent upon browsing to gather vegetation. No cropping mechanism like that of the ungulates appeared, however, for all the incisors except the tusks in both jaws had been lost. Instead, the skin of the nose and upper lip with its underlying musculature elongated to form a trunk or proboscis, that could strip leaves from branches and stuff them into the mouth. As the trunk evolved, the mandible shortened, allowing the trunk to hang downward, and the lower tusks gradually grew smaller. The cheek teeth, meanwhile, increased in surface area by elongating and developing extra rows of large cusps or supplementary smaller ones. The crowns of the teeth became somewhat higher, and delayed eruption of the posterior molars provided the gomphotheres with some additional unworn tooth surface as they grew older. The enlargement of the cheek teeth and, in some forms, the growth of great upper tusks entrained alterations in the shape of the skull. The face became short and deep with much room in the maxillary bones for the development of the tooth buds. The premaxillary bones jutted forward to sheathe the bases of the tusks, and, posteriorly, the skull roof rose high, creating an expanded occipital region for the attachment of the muscles that held up the weighty head.

Deinotheres

Only one group of tusked forms, the deinotheres, deviated significantly from the common course of proboscidean evolution. These animals, all species of *Deinotherium*, could be distinguished from other Miocene and Pliocene members of the order because they lacked upper tusks entirely but produced strong, downwardly curving lower ones at the end of the mandible. From the dorsally oriented nasal openings, it is presumed that *Deinotherium* had evolved a trunk and that it extended forward and downward in front of the lower jaw. Since the animals grew as large as other proboscideans and seem to have browsed with the aid of a trunk, it is difficult to imagine how they could have used the peculiar tusks. Engaging their pointed ends upon anything in the animal's path would have required *Deinotherium* to toss its head upward in an extremely awkward manner. The deinotheres survived, not only with these curious lower incisors, but also with a primitive set of cheek teeth. Instead of becoming longer and higher-crowned, the teeth remained small and developed simple transverse crests. The skull and mandible did not deepen to the same extent as in other proboscideans, and as a result the jaw muscles would have been relatively short and the gape not so wide as in the more progressive forms. In the case of the deinotheres, the features of the skull and teeth selected against in other lines apparently functioned to the animals' advantage, for the last of these peculiar beasts were not extinguished until the general failure of large mammals in the Pleistocene.

Uncertainty concerning relationships among the proboscideans

When the remains of the deinotheres are set aside, paleontologists have left in their collections scores of fossils that they can segregate into separate lines only by remarking upon similarities in tooth structure, in size range, and in the degree of progress in the development of typical proboscidean characters. In the 1930s and 1940s Osborn, Watson, and several other investigators tried, by painstaking comparative studies, to identify the different lineages and to interrelate them, but the material proved so confusing that they ended with a bewildering array of new generic names and evolutionary schemes that varied widely. They differed in their interpretations of the relationship of *Moeritherium, Phiomia,* and *Paleomastodon* to the later proboscideans and could not agree concerning the ancestry of specialized mid-Tertiary gomphotheres like *Rhynchotherium,* in which the lower tusks turned slightly downward, and *Platybelodon,* in which they broadened to form a shovel-like tool. In contrast to the gomphotheres, which slowly evolved a shortened lower jaw and by Pleistocene times developed forms like *Stegomastodon* and *Cuvieronius* that no longer possessed lower tusks, a group of proboscideans was recognized, which were high-skulled, short-jawed, and almost bereft of lower tusks from their first appearance in the Miocene. These animals, which presumably arose from early gomphotheres, were set aside in a separate family, the Mastodontidae, and divided into half a dozen genera, which were traced worldwide to the Pleistocene.

Origin of the elephants

The origin of the elephants remained another point in dispute. Since paleontologists were most familiar with Pleistocene members of the Elephantidae found fossil in Eurasia and North America, they attempted at first to relate the mammoths and elephants to Pliocene mastodons. Searching for proboscideans that showed signs of evolving the high-crowned, ridged molars that allowed the Pleistocene mammoths and elephants to shear grass, investigators singled out *Stegodon* and its older relative *Stegolophodon* as the transitional forms. On the basis of molar structure, it did appear that these animals could have been the ancestors of the elephantids, but new fossils found in sub-Saharan Africa in the 1960s produced evidence that the elephantid line originated, not in Asia in the Pliocene, but much earlier in Africa. Within the last several years, Maglio and other workers have discovered a group of primitive elephants belonging to the genus *Stegotetrabelodon,* which was evidently widespread on the African continent in middle and late Pliocene times. The molars of these animals were low-crowned, like those of gomphotheres (Fig. 9.62), but the heavily enameled cusps had begun to coalesce into knife-like transverse ridges. The stegotetrabelodonts, which may have separated from the gomphotheres as early as the end of the Miocene, apparently gave rise in the middle Pliocene to *Primelephas,* a form in which the molars were still low-crowned but were slightly more advanced in their

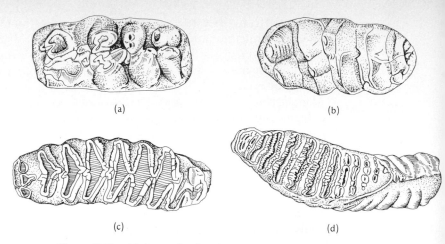

Figure 9.62. Molar teeth of proboscideans in occlusal view. **(a)** *Serridentinus*, a gomphothere. **(b)** *Mastodon rugosidens*, a mastodont. **(c)** *Loxodonta africana*, the modern African elephant. **(d)** *Elephas indicus*, the modern Indian elephant.

pattern of shearing ridges. Since late Pliocene and early Pleistocene deposits in Africa reveal a series of elephantids that were surely ancestral to the Pleistocene forms known from Eurasia, it appears that *Stegolophodon* and *Stegodon* could only have been mastodonts which evolved a molar structure convergent with that of the already well-established elephants.

At the beginning of the Pleistocene, the elephantids were a flourishing group. Their molars had become high-crowned with a large number of dentin-filled, enamel-covered cutting plates held together by cement. Since the teeth erupted successively and gradually moved forward, the animals came into possession of a slow-wearing, self-renewing shearing mechanism that enabled them to eat grass. As in horses, the shift from browsing to grazing prompted the radiation of the stock in which it occurred. Elephants assigned to *Elephas*, *Loxodonta*, *Mammuthus*, or other genera spread throughout the Old World during the Pleistocene, and some species of *Mammuthus* reached North America. Perhaps because the elephants, like the bovid artiodactyls, evolved in Africa and began their migrations late, their colonization of the Western Hemisphere was limited. Whereas more than half a dozen genera of gomphotheres and mastodonts had passed eastward over the Bering bridge by the close of the Pliocene and *Cuvieronius* and two other forms continued through Panama to South America, all the elephants except *Mammuthus* apparently were stopped by the cold in northeastern

Radiation of the elephants

Asia. Of the species of *Mammuthus* that emerged in North America, some became adapted to its warmer, temperate part, but none is known to have penetrated the tropics and invaded the continent to the south. The career of *Mammuthus* and of the advanced gomphotheres and mastodonts everywhere came to an abrupt end as the Pleistocene epoch gave way to Recent time. Some of the animals lived long enough to have been hunted and painted by early man, but, like the modern rhinoceroses, the proboscideans have now been reduced to relict populations, one in India and southeast Asia, consisting of varieties of *Elephas*, and the other in Africa, of *Loxodonta*.

Primates at Fayum   With the proboscideans and other subungulates in the Lower Oligocene strata at Fayum were remains of other mammals, primates, which seem also to have been evolving in isolation on the African continent for some time. The fossils consist for the most part of fragments of jaws and teeth, for the animals were evidently washed into stream channels after death and their bones disarticulated before coming to rest in the mud. Five or six genera have been recognized, all apparently primitive members of the Catarrhini, advanced primates with two incisors, one canine, almost always two premolars, and three molars in each half of the upper and lower jaw. The presence of this assemblage of forms at Fayum and their absence elsewhere in Oligocene deposits suggests that, like the proboscideans, the Old World primates arose in Africa and spread to Eurasia. The principal evidence against an African origin for the Catarrhini, which includes cercopithecoid monkeys, apes, and man, has been the discovery in late Eocene strata in Burma of teeth which have been attributed to two genera of apes. If these teeth have been identified and dated correctly, they imply the existence of catarrhines in Eurasia at the time when the omomyids and other lower primates were diversifying rapidly there. Although it is reasonable to assume that the first catarrhines could have been produced on the northern continent where primitive primates were numerous, the lack of catarrhine remains at other Eurasian Eocene sites has made this possibility seem unlikely. The variety of the Old World forms at Fayum has led Simons to suppose that the center of catarrhine evolution was, in fact, in Africa and that the Burmese apes were migrants, Oligocene rather than Eocene in age. Patterson agrees that the catarrhines probably arose in Africa and strongly doubts that the Burmese animals were apes.

Because of the fragmentary nature of the fossils and the generalized character of the genera they represent, it has proved difficult to link the Fayum primates to other forms. Such cranial elements as are known reveal that their possessors had the forward-turned eyes, bone-encased orbits, short face, and large brain associated with advanced members of the primate order, but there is little basis for diagnosing the relationship of the animals more specifically. *Oligopithecus*, a genus known from part of its mandible and lower teeth, is believed by Simons to be somewhat similar to Eocene omomyids and so supportive of the theory that the catarrhines as well as the New World platyrrhines may have come of omomyid stock. Another Fayum specimen, *Parapithecus*, based upon a dozen fragments, including an almost complete mandible with teeth in place, appears to be tarsioid in its affinities. With so few skeletal remains in hand, both of the Fayum genera and the northern Eocene animals, it is impossible to carry speculations about the interrelationship of these primates any further. Conclusions are equally hard to draw concerning the liaison between the Fayum catarrhines and more advanced members of the group. One form, *Aeolopithecus*, seems to foreshadow the gibbons, the smallest and in some ways the most primitive of the modern apes. *Propliopithecus* and *Aegyptopithecus* cannot be associated with particular living catarrhines but were certainly precursors of the Miocene apes.

Origin of Old World monkeys

The remains at Fayum suggest that the cercopithecoids, or Old World monkeys (Fig. 9.63), had also begun their emergence by early Oligocene time. *Parapithecus* and another small primate, *Apidium*, found at Fayum exhibit the four-cusped, rather high-crowned molars pinched transversely in the middle and the peculiarly shaped last premolar that characterize the cercopithecoid monkeys of today. Paleontologists suspect that the cercopithecoids may have undergone separation into the two subfamilies now extant by the beginning of the Miocene. One of these subfamilies consists of monkeys, like the langurs, which remained entirely arboreal and developed a stomach with complex subdivisions for food storage; the other, consisting of forms with a simple stomach and cheek pouches for food retention, includes small, arboreal species and macaques and baboons that became larger and more terrestrial. The evidence on which the existence of two groups of cercopithecoids in the Miocene is predicated comes from the shape of certain skull and jaw elements and is admittedly tenuous, but discovery of cercopithecoid remains in widely distant regions of the African continent indicates that by Miocene times the Old World monkeys were thriving.

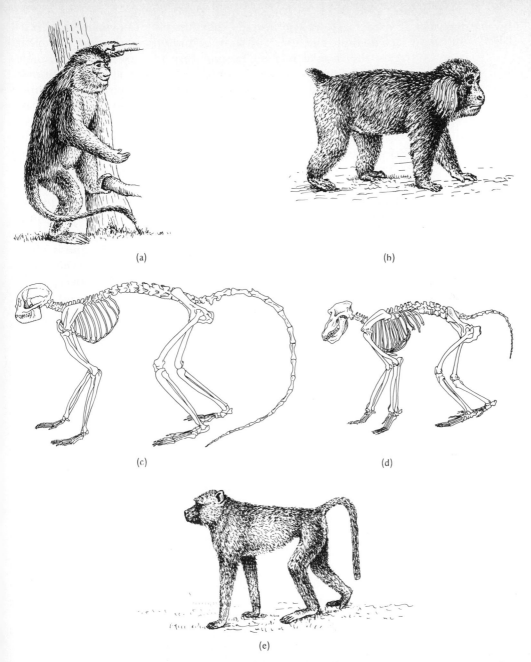

Figure 9.63. Old World monkeys. **(a)** *Semnopithecus*, the langur. **(b)** *Macaca*, the macaque. **(c)** Skeleton of *Cercopithecus*. (From LeGros Clark.) **(d)** Skeleton of *Papio*. **(e)** *Papio*, the baboon.

Since the cercopithecoids split from the catarrhine stock that gave rise to the apes and man before Oligocene time, it is clear that the monkeys never entered into the ancestry of the most advanced primates. The latter forms are described as hominoids because they share a significant complex of characters that man recognizes in himself. What paleontologists have said about the evolution of the Hominoidea has been publicized far beyond scientific circles, for people generally have been curious about the origin of their own kind and, especially in the last 150 years, about the question of their relationship to apes. Unfortunately, the conclusion of the paleontologists, that man did descend from some primitive ape-like stock, has been distorted by the general impression that apes have always been, as they are now, long-armed, brachiating animals with ferocious-looking canine teeth. This narrow but firmly held concept of the ape is often not completely dispelled by the explanation that the older members of the family Pongidae did not exhibit the special traits characteristic of the gibbon, orangutan, chimpanzee, and gorilla of today.

The remains of the apes that lived during the Miocene are extremely fragmentary (Fig. 9.64), consisting of jaws and teeth from Europe and southern Asia ascribed to *Dryopithecus* and rather more complete remains, including limb bones, from Africa, first assigned to a separate genus, *Proconsul,* and then judged by some workers to be identical to the Eurasian form. The arms of the African animals were still slightly shorter than the legs, indicating that the African dryopithecines, at least, were not fully adapted for swinging hand over hand through the limbs of the trees. The skull roof in these apes was domed but without the beetling brow ridges over the orbits or the bony crests developed in modern pongids for the attachment of powerful neck and jaw muscles. Since the anterior teeth were not grossly enlarged nor the jaws especially heavy, the lower part of the face did not protrude markedly. The dryopithecines did show a resemblance to the living apes, however, in having canine teeth that projected beyond the rest of the dentition and in having a first lower premolar with a single cutting edge rather than a bicuspid occlusal surface.

Many investigators believe that the dryopithecines were sufficiently generalized in structure to have given rise both to the later apes and to the hominids,* the group of primates which

---

*"Hominid" denotes a primate in the family of man (Hominidae). "Hominoid" refers collectively to men, pongid apes, and *Oreopithecus*—forms included in the superfamily Hominoidea.

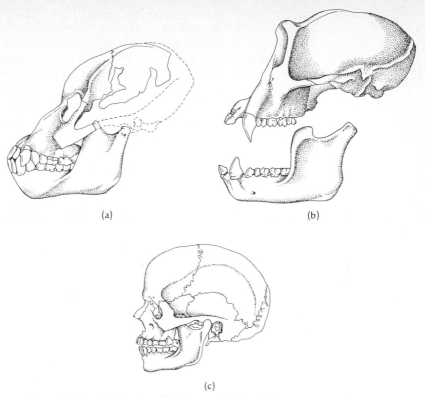

(a)

(b)

(c)

Figure 9.64. Skulls of apes and man. **(a)** *Dryopithecus* (= *Proconsul*), a Miocene and early Pliocene anthropoid of the Old World. (From LeGros Clark and Leakey.) **(b)** *Pan*, the Recent chimpanzee. (From LeGros Clark.) **(c)** *Homo sapiens*, modern man. (From Owen.)

includes modern man. This idea has been questioned by workers who believe that the projecting canines of the dryopithecines represent teeth already undergoing specialization in the pongid direction and that the small canines characteristic of hominids must have been inherited from older, more primitive primate ancestors. Supporters of this hypothesis presume that the hominids existed as a line separate from the dryopithecines during the Miocene but are not preserved in the fossil record because their habitat was one in which conditions were not conducive to preservation. Proponents of the dryopithecine apes as common ancestors of the living hominoids have countered this argument by pointing out that there is in this case no necessity to propose a hypothetical group of Miocene hominids. It is certainly possible, they assert, for a line of primates with small canines to have descended from progenitors with larger ones. Such a change has occurred in the evolution of the modern cats as well as in other groups of mammals, and in the hominid dentition a number of features suggest that reduction in tooth size has, in fact, taken place. Whereas in some specimens of early man the canines are quite prominent and separated by a short space from the lateral incisors in a manner reminiscent of pongids, in living man, *Homo sapiens*, the canine crowns are routinely no bigger than those of the adjacent teeth and form part of a continuous tooth row. Although the pongid diastema never appears in modern man, the canines occasionally erupt with especially sharp summits that project beyond the teeth on either side. The roots are always stout enough to support crowns several times bigger than the existing ones, and the permanent canines still erupt in the fashion of large, heavy teeth, late and slowly.

Origin of the hominid line

Available fossil evidence appears to be in favor of the separation from other dryopithecines, during the early Miocene or even before, of a population of increasingly terrestrial individuals that became the root of the hominid line. Far from being an exceptional evolutionary event, the transition from a wholly arboreal life to one in more open country occurred elsewhere among catarrhines in the emergence of the baboons and their relatives and eventually in the stock leading to the gorillas. An increase in size that made movement through the treetops more difficult may have impelled these Old World primates to adopt new habits, but it seems that a change in the climate may also have contributed to their spread away from the forests. A period of relative aridity that brought about shrinkage of wooded areas and expansion of drier grasslands could very well have promoted the survival of animals able to travel, forage, and protect themselves out from under the forest canopy. That the evolution of terrestrial, even semibipedal, hominoids may have taken place several times during such an interval is suggested by the discovery in early Pliocene deposits in Italy of a form called *Oreopithecus*. This animal, known from extraordinarily complete dental, cranial, and postcranial remains, was neither a hominid nor a pongid but represented an independent and ultimately unsuccessful catarrhine line. It bore a unique mosaic of characters: keeled lumbar vertebrae and molar teeth similar to those of cercopithecoids, elongated arms somewhat gibbon-like in proportions, and small canines followed in the lower jaw by bicuspid premolars like those of men. The long arms and curving fingers of *Oreopithecus* imply that it was adapted for brachiating through the trees, but its broad pelvis could mean only that it stood upright or nearly so part of the time. Since *Oreopithecus* was as large as a chimpanzee, it might have been at first glance hard to distinguish it from semibipedal forms of the early hominid group.

Where the hominids arose is a question as difficult to settle as the time and circumstances of their origin. Since 1891, when Eugene Dubois discovered a human skullcap in Middle Pleistocene sediments in Java, remains of primitive man have been found in many places in Eurasia and Africa. The discovery in Pliocene and earliest Pleistocene strata in South and East Africa of forms that are clearly hominid and of tools dated at about 2.6 million years has caused many investigators to assume that the first hominids appeared in or near that region. The terrain in that part of Africa during late Tertiary times may well have supported the mixed forest and grassland areas that presumably encouraged the evolution of terrestrial catarrhines, and Miocene remains prove the presence in East Africa of genera like *Ramapithecus* (= *Kenyapithecus*) that are thought to have been already diverging in a hominid direction. Within the last decade, Patterson has described part of a humerus and a lower jaw fragment from Kenya that demonstrate the existence of hominids in that region 4 to 5.5 million years ago, late in Pliocene time. Although these remains give no clue to the level of the cranial development of the hominids or to their mode of locomotion, it does seem possible that these forms were close to the base of the line from which *Homo* eventually sprang. While the discovery of human fossils of such great age in Africa argues strongly for the origin of man there, the possibility still exists, of course, that similar or even older remains may someday be found elsewhere. During Miocene and Pliocene times, dryopithecine apes and more advanced forms like *Ramapithecus* were widespread in the warm parts of the Old World, and it is conceivable that the population to which Patterson's hominids belonged could have arisen beyond the borders of Africa in southern Eurasia.

Australopithecine hominids

Whether or not the transition from hominoid to hominid took place in sub-Saharan Africa, the men who lived there beginning at least 5.5 million years ago and who by the early Pleistocene may have been tool users certainly represent a developmental stage intermediate between forms at the level of *Ramapithecus* and the more advanced mid-Pleistocene men. Although R. A. Dart, who described the first of the primitive hominid skulls discovered in South Africa, named his specimen *Australopithecus* or "southern ape," he was quite aware of its hominid structure and possible relationship to modern man. After he published his account of *Australopithecus* in 1925, R. Broom joined the search for additional remains and eventually collected parts of two more skulls. One of them, which seemed distinctive because of its especially heavy jaws and large teeth, Broom called *Paranthropus robustus.* L. S. B. Leakey erected still another genus, *Zinjanthropus*, for the skull he unearthed in Olduvai Gorge in East Africa, since he regarded it as significantly different from the South African specimens. As paleontologists continued to study these fossil hominids and others found later, however, it became apparent that all the individuals discovered had reached the same level of development and that none, except perhaps *Paranthropus*, exhibited traits variant enough to merit generic distinction. *Australopithecus*, as almost all the early Pleistocene hominids are now known, was a small form between 4 and 5 feet tall with teeth and jaws human in character but sufficiently large to make the lower part of the face protrude slightly. Analysis of the skulls and meager postcranial remains available makes it clear that in approaching human status the australopithecines had evolved an almost completely upright stance but progressed more slowly in the enlargement of the brain. The broad iliac bones, the orientation of the articular surfaces at the lower end of the femur, and the position of the foramen magnum beneath rather than at the back of the cranium prove without a doubt that *Australopithecus* carried himself erect and had his hands free. How he used his hands was most probably limited not by structural factors but by the complexity of his brain.

From the volume of the cranium it is known that the brain of *Australopithecus* was not much bigger than that of the modern great apes. It is not necessarily valid to conclude, however, that the actions and behavior of this hominid were therefore similar to those of the pongids. It is entirely possible that although the brain of *Australopithecus* was less than half the size of the brain in modern man, the basic neural connections already established were more like those of *Homo sapiens* than those of the gorilla or the chimpanzee. Evidence to support this contention is scanty and perforce indirect. The shape of the cranium, much higher in front and more rounded in the occipital region than in pongids, suggests that besides merely increasing in size the brain of *Australopithecus* had developed, differentially, centers important in man. Since comparative neuroanatomy is a dark field even among living forms, investigators have not pursued this line of reasoning farther. They have, instead, scrutinized the areas where australopithecine remains were uncovered for artifacts that might serve as clues to the behavior of the primitive hominids. At Olduvai, at Lake Rudolf, and at South African sites, stones were found, some of them split and not all of the same materials as the neighboring rocks, that seem to have been carried to the area and used as tools. A few workers are still not convinced that these so-called pebble tools were really tools or even definitely associated with *Australopithecus* or any more advanced hominid, but others, like Dart and Leakey, believe that these creatures not only employed these stones, perhaps as weapons, but also utilized animal teeth and bones for various tasks. The presence near remains of *Australopithecus* of bones of mammals apparently broken open for their marrow implies that besides differing from the pongids in using tools, the australopithecines had developed the habits of hunting beyond the level usual for large primates. The fractures evident in one *Australopithecus* mandible suggest that the early Pleistocene hominids may have become further distinguished from pongids by indulging in the uniquely human practice of attacking members of their own species. Possibly, if fighting between individuals was common, it influenced the direction of human evolution by fostering the selection of the strongest, most aggressive, and cleverest members of the australopithecine population.

Evolution of Homo

By Middle Pleistocene times, when glacial conditions prevailed intermittently over much of northern Eurasia, the australopithecines had been supplanted by men of the type discovered by Dubois in Java (Fig. 9.65). Although Dubois called his hominid *Pithecanthropus*, or "ape-man," the skull in this form was so similar in structure to that of modern man that paleontologists decided eventually to include it in the genus *Homo* as *Homo erectus*. The Java man, as Dubois' specimen became popularly known, and its fellow Peking man (originally distinguished as *Sinanthropus*) had brains about two-thirds the size of the brain in living man and were capable of making stone tools, of killing fairly large animals, and apparently of building fires. If these men were derived from early Pleistocene or slightly older australopithecine ancestors, it seems that the pace of human cultural development had accelerated appreciably.

The physical changes which underlay the spectacular advance from *Australopithecus* to *Homo* during the Pleistocene epoch are

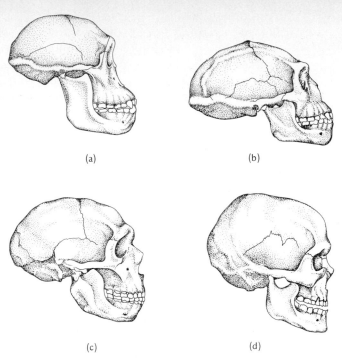

(a) (b)

(c) (d)

Figure 9.65. Hominid skulls. **(a)** *Australopithecus.* **(b)** *Homo erectus* (= *Pithecan-thropus erectus*). **(c)** *Homo sapiens neanderthalensis.* **(d)** *Homo sapiens sapiens* (Cro-Magnon variety.) (From Romer.)

too complex to enumerate, but A. H. Schultz and others have speculated that many of them may have stemmed from a slowing of the rate at which the body develops to maturity. Prolongation of the embryonic stage, during which the cells in the brain multiply and differentiate, for instance, would have resulted in the production of the great number of nerve cells that makes possible the complicated neural circuits responsible for human intelligence and activity. The perfection of the bipedal stance that assured the free use of the hands seems to have come about through the retention of skeletal arrangements that are developed before birth in apes but then abandoned. The nearly hairless condition of *Homo* and the relatively late onset of reproductive function and cessation of growth are further evidence that a retardation of the rate of maturation may have occurred in human evolution. The slowing of development and the preservation of traits characteristic of foetal pongids, if they did occur, were accompanied by mutations that extended the development of old structures in new directions: both the facial muscles and those that control the shaping of sounds must have become increasingly elaborate during this period, creating the possibility of facial expression and speech. The apparent speed with which *Homo* developed the rudiments of Stone Age culture surely depended in great measure upon his growing ability to communicate with his fellows and the formation of stable associations centered about young individuals requiring years of postnatal care. Technological innovations like the flaking of flints soon were not lost with the death of the inventor but were demonstrated persistently to the young and passed to other members of the familial, and later the tribal, group. Techniques employed in daily activities could become complex because for the first time the methods by which individuals solved problems could be transmitted cumulatively and improved upon from generation to generation.

The discovery, in Middle and Upper Pleistocene deposits in Europe, North Africa, and nearby parts of Asia, of neanderthaloid men, forms with a brain equal in size to that of *Homo sapiens* but with a somewhat heavier and differently shaped skull, makes it clear that pithecanthropine men were soon replaced by more intelligent beings presumably descended from them. Since there is evidence that later Pleistocene men continued to prey upon each other as the earlier hominids had done, the ability to fight, and eventually to wage war cooperatively against another group, almost certainly was a more important factor in the triumph of the neanderthalers than any other manifestation of their increased brain power. That these men did employ their enhanced mental abilities in the development of a more complex societal life is, however, implied by evidence of their having buried their dead. The fossil record does not reveal much of the history of the neanderthaloid men—how far they spread or how many descendant races they produced—but it is certain that they were displaced in western Eurasia about 50,000 years ago by the earliest known representatives of *Homo sapiens.*

The replacement of neanderthalers by the Cro-Magnon variety of *Homo sapiens* was so abrupt in Europe that investigators are sure that the more modern man must have evolved elsewhere and then invaded the territory of the less progressive forms. Presumably, Cro-Magnon men and the neanderthalers sprang from common antecedents farther east in Asia or possibly even in Africa because, except for their higher forehead, flatter face, more prominent chin, and straighter stance, the Cro-Magnon men were not far removed structurally from the people they overwhelmed. Only their migration into Europe can be followed in any detail, but rare fossils from other parts of the Old World attest to the rapid spread of the modern form of *Homo sapiens* throughout the Eastern Hemisphere. If, as was probable, the coming of these representatives of *Homo sapiens* to an area inhabited by other varieties of men precipitated the same events that occurred when one group of people invaded the territory of another in historical times, modern men probably subdued neanderthaloid populations in their path and then interbred with them. The result of radiating into different geographical regions, mingling with men already there, and continuing to evolve in isolation was the appearance of the several races of *Homo sapiens* that now exist. Despite the differences they developed, all men of the modern species showed a similar tendency to conceptualize and to express themselves not only verbally but in some artistic form. Either by cave painting, producing ritual objects, or adorning their persons they manifested their consciousness of themselves and their relationship to the world around them. Seminomadic at first, they gathered plant foods, hunted, and fished. When the last of the ice sheets had begun to recede about 30,000 years ago, men from eastern Asia moved over a dry corridor in the Bering region to the New World. Both in the New World and the Old, although agriculture was invented and more permanent settlements sprang up, men continued to hunt with weapons they continued to refine. As well as small animals, they brought down large mammals like mammoths, horses, and bison in great numbers and thus took their place as the most powerful and relentless predators of all time.

Late Pleistocene extinction of the great terrestrial mammals

Many paleontologists believe that increasingly expert hunting by man caused the mass extinction in late Pleistocene and early Recent times of the large mammalian herbivores and the carnivores dependent upon them. The disappearance of the mastodonts and mammoths, of virtually all the remaining perissodactyls and many of the even-toed ungulates, of the ground sloths, and of their enemies among the cats and dog-like forms cannot be explained, these workers say, as part of a general wave of extinction like that at the end of the Paleozoic or the Mesozoic era. When the great terrestrial mammals vanished, no other kinds of land vertebrates went into decline and marine genera continued to thrive. Unlike some of the early Pleistocene mammals, those extinguished between 50,000 and 10,000 years ago seem not to have failed in competition with superior members of their own line or to have died out as the climate worsened. In almost every case, the disappearance of a species left its niche empty. Most of the animals that were lost from the Northern Hemisphere survived the rigors of the earlier Pleistocene Ice Ages and then succumbed as the earth warmed again.

Man is implicated in the destruction of the large mammals, P. S. Martin has pointed out, because his appearance on each continent was followed shortly by the elimination of those animals from the fauna. In Africa and southern Eurasia, where *Homo* established himself first, the major period of mammalian extinction began more than 40,000 years ago. In the colder north, which man reached much later, when his ability to clothe and shelter himself had improved, most of the large animals flourished until about 13,000 years before the present. From that time onward, they vanished, species after species, first in northern Eurasia, then in North America, and finally in South America. The survival of the giant marsupials in Australia some 30,000 years beyond the time when the mammals in nearby southeast Asia started to dwindle was due, presumably, to the fact that man did not enter the island continent until he became skilled enough in sailing to travel considerable distances by sea. In every region where remnants of human hunting and living sites have been found, it is evident from the kinds of animal bones present that men selectively decimated populations of easily accessible, gregarious herbivores. They killed and ate small mammals, too, like rabbits, gophers, and raccoons, but the little animals replaced themselves rapidly and the number caught could never have amounted to a significant percentage of the whole. The larger forms were relatively fewer and reproduced more slowly, however, so that killing them continuously could conceivably have caused their extinction. The way in which men hunted herds of ungulates made their disappearance even more likely. Whereas they were obliged to kill rabbits and rodents one by one, they often drove dozens of the larger animals over a ledge, recovered the carcasses that they could carry, and left the rest to rot. Had men depended entirely upon the flesh of the herbivores for food, a diminution in the number of these animals would have entrained a decrease in the human population and a consequent reduction in hunting pressure. No balance between prey and predator would have come into effect in this case, however, because man, being an omnivore, was able to multiply independently of any other single group of animals.

He had another advantage, also, in his contest with the larger species he hunted. Since he entered most regions as an invader, the native animals had evolved no adequate defense against him. Their legs, horns, hides, and habits of life had been selected for as adaptations for coexistence with the predators that emerged beside them. When men came upon them with wit and weapons, the special form of the animals had already been determined and there was no possibility of any new protective mechanism developing rapidly enough to assure their survival.

Although predation by man is generally acknowledged to be a factor in the disappearance of the great Pleistocene mammals, there are many investigators who do not agree with Martin that it was the primary cause. They argue that a sizable number of the genera that became extinct cannot be demonstrated to have been associated with man and that many that were hunted selectively, like deer, elk, and bison, survived nevertheless. They questioned the ability of men without firearms to vanquish herds of thousands of individuals even by stampeding them over rough terrain and point out that no people still living at the Stone Age level endanger the existence of the species of animals they hunt. They admit that most modern but still primitive men live in warm countries, where the abundance of vegetation makes intensive hunting unnecessary, but they assert that even in colder regions, where men were forced to rely on game for food during many months of the year, the pattern of extinctions cannot be explained solely on the basis of man's hunting activities. Horses, for instance, became extinct in North and South America but not in Eurasia, where they are known to have been killed in great numbers. Herds of red deer remained in Europe while animals less desirable for food, like the cave lion, or less accessible, like the woolly rhinoceros of the far north, disappeared completely.

Instead of considering hunting responsible for the wave of Pleistocene extinctions, J. E. Guilday, R. H. Slaughter, and other workers have attributed the loss of mammalian genera to a deterioration of the climate which operated selectively against large terrestrial forms. Although it might seem that the climate of the Northern Hemisphere was becoming more tolerable as the last ice sheet retreated, it is also possible that certain parts of the temperate region grew excessively arid as the mean temperature rose. Large quadrupeds, all of which were mammals by Pleistocene time, would have felt the stress of changing conditions first, since big animals make the greatest demands upon the environment. Theories of Pleistocene extinction involving alterations in the climate are hard to prove, of course, because no single change could have been responsible for the losses that occurred, and, in any event, it is difficult to demonstrate from fossilized pollen and other clues the short-term climatic fluctuations which can affect the well-being of a species. Typical of the latter are unusual extremes of temperature, which may prevail for several decades in an area, leaving adult animals unaffected but increasing the death rate among the newborn. A population which, when vigorous, was capable of maintaining itself under considerable hunting pressure might, when weakened by such a process, diminish rapidly and disappear.

The possibility of extinction due to climatic change or to climatic change and overpredation was greatest where animals were hemmed in by natural barriers and unable to migrate. In North America, where the temperature gradient from north to south was quite steep and the land constricted to a narrow neck in the lower latitudes, animals moving away from areas no longer suitable for them were limited in the distance that they could go. Guilday and others have argued that, without escape routes, the large North American mammals suffered more from the alteration of environmental conditions than the African forms. In Africa, when the Sahara region became everywhere too dry for the tropical fauna that once inhabited it, many of the animals survived because they were able to follow the grasslands and rain forests southward. The disappearance of a particular kind of environment, instead of its shift, would have doomed the mammals adapted for it even in the absence of human hunters. Thus, although man is known to have hunted the woolly mammoth and the woolly rhinoceros, K. Kowalski believes that the extinction of these animals actually stemmed from their inability to adjust to the forests that replaced their steppe tundra as Eurasia warmed or to move north, where for much of the year the long polar night prevailed and the ground was covered with snow. By the time man's hunting techniques had improved sufficiently to allow him to kill these large and formidable mammals, they may already have been in decline. A similar but slower process of deterioration may have also been taking place among the bison in North America. Before man arrived in the Western Hemisphere, environmental stress had apparently induced a reduction in the size of these animals—the smaller individuals that survived better under less favorable conditions having been selected for—and a restriction in their range. Whether the bison would have provided an example, in historical times, of a large mammal first weakened by climatic change and then extinguished by men using hunting methods little different from those of their Pleistocene ancestors will never be known, because of the intervention of Europeans with firearms.

Future of the class Mammalia

So little time has passed, in the geological sense, since the loss of the great land mammals at the end of the Pleistocene epoch that it is difficult to assess the present status of the class Mammalia with any perspective. It is fair to say that it is now in a comparatively impoverished condition and possibly safe to guess that it will remain so as long as man remains a dominant species. Man's intolerance, except in zoological parks and preserves, of any animal that approaches or exceeds his own size will undoubtedly prevent the resurgence in any environment of familiar forms or the genesis of new ones. If, as all but the most objective or pessimistic of modern men have long assumed, *Homo sapiens* will survive, because of his extraordinary character, as long as the earth is habitable, the class Mammalia may be placed beside the Reptilia as an assemblage of land animals that has passed its peak and produced as its final contribution to the world's fauna numerous small and specialized forms. Should man not survive beyond the term characteristic of other highly successful genera, however, the future history of the Mammalia is not so easy to predict. Possibly, whatever happens to man, the class will continue to diminish in variety. The alternative is, of course, that from some now apparently insignificant stock a new branch of dominating forms will arise. The remnants of the modern fauna will be replaced by animals that solve the old problems of existence in new ways, and a fresh chapter of mammalian evolution will be in the offing.

10
Epilogue:
The Most Successful
Vertebrates

*Definition of an
"advanced" animal*

In discussions of vertebrate evolution, the terms "advanced," "progressive," and "successful" invariably appear, applied narrowly to particular forms or more broadly to whole groups of animals. Although the meaning of the words in each instance is generally clear from the context, their use implies certain assumptions about the direction of change among the vertebrates which ought to be explored. The designation of an animal in any class as an advanced form signifies that its structure is so modified as to give it an advantage over other members of its line under existing environmental conditions. The mackerel is considered a more advanced teleost fish than a herring, for instance, because the fins are shaped and arranged in such a way that it is both faster and more maneuverable and its mouth is built so that the jaws are more effective in grasping prey. When an entire class of vertebrates is said to be more advanced than another, the implication is that its members have become biologically more efficient and structurally sophisticated enough to live in surroundings which place the organism under some new stress. The ancient jawless fishes, the ostracoderms, lowly as they were, in being able to radiate into fresh waters and salt were presumably more advanced than their shadowy ancestors that were apparently confined to one underwater environment or the other. Placoderms and the fishes that replaced them remained water-bound, but their possession of jaws and teeth and increasingly refined paired fins provided a basis for their occupation of niches closed to the older forms. Although the extant fishes are highly adapted for their various ways of life and cannot be considered primitive animals, they are less advanced than the amphibians and reptiles that have succeeded in shifting their habitat from water to air—a medium that offers neither the support nor the moisture necessary for living cells. The status of the most advanced vertebrates is reserved for the birds and mammals, which evolved the accelerated metabolic rate and internal mechanisms of thermoregulation that enabled them to extend their range into regions where temperatures are inimical to unprotected vertebrate tissues.

Progressive forms Animals at any level can be described as progressive if they show changes which improve their ability to function and thus to cope effectively with environmental conditions. As a rule, progressive forms adapt to new surroundings by evolving special structures that fit them physically for overcoming particular obstacles. To this principle man seems to be something of an exception. He has acquired predominance over other vertebrates and established himself everywhere without relinquishing his rather generalized anthropoid structure or becoming extraordinary in size. Instead of depending upon slow alteration of his body through mutation and natural selection, he has lately shortcut the process of adaptation by using the technology his intellect has made possible. Through the development of his brain to its present state, he may have at once escaped the disadvantages of overspecialization and attained an independence of the environment not to be rivaled by any other form.

Criteria for "successful" organisms From this heady and perhaps over-optimistic assessment of man's progress, it might be concluded that *Homo sapiens* is the most successful of the vertebrates. If success is to be measured solely by cerebral development and the phenomena which accompany it, obviously man deserves the crown. However, a more searching examination of what constitutes success among living things makes it plain that the nature of the brain is too narrow a criterion. The test of the success of any organism has always been its ability to survive, and man must ultimately be judged by this standard as well as by his unique qualities of mind. When survival is made the measure, it is hard to argue for man's supremacy, for every kind of vertebrate now in existence represents a population that has come the same distance through time, adapting and readapting to environmental stresses that threatened to annihilate it. On the basis of sheer persistence, furthermore, it has to be admitted that the jawless parasitic lamprey *Petromyzon* has been a more successful form than the woolly rhinoceros, although the latter belonged to a far more advanced class of backboned animals.

Division of the vertebrates into successful and unsuccessful types on grounds of their survival or failure to survive to the present day is, of course, too simple a resolution of the problem. There is validity in terming one form more or less successful than another on other bases, and it is in attempting to define those other bases that the full complexity of the question of success among the vertebrates reveals itself. If animals that have maintained themselves with little change over many millions of years are to be accorded a high place on the scale, then the lungfishes, the coelacanth, and *Sphenodon* have outshone all the mammals. The use of this standard, however, leaves out of the list of highly successful vertebrates many of the most progressive forms. These animals, like the rodent *Paramys* or the perissodactyl *Hyracotherium*, might be considered successful precisely because they did exhibit rapid change and produced descendants which were able to enlarge their range. If success is to be linked to adaptability, there are many groups which should be rated high despite their eventual extinction. The placoderms, in which jaws and paired appendages became established, attained a variety and a distribution in the various niches of the underwater world not equalled by the sharks and bony fishes until many years after their first appearance. The ichthyostegids also seem to have scored a significant success in breaking away from the traditional home of the vertebrates and coming out upon the land.

The placoderms and the ichthyostegids can also be considered among the more successful vertebrates because they were prominent in the fauna of their time. On this basis, the dinosaurs deserve acknowledgement as the most successful terrestrial forms of the Mesozoic era, although the turtles, which were contemporary with them, proved more versatile and survived longer. When prominence, but not necessarily dominance, in the fauna is made the chief criterion, man shares the laurels of success at this time with the rodents and the teleost fishes. Compared to the primates, the latter two groups are greater in variety and in number of individuals, factors which in their own right are deemed aspects of success.

The fact that few of the flourishing modern vertebrates are notable for their large size would make it seem that bulk is not in itself a sign of prosperity in a vertebrate form. Increasing size cannot be entirely dismissed as an adjunct of success, however, because the trait has so often been selected for as new groups established themselves. Animals that are larger than their fellows have the advantage of greater strength and benefit from the increased size of whatever structures they use as weapons. An individual may thus survive longer and produce a greater number of offspring than smaller members of the same species. The disadvantage of evolution of great size, apart from the development of overspecialized structures which may accompany it, lies in the lessened ability of large organisms to adapt to changes in the environment. Since large forms demand a maximum of food and space, they can be cut off by restrictions in the availability of either that are too slight to exert stress on smaller animals. Under unfavorable conditions, the smaller members of a population are often selected for, and dwarf strains emerge, but large animals that require a relatively long time for growth to maturity and, upon reaching it, produce offspring rather slowly generally prove less flexible than small species that mature rapidly and reproduce prolifically. When generations succeed one another more quickly and each mating yields numerous individuals, the chances are simply better that advantageous mutations will appear and spread through the population, enabling it to continue. It might be said, then, that great size, like extreme specialization, affords success in a stable environment, but since no environment persists indefinitely without change, such success is inevitably a prelude to failure.

The future of man　　Whether man's present dominance among terrestrial vertebrates has been attained by a route that will also lead to failure is a question that has only lately received the serious consideration it deserves. Although it has been recognized that the hominid line is a relatively young one and that the evolution of *Homo* is an example of the rapid differentiation of a progressive stock at a new adaptive level, it has been assumed that man's position is somehow more secure than that of less advanced forms. Both scientists and philosophers have pointed out that man's brain is unique and assures his being able to avoid the fate of organisms governed rigidly by instinct unmodified by intellect. Not only has his capacity to think abstractly allowed him to develop technological methods by which he can protect himself in an unfavorable environment, but his ability to probe the events of the past makes it possible for him to learn from his own experience and that of other organisms how to avoid the pitfalls through which a species comes to disaster. The chief vehicle for the continued progress of man, it has been asserted, is the social framework that has grown under his hand. While it favors, theoretically, the kind of orderly and safeguarded existence that fosters the prosperity of any group, it also affords an opportunity for the cooperative study and invention by which man believes he will maintain himself. Men who leave written records have repeatedly stated or implied that through the growth of civilization (and by "civilization" they invariably mean the kind of society to which they themselves belong), human beings would be protected from the forces that threaten them.

Although the significance of trends of the last 5,000 to 10,000 years for the course of history in the next 10 or 100 million is doubtful, men have nevertheless taken their increased accomplishments of the past as a guarantee of their future success. Surely, they reason, since they are able to settle on every continent, live in underwater chambers, and glide around in space, *Homo sapiens* will not be extinguished by the sort of climatological changes that have been responsible for the disappearance of other genera. One by one, the diseases that once reduced whole populations to nearly zero have been contained, and the menace of famine has been mitigated to some degree by the improvement of food plants and farming methods. War, the third of the traditional scourges of mankind, has come to be considered as an evil rather than as an admirable masculine occupation, and, until recently, each international conflict was hopefully declared to be the last. Freed of engagement in disruptive wars, man was going to turn his attention to the arts of peace, not the least of which was to be improvement of the understanding of genetics to the point where man would be able to guide his physical as well as his cultural evolution in a beneficial direction.

Factors endangering man's progress

The events of the twentieth century have caused some people to wonder, however, whether human progress is as real and the human intellect as powerful as hitherto they have seemed to be. As contact among men increases, both because of improvement in methods of communication and because of crowding brought about by overpopulation, man's ancient aggressive instincts are manifested in a less controllable and a more destructive way. Whereas man's readiness to attack, like some new structural trait, was once advantageous to his survival, it seems now, under altered circumstances, to be a characteristic that may imperil his existence. His intellect, which was supposed to govern his instincts, has functioned erratically in that task, exerting its effect far more consistently in the area of external and material invention. As a result, men whose inclination to fight over possessions, territory, or principles is as strong as ever find themselves armed with guns and bombs instead of spears. In addition to the massive physical destruction that is now possible in war, man may undergo increasing degradation in the cities that he has created. Paradoxically, as cities have become centers of human achievement, they have also become traps where great numbers of people are immobilized and made vulnerable to repressive and sometimes deadly forces that grow more dangerous as the machinery through which they operate is further refined. Under prevailing conditions, new technological developments become a two-edged sword. Inventions intended to conserve or enhance human life are consciously or unconsciously used to ill effect. The industrial processes through which men raised themselves far beyond a subsistence economy threaten now to produce such a volume of noxious material that the environment could become chemically unsuitable for man or the organisms upon which he depends. Also, the medical and biological manipulations of which man is now capable bring as much potential for harm as for good. Since man's aggressive spirit still impels him to impose his will over others, the possibility that he will discover a method of altering the genetic complex and so of engineering changes in human beings becomes a nightmare instead of an utopian dream.

Success of man: still an open question

Those who are pessimistic concerning man's future have demonstrated not that *Homo* is on the road to extinction but that he could be. In so doing, they have countered the old assumption that modern man is not subject to the same evolutionary rules as other organisms and made it possible to review the question of man's success from a new perspective. It is still generally agreed that man has been a progressive form and in many respects the most advanced one, but it is no longer maintained that his line is certainly the most successful among the vertebrates. In adaptability, in number and variety, and in prominence in the fauna, there are groups that rival his own and in endurance in their present form greatly surpass it. If man were to weaken himself sufficiently in the next few million years to succumb to forces which would not have extinguished him in his prime, he would rank as another of nature's short-lived, unsuccessful experiments despite his highly developed brain. A lower level of intelligence combined with smaller size, a shorter life cycle, and an even higher reproductive rate might then, once again, be proved a better guarantee of survival for a vertebrate group. Should *Homo sapiens* and his descendants retain their dominant position as long as the dinosaurs did, however, man's place among the most successful vertebrates would be indisputable.

References

Chapter 1. Fossils: Getting the Evidence

Dalrymple, G. B., and M. A. Lanphere: *Potassium-Argon Dating: Principles, Techniques and Applications to Geochronology*, Freeman, San Francisco, 1969.

Eicher, D. L.: *Geologic Time*, Prentice-Hall, Englewood Cliffs, N.J., 1968.

Fenton, C. L., and M. A. Fenton: *The Fossil Book*, Doubleday, Garden City, N.Y., 1958.

Kummel, B., and D. Raup: *Handbook of Paleontological Techniques*, Freeman, San Francisco, 1965.

Chapter 2. The Origin of the Vertebrates

Berrill, N. J.: *The Origin of Vertebrates*, Clarendon Press, Oxford, 1955.

Bryant, W. L.: A Study of the Oldest Known Vertebrates, *Astraspis* and *Eriptychius*, *Proc. Am. Philos. Soc.*, **76**:409–427 (1936).

Denison, R. H.: Evolution and Classification of the Osteostraci: The Exoskeleton of Early Osteostraci, *Fieldiana: Geol.* **11**:155–218 (1951).

———: Early Devonian Fishes from Utah, pt. 1: Osteostraci, *Fieldiana: Geol.*, **11**:263–287 (1952).

———: Early Devonian Fishes from Utah, pt. 2: Heterostraci, *Fieldiana: Geol.*, **11**(7):299–355 (1953).

————: A Review of the Habitat of the Earliest Vertebrates, *Fieldiana: Geol.*, **11**(8):361–457 (1956).

————: Ordovician Vertebrates from Western United States, *Fieldiana: Geol.*, **16**(5):131–192 (1967).

Heintz, A.: Les organes olfactifs des Heterostraci, in Problèmes actuels de paléontologie (Évolution des vertébrés), *Colloq. Int. Cent. Natl. Rech. Sci.*, **104**:13–29 (1962).

Kiaer, J.: The Downtonian Fauna of Norway, pt. 1: Anaspida, *Skr. Videnskaps. Selsk. Kristiania, Mat. Naturvidensk. Kl.*, **1**(6):1–139 (1924).

Ritchie, A.: A New Interpretation of *Jamoytius kerwoodi* White, *Nature (Lond.)*, **188**:647–649 (1960).

————: New Light on the Morphology of the Norwegian Anaspida, *Skr. Nor. Videnskaps.-Acad. Oslo Mat.-Naturvidensk. Kl.*, n.s., **14**:1–35 (1964).

————: New Evidence on *Jamoytius kerwoodi* White, an Important Ostracoderm from the Silurian of Lanarkshire, Scotland, *Palaeontology*, **11**:21–39 (1968).

Robertson, J. D.: The Habitat of the Early Vertebrates, *Biol. Rev.*, **32**:156–187 (1957).

Romer, A. S.: Fish Origins: Fresh or Salt Water?, *Pap. Mar. Biol. Oceanogr., Deep Sea Res.* **3**(suppl.):261–280 (1955).

————and B. H. Grove: Environment of the Earliest Vertebrates, *Am. Midl. Nat.*, **16**:805–856 (1935).

Smith, H. W.: Water Regulation and Its Evolution in the Fishes, *Q. Rev. Biol.*, **7**:1–26 (1932).

Stensiö, E. A.: The Downtonian and Devonian Vertebrates of Spitzbergen, pt. I: Family Cephalaspidac, *Skr. Svalbard Nord.*, 12, 1927.

Westoll, T. S.: The Origin of Continental Vertebrate Faunas, *Trans. Geol. Soc. Glasg.*, **23**:79–105 (1958).

White, E. I.: *Jamoytius kerwoodi*, a New Chordate from the Silurian of Lanarkshire, *Geol. Mag.*, **83**:89–97 (1946).

Zangerl, R., and D. Bardack: First Fossil Lamprey: A Record from the Pennsylvanian of Illinois, *Science*, **162**:1265–1267 (1968).

Chapter 3. Bone and Cartilage in Early Vertebrates

Enlow, D. H., and S. O. Brown: A Comparative Histological Study of Fossil and Recent Bone Tissues, *Tex. J. Sci.*, pt. I, **8**(4):405–443; pt. II, **9**(2):186–214; pt. III, **10**(2):187–230 (1956–1958).

Lamarck, J. B.: *Philosophie zoologique*, 1809, reprinted in English as *Zoological Philosophy*, Hafner, New York, 1963.

Moss, M. L.: The Biology of Acellular Teleost Bone, in Comparative Biology of Calcified Tissue, *Ann. N.Y. Acad. Sci.*, **109**:337–350 (1963).

————: Bone, Dentin, and Enamel and the Evolution of Vertebrates, in P. Person (ed.), *Biology of the Mouth, Am. Assoc. Adv. Sci. Publ.* 89, 1968.

Ørvig, T.: Histologic Studies of Placoderms and Fossil Elasmobranchs, pt. 1: The Endoskeleton, with Remarks on the Hard Tissues of Lower Vertebrates in General, *Ark. Zool. K. Sven. Vetenskapsakad.*, ser. 2,**2**(2):321–454 (1951).

————: Phylogeny of Tooth Tissues: Evolution of Some Calcified Tissues in Early Vertebrates, pp. 45–110 in A. E. W. Miles (ed.), *Structural and Chemical Organization of Teeth*, vol. 1, Academic, New York, 1967.

————: The Dermal Skeleton: General Considerations, pp. 373–397 in T. Ørvig (ed.), *Current Problems of Lower Vertebrate Phylogeny*, Interscience-Wiley, New York, 1968.

Romer, A. S.: The "Ancient History" of Bone, in Comparative Biology of Calcified Tissue, *Ann. N.Y. Acad. Sci.*,, **109**:168–176 (1963).

Smith, H. W.: Studies in the Physiology of the Kidney, *Univ. Kans. Porter Lect. Ser.*, **4**:1–106 (1939).

Chapter 4. The First Fishes with Jaws

De Beer, G. R.: *The Development of the Vertebrate Skull*, Oxford Press, London, 1937.

Denison, R. H.: The Soft Anatomy of *Bothriolepis*, *J. Paleontol.*, **15**:553–561 (1941).

————: Early Devonian Fishes from Utah, pt. 3: Arthrodira, *Fieldiana: Geol.*, **11**:459–551 (1958).

Gray, J.: The Locomotion of Fishes, pp. 1–16 in *Essays on Marine Biology*, Oliver & Boyd, London, 1953.

Gregory, W. K., and H. C. Raven: Studies on the Origin and Early Evolution of Paired Fins and Limbs, *Ann. N.Y. Acad. Sci.*, **42**:275–361 (1941).

Harris, J. E.: The Role of the Fins in the Equilibrium of the Swimming Fish, *J. Exp. Biol.*, **13**:476–493 (1936); **15**:32–47 (1938).

Heyler, D.: Les acanthodiens et le problème de l'aphetohyoidie, in Problèmes actuels de paléontologie (Évolution des vertébrés), *Colloq. Int. Cent. Natl. Rech. Sci.*, **104**:39–47 (1962).

Holmgren, N.: Studies on the Head of Fishes: An Embryological, Morphological, and Phylogenetical Study, pt. 3: The Phylogeny of Elasmobranch Fishes, *Acta Zool. Stockh.*, **23**:129–261 (1942).

Jarvik, E.: On the Visceral Skeleton in *Eusthenopteron* with a Discussion of the Parasphenoid and Palatoquadrate in Fishes, *K. Sven. Vetenskapsakad. Handl.*, **5**:1–104 (1954).

Miles, R. S.: Some Features in the Cranial Morphology of Acanthodians and the Relationships of the Acanthodii, *Acta Zool. Stockh.*, **46**:233–255 (1965).

————: The Cervical Joint and Some Aspects of the Origin of the Placodermi, *Colloq. Int. Cent. Natl. Rech. Sci.*, **163**:49–71 (1967).

————: The Holonematidae (Placoderm Fishes): A Review Based on New Specimens of *Holonema* from the Upper Devonian of Western Australia, *Philos. Trans. R. Soc. Lond.*, **B263**:101–234 (1971).

Moy-Thomas, J. A.: On the Structure and Affinities of the Carboniferous Cochliodont *Helodus simplex, Geol. Mag.,* **73**:488–503 (1936).

———and R. S. Miles: *Palaeozoic Fishes,* 2d ed., Chapman & Hall,
London, 1971.

Ørvig, T.: Y a-t-il une relation directe entre les arthrodires ptyctodon-
tides et les holocéphales?, in Problèmes actuels de paléontologie
(Évolution des vertébrés), *Colloq. Int. Cent. Natl. Rech. Sci.,* **104**:49–61
(1962).

Simpson, G. G.: *Principles of Animal Taxonomy,* Oxford Press, London,
1961.

Stensiö, E.: On the Pectoral Fin and Shoulder Girdle of the Arthrodires, *K. Sven. Vetenskapsakad. Handl.,* ser.(4) **8**(1):5–229 (1959).

———: Anatomical Studies on the Arthrodiran Head, pt. 1: Preface,
Geological and Geographical Distribution, the Organization of the
Arthrodires, the Anatomy of the Head in the Dolichothoraci, Coccos-
teomorphi, and Pachyosteomorphi, Taxonomic Appendix, *K. Sven.
Vetenskapsakad. Handl.,* **9**(2):1–419 (1963).

Tarlo, L. B. Halstead: The Tessellated Pattern of Dermal Armor in the
Heterostraci, *J. Linn. Soc. (Zool.),* **47**:45–54 (1967).

Watson, D. M. S.: The Interpretation of Arthrodires, *Proc. Zool. Soc.
Lond.,* (3)**1934**:437–464.

———: The Acanthodian Fishes, *Philos. Trans. R. Soc. Lond.,* **B228**:49–
146 (1937).

———: Some additions to our Knowledge of Antiarchs, *Paleontology,*
4(2):210–220 (1961).

Westoll, T. S.: The Paired Fins of Placoderms, *Trans. R. Soc. Edinb.,*
63:381–398 (1945).

———: The Lateral Fin-Fold Theory and the Pectoral Fins of Ostracoderms and Early Fishes, pp. 180–211 in T. S. Westoll (ed.),
Studies on Fossil Vertebrates, Athlone Press, London, 1958.

———: The Hyomandibular Problem in Placoderm Fishes, *16th Int.
Cong. Zool. Wash., D.C., Contrib. Pap.,* **1**:176 (1963).

———: *Radotina* and other Tesserate Fishes, *J. Linn. Soc. (Zool.),*
47:83–98 (1967).

White, E. I.: Australian Arthrodires, *Bull. Br. Mus. (Nat. Hist.), Geol.,*
1(9):249–304 (1952).

Chapter 5. The Rise of the Modern Fishes

Agassiz, L.: *Recherches sur les poissons fossiles,* 5 vols, Neuchatel, 1833–
1843.

Andrews, S. M., and T. S. Westoll: The Postcranial Skeleton of
Eusthenopteron foordi Whiteaves, *Trans. R. Soc. Edinb.,* **68**(9):207–329
(1970).

Bendix-Almgreen, S. E.: New Investigations on *Helicoprion* from the
Phosphoria Formation of South-east Idaho, U.S.A., *Biol. Skr. Dan.
Vidensk. Selsk.,* **14**(5):1–54 (1966).

Berg, L. S.: *Classification of Fishes, Both Recent and Fossil,* Edwards, Ann Arbor, Mich., 1947, trans. of Russian original, *Tr. Zool. Inst. Akad. Nauk SSSR,* **5**(2):87–517 (1940); 2d Russ. ed. (1949) trans. into German as *System der rezenten und fossilen Fischartigen and Fische,* Berlin, 1958.

Brough, J.: On the Evolution of Bony Fishes during the Triassic Period, *Biol. Rev.,* **11**:385–405 (1936).

Denison, R. H.: The Evolutionary Significance of the Earliest Known Lungfish, *Uranolophus,* pp. 247–257 in T. Ørvig (ed.), *Current Problems of Lower Vertebrate Phylogeny,* Interscience-Wiley, New York, 1968.

Gardiner, B. G.: Further Notes on Paleoniscoid Fishes with a Classification of the Chondrostei, *Bull. Br. Mus. (Nat. Hist.), Geol.,* **14**(5):143–206 (1967).

Greenwood, P. H., D. E. Rosen, S. H. Weitzman, and G. S. Myers: Phyletic Studies of Teleostean Fishes, with a Provisional Classification of Living Forms, *Bull. Am. Mus. Nat. Hist.,* **131**:339–456 (1966).

Jarvik, E.: The Systematic Position of the Dipnoi, pp. 223–245 in T. Ørvig (ed.), *Current Problems of Lower Vertebrate Phylogeny,* Interscience-Wiley, New York, 1968.

Jessen, H.: The Position of the Struniiformes (*Strunius* and *Onychodus*) among Crossopterygians, in Problèmes actuels de paléontologie (Évolution des vertébrés), *Colloq. Int. Cent. Natl. Rech. Sci.,* **163**:173–180 (1967).

Millot, J., and J. Anthony: Anatomie de *Latimeria chalumnae,* pt. I: Squelette, muscles et formations de soutien, Centre National de la Recherche Scientifique, Paris, 1958.

Patterson, C.: A Review of Mesozoic Acanthopterygian Fishes, with Special Reference to Those of the English Chalk, *Philos. Trans. R. Soc. Lond.,* **B247**:213–482 (1964).

——: The Phylogeny of the Chimaeroids, *Philos. Trans. R. Soc. Lond.,* **B249**:101–219 (1965).

Radinsky, L.: Tooth Histology as a Taxonomic Criterion for Cartilaginous Fishes, *J. Morphol.,* **109**:73–92 (1961).

Schaeffer, B.: The Evidence of the Fresh-Water Fishes, in The Problem of Land Connections across the South Atlantic, with Special Reference to the Mesozoic, *Bull. Am. Mus. Nat. Hist.,* **99**:227–236 (1952).(a)

——: Rates of Evolution in the Coelacanth and Dipnoan Fishes, *Evolution,* **6**(1):101–111 (1952).(b)

——: *Latimeria* and the History of Coelacanth Fishes, *Trans. N.Y. Acad. Sci.,* (2)**15**:170–178 (1953).

——: Comments on Elasmobranch Evolution, in P. W. Gilbert, R. F. Mathewson, and D. P. Rall (eds.), *Sharks, Skates, and Rays,* Johns Hopkins, Baltimore, 1967.

——: The Origin and Basic Radiation of the Osteichthyes, pp. 207–222 in T. Ørvig (ed.), *Current Problems of Lower Vertebrate Phylogeny,* Interscience-Wiley, New York, 1968.

——and D. H. Dunkle: A Semionotid Fish from the Chinle Formation, with Consideration of Its Relationships, *Am. Mus. Novit.*, **1457**:1–29 (1950).

——and D. E. Rosen: Major Adaptive Levels in the Evolution of the Actinopterygian Feeding Mechanism, *Am. Zool.*, **1**(2):187–204 (1961).

Simpson, G. G.: Periodicity in Vertebrate Evolution, in The Distribution of Evolutionary Explosions in Geologic Time (Symposium), *J. Paleontol.*, **26**:298–394 (1952).

Stahl, B. J.: Morphology and Relationships of the Holocephali with Special Reference to the Venous System, *Bull. Mus. Comp. Zool.*, **135**:141–213 (1967).

Thomson, K. S.: The Biology of the Lobe-Finned Fishes, *Biol. Rev.*, **44**:91–154 (1969).

——and K.S.W. Campbell: The Structure and Relationships of the Primitive Devonian Lungfish—*Dipnorhynchus sussmilchi* (Etheridge), *Bull. Peabody Mus. Nat. Hist., Yale Univ.* **38**:1–109 (1971).

Westoll, T. S.: The Permian Fishes *Dorypterus* and *Lekanichthys*, *Proc. Zool. Soc. Lond.*, **B111**:39–58 (1941).

——: The Haplolepidae, a New Family of Late Carboniferous Bony Fishes, *Bull. Am. Mus. Nat. Hist.*, **83**:1–121 (1944).

——: On the Evolution of the Dipnoi, pp. 121–184 in G. L. Jepsen, E. Mayr, and G. G. Simpson (eds.), *Genetics, Paleontology, and Evolution*, Princeton Press, Princeton, N.J., 1949.

White, E. G.: Interrelationships of the Elasmobranchs with a Key to the Order Galea, *Bull. Am. Mus. Nat. Hist.*, **74**:25–138 (1937).

Zangerl, R.: A New Shark of the Family Edestidae, *Ornithoprion hertwigi*, *Fieldiana: Geol.*, **16**(1):1–43 (1966).

——and G. R. Case: Iniopterygia, a New Order of Chondrichthyan Fishes from the Pennsylvanian of North America, *Fieldiana: Geology Mem.*, **6**:1–67 (1973).

Chapter 6. The Amphibians: Gaining the Land

Bolt, J. R.: Lissamphibian Origins: Possible Protolissamphibian from the Lower Permian of Oklahoma, *Science*, **166**:888–891 (1969).

Carroll, R. L.: Early Evolution of the Dissorophid Amphibians, *Bull. Mus. Comp. Zool.*, **131**(7):161–250 (1969).

Estes, R.: Fossil Salamanders and Salamander Origins, *Am. Zool.*, **5**:319–334 (1965).

——and O. A. Reig: The Early Fossil Record of Frogs: A Review of the Evidence, in J. L. Vial (ed.), *Evolutionary Biology of the Anurans, Contemporary Research on Major Problems*, University of Missouri Press, Columbia, Missouri, 1973.

——and M. Wake: The First Fossil Record of Caecilian Amphibians, *Nature*, **239**:228–231 (1972).

Gans, C.: Respiration in Early Tetrapods: The Frog Is a Red Herring, *Evolution*, **24**:740–751 (1970).

Goin, C. J., and O. B. Goin: Further Comments on the Origin of the Tetrapods, *Evolution*, **10**(4):440–441 (1956).

Hecht, M. K.: A Reevaluation of the Early History of the Frogs, *Syst. Zool.*, pt. I, **11**(1):39–44 (1962); pt. II, **12**(1):20–35 (1963).

Holmgren, N.: On the Origin of the Tetrapod Limb, *Acta Zool. Stockh.*, **4**:1–295 (1933).

Hotton, N.: *Mauchchunkia bassa*, gen. et sp. nov., an Anthracosaur (Amphibia, Labyrinthodontia) from the Upper Mississippian, *Kirtlandia, Cleve. Mus. Nat. Hist.*, **12**:1–38 (1970).

Inger, R. F.: Ecological Aspects of the Origins of the Tetrapods, *Evolution*, **11**(3):373–376 (1957).

————: The Development of a Phylogeny of Frogs, *Evolution*, **21**:369–384 (1967).

Jarvik, E.: On the Structure of the Snout of Crossopterygians and Lower Gnathostomes in General, *Zool. Bidr. Upps.*, **21**:235–675 (1942).

————: On the Fish-like Tail in the Ichthyostegid Stegocephalians with Descriptions of a New Stegocephalian and a New Crossopterygian from the Upper Devonian of East Greenland, *Medd. Grønland*, **114**:1–90 (1952).

————: The Oldest Tetrapods and Their Forerunners, *Sci. Mon.*, **80**:141–154 (1955).

————: The Composition of the Intermandibular Division of the Head in Fish and Tetrapods and the Diphyletic Origin of the Tetrapod Tongue, *K. Sven. Vetenskapsakad. Handl.*, **9**(1):1–74 (1963).

Parsons, T. S., and E. E. Williams: The Relationships of the Modern Amphibia: A Re-Examination, *Q. Rev. Biol.*, **38**(1):26–53 (1963).

Romer, A. S.: Notes on the Crossopterygian Hyomandibular and Braincase, *J. Morphol.*, **69**(1):141–160 (1941).

————: Review of the Labyrinthodontia, *Bull. Mus. Comp. Zool.*, **99**:1–368 (1947).

————: Tetrapod Limbs and Early Tetrapod Life, *Evolution*, **12**(3):365–369 (1958).

————: A Temnospondylous Labyrinthodont from the Lower Carboniferous, *Kirtlandia, Cleve. Mus. Nat. Hist.*, **6**:1–20 (1969).

————: A New Anthracosaurian Labyrinthodont, *Proterogyrinus scheelei*, from the Lower Carboniferous, *Kirtlandia, Cleve. Mus. Nat. Hist.*, **10**:1–16 (1970).

————: Skin Breathing: Primary or Secondary? *Respir. Physiol.*, **14**:183–192 (1972).

Schaeffer, B.: The Rhipidistian-Amphibian Transition, *Amer. Zool.*, **5**:267–276 (1965).

Schmalhausen, I. I.: *The Origin of Terrestrial Vertebrates*, Academic, New York, 1968.

Smith, H. W.: *From Fish to Philosopher*, Anchor Books, Doubleday, Garden City, New York, 1961.

Szarski, H.: The Origin of the Amphibia, *Q. Rev. Biol.*, **37**(3):189–241 (1962).

Thomson, K. S.: A Critical Review of the Diphyletic Theory of Rhipidistian-Amphibian Relationships, pp. 285–305 in T. Ørvig, (ed.), *Current Problems of Lower Vertebrate Phylogeny*, Interscience-Wiley, New York, 1967.

Tihen, J. A.: Evolutionary Trends in Frogs, *Am. Zool.*, **5**:309–318 (1965).

Watson, D. M. S.: The Structure, Evolution, and Origin of the Amphibia: The Orders Rhachitomi and Stereospondyli, *Philos. Trans. R. Soc. Lond.*, **B209**:1–73 (1919).

————: The Origin of Frogs, *Philos. Trans. R. Soc. Edinb.*, **60**(1):195–231 (1940).

Westoll. T. S.: The Origin of the Primitive Tetrapod Limb, *Proc. R. Soc.*, **B131**:373–393 (1943). (a)

————: The Origin of the Tetrapods, *Biol. Rev.*, **18**:78–98 (1943). (b)

————: The Origin of Continental Vertebrate Faunas, *Geol. Soc. Glasg.*, **23**:1–27 (1958).

Williams, E. E.: Gadow's Arcualia and the Development of Tetrapod Vertebrae, *Q. Rev. Biol.*, **34**(1):1–32 (1959).

Chapter 7. The Rise and Fall of the Reptiles

Axelrod, D. I., and H. P. Bailey: Cretaceous Dinosaur Extinction, *Evolution*, **22**:595–611 (1968).

Bakker, R. T.: The Superiority of Dinosaurs, *Discovery*, **3**(2):11–22 (1968).

————: Ecology of the Brontosaurs, *Nature*, **229**:172–74 (1971).

————: Anatomical and Ecological Evidence of Endothermy in Dinosaurs, *Nature*, **238**:81–85 (1972).

Bellairs, A. d'A., and G. Underwood: The Origin of Snakes, *Biol. Rev.*, **26**:193–237 (1951).

Carroll, R. L.: Problems of the Origin of Reptiles, *Biol. Rev.*, **44**:393–432 (1969). (a)

————: A Middle Pennsylvanian Captorhinomorph and the Interrelationships of Primitive Reptiles, *J. Paleontol.*, **43**(1):151–170 (1969). (b)

————: The Ancestry of Reptiles, *Philos. Trans. R. Soc. Lond.*, **B257**:267–308 (1970).

———— and D. Baird: The Carboniferous Amphibian *Tuditanus* (*Eosauravus*) and the Distinction between Microsaurs and Reptiles, *Am. Mus. Novit.*, **2337**:1–50 (1968).

Charig, A. J., J. Attridge, and A. W. Crompton: On the Origin of the Sauropods and the Classification of the Saurischia, *Proc. Linn. Soc. Lond.*, **176**(2):197–221 (1965).

Ewer, R. F.: The Anatomy of the Thecodont Reptile *Euparkeria capensis* Broom, *Philos. Trans. R. Soc. Lond.*, **B248**:379–435 (1965).

Goodrich, E. S.: *Studies on the Structure and Development of Vertebrates*, London, 1930; reprinted by Dover, New York, 2 vols., 1958.

Gregory, J. T.: The Genera of Phytosaurs, *Am. J. Sci.*, **260**:652–690 (1962).

Gregory, W. K.: Pareiasaurs versus Placodonts as Near Ancestors to the Turtles, *Bull. Am. Mus. Nat. Hist.*, **86**:279–359 (1946).

McDowell, S. B., and C. M. Bogert: The Systematic Position of *Lanthanotus* and the Affinities of the Anguinomorph Lizards, *Bull. Am. Mus. Nat. Hist.*, **105**:1–142 (1954).

Olson, E. C.: Late Permian Terrestrial Vertebrates, U.S.A. and U.S.S.R., *Trans. Am. Philos. Soc.*, n.s., **52**(2):1–224 (1962).

———: Relationships of *Diadectes*, *Fieldiana: Geol.*, **14**(10):199–227 (1966).

———: The Family Caseidae, *Fieldiana: Geol.*, **17**(3):225–331 (1968).

Ostrom, J. H.: Functional Morphology and Evolution of Ceratopsian Dinosaurs, *Evolution*, **20**:290–308 (1966).

Parrington, F. R.: The Problem of the Classification of Reptiles, *J. Linn. Soc. (Zool.)*, **44**:99–115 (1958).

Parsons, T. S., and E. E. Williams: Two Jurassic Turtle Skulls: A Morphological Study, *Bull. Mus. Comp. Zool.*, **125**(3):43–107 (1961).

Reig, O. A.: The Proterosuchia and the Early Evolution of the Archosaurs: An Essay about the Origin of a Major Taxon, *Bull. Mus. Comp. Zool.*, **139**(5):229–292 (1970).

Robinson, P. L.: The Evolution of the Lacertilia, in Problèmes actuels de paléontologie (Évolution des vértebrés), *Colloq. Int. Cent. Natl. Rech. Sci.*, **163**:395–407 (1967).

Romer, A. S: Ichthyosaur Ancestors, *Am. J. Sci.*, **246**:109–121 (1948).

———: Origin of the Amniote Egg, *Sci. Mon.*, **85**(2):57–63 (1957).

———: *Vertebrate Paleontology*, 3d ed., University of Chicago Press, Chicago, 1966.

———: Early Reptilian Evolution Re-viewed, *Evolution*, **21**(4):821–833 (1967).

———: An Ichthyosaur Skull from the Cretaceous of Wyoming, *Contrib. Geol.*, **7**(1):27–41 (1968).

———: The Chañares (Argentina) Triassic Reptile Fauna, pt. XI: Two New Long-Snouted Thecodonts, *Chanaresuchus* and *Gualosuchus*, *Breviora Mus. Comp. Zool.*, **379**:1–22 (1971). (a)

———: Unorthodoxies in Reptilian Phylogeny, *Evolution*, **25**(1):103–112 (1971). (b)

——— and L. I. Price: Review of the Pelycosauria, *Geol. Soc. Am., Spec. Pap.*, **28**:1–538 (1940).

Seeley, H. G.: *Dragons of the Air*, Appleton, New York, 1901; reprinted by Dover, New York, 1967.

Sill, W. D.: *Proterochampsa barrionuevoi* and the Early Evolution of the Crocodilia, *Bull. Mus. Comp. Zool.*, **135**(8):415–446 (1967).

Szarski, H.: The Origin of Vertebrate Foetal Membranes, *Evolution*, **22**(1):211–214 (1968).

Vaughn, P. P.: The Permian Reptile *Araeoscelis* Restudied, *Bull. Mus. Comp. Zool.*, **113**(5):305–467 (1955).

———: The Paleozoic Microsaurs as Close Relatives of Reptiles, Again, *Am. Midl. Nat.*, **67**(1):79–84 (1962).

Walker, A. D.: Triassic Reptiles from the Elgin Area: *Ornithosuchus* and the Origin of Carnosaurs., *Philos. Trans. R. Soc. Lond.*, **B248**:53–134 (1964).

———: *Protosuchus, Proterochampsa*, and the Origin of Phytosaurs and Crocodiles, *Geol. Mag.*, **105**:1–14 (1968).

Walls, G. L.: The Vertebrate Eye and Its Adaptive Radiation, *Bull. Cranbrook Inst. Sci.*, **19**:1–785 (1942).

Watson, D. M. S.: *Eunotosaurus africanus* Seeley, and the Ancestry of the Chelonia, *Proc. Zool. Soc. Lond.*, **1914**:1011–1020.

———: On *Bolosaurus* and the Origin and Classification of Reptiles, *Bull. Mus. Comp. Zool.*, **111**(9):299–450 (1954).

———: On *Millerosaurus* and the Early History of the . Sauropsid Reptiles, *Philos. Trans. R. Soc. Lond.*, **B240**:325–400 (1957).

Westoll, T. S.: Ancestry of captorhinomorph Reptiles, *Nature*, **149**:667–668 (1942).

Young, J. Z.: *The Life of Vertebrates*, 2d ed., Oxford University Press, New York, 1962.

Chapter 8. The Legacy of the Reptiles: The Birds

De Beer, G. R.: *Archaeopteryx lithographica*, British Museum (Natural History), London, 1954.

———: The Evolution of Ratites, *Bull. Brit. Mus. (Nat. Hist.) Zool.*, **4**(2):57–70 (1956).

Galton, P. M.: Ornithischian Dinosaurs and the Origin of Birds, *Evolution*, **24**:448–462 (1970).

Gregory, J. T.: The Jaws of the Cretaceous Toothed Birds, *Ichthyornis* and *Hesperornis, Condor*, **54**(2):137–145 (1952).

Heilmann, G.: *The Origin of Birds*, Appleton, New York, 1927; reprinted by Dover, New York, 1972.

Howard, H.: Fossil Evidence of Avian Evolution, *Ibis*, **92**:1–21 (1950).

———: Fossil Birds, *Los Ang. Cty. Mus., Sci. Ser.* no. 10, 1955.

Lowe, P. R.: On the Primitive Characteristics of the Penguins and Their Bearing on the Phylogeny of Birds, *Proc. Zool. Soc. Lond.*, **1933**:483–588.

———: On the Relationship of the Struthiones to the Dinosaurs and to the Rest of the Avian Class, with Special Reference to the Position of *Archaeopteryx, Ibis*, (13)**5**:398–432 (1935).

———: An Analysis of the Characteristics of *Archaeopteryx* and *Archaeornis:* Were They Reptiles or Birds?, *Ibis*, **86**:517–543 (1944).

Marsh, O. C.: Birds with Teeth, in Third Annual Report of the Director of the U.S. Geological Survey 1881–1882, U.S. Government Printing Office, Washington, D.C., 1883.

Marshall, A. J. (ed): *Biology and Comparative Physiology of Birds*, 2 vols., Academic, New York, 1960–1961.

Matthew, W. D.: Climate and Evolution, *Ann. N.Y. Acad. Sci.*, **24**:171–318 (1915).

Meillon, B. de: The Fleas of Sea Birds in the Southern Ocean, *Aust. Natl. Antarct. Res. Exped.*, ser. B, vol. 1, 1952.

Menzbier, M.: Vergleichende Osteologie der Pinguine, *Bull. Soc. Imp. Nat., Moscow*, n.s., **1**:483–587 (1887).

Nopcsa, F.: Ideas on the Origin of Flight, *Proc. Zool. Soc. Lond.*, **1907**:223–236.

Ostrom, J. H.: *Archaeopteryx*: Notice of a "New" Specimen, *Science*, **170**:537–538 (1970).

Simpson, G. G.: Fossil Penguins, *Bull. Am. Mus. Nat. Hist.*, **87**:1–99 (1946).

———: Australian Fossil Penguins, with Remarks on Penguin Evolution and Distribution, *Rec. South Aust. Mus.*, **13**(1):51–70 (1957).

Swinton, W. E.: *Fossil Birds*, British Museum (Natural History), London, 1958.

Welty, J. C.: *The Life of Birds*, Saunders, Philadelphia, 1962.

Wetmore, A.: The systematic Position of *Palaeospiza bella*, Allen, with Observations on Other Fossil Birds, *Bull. Mus. Comp. Zool.*, **67**(2):183–193 (1925).

———: Birds of the Pleistocene in North America, *Smithson. Misc. Collect.*, **138**(4):1–24 (1959).

———: A Classification for the Birds of the World, *Smithson. Misc. Collect.*, **139**(11):1–37 (1960).

Chapter 9. The Legacy of the Reptiles: The Mammals

Audley-Charles, M. G.: Mesozoic Palaeogeography of Australasia, *Palaeogeogr., Palaeoclimatol., Palaeoecol.*, **2**:1–25 (1966).

Bakker, R. T.: Dinosaur Physiology and the Origin of Mammals, *Evolution*, **25**(4):636–658 (1971).

Bigalke, R. C.: The Contemporary Mammal Fauna of Africa, *Q. Rev. Biol.*, **43**(3):265–300 (1968).

Brink, A. S.: Note on a Very Tiny Specimen of *Thrinaxodon liorhinus*, *Palaeontol. Afr.*, **3**:73–76 (1955).

Chiarelli, A. B.: *Evolution of the Primates*, Academic, New York, 1973.

Clemens, W. A.: Origin and Early Evolution of Marsupials, *Evolution*, **22**(1):1–18 (1968).

———: Mesozoic Mammalian Evolution, *Ann. Rev. Ecol. Syst.*, **1**:357–390 (1970).

Colbert, E. H.: The Osteology and Relationships of *Archaeomeryx*, an Ancestral Ruminant, *Am. Mus. Novit.*, **1135**:1–24 (1941).

———: *Evolution of the Vertebrates*, 2d ed., Wiley, New York, 1969.

Cooke, H. B. S.: The Fossil Mammal Fauna of Africa, *Q. Rev. Biol.*, **43**(3):234–264 (1968).

Crompton, A. W.: The Evolution of the Mammalian Jaw, *Evolution*, **17**(4):431–439 (1963). (a)

———: On the Lower Jaw of *Diarthrognathus* and the Origin of the Mammalian Lower Jaw, *Proc. Zool. Soc. Lond.*, **140**(4):697–753 (1963). (b)

————: A Preliminary Description of a New Mammal from the Upper Triassic of South Africa, *Proc. Zool. Soc. Lond.*, **142**(3):441–452 (1964).

————: In Search of the Insignificant, *Discovery,* **3**(2):23–32 (1968).

———— and F. A. Jenkins, Jr.: Molar Occlusion in Late Triassic Mammals, *Biol. Rev.*, **43**:427–458 (1968).

Dart, R.: *Australopithecus africanus*: The Man-Ape of South Africa, *Nature*, **115**:195–199 (1925).

Gazin, C. L.: A Review of the Middle and Upper Eocene Primates of North America, *Smithson. Misc. Collect.*, **136**(1):1–112 (1958).

Gregory, W. K.: On the Structure and Relations of *Notharctus*, an American Eocene Primate, *Mem. Am. Mus. Nat. Hist.*, n.s., **3**:49–243 (1920).

————: A Half Century of Trituberculority: The Cope-Osborn Theory of Dental Evolution, with a Revised Summary of Molar Evolution from Fish to Man, *Proc. Am. Philos. Soc.*, **73**:169–317 (1934).

———— and G. G. Simpson: Cretaceous Mammal Skulls from Mongolia, *Am. Mus. Novit.*, **225**:1–20 (1926).

Guilday, J. E.: Differential Extinction during Late-Pleistocene and Recent Times, pp. 121–140 in P. S. Martin and H. E. Wright, Jr. (eds.), *Pleistocene Extinctions: The Search for a Cause*, Yale University Press, New Haven, 1967.

Hopson, J. A.: The Classification of Nontherian Mammals, *J. Mammal.*, **51**(1):1–9 (1970).

———— and A. W. Crompton: Origin of Mammals, pp. 15–72 in T. Dobzhansky, M. K. Hecht, and W. C. Steere (eds.), *Evolutionary Biology*, Appleton-Century-Crofts, New York, 1969.

Jenkins, F. A., Jr.: Cynodont Postcranial Anatomy and the "Prototherian" Level of Mammalian Organization, *Evolution*, **24**:230–252 (1970).

————: The Postcranial Skeleton of African Cynodonts, *Bull. Peabody Mus. Nat. Hist., Yale Univ.* **36**:1–216 (1971).

Keast, A.: Australian Mammals: Zoogeography and Evolution, *Q. Rev. Biol.*, **43**(4):373–408 (1968).

Kellogg, R.: Pinnipeds from Miocene and Pleistocene Deposits of California . . . and a Résumé of Current Theories Regarding the Origin of the Pinnipedia, *Univ. Calif. Publ. Bull., Dept. Geol. Sci.*, **13**:23–132 (1922).

————: A Review of the Archaeoceti, *Publ. Carnegie Inst. Wash. D.C.*, **482**:1–366 (1936).

Kermack, D. M., and K. A. Kermack (eds.): *Early Mammals: Supplement 1 to the Zoological Journal of the Linnean Society, vol. 50*, Academic, New York and London, 1971.

Kermack, D. M., K. A. Kermack, and F. Mussett: The Welsh Pantothere *Kuehneotherium praecursoris, J. Linn. Soc. (Zool.)*, **47**:407–423 (1968).

Kermack, K. A.: The Cranial Structure of the Triconodonts, *Philos. Trans. R. Soc. Lond.*, **B246**:83–103 (1963).

————, P. M. Lees, and F. Mussett: *Aegialodon dawsoni,* a new Trituberculosectorial Tooth from the Lower Wealden, *Proc. R. Soc. Lond.,* **B162**:535–554 (1965).

Kowalski, K.: The Pleistocene Extinction of Mammals in Europe, pp. 349–364 in P. S. Martin and H. E. Wright, Jr. (eds.), *Pleistocene Extinctions: The Search for a Cause,* Yale University Press, New Haven, 1967.

Kühne, W. G.: A Symmetrodont Tooth from the Rhaeto-Lias, *Nature,* **166**:696–697 (1950).

Kurtén, B.: Holarctic Land Connexions in the Early Tertiary, *Commentat. Biol. Soc. Sci. Fenn.,* **29**(5):1–5 (1966).

LeGros Clark, W. E.: The Crucial Evidence for Human Evolution, *Proc. Am. Philos. Soc.,* **103**(2):159–172 (1959). (a)

————: *The Antecedents of Man,* Edinburgh University Press, Edinburgh, 1959; also Harper & Row, New York, 1963.

Lillegraven, J. A.: The Latest Cretaceous Mammals of Upper Part of Edmondton Formation of Alberta, Canada, and a Review of the Marsupial-Placental Dichotomy in Mammalian Evolution, *Univ. Kans. Paleontol. Contrib. Vertebr.,* **12**:art. 50 (1969).

Marsh, O. C.: Fossil Horses in America, *Am. Nat.,* **8**:288–294 (1874).

Martin, P. S.: Prehistoric Overkill, pp. 75–120 in P. S. Martin and H. E. Wright, Jr. (eds.), *Pleistocene Extinctions: The Search for a Cause,* Yale University Press, New Haven, 1967.

Matthew, W. D.: The Phylogeny of the Felidae, *Bull. Am. Mus. Nat. Hist.,* **28**:289–316 (1910).

————: The Phylogeny of Dogs, *J. Mammal.* **11**(2):117–138 (1930).

————: A Review of the Rhinoceroses with a Description of *Aphelops* Material from the Pliocene of Texas, *Univ. Calif. Publ., Dept. Geol. Sci.,* **20**:411–480 (1932).

McKenna, M. C.: Collecting Small Fossils by Washing and Screening, *Curator,* **3**:221–235 (1962).

Osborn, H. F.: The Evolution of Mammalian Molars to and from the Triangular Type, *Am. Nat.,* **22**:1067–1079 (1888).

————: The Extinct Giant Rhinoceros *Baluchitherium* of Western and Central Asia, *Nat. Hist.,* **23**:209–228 (1923).

————: *Proboscidea: A Monograph of the Discovery, Evolution, Migration, and Extinction of the Mastodonts and Elephants of the World,* vol. 1, *Moeritherioidea, Deinotherioidea, Mastodontoidea,* 1936, vol. 2, *Stegodontoidea, Elephantoidea,* 1942, New York.

Parrington, F. R.: The Origins of Mammals, *Adv. Sci. (Lond.),* **24**:1–9 (1967).

Patterson, B.: Early Cretaceous Mammals and the Evolution of Mammalian Molar Teeth, *Fieldiana: Geol.,* **13**:1–105 (1956).

———— and W. W. Howells: Hominid Humeral Fragment from Early Pleistocene of Northwestern Kenya, *Science,* **156**:64–66 (1967).

———— and R. Pascual: The Fossil Mammal Fauna of South America, *Q. Rev. Biol.,* **43**(4):404–451 (1968).

Pilgrim, G. E.: The Dispersal of the Artiodactyla, *Biol. Rev.*, **16**:134–163 (1941).

Radinsky, L.: The Early Evolution of the Perissodactyla, *Evolution*, **23**:308–328 (1969).

Raven, P. H., and D. I. Axelrod: Plate Tectonics and Australasian Paleobiogeography, *Science*, **176**:1379–1386 (1972).

Reed, C. A.: Polyphyletic or Monophyletic Ancestry of Mammals, or: What is a Class?, *Evolution*, **14**(3):314–322 (1960).

Ride, W. D. L.: On the Evolution of Australian Marsupials, pp. 281–306 in *The Evolution of Living Organisms: A Symposium of the Royal Society of Victoria*, Melbourne, 1959.

Romer, A. S.: The Chañares (Argentina) Triassic Reptile Fauna, pt. VI: A Chiniquodontid Cynodont with an Incipient Squamosal-Dentary Jaw Articulation, *Breviora Mus. Comp. Zool.*, **344**:1–18 (1970).

Schaeffer, B.: The Origin of a Mammalian Ordinal Character, *Evolution*, **2**:164–175 (1948).

Schaub, S.: La Trigonodontie des rongeurs simplicidentes, *Ann. Paleont.*, **39**:29–57 (1953).

Schultz, A. H.: Past and Present Views on Man's Specializations, *Irish J. Med. Sci.*, August, 1957, p. 341.

Sharman, G. B.: Reproductive Physiology of Marsupials, *Science*, **167**:1221–1228 (1970).

Simons, E. L.: A Critical Reappraisal of Tertiary Primates, pp. 65–129 in J. Buettner-Janusch (ed.), *Evolutionary and Genetic Biology of Primates*, New York, 1963.

――― and A. E. Wood: Early Cenozoic Mammalian Faunas Fayum Province, Egypt, *Bull. Peabody Mus. Nat. Hist., Yale Univ.* **28**:1–105 (1968).

Simpson, G. G.: *A Catalogue of the Mesozoic Mammalia in the Geological Department of the British Museum*, British Museum (Natural History), London, 1928.

―――: American Mesozoic Mammalia, *Mem. Peabody Mus. Yale Univ.*, **3**(1):1–235 (1929).

―――: *Metacheiromys* and the Edentata, *Bull. Am. Mus. Nat. Hist.*, **59**:295–381 (1931).

―――: Mammals and Land Bridges, *Wash. (D.C.) Acad. Sci. J.*, **30**:137–163 (1940).

―――: The Principles of Classification and a Classification of Mammals, *Bull. Am. Mus. Nat. Hist.*, **85**:1–450 (1945).

―――: Holarctic Mammalian Faunas and Continental Relationships during the Cenozoic, *Bull. Geol. Soc. Am.*, **58**:613–687 (1947).

―――: *Horses*, Oxford University Press, New York, 1951. (a)

―――: History of the Fauna of Latin America, pp. 369–408 in G. A. Baitsell (ed.), *Science in Progress*, 7th ser., Yale University Press, New Haven, 1951. (b)

―――: Evolution and Geography *Oreg. State Syst. Higher Educ. Condon Lect.*, 1953.

————: Mesozoic Mammals and the Polyphyletic Origin of Mammals, *Evolution,* **13**(3):405–414 (1959).

————: Diagnosis of the Classes Reptilia and Mammalia, *Evolution,* **14**(3):388–392 (1960).

————: Historical Zoogeography of Australian Mammals, *Evolution,* **15**(4):431–446 (1961).

Slaughter, B. H.: Animal Ranges as a Clue to Late-Pleistocene Extinction, pp. 154–167 in P. S. Martin and H. E. Wright, Jr. (eds.), *Pleistocene Extinctions: The Search for a Cause,* Yale University Press, New Haven, 1967.

————: Earliest Known Marsupials, *Science,* **162**:254–255 (1968).

Sloan, R. E., and L. Van Valen: Cretaceous Mammals from Montana, *Science,* **148**:220–227 (1965).

Stehlin, H. G., and S. Schaub: Die Trigonodontie der simplicidentaten Nager, *Schweiz. Palaeontol. Abh.,* **67**:1–384 (1951).

Stirton, R. A.: Comments on the Relationships of the Cervoid Family Palaeomerycidae, *Am. J. Sci.,* **242**:633–655 (1944).

Szalay, F. S.: The Beginnings of Primates, *Evolution,* **22**:19–36 (1968).

————: Origin and Evolution of Function of the Mesonychid Condylarth Feeding Mechanism, *Evolution,* **23**:703–720 (1969).

Van Valen, L: Therapsids as Mammals, *Evolution,* **14**(3):304–313 (1960).

————: Deltatheridia, A New Order of Mammals, *Bull. Am. Mus. Nat. Hist.,* **132**(1):1–126 (1966).

————: New Paleocene Insectivores and Insectivore Classification, *Bull. Am. Mus. Nat. Hist.,* **135**(5):217–284 (1967).

————: The Multiple Origins of the Placental Carnivores, *Evolution,* **23**:118–130 (1969).

————: Adaptive Zones and the Orders of Mammals, *Evolution,* **25**(2):420–428 (1971). (a)

————: Toward the Origin of Artiodactyls, *Evolution,* **25**(3):523–529 (1971). (b)

———— and R. E. Sloan: The Earliest Primates, *Science,* **150**:743–745 (1965).

Watson, D. M. S.: The Evolution of the Proboscidea, *Biol. Rev.,* **21**:15–29 (1946).

Wood, A. E.: Rodents—A Study in Evolution, *Evolution,* **1**(3):154–162 (1947).

————: Comments on the Classification of Rodents, *Breviora Mus. Comp. Zool.,* **41**:1–9 (1954).

————: A Revised Classification of the Rodents, *J. Mammal.,* **36**(2):165–187 (1955).

————: The Early Tertiary Rodents of the Family Paramyidae, *Trans. Am. Philos. Soc.,* n.s., **52**(1):1–261 (1962).

————: Grades and Clades among Rodents, *Evolution,* **19**(1):115–130 (1965).

———— and B. Patterson: The Rodents of the Deseadan Oligocene of Patagonia and the Beginnings of South American Rodent Evolution, *Bull. Mus. Comp. Zool.,* **120**(3):282–428 (1959).

Sources of Illustrations

Balinsky, B. I.: *An Introduction to Embryology,* 3d ed., W. B. Saunders Company, Philadelphia, 1970.

Ballard, W. W.: *Comparative Anatomy and Embryology,* The Ronald Press, New York, 1964.

Brough, J.: *Proceedings of the Zoological Society of London,* 1931.

Bystrow, A. P.: *Acta Zoologica,* 1938.

————: *Acta Zoologica,* 1939.

————: *Acta Zoologica,* 1957.

Carroll, R. L.: Problems of the Origin of Reptiles, *Biological Reviews,* 1969; published by Cambridge University Press.

Colbert, E. H.: *American Museum Novitates* (published by the American Museum of Natural History), 1941.

————: Evolution of the Vertebrates, John Wiley & Sons, Inc., New York, 1955.

————: *The Age of Reptiles,* W. W. Norton & Company, Inc., New York, 1966.

———— and C. C. Mook: *Bulletin of the American Museum of Natural History,* 1951.

Crompton, A. W.: Proceedings of the Zoological Society of London, 1963.

DeBeer, G.: *Archaeopteryx lithographica,* Trustees of the British Museum (Natural History), 1954.

Ewer, R. F.: *Philosophical Transactions of the Royal Society of London*, 1965.

Gardiner, B. G.: *Bulletin of the British Museum (Natural History), Geology*, Trustees of the British Museum (Natural History), 1967.

Goodrich, E. S.: *Studies on the Structure and Development of Vertebrates*, Dover Publications, Inc., New York, 1958.

Gregory, W. K.: *Transactions of the American Philosophical Society* (published by the American Philosophical Society), 1933.

————: The Transformation of Organic Designs: A Review of the Origin and Deployment of the Earlier Vertebrates, *Biological Reviews*, 1936; published by Cambridge University Press.

————: *Evolution Emerging*, vol. 2, The Macmillan Company (Crowell-Collier subsidiary), New York, 1951.

———— and C. L. Camp: *Bulletin of the American Museum of Natural History*, 1918.

———— and W. Granger: *Bulletin of the American Museum of Natural History*, 1936.

———— and H. C. Raven: *Annals of the New York Academy of Sciences*, 1941.

Greenwood, Rosen, Weitzman, and Myers: *Bulletin of the American Museum of Natural History*, 1966.

Gross, W.: *Journal of the Linnaean Society (Zoology)*, 1967.

Hall, E. R., and K. R. Kelson: *The Mammals of North America*, vol. 2, The Ronald Press, New York, 1959.

Heilmann, G.: *The Origin of Birds*, Appleton-Century-Crofts, Inc., New York, 1927.

Hopson, J. A., and A. W. Crompton, in T. Dobzhansky, M. K. Hecht, and W. C. Steere (eds.): *Evolutionary Biology*, vol. 3, Appleton-Century-Crofts, Inc., New York, copyright © 1969. By permission of Appleton-Century-Crofts, Educational Division, Meredith Corporation.

Howard, H.: Los Angeles County Museum, Science Series, 1955.

Howell, A. B.: *Aquatic Mammals*, Charles C Thomas, Springfield, Ill., 1930.

Jarvik, E.: *Meddeleser om Grønland*, 1952.

————: *The Scientific Monthly* (now combined with *Science*; published by the American Association for the Advancement of Science), **80**(3):141–154 (1955).

Jenkins, F. A., Jr.: The Postcranial Skeleton of African Cynodonts, *Bulletin 36*, Peabody Museum of Natural History, Yale University, March, 1971.

Jessen, H.: *Arkiv for Zoologi*, 1966.

Jollie, M.: *Chordate Morphology*, Reinhold Publishing Corporation, New York, 1962.

Kellogg, R.: *Publication of the Carnegie Institute* (Washington), 1936.

————: *The Quarterly Review of Biology*, 1928.

Kiaer, J.: *Skrifter Vidensk. Selsk. Kristiana, Mat.-Nat. Kl.*, 1924.

————: *Skrifter om Svalbard og Ishavet*, 1932.

Kurtén, B.: *Commentationes Biologicae Societas Scientiarum Fennica*, 1966.

LeGros Clark, W. E.: *The Antecedents of Man*, Edinburgh University Press, 1959. (Paperback edition published by Harper Torchbooks.) Reprinted by permission of Quadrangle Books, copyright © 1959, 1962, 1971 by W. E. LeGros Clark.

———— and L. S. B. Leakey: The Miocene Hominoidea of East Africa, in Fossil Mammals of Africa, no. 1, Trustees of the British Museum (Natural History), 1951.

Lillegraven, J. A.: *The University of Kansas Paleontological Contributions*, 1969, courtesy of The University of Kansas.

Longwell, C. R., and R. F. Flint: *Introduction to Physical Geology*, 2d ed., John Wiley & Sons, Inc., New York, 1962.

Lull, R. S.: *American Journal of Science*, 1921.

Matthew, W. D.: *Transactions of the American Philosophical Society* (published by the American Philosophical Society), 1937.

McKenna, M. C.: *The Curator* (published by the American Museum of Natural History), 1962.

Miles, R. S.: *Colloques Internationaux du Centre National de la Recherche Scientifique* (Editions du Centre National de la Recherche Scientifique), 1966.

————: *Journal of the Linnaean Society* (Zoology), 1967.

————: The Holonematidae (placoderm fishes), a Review Based on New Specimens of *Holonema* from the Upper Devonian of Western Australia, *Philos. Trans. Roy. Soc. Com.*, **B263**:101–234 (1971).

Millot, J., and J. Anthony: *Colloques Internationaux du Centre National de la Recherche Scientifique* (Editions du Centre National de la Recherche Scientifique), 1958.

Moy-Thomas, J. A.: *Proceedings of the Zoological Society of London*, 1934.

———— and R. S. Miles: *Palaeozoic Fishes*, Chapman & Hall, Ltd., London, 1971.

Nielsen, E.: *Meddeleser om Grønland*, 1932.

Nilsson, T.: *Acta Universitatis Lundensis*, 1946.

Noble, G. K.: *The Biology of the Amphibia*, Dover Publications, Inc., New York, 1954.

Norman, J. R.: *A History of Fishes*, 6th ed., Ernest Benn, Ltd., London, 1960.

Ørvig, T., in A. E. W. Miles (ed.): *Structural and Chemical Organization of Teeth*, vol. 1, Academic Press, Inc., New York, 1967.

Osborn, H. F.: *American Museum Novitates* (published by the American Museum of Natural History), 1924.

Parker, T. J., and W. A. Haswell: *A Textbook of Zoology*, 6th ed., St. Martin's Press, Inc., Macmillan and Company, Ltd., London, 1940.

Patterson, B., and R. Pascual: *The Quarterly Review of Biology*, 1968.

Pearson, H. S.: *Proceedings of the Zoological Society of London*, 1924.

Peyer, B.: *Comparative Odontology*, University of Chicago Press, 1968.

Piveteau, J. (ed.): *Traité de Paléontologie*, Masson et Cie., Paris, vol. 5, 1955; vol. 6, 1961.

Poole, D. F. G., in A. E. W. Miles (ed.): *Structural and Chemical Organization of Teeth*, vol. 1, Academic Press, Inc., New York, 1967.

Rawles, M. E., in A. J. Marshall (ed.): *Biology and Comparative Physiology of Birds*, vol. 1, Academic Press., Inc., New York, 1960–1961.

Raymond, P. E.: *Prehistoric Life*, Harvard University Press, Cambridge, Mass., 1947.

Ritchie, A.: *Palaeontology*, 1968.

Romer, A. S.: *The Osteology of the Reptiles*, University of Chicago Press, 1956.

———: *Vertebrate Paleontology*, 3d ed., University of Chicago Press, 1966.

———: *Harvard Alumni Bulletin*, Harvard Bulletin, Inc., 1967.

———: *Kirtlandia* (published by the Cleveland Museum of Natural History), 1969.

———: *The Vertebrate Body*, 4th ed., W. B. Saunders Company, New York, 1970.

———: *Man and the Vertebrates*, 3d ed., University of Chicago Press, 1941.

——— and R. V. Witter: *Journal of Geology*, 1942.

Sawin, H. J.: *University of Texas Publication*, 1945.

Schaeffer, B.: *Evolution*, 1952.

———: *Bulletin of the American Museum of Natural History*, 1967.

———, in P. W. Gilbert, R. F. Mathewson, and D. P. Rall (eds.): *Sharks, Skates, and Rays*, The Johns Hopkins Press, Baltimore, Md., 1967.

——— and D. Rosen: *American Zoologist* (published by the American Society of Zoologists), 1961.

Scott, W. B., and G. L. Jepsen: *Transactions of the American Philosophical Society* (published by the American Philosophical Society), 1941.

Seeley, H. G.: *Dragons of the Air*, Dover Publications, Inc., New York, 1967.

Simons, E. L.: *American Scientist* (published by The Society of the Sigma Xi), 1960.

Simpson, G. G.: Trustees of the British Museum (Natural History), 1928.

———: *American Mesozoic Mammalia*, Yale University Press, New Haven, Conn., 1929.

Sisson, S., and J. D. Grossman: *The Anatomy of the Domestic Animals*, 4th ed., W. B. Saunders Company, New York, 1953.

Sloan, R. E., and L. Van Valen: *Science* (now combined with *The Scientific Monthly*; published by the American Association for the Advancement of Science), **148**:220–227 (1965).

Steen, M. S.: *Proceedings of the Royal Society of London*, 1938.

Stehlin, H. G., and S. Schaub: *Schweiz. Paläont. Abhandlungen*, 1951.

Stensiö, E.: *Skrifter om Svalbard og Nordishavet*, 1927.

———: *Meddeleser om Grønland*, 1931.

————: Trustees of the British Museum (Natural History), 1932.

————: *Kungl. Svenska Vetenskapsakademiens Handlingar*, 1963.

Storer, T. I.: *General Zoology*, 2d ed., McGraw-Hill Book Company, 1951.

Swinton, W. E.: *Fossil Birds*, 2d ed., Trustees of the British Museum (Natural History), 1965.

Torrey, T. W.: *Morphology of the Vertebrates*, 2d ed., John Wiley & Sons, Inc., New York, 1967.

Travis, D. F.: *Annals of the New York Academy of Sciences*, 1963.

Walker, A. D.: *Philosophical Transactions of the Royal Society of London*, 1964.

Watson, D. M. S.: *Philosophical Transactions of the Royal Society of London*, 1919.

————: *Philosophical Transactions of the Royal Society of London*, 1937.

————: *Transactions of the Royal Society of Edinburgh*, 1938.

————: *Philosophical Transactions of the Royal Society of London*, 1957.

————: *Palaeontology*, 1961.

Weichert, C. K.: *Anatomy of the Chordates*, 3d ed., McGraw-Hill Book Company, 1965.

Welty, J. C.: *The Life of Birds*, W. B. Saunders Company, New York, 1962.

Wendt, H.: *Before the Deluge*, translated by Richard and Clara Winston. Copyright © 1968 by Doubleday & Company, Inc., Garden City, New York. Reproduced by permission of the publisher.

Westoll, T. S.: *Proceedings of the Zoological Society of London*, 1941.

————: *Proceedings of the Royal Society of London*, 1943.

————: *Bulletin of the American Museum of Natural History*, 1944.

————: *Transactions of the Royal Society of Edinburgh*, 1945.

Wood, A. E.: *American Museum Novitates* (published by the American Museum of Natural History), 1949.

————: *Transactions of the American Philosophical Society* (published by the American Philosophical Society), 1962.

Young, J. Z.: *The Life of Vertebrates*, 2d ed., The Clarendon Press, Oxford, 1962.

Zangerl, R.: *Fieldiana: Geology* (published by the Field Museum of Natural History, Chicago), 1966.

———— and D. Bardack: *Science* (now combined with *The Scientific Monthly*; published by the American Association for the Advancement of Science), **162**:1265–1267 (1968).

Index

Page numbers in *italic* indicate illustrations.